Biodegradation and Bioremediation

SECOND EDITION

Biodegradation
and
Bioremediation

SECOND EDITION

Martin Alexander
Department of Soil, Crop, and Atmospheric Sciences
College of Agriculture and Life Sciences
Cornell University
Ithaca, New York

ACADEMIC PRESS

San Diego London Boston New York Sydney Tokyo Toronto

Cover photograph © PhotoDisc.

This book is printed on acid-free paper. ∞

Academic Press
a division of Harcourt Brace & Company
525 B Street, Suite 1900, San Diego, California 92101-4495, USA
http://www.apnet.com

Academic Press
24-28 Oval Road, London NW1 7DX, UK
http://www.hbuk.co.uk/ap/

Library of Congress Catalog Card Number: 98-89646

International Standard Book Number: 0-12-049861-8

PRINTED IN THE UNITED STATES OF AMERICA
99 00 01 02 03 04 BB 9 8 7 6 5 4 3 2 1

CONTENTS

CHAPTER 5
Activation 51

CHAPTER 6
Kinetics 73

CHAPTER 7
Threshold 105

CHAPTER 8
Sorption 117

CHAPTER 9
Nonaqueous-Phase Liquids and Compounds with Low Water Solubility 135

CHAPTER 10
Bioavailability: Aging, Sequestering, and Complexing 157

CHAPTER 11
Effect of Chemical Structure on Biodegradation 177

CHAPTER 12
Predicting Products of Biodegradation 195

PREFACE

Biodegradation of individual compounds has been the subject of active concern for more than 40 years. The initial interest was in the fate and persistence of pesticides in soils; however, the field has expanded enormously in recent years to encompass a wide variety of chemicals and a broad array of issues. Moreover, the technologies that have been developed markedly enhance biodegradation or result in microbial destruction of organic pollutants that otherwise would persist at the sites of contamination. These bioremediation technologies have led to the cleanup of many polluted groundwaters and soils and have fostered the development of a new bioremediation industry.

This book presents the basic principles of biodegradation and shows how those principles relate to bioremediation. It considers some of the microbiological, chemical, environmental, engineering, and technological aspects of biodegradation and bioremediation, but it does not cover all facets. The field is too large and diverse, and its knowledge base is expanding too rapidly to be covered in a single text. Nevertheless, key general principles underlie the science and the technology, and these can be presented within a single volume.

An introduction to biodegradation and bioremediation requires more than knowledge of one or two disciplines; information from many disciplines is needed. The processes are microbiological, the behavior of the compounds follows chemical principles, changes in hazard and exposure are topics of concern in environmental toxicology, the areas containing the pollutants represent environments with unique properties, and the technologies are based on approaches common in environmental engineering. Thus, the book is addressed to—and should be of value to—microbiologists, chemists, toxicologists, environmental scientists, and environmental engineers. Although some topics may not be covered as extensively as others, the references appended to each chapter provide a guide to further information.

In the few years since the first edition of this book was written, much new information about various aspects of biodegradation has been acquired, and many new approaches to practical bioremediation have been devised. That new information has resulted in additions and modifications—sometimes small, sometimes large—to many chapters, and thus numerous recent publications are cited. The earlier review of bioremediation technologies has been altered extensively and expanded to two chapters (Chapters 16 and 17) to show the reader the latest approaches. Our appreciation of the impact and effects of nonaqueous-phase liquids (NAPLs), aging, and sequestration and how they affect bioavailability has grown markedly in only a short time, and this new knowledge is presented in Chapters

9 and 10. Several entirely new chapters have also been included in this edition to give the reader a broader introduction to existing information and to issues of concern to the scientist, the engineer, and the practitioner. These chapters deal with predicting products of biodegradation (Chapter 12), bioremediation of metals and other inorganic pollutants (Chapter 18), and formation and biodegradation of air pollutants (Chapter 20).

Acronyms, abbreviations, and sometimes trade names are used for many of the compounds or groups of compounds. Because these terms are widely accepted and are part of the vocabulary of professionals in biodegradation and bioremediation, they are used here. When two terms apply to the same chemical, the more widely accepted term or the term approved by the relevant technical society (as with pesticides) is adopted. Each term, acronym, or abbreviation is defined or its formula is given in the Appendix.

The book could not have been written without the understanding, patience, and encouragement of my wife, Renee. To her, my expressions of thanks will never suffice.

The book is dedicated to Laura, Jeremy, Anna, Jonathan, Maya, and all other children that they may live in a cleaner and healthier environment.

Martin Alexander

CHAPTER 1

Introduction

The application of highly sensitive analytical techniques to environmental analysis has provided society with disturbing information. The air we breathe, the water we drink and bathe in, the soil in which our crops are grown, and the environments in which populations of animals and plants grow are contaminated with a variety of synthetic chemicals. In agricultural areas and adjacent ground and surface waters, some of these chemicals are pesticides or products generated from pesticides. Many are industrial chemicals that have been deliberately or inadvertently discharged into waters or onto soils following their intended use. Others are by-products of manufacturing operations that do not utilize waste-treatment facilities or by-products that were inadequately treated. Some are probably formed in nature from synthetic compounds, and a few are generated by reaction of natural organic materials with Cl_2 used to treat water for human consumption.

As a rule, these organic compounds are not found individually but rather in simple or complex mixtures. The mixtures may be associated with the release, storage, or transport of many chemicals in surface or groundwaters, waste-treatment systems, soils, or sediments. The number of chemicals found to date is enormous, and the types of mixtures are similarly countless. Moreover, the concentrations of individual compounds vary appreciably, and they may be higher than 1.0 g per liter of water or per kilogram of soil at sites subject to spills from tank cars or trucks, to discharge of industrial waste, or to leakages from storage or disposal facilities for industrial chemicals. In contrast, the concentration may be lower than 1.0 μg per liter of water or per kilogram of soil at some distance from the point of release, spill, or storage. Even at these low concentrations, some chemicals are toxic, or risk analyses suggest that exposure of large populations to the low levels will result in deleterious effects to a few individuals; in addition, some chemicals at low concentrations are subject to biomagnification and may reach levels that have deleterious effects on humans, animals, or plants.

It is not surprising that synthetic chemicals are present in the human environment, for example, in areas used for food and feed production, and environments supporting natural populations of animals and plants. Modern society relies on a striking array of organic chemicals, and the quantities used are staggering. Values for the annual production of organic compounds in the United States alone show the vast tonnages that are part of human activities (Table 1.1). Although many of these chemicals are consumed or destroyed, a high percentage are released into the air, water, and soil. The quantity released varies with the compound and its particular use, but regulatory agencies in industrialized countries have found that significant percentages of the total quantity consumed by industry, agriculture, and domestic pursuits do, indeed, find their way into air, water, and soil. With effective regulation and appropriate enforcement, the quantities released can be dramatically reduced. This is evident from data for emissions in the United States (Table 1.2). Thus, the focus on bioremediation in countries in which regulation and

Table 1.1

Production of Organic Chemicals in the United States in 1996 (in kg × 10^6)[a]

Chemical	Production
Ethylene	22,300
Propylene	11,400
Styrene	5,390
1,2-Dichloroethane	5,140
Ethylbenzene	4,700
Ethylene oxide	3,280
p-Xylene	2,800
Cumene	2,670
1,3-Butadiene	1,740
Acrylonitrile	1,530
Benzene	960
Isopropyl alcohol	628
Aniline	489
o-Xylene	402
2-Ethylhexanol	345

[a] From Anonymous (1997).

Table 1.2

Emissions of Organic Compounds from Manufacturing in the United States in 1995 (in kg × 10^6)[a]

Compound	Emissions
Methanol	111
Toluene	66.2
Xylene (all isomers)	43.4
n-Hexane	35.1
Methyl ethyl ketone	31.8
Dichloromethane	26.0
Glycol ethers	19.9
Styrene	19.0
Ethylene	15.5
Acetonitrile	13.1
n-Butanol	12.6

[a] From Hanson (1997).

enforcement are strict is chiefly on releases in the past, when regulations, enforcement, or both were lax. However, biodegradation and bioremediation of wastes before their discharge remain of considerable interest because they are part of the strategy to keep emissions small.

Predicting the hazards of an organic compound to humans, animals, and plants requires information not only on its toxicity to living organisms but also on the degree of exposure of the organisms to the compound. The mere discharge of a chemical does not, in itself, constitute a hazard: the individual human, animal, or plant must also be exposed to it. In evaluating exposure, the transport of the chemical and its fate must be considered. A molecule that is not subject to environmental transport is not a health or environmental problem except to species at the specific point of release, so that information on dissemination of the chemical from the point of its release to the point where it could have an effect is of great relevancy. However, the chemical may be modified structurally or totally destroyed during its transport, and the fate of the compound during transport, that is, its modification or destruction, is crucial to defining the exposure. A compound that is modified to yield products that are less or more toxic, or that is totally degraded or is biomagnified—these being factors associated with the fate of the molecule—represents greater or lesser hazard to the species that are potentially subject to injury.

At the specific site of discharge or during its transport, the organic molecule may be acted on by abiotic mechanisms. Photochemical transformations occur in the atmosphere and at or very near the surfaces of water, soil, and vegetation, and these processes may totally destroy or appreciably modify a number of different types of organic chemicals. Nonenzymatic, nonphotochemical reactions are also prominent in soil, sediment, and surface and groundwater, and these may bring about significant changes; however, such processes rarely if ever totally convert organic compounds to inorganic products in nature, and many of these nonenzymatic reactions only bring about a slight modification of the molecule so that the product is frequently similar in structure, and often in toxicity, to the precursor compound.

However, biological processes may modify organic molecules at the site of their discharge or during their transport. Such biological transformations, which involve enzymes as catalysts, frequently bring about extensive modification in the structure and toxicological properties of pollutants or potential pollutants. These biotic processes may result in the complete conversion of the organic molecule to inorganic products, cause major changes that result in new organic products, or occasionally lead to only minor modifications. Plants and, to a lesser extent, animals in natural or man-modified environments may cause a number of changes in a wide array of chemicals, and these are of enormous importance in reducing or in sometimes increasing the toxicity of the chemical to the plant or animal that is exposed. Nevertheless, the available body of information suggests that the major agents causing the biological transformations in soil, sediment, wastewater, surface and groundwater, and many other sites are the microorganisms that inhabit these environments. These microfloras are thus frequently the major agents affecting the fate of chemicals at the sites of their release or in the environments through which they pass.

Biodegradation can be defined as the biologically catalyzed reduction in complexity of chemicals. In the case of organic compounds, biodegradation frequently, although not necessarily, leads to the conversion of much of the C, N, P, S, and other elements in the original compound to inorganic products. Such a conversion of an organic substrate to inorganic products is known as *mineralization*. *Ultimate* biodegradation is a term sometimes used as a synonym for mineralization. Thus, in the mineralization of organic C, N, P, S, or other elements, CO_2 or inorganic forms of N, P, S, or other elements are released by the organism and enter the surrounding environment. Plant and animal respiration are mineralization processes that destroy numerous organic molecules of living organisms, but the mineralization of synthetic chemicals by biological processes appears to result largely or, in some environments, entirely from microbial activity. Indeed, microorganisms are frequently the sole means, biological or nonbiological, of converting synthetic chemicals to inorganic products. Few nonbiological reactions in nature bring about comparable changes. It is because of their ability to mineralize anthropogenic compounds that microorganisms play a large role in soils, waters, and sediments.

Many synthetic molecules discharged into these environments are directly toxic or become hazardous following biomagnification. Because mineralization results in the total destruction of the parent compound and its conversion to inorganic products, such processes are beneficial. In contrast, nonbiological and many biological processes, although degrading organic compounds, convert them to other organic products. Some of these products are toxic, but others evoke no untoward response. Nevertheless, the accumulation in nature of an organic product is still cause for some concern inasmuch as a material not currently known to be harmful, may, with new techniques or measurements of new toxicological manifestations, reveal itself to be undesirable. The literature of toxicology contains examples in which the increasing base of knowledge or new procedures or approaches have revealed that chemicals previously deemed to be safe were in fact harmful. Thus, mineralization is especially important in ridding natural environments of actual or possible hazards to humans, animals, and plants.

Microorganisms carry out biodegradation in many different types of environments. Of particular relevance for pollutants or potential pollutants are sewage-treatment systems, soils, underground sites for the disposal of chemical wastes, groundwater, surface waters, oceans, sediments, and estuaries. Microbial processes in the various kinds of aerobic and anaerobic systems for treating industrial, agricultural, and municipal wastes are extremely important because these treatment systems represent the first point of the discharge of many chemicals into environments of importance to humans or other living organisms. Microbial processes have long been known to be important in sewage and wastewater for the destruction of a large number of synthetic compounds. Soils also receive countless synthetic molecules from farming operations, land spreading of industrial wastes, accidental spills, or sludge disposal, and the degradation of natural materials in soils was recognized even in prehistoric times. In this century, the disposal of industrial wastes on or below the surface of the land became widespread before the evidence of groundwater pollution became prominent, but the sites adjacent to these points of chemical disposal contain microbial communities that, should they not be directly affected by the toxicity of the wastes, destroy many of the organic compounds. Groundwater adjacent to these waste-disposal sites, lakes and rivers that receive inadvertent or deliberate discharges of chemicals, and the oceans and estuaries similarly contain highly diverse and often very active communities of bacteria, fungi, and protozoa that, directly or indirectly, destroy many natural products as well as various synthetics. In addition, a variety of pollutants are retained by the sediments below fresh or marine waters, and these sediments also contain large and metabolically active communities of heterotrophic microorganisms.

Natural communities of microorganisms in these various habitats have an amazing physiological versatility. They are able to metabolize and often mineralize an enormous number of organic molecules. Probably every natural product, regardless of its complexity, is degraded by one or another species in some particular environment; if not, such compounds would, this long after the appearance of life on earth, have accumulated in enormous amounts. The lack of significant accumulation of natural products in oxygen-containing ecosystems, in itself, is an indication that the indigenous microfloras act on an astounding array of natural products. A particular species metabolizes only a small number from this array, but another species in the same habitat is able to make up for the deficiencies of its neighbor. Although certain bacteria and fungi act on a broad range of organic compounds, no organism known to date is sufficiently omnivorous to destroy a very large percentage of the natural chemicals that are formed by plants, animals, and other microorganisms.

Similarly, communities of bacteria and fungi metabolize a multitude of synthetic chemicals. The number of such molecules that can be degraded has yet to be counted, but literally thousands are known to be destroyed as a result of microbial activity in one or another environment. It is not clear how many of the millions of known organic molecules synthesized in the laboratory or made industrially can be modified in these ways, but of the list of chemicals currently regarded as pollutants and that are derived from the activities of human society, many clearly can be modified and often are mineralized by actions of these natural communities. Because too few of the known organic compounds have been tested, however, it is not yet certain to what degree the impressive microbial versatility applies to all organic compounds, but at least this versatility has been amply demonstrated with regard to many of the environmental pollutants of current concern.

Several conditions must be satisfied for biodegradation to take place in an environment. These include the following: (a) An organism that has the necessary enzymes to bring about the biodegradation must exist. The mere existence of an organism with the appropriate catabolic potential is necessary but not sufficient for biodegradation to occur. (b) That organism must be present in the environment containing the chemical. Although some microorganisms are present in essentially every environment near the earth's surface, particular environments may not contain an organism with the necessary enzymes. (c) The chemical must be accessible to the organism having the requisite enzymes. Many chemicals persist even in environments containing the biodegrading species simply because the organism does not have access to the compound that it would otherwise metabolize. Inaccessibility may result from the substrate being in a different microenvironment from the organism, in a solvent not miscible with water, or sorbed to solid surfaces. (d) If the initial enzyme bringing about the degradation is extracellular, the bonds acted upon by that enzyme must be exposed for the catalyst to function. This is not always the case because of sorption of many organic molecules. (e) Should the enzymes catalyzing the initial degradation be intracellular, that molecule must penetrate the surface of the cell to the internal sites where the enzyme acts. Alternatively, the products of an extracellular reaction must penetrate the cell for the transformation to proceed further. (f) Because the population or biomass of bacteria or fungi acting on many synthetic compounds is initially small, conditions in the environment must be conducive to allow for proliferation of the potentially active microorganisms (Alexander, 1973).

Because microorganisms are frequently the major and occasionally the sole means for degradation of particular compounds, the absence of a microorganism from a particular environment, or its inability to function, often means that the compound disappears very slowly. If microorganisms are the sole agents of destruction, the chemical will not be destroyed at all. If any of the conditions mentioned are not met, the chemical similarly will be long-lived. Hence, the frequent finding that organic pollutants are persisting is evidence that microorganisms are not functioning, they are acting very slowly, or no microorganism exists with the capacity to modify the molecule. It is not certain at present how many compounds persist in one or another environment because of the absence of microorganisms in that site, the occurrence of conditions not conducive for microbial biodegradation, or the complete absence in nature of species having the capacity to bring about the transformation. Monitoring programs have revealed that many chlorinated hydrocarbons used in industry and agriculture, compounds containing substituents other than halogens, and other categories of materials endure for long periods, but this very persistence shows either that microorganisms are not omnipotent or that particular environmental conditions prevent appreciable biological activity. Microbial successes are clearly evident because the organic molecule is destroyed; in contrast, their failings are also evident because the chemical endures.

REFERENCES

Alexander, M., *Biotechnol. Bioeng.* **15,** 611–647 (1973).
Anonymous, *Chem. Eng. News* **75**(25), 38–79 (1997).
Hanson, D. J. *Chem. Eng. News* **75**(23), 22–23 (1997).

CHAPTER 2

Growth-Linked Biodegradation

Microorganisms use naturally occurring and many synthetic chemicals for their growth. They use these molecules as a source of C, energy, N, P, S, or another element needed by the cells. Most attention has been focused on the acquisition of C and energy to sustain the growth of bacteria and fungi. For the synthetic substrates that are extensively degraded, the molecule is simply another organic substrate from which the population can obtain the needed elements or the energy required for biosynthetic reactions.

A common research procedure that relies on the ability of microorganisms to use organic compounds as sources of C and energy for growth is known as the enrichment-culture technique. The method is based on the selective advantage gained by an organism that is able to use a particular test compound as a C and energy source in a medium containing inorganic nutrients but no other sources of C and energy. Under these conditions, a species that is able to grow by utilizing that chemical will multiply. Few other bacteria and fungi will proliferate in this medium. However, species that use products excreted by the populations acting on the added organic nutrient will also flourish, and thus the final isolation of a microorganism in pure culture requires plating on an agar medium so that individual colonies can be selected. That agar medium is also made selective by having a single source of C and energy. Repeated transfer of the enrichment through solutions that contain the test compound and inorganic nutrients further increases the degree of selectivity before plating because organic materials and unwanted species from the original environmental sample are diluted by the serial transfers.

The enrichment-culture technique has been the basis for the isolation of pure cultures of bacteria and fungi that are able to use a large number of organic molecules as C and energy sources. However, attempts to obtain microorganisms that are able to grow on a variety of other organic compounds have met with failure. Undoubtedly, many of the failures can be attributed to misuse of the technique or errors in the approach of the investigator; for example, sometimes the concentration of the organic nutrient may be too low to give detectable turbidity in the enrichment solution or too high so that the microorganisms fail to develop because of the toxicity. In other instances, the failure results from the absence from the selective medium of the growth factors essential for the organisms degrading the compound. Nevertheless, when the failure to isolate a microorganism by enrichment culture agrees with the prolonged persistence of the chemical in nature, it is likely that the compound is not used by microorganisms as a source of C and energy.

Members of a large number of genera of bacteria and fungi have been isolated that grow on one or more synthetic compounds. Much of the early literature deals with sugars, amino acids, other organic acids, and other cellular or tissue constituents of living organisms, but a variety of pesticides have also been shown to support the growth of one or another bacterium or fungus. Under these conditions, bacteria increase in numbers and fungi increase in biomass in culture media. At the same time, the chemical disappears, typically at a rate that parallels the increase in cell number or biomass. As the concentration of the C source declines, the rate of cell or biomass increase diminishes until, when all the substrate is consumed, the population rise ends.

As a rule, mineralization of organic compounds is characteristic of growth-linked biodegradation, in which the organism converts the substrate to CO_2, cell components, and products typical of the usual catabolic pathways. It is likely, however, that mineralization in nature occasionally may not be linked to growth but instead results from nonproliferating populations. Conversely, some species growing at the expense of a C compound may still not mineralize and produce CO_2 from the substrate; however, if O_2 is present, the organic products excreted by one species probably will be converted to CO_2 by another species, so that even if the initial population does not produce CO_2, the second species will. The net effect is still one of mineralization.

A compound, such as many environmental pollutants, that represents a novel C and energy source for a particular population still is transformed by the metabolic pathways that are characteristic of heterotrophic microorganisms. For the organism to grow on the compound, it must thus be converted to the intermediates that characterize these major metabolic sequences. If the compound cannot be modified enzymatically to yield such intermediates, it will not serve as a C and energy source because the energy-yielding and biosynthetic processes cannot function. The initial phases of the biodegradation thus involve modification of the novel substrate to yield a product that is itself an intermediate or, following further metabolism, is converted to an intermediate in these ubiquitous metabolic sequences. This need to convert the synthetic molecule to intermediates is characteristic of both aerobes and anaerobes as they derive C and energy from the substrate.

It should be stressed, however, that an organic compound need not be a substrate for growth in order for it to be metabolized by microorganisms. Two categories of transformations exist. In the first, the biodegradation provides C and energy to support growth, and the process therefore is growth-linked. In the second, the biodegradation is not linked to multiplication; the reasons will be considered in the following.

Several studies have demonstrated that the number of microbial cells or the biomass of the species acting on the chemical of interest increases as degradation proceeds. During a typical growth-linked mineralization brought about by bacteria, the cells use some of the energy and C of their organic substrate to make new cells, and this increasingly large population causes an increasingly rapid mineralization. In these instances, the mineralization reflects the population changes. During the decomposition of 2-, 3-, or 4-chlorobenzoate or 3,4-dichlorobenzoate in sewage, for example, bacteria acting on these compounds multiply, and the increase in cell numbers parallels the destruction of the chemicals that serve as their source of C (DiGeronimo *et al.*, 1979). Similarly, bacteria capable of metabolizing 4-nitrophenol proliferate in sewage samples as the chemical disappears from the water phase (Fig. 2.1). Bacteria using 2,4-D similarly increase in numbers as the microbial community of soil destroys this herbicide (Kunc and Rybarova, 1983). Many observations have been made that pure cultures grow as they destroy synthetic chemicals, for example, during the decomposition of the herbicide IPC by *Arthrobacter* sp. (Clark and Wright, 1970). It also has been reported frequently that synthetic molecules, such as the herbicide endothal (Sikka and Saxena, 1973), are converted to typical constituents of microbial cells as the chemicals are used as C and energy sources.

ASSIMILATION OF CARBON

Many measurements have been made of the percentage of the C in the organic substrate that is converted into the cells that are carrying out the biodegradation. The values reflect the biological efficiency of converting the substrate into biomass, with the higher values characterizing the more efficient organisms. Such measurements are simple and straightforward in liquid media with water-soluble substrates since the biomass is particulate and thus can be readily distinguished from C in solution. Measurements in soils, wastewater, sewage, or sediments, in contrast, are complicated because other particulate matter is present in addition to the cells and because complex water-insoluble products are often formed that must be

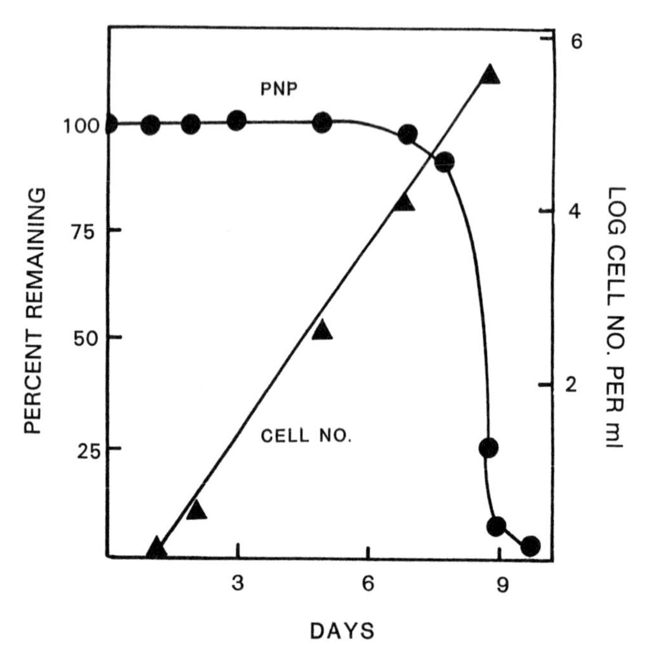

Figure 2.1 Growth of bacteria degrading 4-nitrophenol (PNP) in sewage amended with 2 mg of 4-nitrophenol per liter. (From Wiggins *et al.*, 1987. Reprinted with permission from the American Society of Microbiology.)

distinguished from the cell material. In samples of such environments, therefore, C assimilation is estimated as

$$C_{assimilated} = C_{substrate} - C_{mineralized}.$$

The C that is assimilated is further mineralized as the cells metabolizing the original substrate are themselves decomposed or consumed by protozoa or other predators.

The values from the measurements in pure cultures of microorganisms are often expressed as *growth yield,* which is the weight of biomass formed divided by the weight of substrate used. The values for pure cultures may also be given as a *molar growth yield,* which is the weight of biomass formed divided by moles of substrate metabolized.

The values for such estimates of the efficiency of biomass production vary appreciably, for both aerobes and anaerobes. Some species are efficient in capturing the energy in the organic substrate and converting the C to cells, but others are notably inefficient. Typical values are presented in Table 2.1. Some of the values listed are not actual estimates of C assimilated but rather are quantities of substrate-C not mineralized: such figures represent the total C in biomass and in products.

Table 2.1

Percentages of Substrate-C Converted to Cells or Mineralized

Organisms or environmental sample	Substrate	% of Substrate-C converted to cells	Reference
Bacillus acidocaldarius	Glucose	15–47	Farrand *et al.* (1983)
Candida utilis	Glucose	39[a]	Johnson (1967)
Candida utilis	Acetate	56[a]	Johnson (1967)
Arthrobacter sp.	Glucose	21–28[a]	Cacciari *et al.* (1983)
River water	Biphenyl	<20–40[b]	Bailey *et al.* (1983)
Pond water	Phenol	20–25	Chesney *et al.* (1985)
Lake water	Aniline	40–60[b]	Hoover *et al.* (1986)
Lake water	4-Nitrophenol	<10	Hoover *et al.* (1986)
Soil	Maleic hydrazide	44[b]	Helweg (1975)
Soil	Phenthoate	39[b]	Iwata *et al.* (1977)
Soil	Several	20–40	Kassim *et al.* (1981)
Soil	Glucose	17–53[b]	Martin and Haider (1979)
Soil	Several	>50[b]	Scow *et al.* (1986)
Soil	Acetate	>70[b]	Stevenson and Ivarson (1964)
Soil	2,4-D	19–92[b]	Stott *et al.* (1983)
Sewage inoculum	Phthalate esters	2–44	Sugatt *et al.* (1984)

[a] Assumes cells contain 50% C.

[b] The value represents substrate-C converted to cells and organic products (i.e., it is the percentage of C not mineralized).

It is immediately evident that the percentages of substrate-C converted to cells, the amount not mineralized, or both vary enormously. In some instances, little biomass is formed. In others, the yield of cells plus products is very high.

Under certain conditions in fresh and wastewaters, and possibly in other environments as well, essentially all the C is mineralized, and little or none accumulates in the biomass. This is surprising and as yet unexplained because mineralization generates energy, and the metabolic pathways leading to the formation of CO_2 are assumed to involve biochemical sequences that result in C assimilation. In one study, for example, 93 to 98% of benzoate, benzylamine, aniline, phenol, and 2,4-D added to samples of lake water or sewage at levels below 300 μg per liter was converted to CO_2, and direct measurements revealed no C assimilation during the mineralization of 24 ng to 250 μg of benzylamine per liter (Subba-Rao *et al.*, 1982). Similarly, only 1.2% of the C of 2,4-D added to stream water was converted to particulate form, the particle fraction in waters containing the microbial cells

(Boethling and Alexander, 1979). This lack of significant C assimilation may be a result of the inability of the organisms to obtain C and energy for biosynthetic purposes at these low concentrations, the immediate use of the C for respiration in order for the cells to maintain their viability (i.e., for maintenance energy), or the rapid decomposition and mineralization of the cells and their constituents.

In contrast, a high percentage of the C in other compounds or in similar compounds in different environments is incorporated and accumulates in the biomass, even at low substrate concentrations. With some bacteria, moreover, the efficiency of incorporation of substrate-C into cells is essentially the same from 43 ng to 100 mg of glucose-C per liter. This constancy is especially surprising at substrate concentrations so low that presumably all the C is being diverted to respiration for the organisms to maintain their viability (maintenance metabolism), although it is possible that bacteria use other organic molecules in their environment for maintenance and not the compound whose biodegradation is being determined (Seto and Alexander, 1985).

The percentage of the substrate that is either mineralized or incorporated depends on the species carrying out the transformation, the identity of the substrate, its concentration, temperature, and probably other environmental factors. In one investigation, only 15% of NTA was found to be mineralized in lake water at 100 ng/liter, but the value was more than 90% at 10 μg or 1.0 mg/liter; however, the values were 59, 78, and 12% for IPC at 400 ng, 10 μg, and 1.0 mg/liter, respectively, so it is not possible to generalize that the percentage of mineralization increases or decreases with increasing concentration (Hoover *et al.*, 1986). An effect of concentration on the percentage mineralized is also evident in soil (Sielicki *et al.*, 1978). However, the percentage of substrate mineralized by some bacteria may not change over enormous ranges of substrate concentration (Seto and Alexander, 1985). Temperature also affects the percentage of substrate-C that is incorporated into biomass or mineralized by sediment microfloras and cultures of individual bacterial species (Tison and Pope, 1980).

In natural communities, the cells that grow on the chemical of interest themselves are decomposed by other species, or the cells are grazed upon and the C respired as CO_2 by protozoa or other predators. Hence, the percentage of substrate-C incorporated into the biomass of natural communities declines and the percentage mineralized increases with time, at least in the presence of O_2, and the values initially reflect the populations active on the organic compound but, with time, reflect the activities of the community of microorganisms. Thus, patterns of mineralization have a characteristic initial phase that, to a significant degree, represents the species acting on the parent molecule. Thereafter, a slower phase of mineralization is evident as the original cells, as well as their excretions, are destroyed and converted to CO_2 and other products.

In soil and undoubtedly other environments, a small or large part of the substrate-C is also converted to high-molecular-weight complexes that are resistant to rapid biodegradation. Such humic substances may contain much of the C originally added to that environment, and this organic matter is only very slowly converted to CO_2 (Stott *et al.,* 1983).

ASSIMILATION OF OTHER ELEMENTS

Synthetic molecules may be used as sources of required elements other than C. Microorganisms need N, P, S, and a variety of other elements, and these nutrient requirements may be satisfied as the responsible species degrade the compound of interest. It is common for the element that is in organic complex to be converted to the inorganic form before it becomes incorporated into cell components. For example, *Klebsiella pneumoniae* uses bromoxynil as a N source, but it does so only after converting the nitrile to NH_3, which is then assimilated (McBride *et al.,* 1986). Similarly, a strain of *Pseudomonas* that uses 2,6-dinitrophenol as a N source for growth first cleaves the nitro groups to free nitrite that, presumably after reduction of the nitrite to NH_3, sustains multiplication of the bacteria (Bruhn *et al.,* 1987). Bacteria are also able to use a large number of organophosphorus insecticides (Rosenberg and Alexander, 1979), alkyl phosphates and phosphonates (Cook *et al.,* 1978), and the herbicide glyphosate (Balthazor and Hallas, 1986) as P sources. Sulfur may also be extracted from organic molecules and then support multiplication, as shown by the use of *O,O*-diethylphosphorothioate and *O,O*-diethylphosphorodithioate as S sources by *Pseudomonas acidovorans* (Cook *et al.,* 1980). Although organic substrates may contain more than one of the elements needed for growth, the organism frequently is able to use the chemical as a source of only one of its constituent elements.

For heterotrophic microorganisms in most natural ecosystems, the limiting element is generally C, and usually sufficient N, P, S, and other nutrient elements are present to satisfy the microbial demand. Because C is limiting and because it is the element for which there is intense competition, a species with the unique ability to grow on synthetic molecules has a selective advantage. No such selective advantage exists for an organism using an organic compound as the source of an element that already is available in abundant supply. Hence, it is unlikely that microorganisms obtaining other nutrient elements from synthetic molecules are selectively enhanced in such environments. Nevertheless, as the organisms use the molecules as C or energy sources, the biodegradative process usually will still lead to the mineralization of the other elements in the chemical.

REFERENCES

Bailey, R. E., Gonsior, S. J., and Rhinehart, W. L., *Environ. Sci. Technol.* **17,** 617–621 (1983).
Balthazor, T. M., and Hallas, L. E., *Appl. Environ. Microbiol.* **51,** 432–434 (1986).

Boethling, R. S., and Alexander, M., *Appl. Environ. Microbiol.* **37,** 1211–1216 (1979).
Bruhn, C., Lenke H., and Knackmuss, H.-J., *Appl. Environ. Microbiol.* **53,** 208–210 (1987).
Cacciari, I., Lippi, D., Ippoliti, S., and Pietrosanti, W., *Can. J. Microbiol.* **29,** 1136–1140 (1983).
Chesney, R. H., Sollitti, P., and Rubin, H. E., *Appl. Environ. Microbiol.* **49,** 15–18 (1985).
Clark, C. G., and Wright, S. J. L., *Soil Biol. Biochem.* **2,** 19–26 (1970).
Cook, A. M., Daughton, C. G., and Alexander, M., *Appl Environ. Microbiol.* **36,** 668–672 (1978).
Cook, A. M., Daughton, C. G., and Alexander, M., *Appl Environ. Microbiol.* **39,** 463–465 (1980).
DiGeronimo, M. J., Nikaido, N., and Alexander, M., *Appl. Environ. Microbiol.* **37,** 619–625 (1979).
Farrand, S. G., Jones, C. W., Linton, J. D., and Stephenson, R. J., *Arch. Microbiol.* **135,** 276–283 (1983).
Helweg, A., *Weed Res.* **15,** 53–58 (1975).
Hoover, D. G., Borgonovi, G. E., Jones, S. H., and Alexander, M., *Appl. Environ. Microbiol.* **51,** 226–232 (1986).
Iwata, Y., Ittig, M., and Gunther, F. A., *Arch. Environ. Contam. Toxicol.* **6,** 1–12 (1977).
Johnson, M. J., *Science* **155,** 1515–1519 (1967).
Kassim, G., Martin, J. P., and Haider, K., *Soil Sci. Soc. Am. J.* **45,** 1106–1112 (1981).
Kunc, F., and Rybarova, J., *Soil Biol. Biochem.* **15,** 141–144 (1983).
Martin, J. P., and Haider, K., *Soil Sci. Soc. Am. J.* **43,** 917–920 (1979).
McBride, K. E., Kenny, J. W., and Stalker, D. M., *Appl. Environ. Microbiol.* **52,** 325–330 (1986).
Rosenberg, A., and Alexander, M., *Appl. Environ. Microbiol.* **37,** 886–891 (1979).
Scow, K. M., Simkins, S., and Alexander, M., *Appl. Environ. Microbiol.* **51,** 1028–1035 (1986).
Seto, M., and Alexander, M., *Appl. Environ. Microbiol.* **50,** 1132–1136 (1985).
Sielicki, M., Focht, D. D., and Martin, J. P., *Appl. Environ. Microbiol.* **35,** 124–128 (1978).
Sikka, H. C., and Saxena, J., *J. Agric. Food Chem.* **21,** 402–406 (1973).
Stevenson, I. L., and Ivarson, K. C., *Can. J. Microbiol.* **10,** 139–142 (1964).
Stott, D. E., Martin, J. P., Focht, D. D., and Haider, K., *Soil Sci. Soc. Am. J.* **47,** 66–70 (1983).
Subba-Rao, R. V., Rubin, H. E., and Alexander, M., *Appl. Environ. Microbiol.* **43,** 1139–1150 (1982).
Sugatt, R. H., O'Grady, D. P., Banerjee, S., Howard, P. H., and Gledhill, W. E., *Appl. Environ. Microbiol.* **47,** 601–606 (1984).
Tison, D. L., and Pope, D. H., *Appl. Environ. Microbiol.* **39,** 584–587 (1980).
Wiggins, B. A., Jones, S. H., and Alexander, M., *Appl. Environ. Microbiol.* **53,** 791–796 (1987).

CHAPTER 3

Acclimation

Prior to the degradation of many organic compounds, a period is noted in which no destruction of the chemical is evident. This time interval is designated an *acclimation period* or, sometimes, an adaptation or lag period. It may be defined as the length of time between the addition or entry of the chemical into an environment and evidence of its detectable loss. During this interval, no change in concentration is noted, but then the disappearance becomes evident and the rate of destruction often becomes rapid (Fig. 3.1).

This acclimation phase may be of considerable public health or ecological significance because the chemical is not destroyed. Hence, the period of exposure of humans, animals, and plants is prolonged, and the possibility of an undesirable effect is increased. Furthermore, if the chemical is present in flowing waters above or below ground, it may be widely disseminated laterally or vertically because of the absence of detectable biodegradation. In the case of toxicants, such increased dispersal may result in the exposure of susceptible species at distant sites before the harmful substance is destroyed.

Acclimation periods have been reported for many compounds that are introduced into soil, fresh water, sediment, and sewage. Among the chemicals for which such a phase has been described, either aerobically or anaerobically, are the following.

(a) Herbicides: 2,4-D, MCPA, mecoprop, 4-(2,4-DB), TCA, amitrole, dalapon, monuron, chlorpropham, IPC, napropamide, endothal, pyrazon, and DNOC.

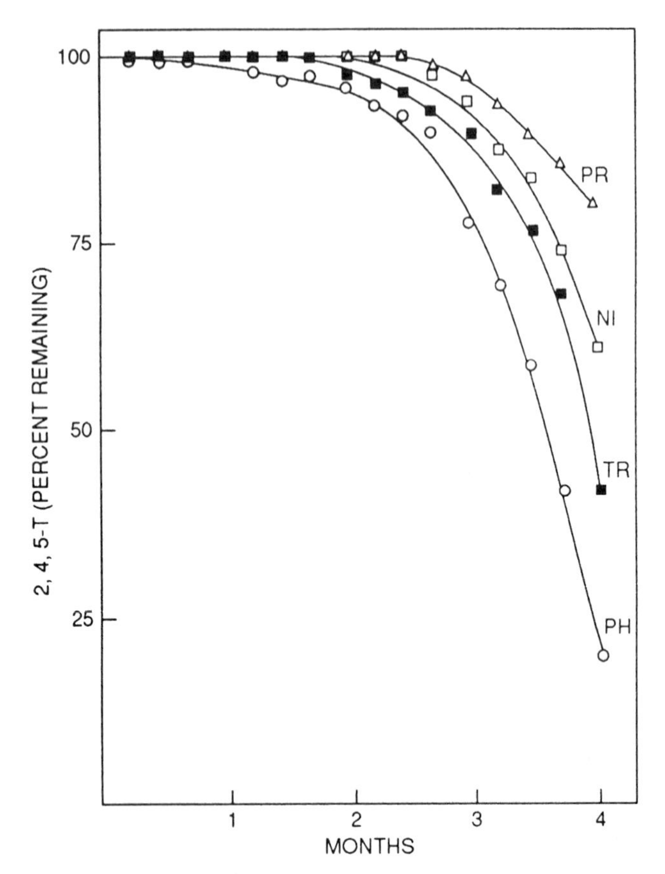

Figure 3.1 Disappearance of 2,4,5-T in soil from the Philippines (PH), Trinidad (TR), Nigeria (NI), and Puerto Rico (PR). (From Rosenberg and Alexander, 1980. Reprinted with permission from the American Chemical Society.)

(b) Insecticides: methyl parathion and azinphosmethyl.
(c) Fungicides: vinclozolin, and iprodione.
(d) Quarternary ammonium compounds: dodecyltrimethylammonium chloride.
(e) Polycyclic aromatic hydrocarbons: naphthalene and anthracene.
(f) Others: phenol, 4-chlorophenol, 4-nitrophenol, chlorobenzene, 1,2- and 1,4-dichlorobenzene, 3,5-dichlorobenzoic acid, PCP, diphenylmethane, and NTA.

The length of the acclimation period varies enormously. It may be less than 1 h or many months. The duration varies among chemicals and environments, and it also depends on the concentration of the compound and a number of environmental conditions. Some typical values are given in Table 3.1. The values shown are not fixed, and the acclimation period for any one of the chemicals may be longer or shorter than the times shown, depending on the concentration, the environment, the temperature, the aeration status, and other, often undefined factors. The time period may be especially long in anaerobic environments for some compounds, such as chlorinated molecules (Linkfield *et al.*, 1989). Especially disturbing is the inability to predict accurately the duration of the acclimation phase for most chemicals in nearly all environments.

The acclimation phase is considered to end at the onset of the period of detectable biodegradation. After the acclimation, the rate of metabolism of the chemical may be slow or rapid, but if a second addition of the chemical is made during this time of active metabolism, the loss of the second increment characteristically occurs with little or no acclimation (Fig. 3.2). The disappearance or marked reduction in the acclimation period has been noted in soils amended with napro-

Table 3.1

Lengths of Acclimation Phases for Several Organic Compounds

Chemical	Environment	Length of acclimation phase	Reference
Several aromatics	Soil	10–30 h	Kunc and Macura (1966)
Dodecyltrimethyl-ammonium chloride	Fresh water	24 h	Ventullo and Larson (1986)
4-Nitrophenol	Water-sediment	40–80 h	Spain and Van Veld (1983)
IPC	Soil	20 days	Robertson and Alexander (1994)
Amitrole	Soil	7 days	Riepma (1962)
Chlorinated benzenes	Biofilm	10 days–5 months	Bouwer and McCarty (1984)
DNOC	Soil	16 days	Hurle and Rademacher (1970)
PCP	Stream water	21–35 days	Pignatello *et al.* (1986)
Mecoprop	Enrichments	30–37 days	Lappin-Scott *et al.* (1986)
NTA	Estuary	50 days	Pfaender *et al.* (1985)
Halobenzoates	Sediment (anaerobic)	3 weeks–6 months	Linkfield *et al.* (1989)
2,4,5-T	Soil	4–10 weeks	Rosenberg and Alexander (1980)
Several	Groundwater	>16 weeks	Wilson *et al.* (1986)

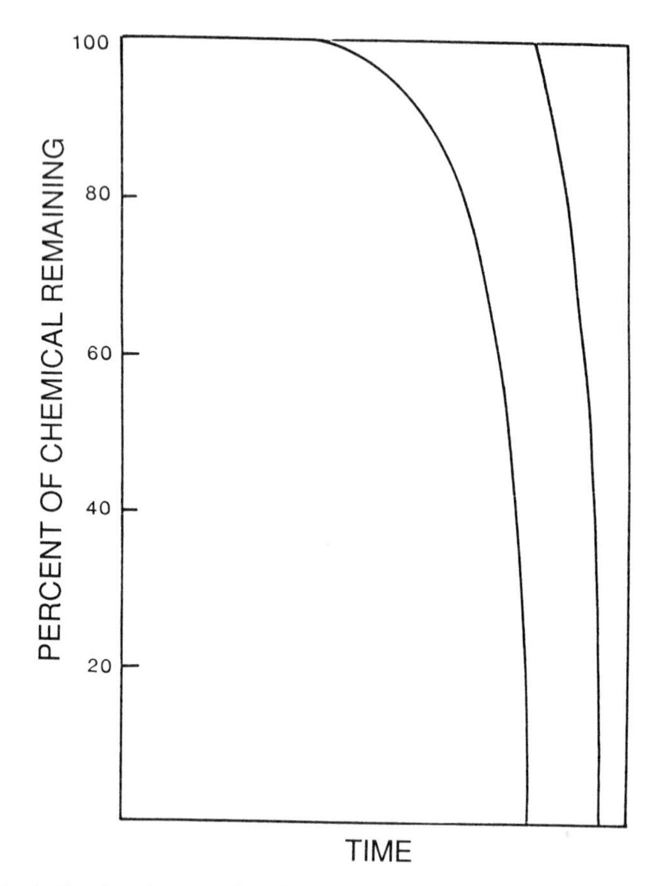

Figure 3.2 Acclimation phase preceding the microbial destruction of an organic chemical after its first addition, and the absence of a delay period and rapid metabolism of the chemical following its second addition.

pamide (Walker *et al.*, 1993), IPC (Robertson and Alexander, 1994), 2,4-D (Audus, 1949), DNOC (Hurle and Pfefferkorn, 1972), amitrole (Riepma, 1962), methomyl (Fung and Uren, 1977), and CIPC (Kaufman and Kearney, 1965), soil suspensions amended with 4-(2,4-DB) (Whiteside and Alexander, 1960), river water supplemented with 4-nitrophenol (Spain *et al.*, 1980), and marine waters containing 4-chlorophenol (Kuiper and Hanstveit, 1984). It is generally assumed that biodegradation is detected immediately following the second introduction of the chemical because the organisms responsible for the transformation became numerous as they grew on the organic chemical following its first introduction.

The rate of biodegradation of the second addition may be the same as the final rate evident during the active phase of breakdown of the first addition

(Kaufman and Kearney, 1965). However, it is far more common to have a greater rate of biodegradation, which is usually measured as the loss of parent compound or the formation of $^{14}CO_2$ form labeled compound, following the second than after the first application. The rate is further enhanced with still more additions. This enhancement of rate upon repeated additions of chemical has been reported frequently for pesticides added to soil. For example, the rate of parathion loss and its conversion to CO_2 rises as soil receives additional monthly treatments with the insecticide (de Andrea *et al.*, 1982). The degradation of iprodione and vinclozolin similarly becomes more rapid as a result of prior additions of these fungicides to soil (Walker *et al.*, 1986). In soil to which EPTC or butylate is applied, the rate of mineralization increases as a result of prior treatments with these herbicides (Obrigawitch *et al.*, 1982, 1983). Greater rates of disappearance of the nematicide enthoprop and diphenamid are evident following the second than after the first introduction into soil (Kaufman *et al.*, 1985). Not all of these instances show an acclimation phase, but such acclimations are common prior to the period of rapid pesticide breakdown in soil. Similar changes occur in water. Thus, dodecyltrimethylammonium chloride is rapidly mineralized in fresh water after an acclimation period, and the rate is faster following the second than after the first addition of the quaternary ammonium compound (Ventullo and Larson, 1986).

The greater rate on subsequent additions probably results from increases in numbers of degrading organisms following repeated treatment with the chemical. Consider a simple illustration. Assume that a chemical is added repeatedly at a concentration of 1.0 μg per unit volume, that its mineralization is a result of bacterial action, and that each bacterial cell that is formed destroys 1 pg (10^{-12} g) of the organic molecule. A considerable amount of time would elapse until 1.0×10^6 cells appear to destroy the first increment. However, far less time would be required for those 1.0×10^6 cells to destroy the next 1.0 μg, but the population size would then grow to 2.0×10^6. Still less time would elapse for those 2.0×10^6 bacteria to destroy the following 1.0 μg, but they would then grow to give 3.0×10^6 cells, etc. The example is an oversimplification in many ways, one of which is that not all cells produced would survive because they might die or be destroyed by protozoa or other predators before the next addition is made, but it does illustrate the greater rate to be expected of the larger populations that may be produced with successive introductions of a C source they use for growth.

Once the indigenous community of microorganisms has become acclimated to the degradation of a chemical and the activity becomes marked, the community may retain its active state for some time, that is, the potential for activity may continue to remain higher than in comparable soils, waters, or sewage-treatment systems that have not acquired this capacity. For example, 2-4-D disappears more quickly from soils treated a year previously with the herbicide than from untreated soils (Newman and Thomas, 1949), and the rate of degradation of isofenphos is also more rapid in soil treated with this insecticide a year earlier than in untreated soil (Chapman *et al.*, 1986b). An effect of pretreatment with TCA is evident even after about 3 years (McGrath, 1976), and the microorganisms in soil can sometimes metabolize 2,4-D for a period of at least 4 years after its last addition (Smith and Aubin, 1994). Conversely, the effect of prior treatment to acclimate the microbial community may be short-lived, as indicated by the loss of the higher activity on 4-nitrophenol after 7 weeks in a fresh water-sediment mixture (Spain and Van Veld, 1983). Too little information is presently available to permit generalizations among chemicals on the duration of the beneficial influence of prior additions of the compound. It is not presently clear why a microbial community that has acclimated to a particular substrate loses that activity; it could result from a decline in numbers or biomass of the responsible microorganisms or a loss of the metabolic activity in the absence of the specific chemical.

FACTORS AFFECTING ACCLIMATION

Acclimation of a microbial community to one substrate frequently results in the simultaneous acclimation to some, but not all, structurally related molecules. Because individual species often act on several structurally similar substrates, the species favored by the first addition may then quickly destroy the analogues. For example, when the microbial community of soil becomes acclimated to destroy 2,4-dichloro- or 4-chloro-2-methylphenoxyacetic acid, it simultaneously acquires the capacity for more rapid destruction of the other herbicide (Soulas *et al.*, 1983); if the soil becomes enriched with organisms that bring about the rapid degradation of EPTC, a more rapid degradation of the structurally similar herbicide butylate will occur (Obrigawitch *et al.*, 1983). An analogous effect is evident with two structurally related dicarboximide fungicides (Mitchell and Cain, 1996). Similarly, stimulation of the organisms that metabolize phenol in water, following acclimation, will result in enhanced metabolism of 4-chlorophenol, 3-aminophenol, and *m*-cresol (Shimp and Pfaender, 1987). Analogous simultaneous acclimations for transformation of several polycyclic aromatic hydrocarbons occur in slurries of marine sediments following acclimation to other polycyclics or benzene (Bauer and Capone, 1988).

The length of the acclimation is affected by several environmental factors. Temperature has a major impact on the duration of the period before the active phase, as indicated by the longer interval before the onset of rapid oil biodegradation at lower than at higher temperatures (Atlas and Bartha, 1972). The pH and aeration status of an environment also appear to affect the duration of the acclimation for some compounds. The concentration of N, P, or both may be important in some environments, as in some natural waters, in which their concentrations are so low that they may limit microbial growth (Lewis *et al.*, 1986; Wiggins *et al.*, 1987). Conversely, the acclimation prior to the biodegradation of P- or N-containing organic compounds could be extended because of high levels of P or N if the responsible organisms use the inorganic phosphate or N from the environment in preference to that which would be released as a result of cleavage of the organic molecule (Daughton *et al.,* 1979).

The concentration of the compound that is being metabolized greatly affects the length of time before one can detect a decline in its concentration. The rate of biodegradation of chemicals at trace levels increases with concentration, but because chemical loss is usually determined and not CO_2 or product formation, the low precision of analyses leads to data indicating a longer acclimation at higher concentration. Thus, based on measurements of loss of the test chemical, the acclimation period lengthens as the concentration of mecoprop in soil increases from 1 to 40 mg/kg (Amrein *et al.,* 1982), the level of picloram rises from 0.25 to 1.0 mg/kg (Grover, 1967), and the concentration of 4-nitrophenol in sewage effluent increases from 1 to 25 mg/liter (Nyholm *et al.,* 1984). However, the length of the acclimation period before degradation of 4-nitrophenol by *Pseudomonas putida* was unaffected by concentrations of the chemical ranging from 15 to 5000 μg/liter (Nishino and Spain, 1993).

The duration of the acclimation period is not fixed even at a single concentration but varies from site to site, and some microbial communities acclimate to a particular chemical whereas others do not. Such variation in the occurrence of acclimation has been noted for 4-nitrophenol added to fresh and marine waters (Spain and Van Veld, 1983) and IPC and 2,4-D added to lake waters (Hoover *et al.,* 1986).

As stated earlier, many chemicals are rapidly degraded only after an acclimation period, and the second addition is destroyed more readily than the first. However, the destruction of some compounds by microorganisms does not show such patterns. For example, NTA mineralization in estuarine waters is occasionally slow and does not become more rapid with time (Pfaender *et al.,* 1985). Similarly, although many thiocarbamate herbicides are destroyed more readily following their second than after their first addition to soil, not all thiocarbamates behave in this fashion (Gray and Joo, 1985). Other herbicides also are not destroyed at rates that become faster following two or more applications to soil, for example, monolinuron and simazine (Paeschke *et al.,* 1978).

There appear to be concentrations of some chemicals below which no acclimation occurs. A typical case is 4-nitrophenol, which is destroyed in samples containing sediments and natural waters at concentrations above but not below 10 μg/liter (Spain and Van Veld, 1983). Similarly, second applications of 2,4-D to soil are mineralized more rapidly than the first if the two additions are at 3.3 or 33 but not at 0.33 mg/kg (Fournier *et al.,* 1981), and analogous enhanced decomposition of carbofuran occurs in soil treated with 1.0 or 10 mg/kg but not at 0.01 or 0.1 mg/kg (Chapman *et al.,* 1986a). However, microorganisms in fresh or marine waters may acclimate to destroy compounds at levels below which they can use single compounds as sole carbon sources for growth (i.e., below the threshold), for example, at 1.0 μg of IPC per liter (Hoover *et al.*, 1986), below 2 μg/liter for dodecyltrimethylammonium chloride in stream water (Shimp *et al.*, 1989), and 0.7 μg/liter for toluene in seawater (Button and Robertson, 1985).

Acclimation for the metabolism of a particular compound may be needed in some environments but not in others. Thus, the mineralization of 1 to 50 μg of 4-nitrophenol, 4 μg of 2,4-dichlorophenol, and 20 μg of NTA per kilogram in soil may proceed with little or no acclimation period (Scow *et al.*, 1986), but acclimations are characteristic of the decomposition of such chemicals at higher concentrations or in other soils.

ACCELERATED PESTICIDE BIODEGRADATION

Farmers commonly grow the same crops in particular areas either continuously or at regular intervals in a crop-rotation sequence. This often results in the reappearance of the same pests each time the individual crop is grown. These pests may be insects, weeds, or plant pathogens. To reduce the severity or prevent the occurrence of large pest populations, the farmer, each year or several times in a single growing season, typically applies pesticides—insecticides for insects, herbicides for weeds, and fungicides for many plant pathogens. Many of the pesticides are applied to the soil, and they may be added before planting and subsequent to emergence of the crop plants (preemergent pesticides) or after the plants emerge from the soil (postemergent pesticides). If the pest-control chemical is added before planting, it must persist sufficiently long to be present at concentrations high enough to control the insects or plant pathogens that harm the plants that later appear above ground or to prevent growth of the weeds that appear some time after the pesticide is applied. Obviously, no soil-applied chemical would be used unless it persisted for the requisite time.

With some useful pesticides, however, a change in their persistence and their consequent suppression of pests occurs with the passage of time. In the years immediately after the introduction and widespread use of these compounds, the control of harmful insects, weeds, and plant pathogens is adequate. Therefore, farmers continue to apply the chemicals as part of their usual field operations. With repeated use of the pesticides, however, the pests unexpectedly are no longer controlled and cause a marked reduction in crop yield. The pesticides appear to be losing effectiveness with time. Sometimes the inability of the original product to continue to control pests is a result of species acquiring resistance to the pesticides. In other cases, the loss of effectiveness is directly attributable to the more rapid degradation of the chemical as a result of its use season after season. This problem, often termed accelerated pesticide degradation or enhanced microbial degradation, is a consequence of a change in the length of the acclimation phase, the rate of biodegradation, or both as a direct consequence of the repeated use. The problem was initially recognized because of the loss of control by carbofuran of the rootworm that affects corn, whose eggs hatch 3 to 5 weeks after planting (Felsot *et al.*, 1981), and the declining effectiveness of EPTC applied as a preemergence herbicide to control weeds growing in corn fields (Kaufman *et al.*, 1985). Enhanced microbial degradation of carbofuran may also result in the inability to control other insects, including those affecting a variety of crops (Wilde and Mize, 1984). The accelerated degradation of EPTC has a spotty distribution, being a problem in some fields but not in nearby locations (Roeth *et al.*, 1989).

A major practical impact of the inability to control certain insects with carbofuran and many weeds with EPTC has been to stimulate considerable research on their enhanced degradation. Loss of insect control by carbofuran is noted if the compound has been used for the previous 2 to 4 years (Felsot *et al.*, 1981). In laboratory studies, even a single application of EPTC to soil leads to its more rapid mineralization than in soil not previously treated with the herbicide. In this instance, it appears that the differences are in the rate of mineralization and not the lengths of the acclimation phase associated with the two treatments (Obrigawitch *et al.*, 1982).

Apart from the practical implications, which are particularly pronounced only with a few pesticides, the phenomenon of accelerated pesticide degradation is characteristic of many pest-control agents, It was early recognized for such herbicides as 2,4-D, MCPA (Torstensson *et al.*, 1975), and DNOC (Hurle and Pfefferkorn, 1972), but the list of affected compounds includes the herbicides napropamide (Walker *et al.*, 1993), IPC (Robertson and Alexander, 1994), TCA (McGrath, 1976), vernolate, and butylate (Gray and Joo, 1985), the insecticides chlorfenvinphos (Hommes and Pestemer, (1985), aldicarb (Bromilow *et al.*, 1996), and fensulfothion (Read, 1983), the fungicides vinclozolin and iprodione (Mitchell and Cain, 1996), and a variety of other important pesticides (Gray and Joo, 1985; Avidov *et al.*, 1988; Slade *et al.*, 1992). Moreover, a soil treated with one pesticide may show enhanced biodegradation and thus shorter periods of effectiveness of structurally related chemicals, for example, among the thiocarbamate herbicides EPTC, butylate, and vernolate (Wilson, 1984).

The enhanced degradation is the result of microbial action on many if not all of these compounds, as evidenced by findings that little or no mineralization or loss of pesticidal activity or of the chemical itself occurred in sterile soil, at least with those substances tested. Not all pesticides are subject to accelerated transformation following their regular use for control of insects, weeds, or plant pathogens, for example, atrazine, simazine, isoproturon, and chlorpyrifos (Racke *et al.*, 1990; Walker and Welch, 1991).

The enhanced or accelerated pesticide biodegradation is not surprising. Considerable research on the acclimation phase and on enhanced rates of transformation following repeated treatments had been undertaken. What is surprising is the many years that elapsed before it was reported to be a major practical problem. The reason for the enhancement may be an increase, following the first addition, in population or biomass of the microorganisms able to degrade the compound and use it as a source of C and energy; if the population or biomass is still large when the soil again receives the pesticide, the disappearance will then occur without an acclimation period. If the population size has diminished by the time of the second addition or if the compound is metabolized but does not serve as a C source, no acceleration should be evident (Robertson and Alexander, 1994). However, evidence exists that the accelerated degradation of EPTC and carbofuran in soil is not attributable to increased numbers of microorganisms but rather to a greater activity per cell (Moorman, 1988; Scow *et al.*, 1990).

To enhance the persistence of the chemicals and thus to allow for the control of pests that would no longer be suppressed, another chemical may be added together with the pest-control agent. The second compound, sometimes called an extender because it extends the life of the pesticide, acts by inhibiting the biodegrading populations. Extenders to increase the duration of effectiveness of EPTC and butylate are fonofos and dietholate (Rudyanski *et al.*, 1987).

EXPLANATIONS FOR THE ACCLIMATION PHASE

Many explanations have been proposed for the acclimation of microbial communities to the biodegradation of organic compounds in natural waters, soils, or wastewaters. Many of these were proposed based on early studies of pure cultures of bacteria growing in media containing single organic substrates, often at cell densities far higher than is common for individual species of bacteria in nature. Some were based on investigations of the biochemistry or genetics of individual species acting in pure culture on very high concentrations of sugars, amino acids, or other natural products that can be metabolized by a diverse array of microbial species. Few of the explanations, however, were derived from studies of natural microbial communities acting on synthetic compounds at environmentally relevant concentrations, and hence the original emphasis placed on certain of these hypotheses must be considered with skepticism. However, more recent studies have been designed to evaluate these hypotheses as they relate to natural communities as contrasted to pure cultures, to cell densities more characteristic of natural ecosystems than those bacterial densities commonly used in tests of pure cultures, to synthetic compounds acted on by only a few rather than a diversity of microbial genera or species, and to concentrations that are characteristic of environmental pollutants rather than of organic nutrients included in culture media.

These inquiries have often led not to the rejection of all the earlier hypotheses but rather to the establishment of somewhat different mechanisms as the most common causes of the acclimation. These explanations are related to (a) proliferation of small populations; (b) presence of toxins; (c) predation by protozoa; (d) appearance of new genotypes; and (e) diauxie.

Proliferation of Small Populations

Soils, natural waters, sewage, and wastewaters typically contain small populations of microorganisms acting on many of the synthetic organic compounds that are capable of supporting growth. The population is so small that one, two, three, or several more doublings of the cell number would not bring about an appreciable loss of the chemical. For example, if the initial density of bacteria is 10^2 cells per unit volume and each bacterium, as it divides, destroys 1.0 pg of organic substrate, a decline in concentration of a compound initially present at 0.1 μg per unit volume would not be detected initially even as the bacteria grow (Fig. 3.3). There would be an apparent acclimation phase simply because the precision of analyses would not detect the loss of the 100 pg destroyed by 10^2 cells (0.1 μg or 10^5 pg less the 10^2 pg metabolized) or 10^3 cells (10^5 pg less the 10^3 pg metabolized). Only when the bacteria have undergone many cell divisions would a decline in concentration of parent chemical be detected. From Fig. 3.3, it is evident that even longer apparent acclimation periods would be evident as a population of such initial size is exposed to concentrations of 10 μg, 1 mg, and 100 mg of organic chemical per unit volume; bacteria growing exponentially must reach even greater abundance, and hence more time elapses, before a decline in concentration is evident.

In the circumstances in which this mechanism applies, the acclimation would not appear to be as long, or would not even exist, if analyses for product formation were performed (rather than loss of substrate). Thus, if one could detect 100 pg of product (e..g, $^{14}CO_2$ from a ^{14}C-labeled substrate), then the activity of 100 cells (each of which might form ca. 1 pg of CO_2) might be detected, although the loss of 100 or 1000 pg of substrate might not be measurable when the initial substrate concentration is 10^5 pg per unit volume. From Fig. 3.3, it would seem that the amount of substrate-C oxidized (equivalent to product-C generated) would be detectable by *sensitive* methods long before the loss of parent compound by *precise* techniques would be noted.

It is not presently clear how often acclimation, which in this case is more apparent than real, results simply from the time required for a small population to become sufficiently large to give a detectable loss of the organic substrate. Certainly many investigations have suggested its importance without providing supporting data. Yet some research has in fact verified that the numbers of bacteria growing at the expense of specific organic chemicals rise as those compounds are degraded, for example, dodecyltrimethylammonium chloride and phenol in aquatic environments (Ventullo and Larson, 1986; Shimp and Pfaender, 1987), 2,4-D in lake

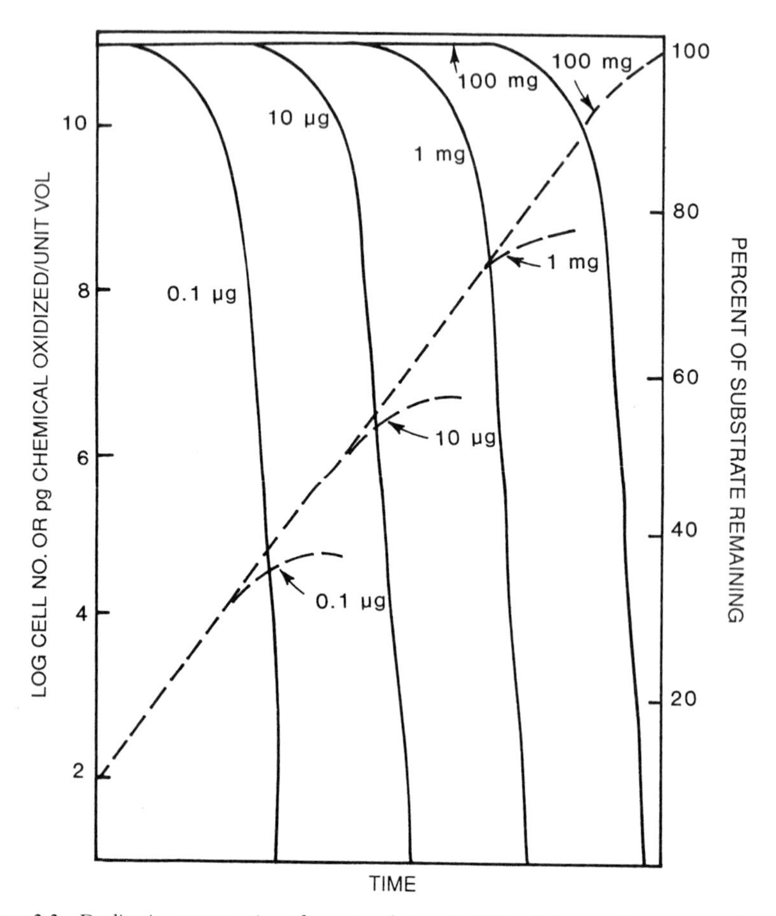

Figure 3.3 Decline in concentration of organic substrate (solid lines), increase in amount of chemical oxidized (or product formed) on a logarithmic scale (dashed lines), and increase in logarithm of cell density (also dashed lines) per unit volume. The estimates shown are based on the assumptions that bacteria are growing exponentially by using the substrate as the sole C source, that each cell consumes 1 pg of substrate and forms ca. 1 pg of CO_2 as it multiplies, and that the rate of growth does not change markedly as the substrate concentration declines.

water (Chen and Alexander, 1989), 4-nitrophenol in natural water plus sediment (Spain *et al.*, 1980), and isofenphos in soil (Racke and Coats, 1987). In lake water amended with 2.0 mg of 4-nitrophenol per liter, for example, the number of cells able to metabolize the compound increases shortly after the addition, but loss of the chemical can only be detected at about 8 days, at which time the density of cells acting on 4-nitrophenol has reached ca. 10^5 per milliliter (Fig. 2.1). In many of these studies, the cell counts are less than those that might be expected based on the initial chemical concentration, but many of the cells that are formed undoubtedly are dying continuously or are consumed by predators such as protozoa. If a metabolized compound that does not support microbial growth has an acclimation phase, this explanation obviously does not apply.

In instances in which acclimation is solely a reflection of the time for the population size to become large enough to effect a detectable change, any factor that enhances (or diminishes) the growth rate would shorten (or lengthen) the acclimation. Thus, in environments in which the concentrations of N, P, or possibly other inorganic nutrients are low, the acclimation phase may be longer than in similar environments having higher levels, and the addition of N to N-poor environments and P to P-deficient environments may shorten the acclimation phase. Such effects have been noted during the degradation of *p*-cresol by a mixture of aquatic microorganisms (Lewis *et al.,* 1986) and during the mineralization of 4-nitrophenol in sewage and lake water (Wiggins *et al.,* 1987; Jones and Alexander, 1988).

From an examination of Fig. 3.3, it is evident why the acclimation may appear to be longer at higher substrate concentrations: more time is required for the cell density to become large enough to give a detectable loss at higher than at lower substrate concentrations. Other reasons may explain the longer acclimation at high than at low concentrations (e.g., toxicity at the higher levels), but difference in time to give the necessary cell number is surely one likely cause. From Fig. 3.3 it is also apparent why second additions of a chemical may often be destroyed with little detectable delay: the population became sufficiently large as a result of the first addition. However, if appreciable time elapsed between the day all the chemical was destroyed and the day when an additional increment is introduced, an acclimation may be evident; many of the cells may have died or been consumed by predators or possibly parasites once the unique C supply for that population was exhausted.

The phenomenon of accelerated pesticide degradation may have a similar explanation, as indicated earlier. The population size rises following the first treatment of soil with the pesticide. If the cell numbers or biomass remains large from season to season or between pesticide applications in a single season, the second increment disappears more readily than the first. If the population declines to about its original size before the next addition, a long acclimation will again be evident. Because many pesticides are added annually to soil, accelerated pesticide degradation may become evident with time if the repeated annual use of the chemical results, at the start of each growing season, in a population size that is larger than that at the start of the previous growing season (even if many cells die after all the chemical is metabolized); this ever larger number of cells would result in an even shorter acclimation phase.

Presence of Toxins

Two circumstances are common in which the presence of inhibitors may affect the length of the period prior to the onset of rapid biodegradation. First, the chemical of interest may be present at such high levels that few of the biodegrading microorganisms present may be able to grow or metabolize, and biodegradation will not be detectable until the rare species, either present or transported to the site, is able to multiply to reach a biomass sufficient to cause appreciable chemical loss. Such high concentrations are known in waste discharges from chemical manufacturing or in accidental spills. Second, many sites containing toxic chemicals have a mixture of compounds, and one or more of the mixture may be inhibitory to the organisms destroying the test substance. When that toxicant disappears by biodegradation, nonenzymatic destruction, sorption, or volatilization, the period in which no destruction of the test substance is evident is replaced by a period in which the destruction becomes marked (Fig. 3.4). The inhibition in many hazardous-waste sites may be complete so that even readily utilizable substrates are not metabolized, but degradation will occur as a pollutant plume in the groundwater moves away from the source and the inhibitors become diluted.

The toxicant may act in several ways. (a) In the first and second circumstances just indicated, it may merely act to slow the growth rate of the degrading species and hence lengthen the period during which no loss of the chemical being measured is evident. Such toxicants may be organic molecules, or they may be inorganic. The role of the latter is shown by the finding that the acclimation phase prior to rapid NTA biodegradation in activated sludge is lengthened if the sludge is rich in heavy metals (Stephenson *et al.,* 1984). (b) The toxicant may be eliminated so that the degraders are able to proliferate, and the acclimation period then represents the sum of the time for lowering the level of the antimicrobial agent to a noninhibitory concentration plus the subsequent time for multiplication of the degrading species to a density adequate for significant chemical loss. For example, in wastewater containing both 4-nitrophenol and 2,4-dinitrophenol, the acclimation phase for the mineralization of the first compound resulted from the toxicity of the second to the 4-nitrophenol-degrading species, but when the dinitro compound was biodegraded, the 4-nitrophenol utilizers proliferated and the acclimation phase soon ended (Wiggins and Alexander, 1988b). Evidence that part of the acclimation period for oil biodegradation reflects the time for volatilization of its antimicrobial constituents has been obtained by Atlas and Bartha (1972). Conversely, the acclimation period for 4-nitrophenol mineralization in the presence of high phenol concentrations is a result of the antimicrobial effect of the second chemical, but this period is markedly shortened as phenol is mineralized (Murakami and Alexander, 1989). (c) The toxicant may suppress the faster-growing species that usually predominate in a mixture of species capable of metabolizing the contaminants, but a resistant and slower-growing species will then have a selective advantage that it did not previously have—and the longer acclimation is simply a reflection of the longer time needed for the appearance of the large biomass of the slow-growing organisms. (d) The toxins may not be present initially, but they may be generated during biodegradation. This possibility is suggested by the observation that phenol min-

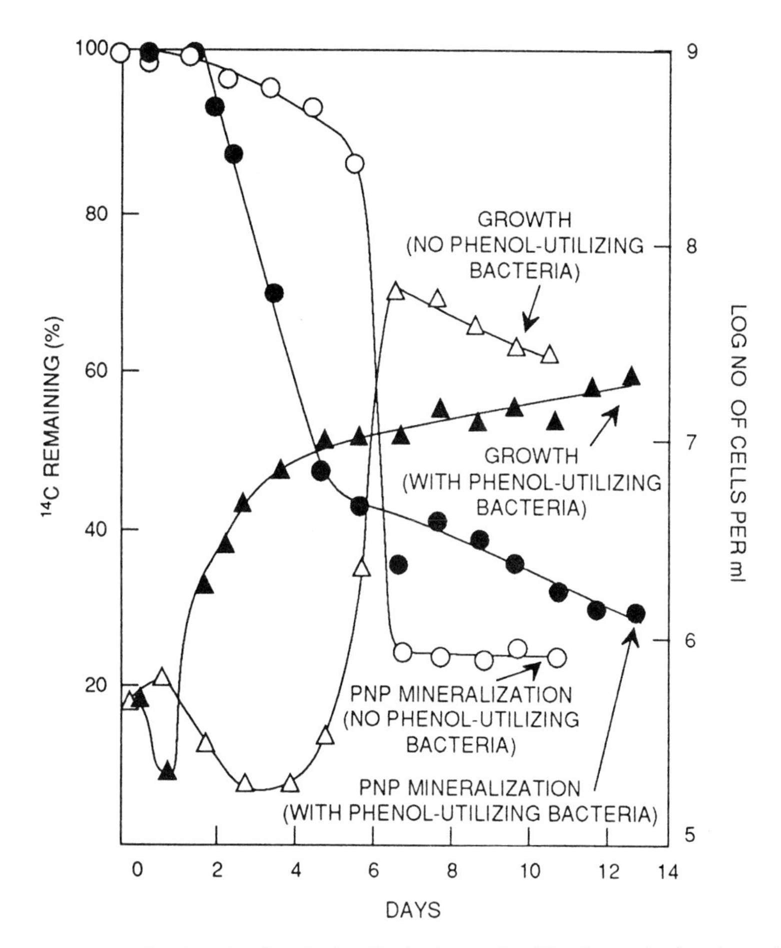

Figure 3.4 Growth and 4-nitrophenol mineralization by a strain of *Pseudomonas* in phenol-containing media in the presence or absence of phenol-degrading bacteria. (From Murakami and Alexander, 1989. Copyright by John Wiley & Sons and reprinted with permission by the copyright holder.)

eralization may be suppressed by products generated microbiologically from 4-nitrophenol that is present together with phenol (Murakami and Alexander, 1989).

Predation by Protozoa

A number of natural ecosystems and aerated waste-treatment systems are characterized by large and active populations of protozoa. These microscopic animals feed and multiply because they prey on the bacteria in these environments.

This feeding reduces the abundance of bacteria if their densities become especially high and probably keeps the bacterial density lower than might otherwise be expected based on the supply of readily utilizable organic nutrients.

Although little attention has been given to the role that protozoa play in governing acclimation, it appears that in sewage, and probably other environments in which bacteria are actively destroying synthetic molecules, these unicellular animals are quite important. The evidence comes from studies in which comparisons were made of the mineralization of 4-nitrophenol in samples of sewage containing protozoa and those in which protozoa were suppressed by additions of inhibitors selective for eucaryotes (Fig. 3.5). Because protozoa are the chief eucaryotes present in many wastewaters, the use of such selective inhibitors largely eliminates protozoan grazing on bacteria. When the predatory activities are thus suppressed, the normally long acclimation interval is markedly shortened; that is, the protozoan consumption of bacteria is directly related to the longer acclimation (Wiggins and Alexander, 1988a). Presumably, the protozoa act to keep the density of bacteria responsible for the degradation so low that no appreciable chemical loss is detectable. With time, the density of total bacteria falls to a low level, because of both this grazing as well as the reduced supply of readily utilizable organic matter, and then the protozoa become less active. As a result, bacteria growing on synthetic compounds proliferate by using the synthetic molecules whose presence gives these

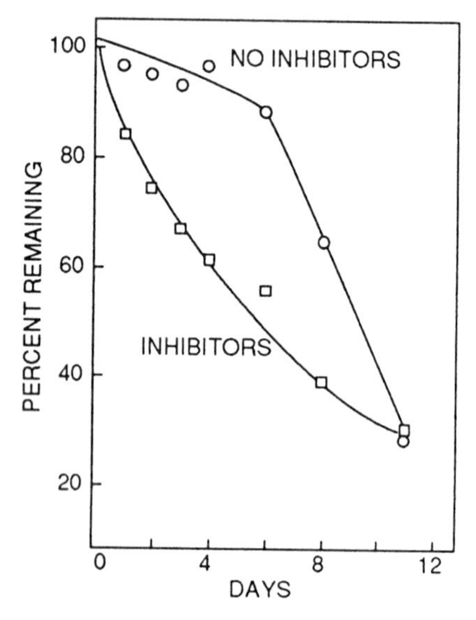

Figure 3.5 Effect of protozoa on the biodegradation of 2 mg of 4-nitrophenol per liter of sewage. Protozoa were suppressed in half of the flasks by addition of the eucaryotic inhibitors cycloheximide and nystatin. (Reprinted with permission from Wiggins and Alexander, 1988a.)

species a selective advantage; at this time, the decline in concentration of the compound becomes evident, and the acclimation phase ends (Fig. 3.6). Acclimation is less likely to be attributable to protozoa in environments in which predation is not marked, such as many natural waters (Wiggins *et al.,* 1987).

Appearance of New Genotypes

Bacteria and fungi may undergo genetic change as a result of a mutant appearing in the population or the transfer of genetic information from one species to another. Such events occur at low frequency, so only a few cells in a population

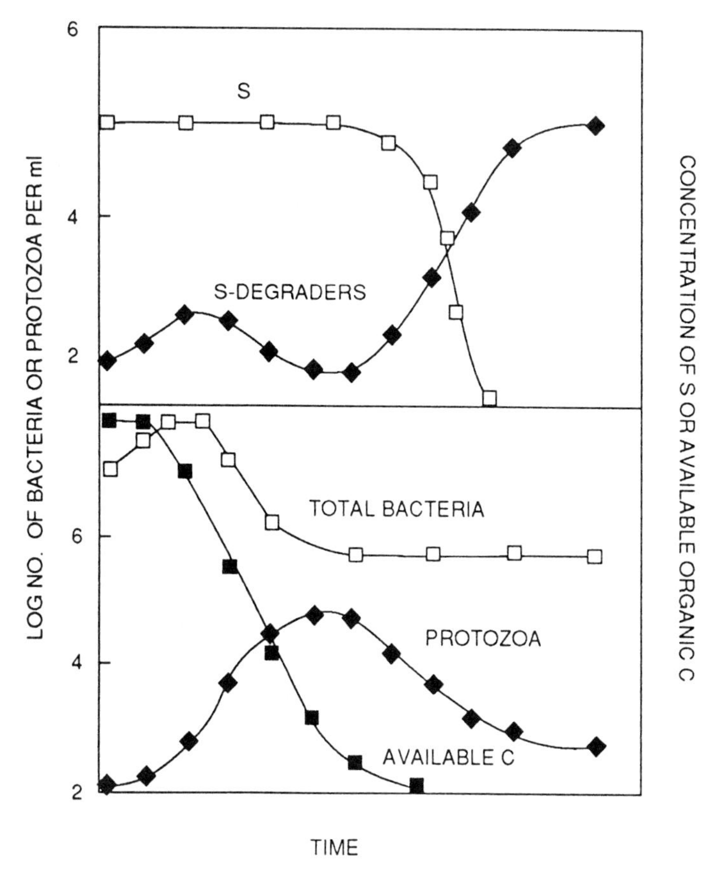

Figure 3.6 Changes in populations of total bacteria, a bacterium degrading a test compound (S), and protozoa and the disappearance of total available C and the test compound in wastewater.

represent a new genotype with a particular set of new phenotypic traits. However, if the new genotype possesses physiological characteristics that give it a selective advantage, it will multiply. The possession of enzymes that degrade a novel substrate, like a synthetic molecule, and that give energy and C to the cells synthesizing these enzymes is clearly a selective advantage if other members of the microbial community are unable to grow at the expense of that molecule. Thus, acclimation could reflect the sum of the times for the mutation or gene transfer to occur and for the resulting organism to multiply to reach the requisite high population density.

The ease of showing the occurrence of mutations for acquisition of certain traits in pure cultures of bacteria and of gene transfer between large populations of two bacterial species has prompted frequent suggestions that acclimation is often a result of the appearance of new genotypes. However, little information exists to support these contentions. This is not to say that genetic changes followed by population growth are not involved in the acclimation prior to biodegradation of some chemicals, but only that the view is not well supported by experimental observations.

One line of evidence that mutations are not the cause of many acclimations is based on the random occurrence of mutations. An event that occurs randomly should occur sporadically and only in occasional replicate samples of the same environment collected at different times, and the acclimation phase that results from the time for a mutation to take place should vary in length among replicate samples from the environment; hence, the observations that the acclimation periods for the mineralization of 4-nitrophenol in replicate samples of sewage, 2,4-D in lake water, and halobenzoates in anaerobic sediments are essentially the same for each compound suggest that a mutation (followed by multiplication) does not account for the acclimation phase with these compounds (Wiggins *et al.,* 1987; Chen and Alexander, 1989; Linkfield *et al.,* 1989). The reproducible duration of the acclimation period prior to the degradation of other chemicals or in samples from several environments also has been used to argue against the significance of mutation (Spain *et al.,* 1980; Fournier *et al.,* 1981). Conversely, the marked difference in the times for onset of biodegradation of 3- and 5-nitrosalicylates and 4-nitroaniline in sewage and lake water suggests that mutations had occurred and that the acclimation reflected the time for the mutant to arise and for the population of the new genotype to multiply (Wiggins and Alexander, 1988b). Mutants of *Pseudomonas putida* may also have appeared in a microbial mixture growing in a medium containing the herbicide dalapon (Senior *et al.,* 1976).

The transfer of genes involved in biodegradation has been shown to take place in media containing two different bacteria. For example, the capacity to degrade chlorocatechols can be transferred from a strain of *Pseudomonas* to a strain of *Alcaligenes,* and the resulting new genotype is able to metabolize 2-, 3-, and 4-chlorophenols, a property possessed by neither of the parent bacteria (Schwien and Schmidt, 1982). Similarly, the transfer of genes coding for steps in the metabolism of 3-chlorobenzoate may take place between dissimilar strains of *Pseudomonas* (Rubio *et al.*, 1986). However, gene transfer leading to the evolution of new genotypes involved in biodegradation in natural environments or samples of such environments brought to the laboratory has yet to be confirmed.

Diauxie

Pure cultures of bacteria growing in media containing relatively high concentrations of two C sources often do not show the single exponential phase that is characteristic of the same organisms multiplying in media with a single C source. Instead, they have two exponential growth phases, which are separated by an interval with little or no growth (Fig. 3.7). During the first exponential period, only one of the substrates is metabolized by the organism to support growth, and the second exponential period corresponds to growth on and degradation of the second organic compound. This biphasic growth and utilization of two substrates in sequence is known as *diauxie*. The C source that permits faster growth usually is metabolized first. Diauxie is characterized by repression of synthesis of the enzymes concerned with initial steps in the metabolism of the second C source as the bacterium uses the first (Harder *et al.*, 1984).

This utilization of one organic compound in preference to a second has been advanced as an explanation for acclimation; that is, the chemical of interest is the second substrate, and its loss does not begin until the supply of the first is depleted. For example, it has been suggested that bacteria in seawater only attack 4-chlorophenol after certain naturally occurring organic constituents of the water are consumed by the bacteria (Kuiper and Hanstveit, 1984). Enrichment cultures

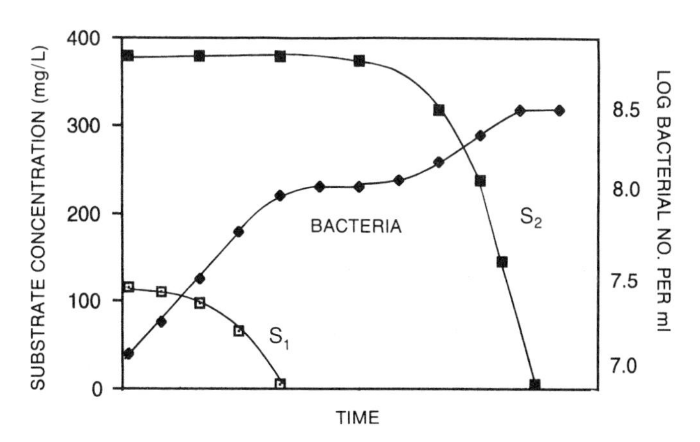

Figure 3.7 Biphasic growth of a bacterium in pure culture with S_1 and S_2 as carbon sources.

that are probably dominated by a single bacterial type, and thus behave much like a pure culture, also show sequential destruction of organic substrates and diauxic growth (Stumm-Zollinger, 1966). Diauxic utilization of substrates, and thus an apparent acclimation before use of the second substrates, may also occur with P compounds, as suggested by a report that *Pseudomonas testosteroni* uses inorganic phosphate before methylphosphonate as a P source for growth, resulting in an initial period when the phosphonate is not being degraded. Only when the inorganic phosphate is consumed does the disappearance of the second P source begin (Daughton *et al.,* 1979).

Direct tests for the occurrence of diauxie in biodegradation or its role in acclimation in natural ecosystems or in waste-treatment systems are scarce. Diauxie does not appear to account for the acclimation prior to mineralization of low concentrations of 4-nitrophenol in sewage or lake water (Wiggins *et al.,* 1987), but it may be important in other environments, especially those in which the one or both of the compounds needed for diauxic biodegradation are present at high concentrations. However, in a heterogeneous microbial community, it is likely that two readily available compounds will be used by different species, rather than one of the molecules persisting while the bacteria active on the first are growing and using their preferred nutrient.

ENZYME INDUCTION AND LAG PHASE

Microorganisms produce many enzymes regardless of whether the substrates for those enzymes are present. These are known as *constitutive enzymes.* In contrast, *inducible enzymes* are formed in appreciable amounts only when the substrate, or sometimes a structurally related chemical or metabolite, is present. The inducer is the specific molecule that, when provided to the cells, is involved in the process of induction. The inducible enzyme may be detectable in the absence of the inducer, but the level is not high. The process of induction has been extensively studied and is known to be a complex process, which typically involves an increase in the rate of formation of the degradative enzymes. The enzymatic activity of a population may also be controlled by *catabolite repression,* in which products generated during the catabolism of one substrate repress the synthesis of enzymes concerned with the degradation of a second substrate that itself would be converted to the same products.

Many of the enzymes involved in one or more of the early steps in the breakdown of synthetic compounds are inducible, for example, many of the dehalogenases that remove chlorine from halogenated molecules and release ionic chloride. However, because enzyme induction is usually largely complete in minutes or hours (Richmond, 1968) and acclimation phases often are weeks in duration (Table 3.1), only the very early portion of the usually far longer acclimation period would involve the time for induction of catabolic enzymes. An acclimation period associated with the time for induction of the enzymes metabolizing 2,4-D by a bacterium is depicted in Fig. 3.8. A delayed induction is also the cause of the acclimation period prior to the rapid degradation of 4-nitrophenol by *Pseudomonas putida;* in this instance, moreover, the bacterium first converts the substrate to hydroquinone, which in turn is apparently transformed to the actual inducer of 4-nitrophenol metabolism (Nishino and Spain, 1993). Except as it may be implicated in diauxie (Stumm-Zollinger, 1966), moreover, catabolite repression also does not seem to be a factor governing much of the time entailed for acclimation.

If the first steps in the metabolism of a compound require the biosynthesis of inducible enzymes and the conditions preclude such synthesis, the compound will not be degraded. In this light, it is noteworthy that a threshold appears to exist for the induction of certain enzymes, at least by some bacteria. For example, induction of the enzymes concerned with early steps in the catabolism of 3- and 4-chlorobenzoates by *Acinetobacter calcoaceticus* occurs at concentrations above but not below 1 μM (Reber, 1982), and the amidase that is necessary for a gram

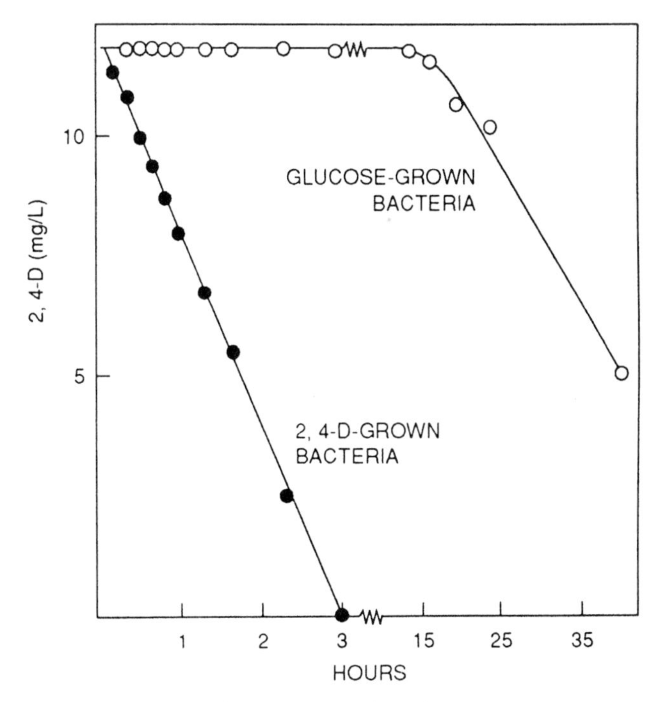

Figure 3.8 Metabolism of 2,4-D by cell suspensions of a bacterium grown on 2,4-D or glucose. (Reprinted with permission from Chen and Alexander, 1989.)

negative bacterium to cleave certain phenylurea herbicides is induced at 50 but not 10 μM linuron (Lechner and Straube, 1984).

The lag phase in the bacterial growth cycle is evident when a bacterium is transferred into fresh medium, even in a medium identical to that in which it previously grew. During this period, the organism does not multiply. Subsequently, the organism initiates rapid growth and, in rich media, enters into the exponential phase of growth. The lag phase is characteristic, or at least well studied, of inocula containing low cell densities, and the small population size coupled with the absence of multiplication denote little or no substrate loss. Because the lag phase in the bacteria studied to date lasts several hours at most and acclimation periods are often far longer, the lag phase per se does not explain the longer acclimations. However, as with enzyme induction, the very initial part of the acclimation may be associated with the true bacterial lag.

REFERENCES

Amrein, J., Hurle, K., and Kirchhoff, J., *Z. Pflanzenkr. Pflanzenschutz, Sonderh.* **9,** 329–341 (1981).

Atlas, R. M., and Bartha, R., *Can. J. Microbiol.* **18,** 1851–1855 (1972).

Audus, L. J., *Plant Soil* **2,** 31–36 (1949).

Avidov, E., Aharonson, N., and Katan, J., *Weed Sci.* **36,** 519–523 (1988).

Bauer, E. J., and Capone, D. G., *Appl. Environ. Microbiol.* **54,** 1649–1655 (1988).

Bouwer, E. J., and McCarty, P. L., *Ground Water* **22,** 433–440 (1984).

Bromilow, R. H., Evans, A. A., Nicholls, P. H., Todd, A. D., and Briggs, G. G., *Pestic. Sci.* **48,** 63–72 (1996).

Button, D. K., and Robertson, B. R., *Mar. Ecol.: Prog. Ser.* **26,** 187–193 (1985).

Chapman R. A., Harris, C. R., and Harris, C., *J. Environ. Sci. Health, Part B* **B21,** 125–141 (1986a).

Chapman, R. A., Harris, C. R., Moy, P., and Henning, K., *J. Environ. Sci. Health, Part B* **B21,** 269–276 (1986b).

Chen, S., and Alexander, M., *J. Environ. Qual.* **18,** 153–156 (1989).

Daughton, C. G., Cook, A. M., and Alexander, M., *Appl. Environ. Microbiol.* **37,** 605–609 (1979).

de Andrea, M. M., Lord, K. A., Bromilow, R. H., and Ruegg, E. F., *Environ. Pollut., Ser. A* **27,** 167–177 (1982).

Felsot, A., Maddox, J. V., and Bruce, W., *Bull. Environ. Contam. Toxicol.* **26,** 781–788 (1981).

Fournier, J. C., Codaccioni, P., and Soulas, G., *Chemosphere* **10,** 977–984 (1981).

Fung, K. K. H., and Uren, N. C., *J. Agric. Food Chem.* **25,** 966–969 (1977).

Gray, R. A., and Joo, G. K., *Weed Sci.* **33,** 698–702 (1985).

Grover, R., *Weed Res.* **7,** 61–67 (1967).

Harder, W., Dijkhuisen, L., and Veldkamp, H., *in* "The Microbe" (D. P. Kelly and N. G. Carr, eds.), Part II, pp. 51–95. Cambridge Univ. Press, Cambridge, UK, 1984.

Hommes, M., and Pestemer, W., *Meded. Fac. Landbouwwet., Rijksuniv. Gent* **50**(2B), 643–650 (1985).

Hoover, D. G., Borgonovi, G. E., Jones, S. H., and Alexander, M., *Appl. Environ. Microbiol.* **51,** 226–232 (1986).

Hurle, K., and Pfefferkorn, V., *Proc. Br. Weed Control Conf., 11th, 1972,* Vol. 2, pp. 806–810 (1972).

Hurle, K., and Rademacher, B., *Weed Res.* **10,** 159–164 (1970).

Jones, S., and Alexander, M., *Appl. Environ. Microbiol.* **54,** 3177–3179 (1988).

Kaufman, D. D., and Kearney, P. C., *Appl. Microbiol.* **13,** 443–446 (1965).

Kaufman, D. D., Katan, Y., Edwards, D. F., and Jordan, E. G., *in* "Agricultural Chemicals of the Future" (J. L. Hilton, ed.), pp. 437–451. Rowman & Allenheld, Totowa, NJ, 1985.
Kuiper, J., and Hanstveit, A. O., *Ecotoxicol. Environ. Saf.* **8,** 15–33 (1984).
Kunc, F., and Macura, J., *Folia Microbiol. (Prague)* **11,** 248–256 (1966).
Lappin-Scott, H. M., Greaves, M. P., and Slater, J. G., *in* "Microbial Communities in Soil" (V. Jensen, A. Kjøller, and L. H. Sørensen, eds.), pp. 211–217. Elsevier Applied Science, London, 1986.
Lechner, U., and Straube, G., *Z. Allg. Mikrobiol.* **24,** 581–584 (1984).
Lewis, D. L., Kollig, H. P., and Hodson, R. E., *Appl. Environ. Microbiol.* **51,** 598–603 (1986).
Linkfield, T. G., Suflita, J. M., and Tiedje, J. M., *Appl. Environ. Microbiol.* **55,** 2773–2778 (1989).
McGrath, D., *Weed Res.* **16,** 131–137 (1976).
Mitchell, J. A., and Cain, R. B., *Pestic. Sci.* **48,** 1–11 (1996).
Moorman, T. B., *Weed Sci.* **36,** 96–101 (1988).
Murakami, Y., and Alexander, M., *Biotechnol. Bioeng.* **33,** 832–838 (1989).
Newman, A. S., and Thomas, J. R., *Soil Sci. Soc. Am. Proc.* **14,** 160–164 (1949).
Nishino, S. F., and Spain, J. C., *Environ. Sci. Technol.* **28,** 489–494 (1993).
Nyholm, N., Lindgaard-Jørgensen, P., and Hansen, N., *Ecotoxicol. Environ. Saf.* **8,** 451–470 (1984).
Obrigawitch, T., Wilson, R. G., Martin, A. R., and Roeth, F. W., *Weed Sci.* **30,** 175–181 (1982).
Obrigawitch, T., Martin, A. R., and Roeth, F. W., *Weed Sci.* **31,** 187–192 (1983).
Paeschke, R. R., Ebing, W., and Heitefuss, R., *Z. Pflanzenkr. Pflanzenschutz* **85,** 280–297 (1978).
Pfaender, F. K., Shimp, R. J., and Larson, R. J., *Environ. Toxicol. Chem.* **4,** 587–593 (1985).
Pignatello, J. J., Johnson, L. K., Martinson, M. M., Carlson, R. E., and Crawford, R. L., *Can. J. Microbiol.* **32,** 38–46 (1986).
Racke, K. D., and Coats, J. R., *J. Agric. Food Chem.* **35,** 94–99 (1987).
Racke, K. D., Laskowski, D. A., and Schultz, M. R., *J. Agric. Food Chem.* **38,** 1430–1436 (1990).
Read, D. C., *Agric. Ecosyst. Environ.* **10,** 37–46 (1983).
Reber, H. H. *Eur. J. Appl. Microbiol. Biotechnol.* **15,** 138–140 (1982).
Richmond, M. H., *Essays Biochem.* **4,** 105–154 (1968).
Riepma, P., *Weed Res.* **2,** 41–50 (1962).
Robertson, B. K., and Alexander, M., *Pestic. Sci.* **41,** 311–318 (1994).
Roeth, F. W., Wilson, R. G., Martin, A. R., and Shea, P. J., *Weed Technol.* **3,** 24–29 (1989).
Rosenberg, A., and Alexander, M., *J. Agric. Food Chem.* **28,** 705–709 (1980).
Rubio, M. A., Engesser, K.-H., and Knackmuss, H.-J., *Arch. Microbiol.* **145,** 116–122 (1986).
Rudyanski, W. J., Fawcett, R. S., and McAllister, R. S., *Weed Sci.* **35,** 68–74 (1987).
Schwien, U., and Schmidt, E., *Appl. Environ. Microbiol.* **44,** 33–39 (1982).
Scow, K. M., Simkins, S., and Alexander, M., *Appl. Environ. Microbiol.* **51,** 1028–1035 (1986).
Scow, K. M., Merica, R. R., and Alexander, M., *J. Agric. Food Chem.* **38,** 908–912 (1990).
Senior, E., Bull, A. T., and Slater, J. H., *Nature (London)* **263,** 476–479 (1976).
Shimp, R. J., and Pfaender, F. K., *Appl. Environ. Microbiol.* **53,** 1496–1499 (1987).
Shimp, R. J., Schwab, B. S., and Larson, R. J., *Environ. Toxicol. Chem.* **8,** 723–730 (1989).
Slade, E. A., Fullerton, R. A., Stewart, A., and Young, H., *Pestic. Sci.* **35,** 95–100 (1992).
Smith, A. E., and Aubin, A. J., *Bull. Environ. Contam. Toxicol.* **53,** 7–11 (1994).
Soulas, G., Codaccioni, P., and Fournier, J. C., *Chemosphere* **12,** 1101–1106 (1983).
Spain, J. C., and Van Veld, P. A., *Appl. Environ. Microbiol.* **45,** 428–435 (1983).
Spain, J. C., Pritchard, P. H., and Bourquin, A. W., *Appl. Environ. Microbiol.* **40,** 726–734 (1980).
Stephenson, T., Lester, J. N., and Perry, R., *Chemosphere* **13,** 1033–1040 (1984).
Stumm-Zollinger, E., *Appl. Microbiol.* **14,** 654–664 (1966).
Torstensson, N. T. L., Stark, J., and Goransson, B., *Weed Res.* **15,** 159–164 (1975).
Ventullo, R. M., and Larson, R. J., *Appl. Environ. Microbiol.* **51,** 356–361 (1986).
Walker, A., and Welch, S. J., *Weed Res.* **31,** 49–57 (1991).
Walker, A., Brown, P. A., and Entwistle, A. R., *Pestic. Sci.* **17,** 183–193 (1986).

Walker, A., Parekh, N. R., Roberts, S. J., and Welch, S. J., *Pestic. Sci.* **39,** 55–60 (1993).
Whiteside, J. S., and Alexander, M., *Weeds* **8,** 204–213 (1960).
Wiggins, B. A., and Alexander, M., *Can. J. Microbiol.* **34,** 661–666 (1988a).
Wiggins, B. A., and Alexander, M., *Appl. Environ. Microbiol.* **54,** 2803–2807 (1988b).
Wiggins, B. A., Jones, S. H., and Alexander, M., *Appl. Environ. Microbiol.* **53,** 791–796 (1987).
Wilde, G., and Mize, T., *Environ. Entomol.* **13,** 1079–1082 (1984).
Wilson, B. H., Smith, G. B., and Rees, J. F., *Environ. Sci. Technol.* **20,** 997–1002 (1986).
Wilson, R. G., *Weed Sci.* **32,** 264–268 (1984).

CHAPTER 4

Detoxication

The most important role of microorganisms in the transformation of pollutants is their ability to bring about detoxication. Detoxication (sometimes designated detoxification) refers to the change in a molecule that renders it less harmful to one or more susceptible species. The susceptible species may be humans, animals, plants, other microorganisms, or the detoxifying population itself. Particular attention, needless to say, is given to detoxications that make organic compounds less injurious to humans, but an abundant body of information also exists on detoxication reactions that alter the toxicity to animals and plants. In studies of environmental pollution, detoxications that reduce the harm to microorganisms have received little inquiry.

Detoxications result in inactivation, with the toxicologically active substance being converted to an inactive product. Because toxicological activity is associated with many chemical entities, substituents, and modes of action, detoxications similarly include a large array of different types of reactions.

A simple way of demonstrating detoxication is to measure the effect of environmental samples on the behavior, growth, or viability of susceptible species. This clearly would not be acceptable when humans are the suscepts, but it is a common procedure when destruction of compounds affecting plants or lower animals is of concern. For example, seeds introduced into soil containing a herbicide or other phytotoxin (Hill *et al.,* 1955) or insects added to a soil containing an insecticidal agent (Thompson, 1973) will not grow, and they often die. However, as detoxication occurs, seeds or insects introduced at progressively later periods into the amended soil will grow but poorly, or they will develop normally. If the tests are done with identical environmental samples that differ only in that they are sterile, the more rapid dissipation of the injurious effect in the nonsterile samples will show that the detoxication results from biological activity (Fig. 4.1). Bioassays are especially useful inasmuch as they reflect the loss of biological activity of a chemical, but they are frequently replaced by direct chemical analysis showing the loss of the parent compound or the formation of products.

Detoxication is advantageous to the microorganisms carrying out the transformation if the concentration of the chemical is in the range that suppresses these species. If the reaction is the first step in a process by which organisms use the molecule as a C source, the reaction is also beneficial, not because it inactivates the molecule but by virtue of its helping the cell to acquire C. For the many microbial detoxication reactions involving substances that are toxic to humans, plants, and animals and provide no nutritional benefit to the microflora, however, the transformations are important in public health, agriculture, or natural biological communities but not for the microorganisms responsible for the conversion.

The enzymatic step or sequence that results in the conversion of the active molecule into the innocuous product usually occurs within the cell. The product may then undergo one of three fates: (a) it may be excreted; (b) after one or more additional enzymatic steps, it may be changed to a compound that enters the normal metabolic pathways within the cell and ultimately the C is excreted as an organic waste; or (c) it may be modified to a new molecule that becomes subject to these normal reaction sequences, and finally the C is released as CO_2 (Fig. 4.2).

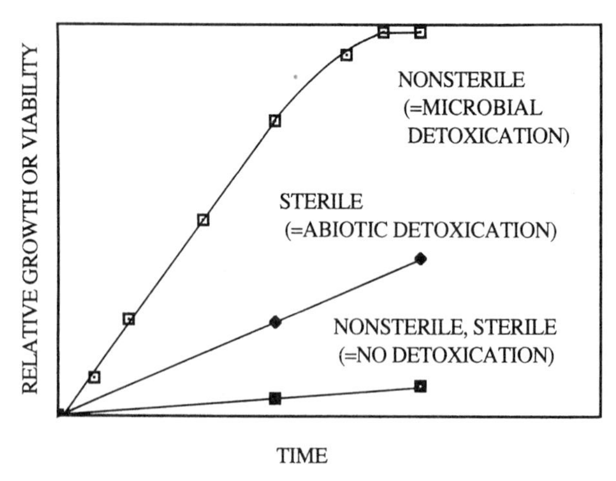

Figure 4.1 Bioassays for destruction of a toxicant in an environmental sample by microorganisms or by an abiotic mechanism.

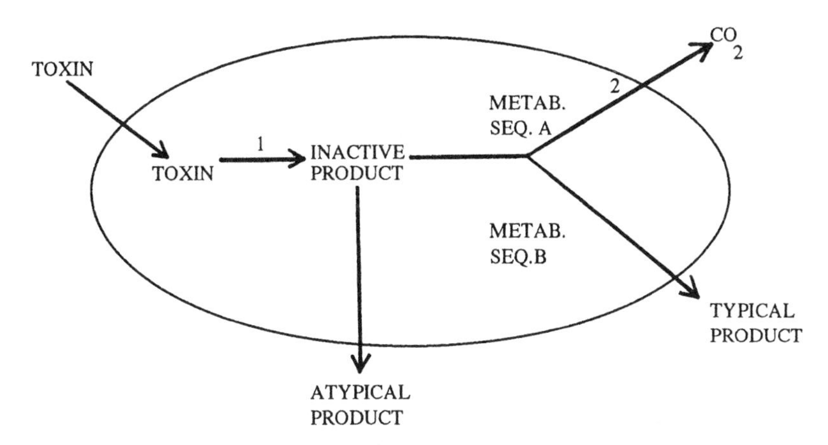

1. DETOXICATION REACTION
2. MINERALIZATION

Figure 4.2 Fate of chemicals that are detoxified.

The product in the first case is structurally similar to the toxin, but it is harmless at the prevailing concentration. The last fate is mineralization, the mineralization of inhibitors being detoxications, but the actual detoxication step occurs at some early step in the catabolic sequences that finally yields CO_2. Cometabolic processes (which will be discussed later) often are detoxications, but the products of the transformation are structurally similar to the original substrate.

Several processes may result in detoxication. These processes represent only the first step in Fig. 4.2. These include (a) hydrolysis, (b) hydroxylation, (c) dehalogenation, (d) demethylation or other dealkylations, (e) methylation, (f) nitro reduction, (g) deamination, (h) ether cleavage, (i) conversion of a nitrile to an amide, and (j) conjugation.

a. Hydrolysis. Cleavage of a bond by the addition of water is a common means by which microorganisms inactivate toxicants. Such reactions may involve a simple hydrolysis of an ester bond, as with the insecticide malathion (Walker and Oesch, 1983) by a carboxyesterase:

$$\mathrm{R\overset{\overset{\displaystyle O}{\|}}{C}OR' + H_2O \rightarrow R\overset{\overset{\displaystyle O}{\|}}{C}OH + HOR'}$$

This hydrolysis of malathion is effected microbiologically (Paris *et al.*, 1975). Such reactions may entail hydrolytic cleavage of anilides by an amidase (Munnecke, 1981):

$$\mathrm{RNH\overset{\overset{\displaystyle O}{\|}}{C}CH_2R' + H_2O \rightarrow RNH_2 + HO\overset{\overset{\displaystyle O}{\|}}{C}CH_2R'}$$

Other hydrolyses resulting in inactivation have been described.

b. Hydroxylation. The addition of OH to an aromatic or aliphatic molecule often makes it less harmful. Thus, the simple replacement of H by OH inactivates the fungicide MBC (Davidse, 1976).

$$\mathrm{RH \rightarrow ROH}$$

The hydroxylation of the ring of 2.4-D similarly converts the parent herbicide to a nontoxic product (Owen, 1989; Jensen, 1982). Microorganisms may bring about such a detoxication when they hydroxylate the ring in the 4-position, a process that leads to a migration of the chlorine to give 2,5-dichloro-4-hydroxyphenoxyacetic acid (Faulkner and Woodcock, 1964).

c. Dehalogenation. Many pesticides and hazardous industrial wastes contain Cl or other halogens, and removal of the halogen often converts the toxicant to an innocuous product. The enzymes are designated dehalogenases. These dehalogenations may involve replacement of the halogen by H (reductive dehalogenation)

$$\mathrm{RCl \rightarrow RH}$$

or by OH (hydrolytic dehalogenation),

$$\mathrm{RCl \rightarrow ROH}$$

or it may result in removal of the halogen and an adjacent H (dehydrodehalogenation).

$$\mathrm{RCH_2CHClR' \rightarrow RCH{=}CHR'}$$

The halogen is released in these reactions as inorganic chloride, fluoride, bromide, or iodide. Three dehalogenations of pesticides are illustrated in Fig. 4.3. The dehydrodechlorinase that acts on DDT converts this insecticide to a nontoxic product (Walker and Oesch, 1983). Similarly, the microbial conversion of lindane to 2,3,4,5,6-pentachloro-1-cyclohexene detoxifies the insecticide (Francis *et al.*, 1975; Yule *et al.*, 1967), as does the conversion of the herbicide dalapon to pyruvic acid (Berry *et al.*, 1979). The cleavage of some of these C—halogen bonds is most surprising because of the strength of the bonds. The strength of a chemical bond is the amount of energy required to break that bond:

$$\mathrm{C{-}F + energy \rightarrow C\cdot + \cdot F}$$

To break the C—F bond, for example, requires considerable energy because the bond energy for the bond between C and F is 116 kcal/mole (Speier, 1964).

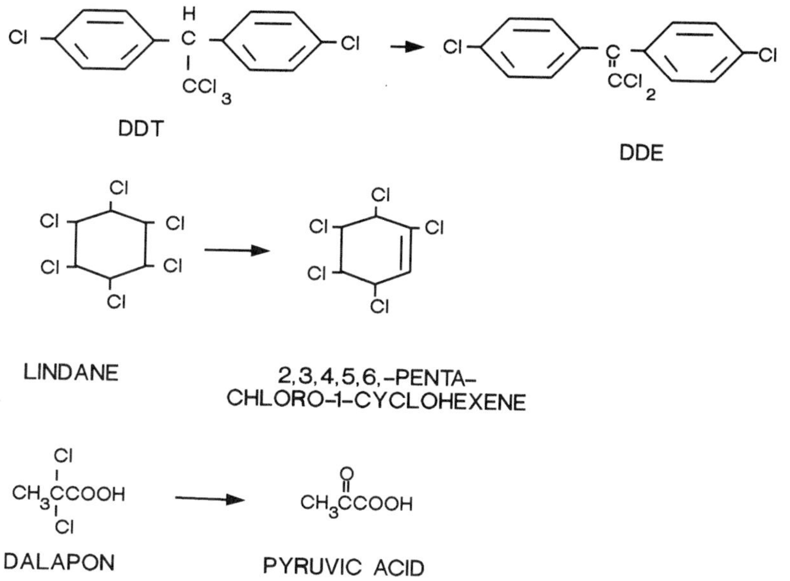

DDT → DDE

LINDANE → 2,3,4,5,6,-PENTA-CHLORO-1-CYCLOHEXENE

DALAPON → PYRUVIC ACID

Figure 4.3 Dehalogenations that represent detoxications.

d. Demethylation or Other Dealkylations. Many pesticides contain methyl or other alkyl substituents. These may be linked to N or O (*N*- or *O*-alkyl substitution). An *N*- or *O*-dealkylation catalyzed by microorganisms frequently results in loss of pesticidal activity. A number of herbicides that are structurally related to phenylurea become less active when microorganisms *N*-demethylate the molecules, as in the case of the conversion of diuron to the monomethyl derivative (Fig. 4.4). The subsequent removal of the second *N*-methyl group renders the molecule wholly nontoxic (Ellis and Camper, 1982; Jensen, 1982; Hathaway, 1986). Similar reactions leading to detoxications occur when an *s*-triazine herbicide like atrazine is dealkylated as it loses its *N*-ethyl or *N*-isopropyl groups (Jensen, 1982), a reaction that occurs in soil (Khan and Marriage, 1977), presumably as a result of microbial action. The microbial *O*-demethylation of chloroneb creates a nontoxic product, 2,5-dichloro-4-methoxyphenol (Fig. 4.4) (Hock and Sisler, 1969).

e. Methylation. The reverse reaction—the addition of a methyl group—may inactivate toxic phenols. Thus, penta- and tetrachlorophenols are fungicides, the former being especially widely used, and these can be detoxified microbiologically by addition of a methyl group in a reaction that represents an *O*-methylation (Cserjesi, 1972; Cserjesi and Johnson, 1972):

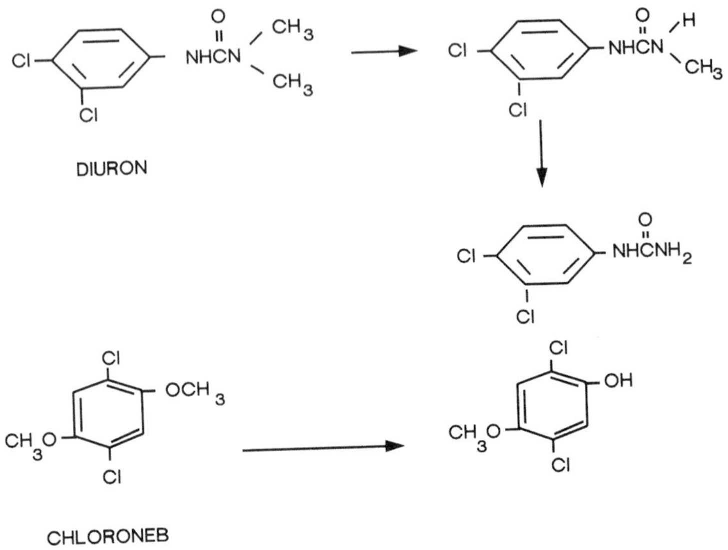

DIURON

CHLORONEB

Figure 4.4 Dealkylations that detoxify pesticides.

$$ROH \rightarrow ROCH_3$$

f. Nitro Reduction. Nitro compounds are harmful to many types of organisms, both higher and lower. These may be rendered less toxic by reduction of the nitro to an amino group:

$$RNO_2 \rightarrow RNH_2$$

Such reductions may result in loss or diminution of the harmful effects as microorganisms convert the broad-spectrum poison 2,4-dinitrophenol to 2-amino-4- and 4-amino-2-nitrophenol (Madhosingh, 1961), the fungicide pentachloronitrobenzene to pentachloroaniline (Nakanishi and Oku, 1969), and the insecticide parathion to aminoparathion (Mick and Dahm, 1970).

g. Deamination. The herbicide known as metamitron can be transformed microbiologically to yield a deaminated product (Engelhardt and Wallnöfer, 1978) that is nontoxic to plants (Hathaway, 1986) (Fig. 4.5).

h. Ether Cleavage. Phenoxy herbicides contain ether linkages (C—O—C), and the cleavage of these linkages destroys the phytotoxicity of the molecule. This is illustrated by the cleavage of the ether bond in 2.4-D (Loos *et al.*,

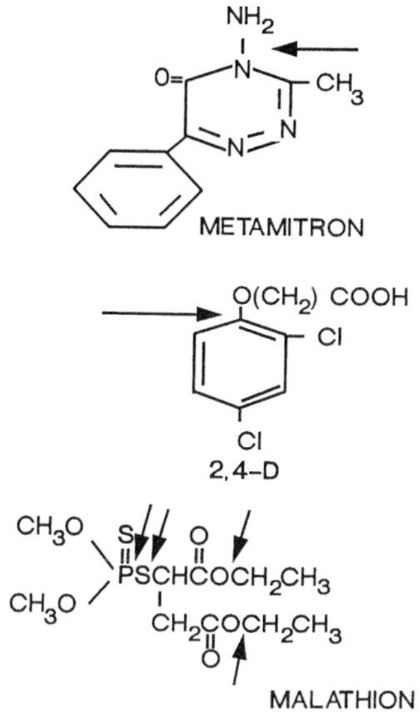

Figure 4.5 Initial reactions that result in detoxications. Arrows indicate cleavage sites.

1967) (Fig. 4.5). This microbial conversion is somewhat surprising because of the bond energy between C and O, which is 85.5 kcal/mole (Speier, 1964), and thus the need of the microorganism to provide the energy to cleave the bond.

i. Conversion of Nitrile to Amide. A potent inhibitor of the growth of certain plants is 2-6-dichlorobenzonitrile, which is sold under the name of dichlobenil. However, when it is converted to 2,6-dichlorobenzamide, the molecule is rendered inactive (Ashton and Crafts, 1981):

$$R-C \equiv N \rightarrow R-\overset{\overset{\displaystyle O}{\|}}{C}-NH_2$$

This detoxication reaction is brought about by microorganisms in soil (Verloop, 1972).

j. Conjugation. A conjugation involves a reaction between a common intermediate in some natural metabolic pathway with a synthetic molecule. Products of the combination of a normal metabolite with a toxicant frequently are harmless. Considerable effort has been directed to animal and plant conjugations that involve sugars, glutathione, and amino acids, but the possible role of microorganisms in conjugations leading to detoxication has received little attention. Moreover, not all of the microbial processes in which toxicants are conjugated have been shown to actually result in loss of toxicity, although by analogy to similar metabolic reactions in animals and plants, many probably do indeed lead to the loss or elimination of harmful effects if the parent molecule is harmful.

Several types of microbiologically effected conjugations have been described. (a) Glucose conjugates. For example, *Cunninghamella elegans* conjugates pyrene with glucose to yield nontoxic glucose conjugates (Cerniglia *et al.*, 1986), and the same fungus forms glucose-containing derivatives of phenanthrene (Casillas *et al.*, 1996). (b) Glucuronide conjugates. Several fungi in culture metabolize phenanthrene with the ultimate formation of glucuronide derivatives (Casillas *et al.*, 1996). (c) Glutathione conjugates. Bacteria are able to generate such complexes from the herbicides EPTC (Tal *et al.*, 1993) and barban (Marty and Bastide, 1988), and a similar process may occur during the metabolism of the herbicide metolachlor in soil (Aga *et al.*, 1996). (d) Amino acid conjugates. Several microbial species detoxify the fungicide sodium dimethyldithiocarbamate by converting it to the less toxic conjugate γ-(dimethylthiocarbamoylthio)-γ-aminobutyric acid (Kaars Sijpesteijn *et al.*, 1962). Cysteine can be conjugated with EPTC by some bacteria (Tal and Rubin, 1993), and other bacteria can conjugate alanine with the herbicide amitrole (Williams *et al.*, 1965). (e) Acyl conjugates. Acylation by the addition of acetyl or sometimes propionyl or formyl groups is common to many microorganisms. An example is the acetylation of aromatic amines formed from the products of TNT breakdown (Noguera and Freedman, 1996). (f) Sulfonate and sulfate conjugates. Metolachor is converted in soil to a corresponding ethanesulfonate derivative (Aga *et al.*, 1996), and phenanthrene and 2-nitrofluorene are metabolized by fungi to yield sulfate conjugates (Casillas *et al.*, 1996; Pothuluri *et al.*, 1996). (g) Dicarboxylic acid conjugates. Such products are generated as a result of the conversion of toluene to benzylsuccinic and benzylfumaric acids by anaerobes (Beller *et al.*, 1992; Biegert *et al.*, 1996). Toxicological assessments are required to establish which of these conjugations also represent detoxications.

A particular microorganism or a microbial community may detoxify a single toxicant in several ways. This is illustrated for the insecticide malathion in Fig. 4.5 (Walker and Stojanovic, 1974). Such multiple pathways are initiated by entirely different enzymes. Other pesticides are also acted on by several dissimilar enzymes, which may thus yield several inactive products.

The 10 reaction types given here are not always detoxications, however. A compound acted on by one or another mechanism may yield a product no less toxic than its precursor. Indeed, several such reactions may yield products far more toxic than the original substrates. Furthermore, a reaction or a sequence that yields a product not injurious to one species may not result in detoxication for a second species. Thus, one cannot consider detoxication in a general sense: the sensitive species that is protected must be defined.

REFERENCES

Aga, D. S., Thurman, E. M., Yockel, M. E., Zimmerman, L. R., and Williams, T. D., *Environ. Sci. Technol.* **30,** 592–597 (1996).

Ashton, F. M., and Crafts, A. S., "Mode of Action of Herbicides." Wiley, New York, 1981.

Beller, H. R., Reinhard, M., and Grbić-Galić, D., *Appl. Environ. Microbiol.* **58,** 3192–3195 (1992).

Berry, E. K. M., Allison, N., Skinner, A. J., and Cooper, R. A., *J. Gen. Microbiol.* **110,** 39–45 (1979).

Biegert, T., Fuchs, G., and Heider, J., *Eur. J. Biochem.* **238,** 661–668 (1996).

Casillas, R. P., Crow, S. A., Jr., Heinze, T. M., Deck, J., and Cerniglia, C. E., *J. Ind. Microbiol.* **16,** 205–215 (1996).

Cerniglia, C. E., Kelly, D. W., Freeman, J. P., and Miller, D. W., *Chem.-Biol. Interact.* **57,** 203–216 (1986).

Cserjesi, A. J., *Int. Biodeterior. Bull.* **8,** 135–138 (1972).

Cserjesi, A. J., and Johnson, E. L., *Can. J. Microbiol.* **18,** 45–49 (1972).

Davidse, L. C., *Pestic. Biochem. Physiol.* **6,** 538–546 (1976).

Ellis, P. A., and Camper, N. D., *J. Environ. Sci. Health, Part B* **B17,** 277–289 (1982).

Engelhardt, G., and Wallnöfer, P. R., *Chemosphere* **7,** 463–466 (1978).

Faulkner, J. K., and Woodcock, D., *Nature (London)* **203,** 865 (1964).

Francis, A. J., Spanggord, R. J., and Ouchi, G. I., *Appl. Microbiol.* **29,** 567–568 (1975).

Hathaway, D. E., *Biol. Rev. Cambridge Philos. Soc.* **61,** 435–486 (1986).

Hill, G. D., McGahen, J. W., Baker, H. M., Finnerty, D. W., and Bingeman, C. W., *Agron. J.* **47,** 93–104 (1955).

Hock, W. K., and Sisler, H. D., *J. Agric. Food Chem.* **17,** 123–128 (1969).

Jensen, K. I. N., *in* "Herbicide Resistance in Plants" (H. M. LeBaron and J. Gressel, eds.), pp. 133–162. Wiley (Interscience), New York, 1982.

Kaars Sijpesteijn, A., Kaslander, J., and van der Kerk, G. J. M., *Biochim. Biophys. Acta* **62,** 587–589 (1962).

Khan, S. U., and Marriage, P. B., *J. Agric. Food Chem.* **25,** 1408–1413 (1977).

Loos, M. A., Bollag, J. M., and Alexander, M., *J. Agric. Food Chem.* **15,** 858–860 (1967).

Madhosingh, C., *Can. J. Microbiol.* **7,** 553–567 (1961).

Marty, J.-L., and Bastide, J., *Pestic. Sci.* **24,** 221–230 (1988).

Mick, D. L., and Dahm, P. A., *J. Econ. Entomol.* **63,** 1155–1159 (1970).

Munnecke, D. M., *in* "Microbial Degradation of Xenobiotics and Recalcitrant Compounds" (T. Leisinger, A. M. Cook, R. Hütter, and J. Nüesch, eds.), pp. 251–269. Academic Press, London, 1981.

Nakanishi, T., and Oku, H., *Phytopathology* **59,** 1761–1762 (1969).

Noguera, D. R., and Freedman, D. L., *Appl. Environ. Microbiol.* **62,** 2257–2263 (1996).

Owen, W. J., *in* "Herbicides and Plant Metabolism" (A. D. Dodge, ed.), pp. 171–198. Cambridge Univ. Press, Cambridge, UK, 1989.

Paris, D. F., Lewis, D. L., and Wolfe, N. L., *Environ. Sci. Technol.* **9,** 135–138 (1975).

Pothuluri, J. V., Evans, F. E., Heinze, T. M. Fu, P. P., and Cerniglia, C. E., *J. Toxicol. Environ. Health* **47,** 587–599 (1996).

Speier, J. L., *Chemistry* **37**(7), 6–11 (1964).

Tal, A., and Rubin, B., *Pestic. Sci.* **39,** 207–212 (1993).

Thompson, A. R., *J. Econ. Entomol.* **66,** 855–857 (1973).

Verloop, A., *Residue Rev.* **43,** 55–103 (1972).

Walker, C. H., and Oesch, F., *in* "Biological Basis of Detoxication" (J. Caldwell and W. B. Jakoby, eds.), pp. 349–368. Academic Press, New York, 1983.

Walker, W. W., and Stojanovic, B. J., *J. Environ. Qual.* **3,** 4–10 (1974).

Williams, A. K., Cox, S. T., and Eagon, R. G., *Biochem. Biophys. Res. Commun.* **18,** 250–254 (1965).

Yule, W. N., Chiba, M., and Morley, H. V., *J. Agric. Food Chem.* **15,** 1000–1004 (1967).

CHAPTER 5

Activation

One of the more surprising, and possibly the most undesirable, aspects of microbial transformations in nature is the formation of toxicants. A large number of chemicals that are themselves innocuous can be, and often are, converted to products that may be harmful to humans, animals, plants, or microorganisms. By such means, the resident microflora creates pollutants where none was present previously. Hence, it is not sufficient to know that the parent compound has disappeared—the products may be the problem, not the parent. The process of forming toxic products from innocuous precursors is known as *activation*.

Activation is a major reason to study the pathways and products of breakdown of organic molecules both in natural ecosystems and in waste-disposal systems that lead to environmental discharges. Because the molecules thus synthesized may pose a problem where the precursors were benign, those metabolites must be identified. Activation occurs in soil, water, wastewater, and other environments in which microorganisms are active, and the products created thereby may have a short residence time or persist for long periods (Fig. 5.1). The conversion may represent a single reaction or a simple sequence in a cometabolic process. Alternatively, the harmful product may be an intermediate in mineralization, yet it may persist long enough to create a pollution problem. The consequences of activation include the biosynthesis of carcinogens, mutagens, teratogens, neurotoxins, phytotoxins, and insecticidal and fungicidal agents. Moreover, the mobility of the product of activation is sometimes far different from that of its precursor, so

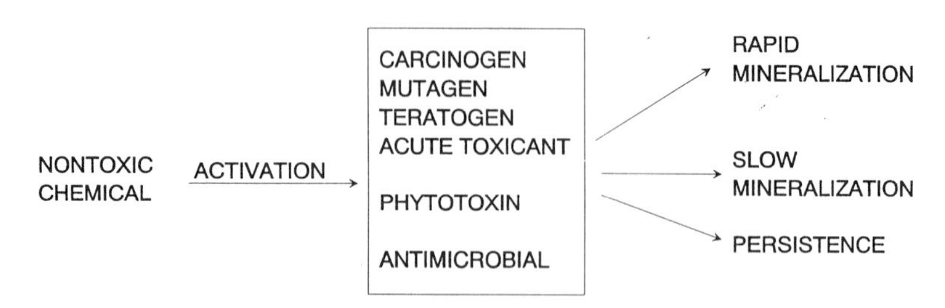

Figure 5.1 General processes associated with activation.

that the product may be transported to distant sites to a far greater or to a significantly smaller extent than the molecule from which it was formed.

Many different pathways, mechanisms, and enzymes are associated with activation. This should not be surprising in view of the differences in structure of chemicals that are toxic to one species or another, or that affect one or another physiological process or target site in the susceptible species. Therefore, a consideration of activation requires a review of a variety of dissimilar mechanisms. The only common attribute among these mechanisms is the greater hazard of the product compared to the starting material.

Particular attention in research on activation has been given to pesticides. This emphasis arose naturally when it became evident that certain pesticides themselves were not especially harmful to the insects, weeds, or sometimes plant pathogens they were designed to control, but rather they were modified within the pest to the molecule that caused injury or death. Subsequently, it was learned that similar conversions were carried out by microorganisms in soils, natural waters, and other environments.

Even before the products of activation were identified, bioassays revealed that a number of pesticides were activated in soil. This became clear when the low level of toxicity in soil receiving the chemical increased with time, with the mortality of sensitive species rising as the molecule underwent some then unspecified transformation. This is depicted in Fig. 5.2 for soils amended with the insecticides zinophos, trichloronat, and carbofuran. Similar increases with time in toxicity to insects have been noted with phorate, diazinon, dasanit, dyfonate, and chlorfenvinphos (Read, 1969, 1971a,b). An analogous instance has been observed for phytotoxin formation; thus, red pine growing in soil treated with ipazine showed no symptoms of toxicity for 80 days, but symptoms of injury developed rapidly thereafter (Kozlowski, 1965). In addition, the herbicide chlornitrofen is converted to its amino derivative by the microflora in river water, and this amino derivative is a mutagen (Tanaka *et al.*, 1996). Such bioassay data indicate that several chemical classes may be activated. Some of the mechanisms of activation of these and of a

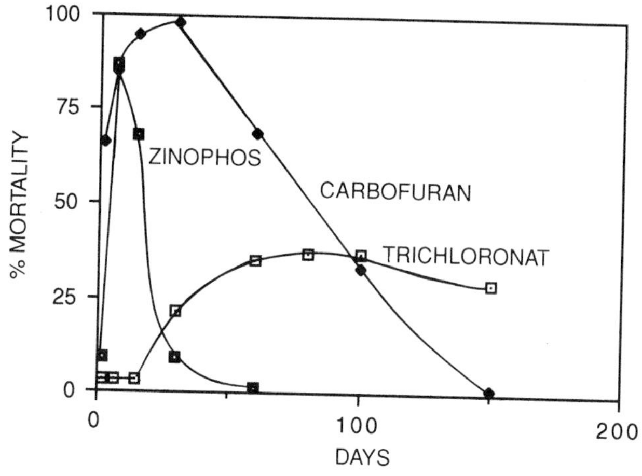

Figure 5.2 Activation of zinophos, trichloronat, and carbofuran added to soil at 3 mg/kg. (Reprinted with permission from Read, 1971a).

variety of other substrates for microorganisms are considered in the following. Except as noted, not all the reactions have been confirmed to result from microbial metabolism, although most are probably a consequence of their activities.

Activations may also occur during bioremediation, particularly if the microbial conversions are incomplete. For example, uncharacterized mutagens were found to be generated during a laboratory study of the bioremediation of PAHs in soil (Belkin *et al.*, 1994). Hence, a bioremediation that destroys the target compounds does not necessarily eliminate the hazard.

MECHANISMS OF ACTIVATION

Dehalogenation

A major activation occurs during the microbial metabolism of trichloroethylene (TCE). This compound was once widely used and now represents a major contaminant of many aquifers. Because TCE is metabolized by many bacteria, its elimination by bioremediation is being actively pursued. Unfortunately, a major product that is frequently encountered is vinyl chloride, which is a potent carcinogen:

$$Cl_2C{=}CHCl \rightarrow ClHC{=}CH_2$$

The same carcinogen can also be formed during the anaerobic metabolism of 1,1- and *trans*-1,2-dichloroethylene (Wilson *et al.*, 1986). Vinyl chloride has been found frequently in TCE-contaminated groundwaters and during tests of technologies to bring about TCE bioremediation by anaerobic bacteria.

TCE can also be converted in cultures of methanotrophs to 2,2,2-trichloro-acetaldehyde, which is often called chloral hydrate (Newman and Wackett, 1991):

$$Cl_2C{=}CHCl \rightarrow Cl_3C{-}CHO$$

Chloral hydrate is a mutagen and is also acutely toxic, and if it is consumed together with an alcoholic beverage, it will lead to unconsciousness. The microbial conversion is not really a dehalogenation but rather causes a migration of the chlorine to the adjacent C. Differing from vinyl chloride, chloral hydrate is not known to be a problem in the field.

HALOGENATION

Fungi, algae, and bacteria possess enzymes that catalyze the addition of chlorine and bromine to organic compounds. Some of these reactions have been well characterized, and the enzymes, substrates, and products are known. Although the enzymes and substrates resulting in other halogenated metabolites have not been identified, the products have been clearly identified. For example, a number of algae form trichloroethylene and tetrachloroethylene (PCE) (Abrahamsson *et al.*, 1995). Some algae generate chloroform, bromoform, and other halogenated methane derivatives (Nightingale *et al.*, 1995) as well as bromo- and iodoethanes (Laturnus, 1995). Trichloroacetic acid also appears to be produced by natural processes (Haiber *et al.*, 1996).

A variety of chlorine-containing aromatic compounds have been detected in decaying plant materials and in forest soils. Some of the substances have been identified, and they include chlorinated benzoic acids and chloroanisoles (Flodin *et al.*, 1997; de Jong *et al.*, 1994; Hjelm *et al.*, 1995). The toxicity of few of these compounds has been established, however. Because such organohalogens are known to be produced by Basidiomycetes in culture (Field *et al.*, 1995; Oberg *et al.*, 1997), they may be the responsible organisms in nature.

Many chlorophenols are harmful and many are persistent. It is possible that these may be produced microbiologically in nature in view of the finding that a fungal chloroperoxidase halogenates phenol to yield monochlorophenols, the latter to give dichlorophenols, the dichlorophenols to produce trichlorophenols, the trichloro compounds to give rise to tetrachlorophenols, and even the latter to generate pentachlorophenol (Wannstedt *et al.*, 1990). Fungal peroxidases may also dimerize 3,4-dichloroaniline to 3,4,3′,4′-tetrachloroazobenzene, a compound similar in toxicity to 2,3,7,8-tetrachloro-*p*-dibenzodioxin (TCDD) (Pieper *et al.*, 1992). It is not known whether any such processes occur in nature, but the enzymatic precedent is cause for further inquiry.

N-Nitrosation of Secondary Amines (Nitrosamine Formation)

Many activations involve chemicals that are used as pesticides, and coinciding with their widespread use, the activation reaction was described. In the instance of *N*-nitrosation, in contrast, the precursors for the toxicants had been widely used for long periods before the activation reaction was discovered. The precursors are secondary amines and nitrate. The former are common synthetic chemicals whose annual production is many millions of kilograms. The latter is an anion that is found in nearly all soils and in most natural waters. Furthermore, secondary amines are natural products that are present, sometimes in reasonable amounts, in plants, fish, decaying products, and other materials.

A secondary amine can be written as RNHR′. The *N*-nitrosation of such a secondary amine occurs in the presence of nitrite, which is formed microbiologically from nitrate. The product is an *N*-nitroso compound or, as it is commonly called, a nitrosamine:

$$\begin{matrix} R \diagdown \\ \quad N{-}N{=}O \\ R' \diagup \end{matrix}$$

In some instances, R and R′ are identical. The reason for concern with nitrosamines is their potency: they are active at very low concentrations as carcinogens, teratogens, and mutagens.

Many secondary and tertiary amines are used in industry. The tertiary amine has the structure:

$$\begin{matrix} R \diagdown \\ \quad N{-}R'' \\ R' \diagup \end{matrix}$$

Secondary or tertiary amines are found in common household products, and a number of pesticides are also secondary or tertiary amines. As a result of the wide use of synthetic amines and their occurrence in living organisms, secondary amines are present in river waters, wastewaters even following treatment (Sander *et al.*, 1974; Neurath *et al.*, 1977), as well as in soil (Vlasenko *et al.*, 1981). Many secondary amines may exist in some rivers (Neurath *et al.*, 1977). Secondary and tertiary amines also appear during the decay of plant residues (Fujii *et al.*, 1972) as well as in the decomposition of creatinine, choline, and phosphatidylcholine in sewage. Certain pesticides are converted to secondary amines in soil (Tate and Alexander, 1974; Mallik *et al.*, 1981). In turn, tertiary amines are converted in soil, sewage, and microbial cultures to secondary amines (Ayanaba and Alexander, 1973, 1974; Tate and Alexander, 1976). This conversion can be depicted as:

$$\begin{matrix} R \\ R' \end{matrix}\!\!>\!N{-}R'' \longrightarrow \begin{matrix} R \\ R' \end{matrix}\!\!>\!NH$$

A well-studied example is the microbial transformation of trimethylamine to dimethylamine:

$$(CH_3)_3N \rightarrow (CH_3)_2NH$$

Such conversions of tertiary to secondary amines in nature are microbial.

Simple secondary and tertiary amines, and their complex nitrogenous precursors, rarely are hazardous. However, the stage is set for activation either once the secondary amine is added or as it is formed. The first reactant for activation is now present; the second reactant needed for activation is nitrite. Although nitrite is rarely found in significant amounts in nature, indeed its presence often goes undetected, it is generated by microorganisms when nitrate is present:

$$NO_3^- \rightarrow NO_2^-$$

The yield is often low, but even a small yield apparently is adequate for activation. The activation is the *N*-nitrosation of the secondary amine to give the highly toxic *N*-nitroso compound:

$$\begin{matrix} R \\ R' \end{matrix}\!\!>\!NH + NO_2^- \longrightarrow \begin{matrix} R \\ R' \end{matrix}\!\!>\!N{-}N{=}O + OH^-$$

Such nitrosations may occur in sewage, lake water (Ayanaba and Alexander, 1974; Yordy and Alexander, 1981), soil (Ayanaba *et al.,* 1973), and wastewater (Greene *et al.,* 1981), and the precursors may be such secondary amines as dimethylamine and diethanolamine.

Microorganisms can convert a number of secondary amines to the corresponding *N*-nitroso compounds (Ayanaba and Alexander, 1973; Hawksworth and Hill, 1971; Kunisaki and Hayashi, 1979). Moreover, microbial enzymes can also *N*-nitrosate several amines in the presence of nitrite (Ayanaba and Alexander, 1973). Nevertheless, the actual nitrosation step may be, to a lesser or greater extent, nonenzymatic and may result from a spontaneous reaction of the amine and nitrite with some metabolic product (Collins-Thompson *et al.,* 1972) or cell constituents (Mills and Alexander, 1976). The extent of conversion of the amine to the nitrosamine is nearly always small at the pH values common in nature, although the yield can be high in artificially acidified solutions. Hence, the microbial role in activation is the enzymatic formation of the secondary amine and nitrite and, enzymatically or otherwise, the actual *N*-nitrosation.

Such activations of secondary amines do occur. For example, a pesticide-manufacturing company in Elmira, Ontario, disposed of dimethylamine in an adjacent uncontrolled waste-disposal site. Several years later, the entire water supply of Elmira, which came from groundwaters, was found to contain *N*-nitrosodimethylamine, $(CH_3)_2N{-}N{=}O$, at levels far in excess of the regulatory level, which, because of the extreme potency of this carcinogen, was several parts per trillion. Although it is not certain whether the formation of *N*-nitrosodimethylamine was microbial or abiotic, the circumstances were suitable for microbial nitrosation and the nitrite precursor was likely of microbial origin.

The nitrosation reaction serves as a good illustration of the fallacy of assuming that dangers reside in synthetic chemicals and not in "natural" compounds. As pointed out earlier, plants and fish contain secondary or tertiary amines. Fish is a notable source of trimethylamine and trimethylamine *N*-oxide. Microbial transformations in the alimentary tract will convert the tertiary amines to secondary amines and the nitrate in drinking water and many vegetables to nitrite. Nitrosation is then a likely consequence, although the reaction may either be microbial or result from an abiotic conversion at the low pH in the stomach. The issue is not whether the chemical is made naturally or by industry, but rather the identity of the molecule itself.

Other examples of such activations are the detection of *N*-nitrosodiethylamine and *N*-nitrosodimethylamine in municipal sludge (Brewer *et al.*, 1980). These two carcinogens and *N*-nitrosomorpholine have been found in sewage-treatment operations, and *N*-nitrosodiethanolamine has been detected in the outlet of a cutting-fluid recovery plant (Richardson *et al.*, 1980). The latter nitrosamine probably is a result of the microbial metabolism of diethanolamine (Yordy and Alexander, 1981), which is a common constituent of cutting fluids and many other products. In addition, the nitroso derivative of cyanazine has also been found in soil treated with this herbicide (Zwickenpflug and Richter, 1994). Some of these nitrosations may result from nonbiological processes, which can also lead to nitrosations.

Epoxidation

Microorganisms are able to form epoxides from several compounds having double bonds:

$$-HC{=}CH- \longrightarrow -\overset{}{HC}\overset{O}{\frown}CH-$$

In the case of several chemicals marketed as insecticides, this oxidation converts the precursor to a product that is more toxic to animals. This type of transformation is illustrated by the conversion of heptachlor to heptachlor epoxide in soil (Duffy

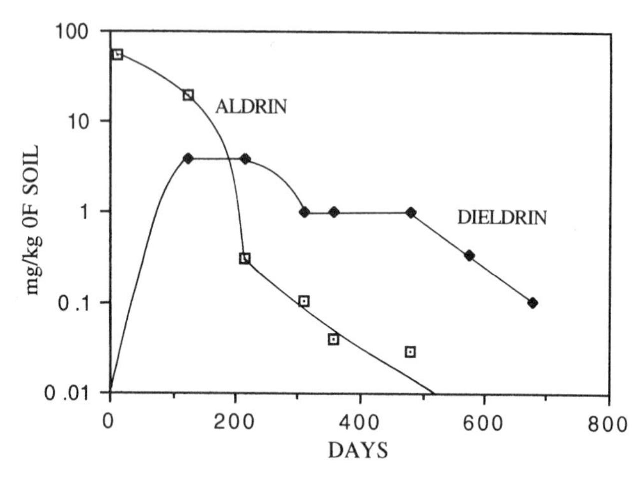

Figure 5.3 Epoxidation of aldrin in soil leading to dieldrin. (Reprinted with permission from Singh *et al.*, 1991.)

and Wong, 1967) and in culture (Elsner *et al.*, 1972) and the analogous oxidation of aldrin to its epoxide, which is called dieldrin, in soil as a result of microbial action (Lichtenstein and Schulz, 1960) and in culture (Korte *et al.*, 1962).

The activation of aldrin and the consequent formation of dieldrin is at the heart of an environmental problem of enormous magnitude. The reason is because dieldrin is not only more toxic than aldrin but it is far more persistent. The disappearance of aldrin and the microbial synthesis and persistence of dieldrin are illustrated in Fig. 5.3. Dieldrin is known to be present in some soils more than 15 years after aldrin or dieldrin itself was introduced into soil to control soil-borne insects harmful to crops. The especially dramatic example involves a site near Denver, Colorado, known as the Rocky Mountain Arsenal. Here, in addition to chemicals being manufactured by the U.S. Army, a private company made aldrin and dieldrin for pest control. Unfortunately, that company disposed of residual insecticide at the site, and the chemical concentration became high at the disposal site but also became disseminated across the large land area of the arsenal grounds. Dieldrin remains there more than 20 years after its disposal. Now, the cost of cleaning up the soil contaminated with dieldrin, as well as other chemicals in the groundwater, could exceed one billion U.S. dollars.

Conversion of Phosphorothionate to Phosphate

A widely used group of insecticides are the phosphorothionates. These have the common structure:

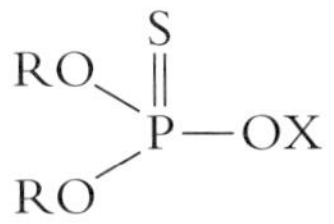

$$\begin{array}{l} \quad\quad\;\; S \\ RO\diagdown\;\| \\ \quad\quad\;\; P-OX \\ RO\diagup \end{array}$$

in which R is a short alkyl chain, typically CH_3 or CH_3CH_2, and X may be one of a variety of different structures. These molecules have little toxicity, but when they are converted to the corresponding phosphates,

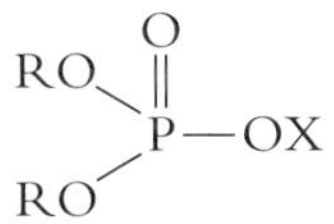

$$\begin{array}{l} \quad\quad\;\; O \\ RO\diagdown\;\| \\ \quad\quad\;\; P-OX \\ RO\diagup \end{array}$$

they become potent insecticides. Many are also highly toxic to humans and other mammals. Although such activations occur within the animal, they can also be carried out by microorganisms in natural environments and in agricultural soils. These activations can take place, for example, in soil treated with chlorfenvinphos (Read, 1971b). This type of reaction is an oxidative desulfuration that leads to the creation of a potent inhibitor of cholinesterase activity. Anticholinesterase potency may increase by a factor of about 10,000-fold (O'Brien 1960). A typical example is the conversion of parathion to its oxygen analogue (known as paraoxon) in soil (Sethunathan and Yoshida, 1973a) and microbial cultures (Munnecke and Hsieh, 1976).

A similar type of activation occurs with a group of compounds known as phosphorodithoates:

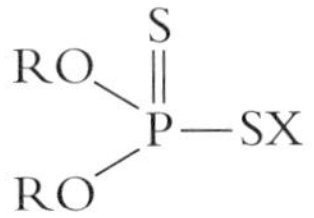

$$\begin{array}{l} \quad\quad\;\; S \\ RO\diagdown\;\| \\ \quad\quad\;\; P-SX \\ RO\diagup \end{array}$$

They also become active when $P{=}S$ is converted to $P{=}O$. A well-described case is the conversion of dimethoate to its oxygen analogue in soil (Duff and Menzer, 1973). This type of activation is shown in Fig. 5.4.

Metabolism of Phenoxyalkanoic Acids

A prominent herbicide is 2,4-D, which is itself a potent phytotoxin. However, a number of structurally related but inactive compounds may be converted in plants to 2,4-D, and thus they act as herbicides following the activation process. Similar conversions occur in soil. Collectively, these phenoxyalkanoic acids are

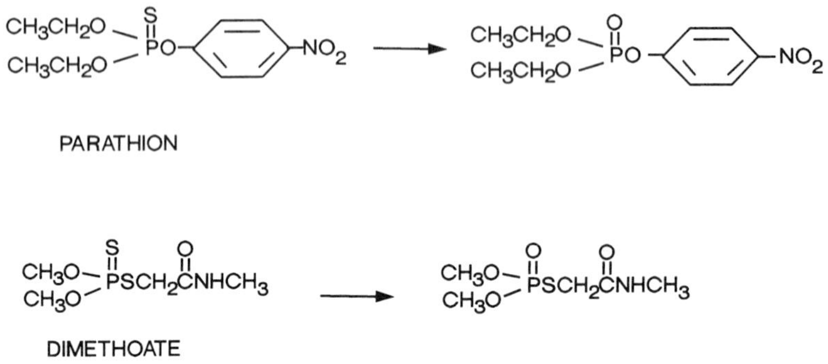

$(CH_3CH_2O)_2P(S)O{-}C_6H_4{-}NO_2 \longrightarrow (CH_3CH_2O)_2P(O)O{-}C_6H_4{-}NO_2$

PARATHION

$(CH_3O)_2P(S)SCH_2C(O)NHCH_3 \longrightarrow (CH_3O)_2P(O)SCH_2C(O)NHCH_3$

DIMETHOATE

Figure 5.4 Conversion of parathion and dimethoate to their oxygen analogues.

ω-(2,4-dichlorophenoxy)alkanoic acids, the ω-linkage referring to the attachment of the last (ω) C of the fatty (alkanoic) acid to the 2,4-dichlorophenoxy moiety through the O. The transformation may thus be viewed as shown in Fig. 5.5 for 6-(2,4-dichlorophenoxy)hexanoic acid as the parent chemical (Gutenmann *et al.*, 1964). The sequence is called β-oxidation because the steps in which each of the two carbons are removed initially involve the oxidation of the β-carbon of the aliphatic acid moiety:

$$RCH_2CH_2COOH \longrightarrow RCH{=}CHCOOH \longrightarrow RCH(OH)CH_2COOH$$
$$\longrightarrow RC(O)CH_2COOH \longrightarrow RC(O)OH + CH_3COOH$$

Because similar reactions occur in bacterial cultures (Taylor and Wain, 1962) and the chemicals probably do not undergo such reactions in nature by abiotic mechanisms, this type of activation in soil appears to be microbial.

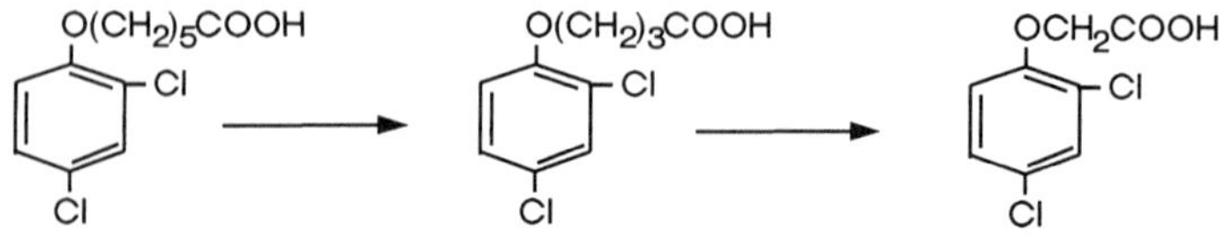

$O(CH_2)_5COOH$ (2,4-dichlorophenyl) $\longrightarrow$ $O(CH_2)_3COOH$ (2,4-dichlorophenyl) $\longrightarrow$ OCH_2COOH (2,4-dichlorophenyl)

Figure 5.5 Transformation of 6-(2,4-dichlorophenoxy)hexanoic acid to 4-(2,4-DB) and finally to the actual phytotoxin, 2,4-D.

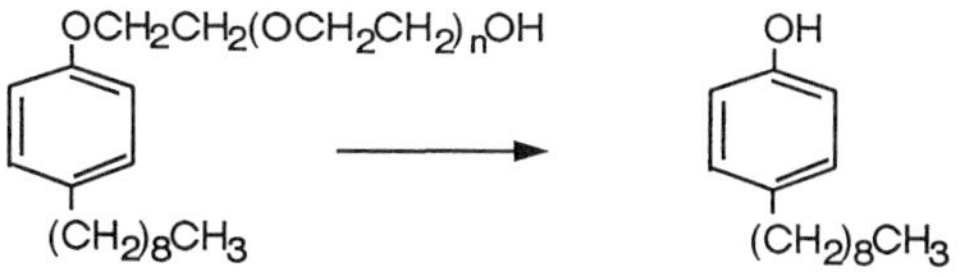

$$\text{4-}(CH_2)_8CH_3\text{-}C_6H_4\text{-}OCH_2CH_2(OCH_2CH_2)_nOH \longrightarrow \text{4-}(CH_2)_8CH_3\text{-}C_6H_4\text{-}OH$$

Figure 5.6 Microbial conversion of nonylphenol polyethoxylates to 4-nonylphenol.

A related activation is evident in the removal of the side chain of nonylphenol polyethoxylate surfactants by microorganisms in sewage sludge. The product is 4-nonylphenol (Fig. 5.6). The latter is important because it is highly toxic to fish and other aquatic organisms (Giger *et al.*, 1984) and it also is a weak estrogen (Routledge and Sumpter, 1996). An estrogen is a substance that enhances the development of secondary sex characteristics in females. It has been found in river water, sometimes at concentrations as high as 0.60 μg/liter, and in many aquatic sediments (Naylor, 1996). In this way, a component of detergents, which are widely used in the home and in commerce, is ultimately released to surface waters, where it may cause harm.

Oxidation of Thioethers

A number of chemicals containing thioether linkages (—C—S—C—) are sold as insecticides yet have only modest toxicity to insects, but they are activated and become more potent as they are oxidized to the corresponding sulfoxides and sulfones:

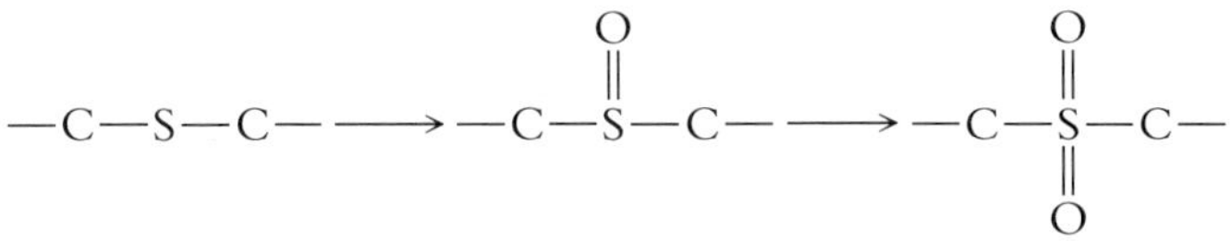

$$-C-S-C- \longrightarrow -C-\overset{\overset{\displaystyle O}{\|}}{S}-C- \longrightarrow -C-\underset{\underset{\displaystyle O}{\|}}{\overset{\overset{\displaystyle O}{\|}}{S}}-C-$$

Similar reactions occur in soil, presumably largely by microbial action, and in pure cultures of certain microorganisms. The chief toxicants in nature appear to be the sulfoxide, the sulfone, or both. Three compounds marketed as insecticides have been widely studied in this regard, namely, aldicarb (synonym Temik), phorate (synonym Thimet), and disulfoton. Aldicarb is also used as a miticide and nematicide. The structures of the compounds are:

aldicarb:

$$CH_3SC(CH_3)_2CH{=}NOC(=O)NHCH_3$$

phorate:

$$CH_3CH_2SCH_2SP(=S)(OCH_2CH_3)_2$$

disulfoton:

$$CH_3CH_2SCH_2CH_2SP(=S)(OCH_2CH_3)_2$$

In each instance, the sulfur adjacent to the terminal CH_3 or CH_3CH_2 can be oxidized to the corresponding sulfoxide and sulfone. The formation of toxic analogues occurs in soil treated with aldicarb (Richey *et al.*, 1977), phorate (Getzin and Shanks, 1970), and disulfoton (Clapp *et al.*, 1976). Microorganisms also form sulfoxides, sulfones, or both in cultures incubated with aldicarb (Jones, 1976) and phorate (Le Patourel and Wright, 1976). Other putative pesticides are also converted to the more toxic sulfoxide and sulfone derivatives in soil (Whitten and Bull, 1974).

Hydrolysis of Esters

Several esters marketed as herbicides are activated by hydrolysis to give the actual phytotoxin, which is the free acid (Hatzois and Penner, 1982; Kerr, 1989):

$$RC(=O)OR' \longrightarrow RC(=O)OH$$

This reaction is carried out in soils amended with flamprop-methyl (Roberts and Standen, 1978), benzoylprop-ethyl (Beynon *et al.*, 1974), and dichlorfop-methyl (Gaynor, 1984). As the names of these pesticides indicate, R′ is CH_3 or CH_2CH_3. The second product of the conversion is presumably the nontoxic alcohol HOR′.

Other Activations

A number of other microbial activities may lead to the formation of toxic products. Insofar as is presently known, these activations are not widespread and may be unique to the individual compound.

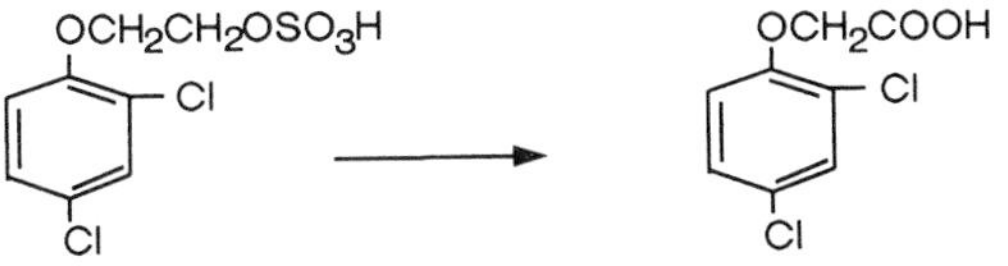

$$\text{2,4-Cl}_2\text{C}_6\text{H}_3\text{OCH}_2\text{CH}_2\text{OSO}_3\text{H} \longrightarrow \text{2,4-Cl}_2\text{C}_6\text{H}_3\text{OCH}_2\text{COOH}$$

Figure 5.7 Activation of 2,4-dichlorophenoxyethyl sulfate.

An early example of activation is the conversion in soil, apparently by microorganisms, of 2,4-dichlorophenoxyethyl sulfate to 2,4-D. The former is innocuous, but the latter is a potent phytotoxin (Fig. 5.7). The reaction also is carried out by *Bacillus cereus* (Audus, 1952; Vlitos and King, 1953). The presumed intermediate is 2,4-dichlorophenoxyethanol. Although microorganisms cleave other sulfate esters, the products are not generally toxic.

Microorganisms are able to convert the fungicide benomyl (synonym Benlate) to benzimidazole carbamic acid methyl ester. Both the parent compound and the product are fungicidal (Clemons and Sisler, 1969). However, for some fungi, the transformation is an activation because they are more sensitive to the product (Felsot and Pedersen, 1991).

Although benzene and ethylbenzene are significant pollutants that enter groundwaters and soils from petroleum and its products, both can be formed by microbial processes in soil (Wheatley *et al.*, 1996). Neither the precursors nor the organisms responsible have been established.

Chlorinated dibenzo-*p*-dioxins and dibenzofurans are among the most toxic substances known, especially TCDD. These extremely hazardous compounds can be produced from 3,4,5- and 2,4,5-trichlorophenols by peroxidases (Öberg *et al.*, 1990; Svenson *et al.*, 1989). Peroxidases may also convert pentachlorophenol to an octachlorodibenzo-*p*-dioxin (Morimoto and Tatsumi, 1997). However, the biological formation of such toxicants in nature or by microorganisms has not been described.

Several other types of activation occur in microbial cultures, but their significance in nature is uncertain. For example, the fungus *Cunninghamella elegans* converts pyrene to 1,6- and 1,8-dihydroxypyrenes, which probably can be injurious to higher organisms (Cerniglia *et al.*, 1986). Several microorganisms convert the fungicide triadimefon to triadimenol, and since some fungi are resistant to the first but sensitive to the second compound, this reaction represents an activation (Deas *et al.*, 1986); the process involves the reaction

$$RR'C{=}O \longrightarrow RR'CHOH$$

Some bacteria, in culture, are able to *O*-methylate chloroguaiacols:

$$ROH \rightarrow ROCH_3$$

The products of such a methylation are toxic to fish (Allard *et al.*, 1985).

Activation by microbial methylation is well known in the cases of mercury, arsenic, and tin. The methylation of mercury has received considerable attention because it has resulted in the contamination of fish and bans on the use of several bodies of water for commercial or sports fishing. The concentrations of methylmercury in aquatic animals may be several orders of magnitude greater than that in the ambient water because the methylation of cationic mercury results in bioconcentration of the element. The transformation occurs in aerobic and anaerobic environments, and both monomethyl- (CH_3Hg^+) and dimethylmercury (CH_3HgCH_3) may be formed (Bisogni and Lawrence, 1975). The process takes place in fresh water, estuarine and marine sediments, and soils, but the focus of interest has been in sediments because of the uptake of methylmercury by fish. A variety of bacteria are able to carry out this metabolic step in culture.

Inorganic arsenic is also subject to methylation. Although arsenite and arsenate are themselves toxic, and indeed are more toxic than some methylarsenic compounds, so that a true activation is not involved, methylation of this element is of special interest because some of the methylated species are volatile, and thus respiratory exposures may occur with harmful outcomes. Indeed, arsenic poisoning has occurred because of human inhalation of methylarsenic released in houses by the activity of microorganisms growing on As-containing wallpapers. Monomethylarsine (CH_3AsH_2), dimethylarsine [$(CH_3)_2AsH$], and arsine (AsH_3) are formed in soil and appear as volatile metabolites (Cheng and Focht, 1979). In addition, the following compounds have been found in marine and fresh waters as well as in microbial cultures (Braman, 1975; Cox and Alexander, 1973):

$CH_3AsO(OH)_2$	methylarsonic acid
$(CH_3)_2AsO(OH)$	dimethylarsinic acid
$(CH_3)_3As$ (or the oxide)	trimethylarsine

Although inorganic tin is of little toxicological significance, methylated forms of this element are highly toxic. Trimethyltin, for example, at low concentrations produces irreversible neuronal damage and neuronal necrosis in the brain, and it is absorbed into the body from the stomach, intestine, and even the skin (Aldridge and Brown, 1988). Microorganisms in sediments and a number of species in pure culture are able to methylate inorganic tin and form mono-, di-, and trimethyltin (Gilmour *et al.*, 1987; Hallas *et al.*, 1982). The fact that sterilized sediments do not form methyltin compounds confirms that the process is microbial. Inasmuch as mono-, di-, and trimethyltin are found in natural waters, albeit at low concentrations (Byrd and Andrea, 1982), the process is not merely a laboratory phenomenon. Nevertheless, it is not clear whether the microbial transformation constitutes a substantive threat to human health.

The reverse reaction—demethylation—may also result in activation, witnessed by the fact that the demethylation of diphenamid (*N,N*-dimethyl-2,2-diphenylacetamide) to yield the monomethyl and the unmethylated 2,2-diphenylacetamide by *Trichoderma viride* and *Aspergillus candidus* converts a nontoxic precursor to two phytotoxins (Kesner and Ries, 1967); a similar conversion has been noted in soil treated with diphenamid (Golab *et al.*, 1968):

$$RN(CH_3)_2 \longrightarrow RN(CH_3)H \longrightarrow RNH_2$$

Another activation to yield a phytotoxin from an innocuous precursor has been noted both in soil as well as in cultures of *Pseudomonas putrefaciens*. The reaction is the conversion of α-amino-2,6-dichlorobenzaldoxime to 2,6-dichlorobenzonitrile (Milbarrow, 1963).

Several other carcinogens are generated microbiologically. For example, the carcinogens 1,1- and 1,2-dimethylhydrazine are produced during anaerobic biodegradation of the explosive known as RDX (Fig. 5.8) (McCormick *et al.*, 1981).

In some instances, the microbial transformation itself does not create a toxicant. Nevertheless, the product can be of public health or ecological importance because it undergoes a nonbiological reaction that leads to a hazardous substance. This is particularly true of compounds that are chlorinated as a result of Cl_2 treatment of water for human consumption. An example of such a problem is the conversion by sewage bacteria of dodecylsulfate to acetoacetic acid; the latter is converted in high yield to chloroform, a toxicant, when the water is treated with Cl_2 (Itoh *et al.*, 1985).

The examples cited here represent activations in a strict sense, that is, the product is more toxic than the precursor. However, microorganisms may increase the exposure of sensitive organisms to toxicants not only because they form more harmful metabolites but also because they alter the *mobility* or *persistence* of toxic compounds. The microbial product is often more mobile and sometimes more persistent than the parent molecule, and the product but not the parent may be

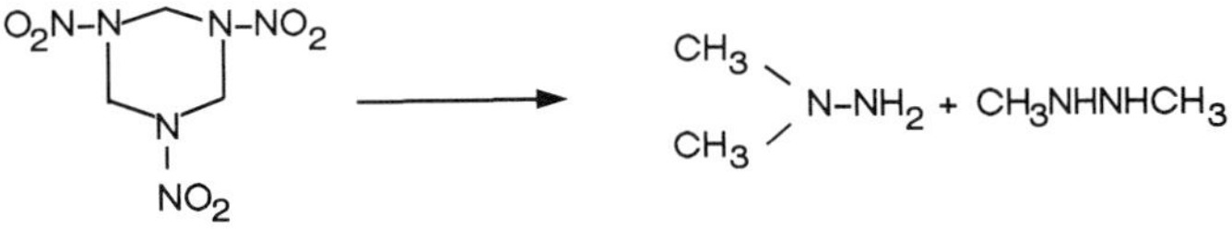

$$\text{RDX } (O_2N\text{-}N\text{, }N\text{-}NO_2\text{, }N\text{-}NO_2 \text{ ring}) \longrightarrow (CH_3)_2N\text{-}NH_2 + CH_3NHNHCH_3$$

Figure 5.8 Biodegradation of RDX.

found in the groundwater underlying soil receiving the precursor or in soil long after the original chemical has disappeared. In this way, the exposure can be enhanced. Thus, dieldrin persists long after aldrin applied to soil has been metabolized and constitutes a long-term source of pollution. With the *N*-nitrosodimethylamine formed from dimethylamine, not only does the conversion result in the appearance of a highly potent carcinogen but the nitroso compound moves through soil into groundwaters, is carried with the flow of the groundwater, and is highly persistent; this contrasts with the lack of appreciable injury, poor mobility, and the frequently rapid biodegradation of the precursor amine.

DEFUSING

A compound that is potentially activated will pose a health or environmental hazard if it undergoes that type of reaction. However, if the microflora converts that substrate to a different metabolite that itself is both harmless and not subject to activation, the potential problem posed by the initial substrate does not arise. Thus, A is converted to C rather than to B:

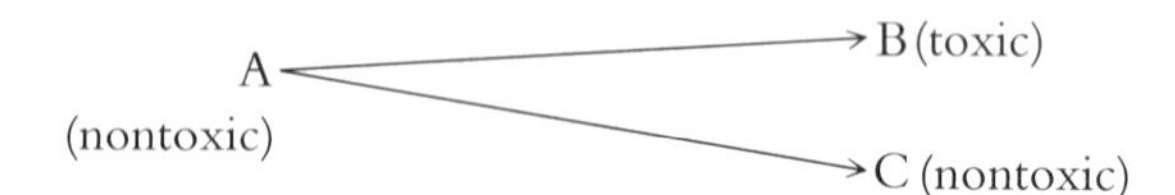

By analogy to a bomb that only explodes after its fuse is lit, this phenomenon has been termed *defusing*. Just as a bomb loses its capacity to do injury by having the fuse removed, the chemical is not activated by the prior occurrence of the defusing reaction.

Defusing is best illustrated by some pesticides that undergo activation. Among the phenoxy herbicides, 4-(2,4-DB) is activated when it undergoes β-oxidation to yield 2,4-D. Hence, bacteria that cleave the molecule by removing butyric acid from the side chain to release 2,4-dichlorophenol in culture (MacRae *et al.*, 1963) are defusing the molecule (Fig. 5.9). Diazinon is activated when the P=S is converted to P=O, but its cleavage to 2-isopropyl-4-methyl-6-hydroxypyrimidine and diethylthiophosphate in soil (Konrad *et al.*, 1967) and in culture (Sethunathan and Yoshida, 1973b) similarly represents a defusing (Fig. 5.9). Defusing has also been reported for a number of other insecticides that are activated by the conversion of P=S to P=O, for example, when part or all of the parathion (Munnecke and Hsieh, 1976) or malathion (Rosenberg and Alexander, 1979) is cleaved in bacterial cultures or dimethoate is cleaved in soil (El Beit *et al.*, 1978) before the chemical is activated.

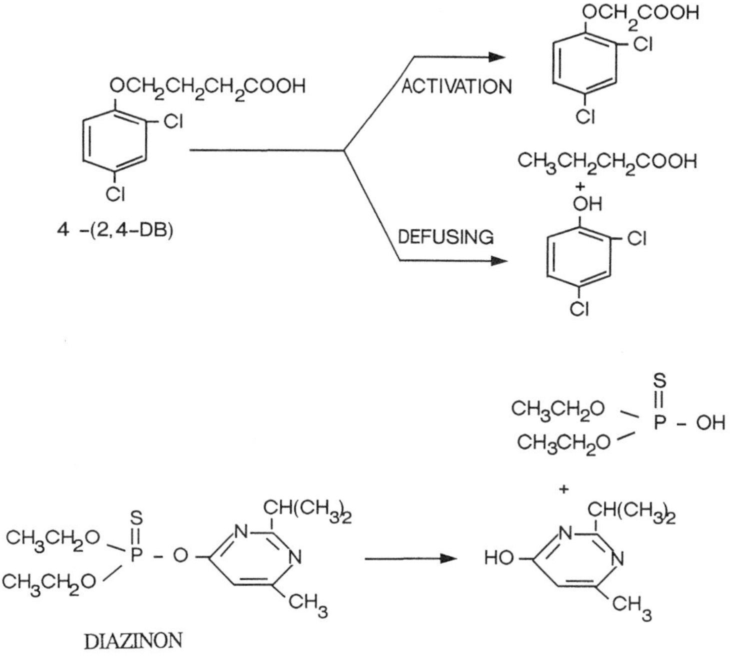

Figure 5.9 Defusing of two pesticides.

CHANGE IN TOXICITY SPECTRUM

On occasion, a compound harmful to one group of organisms is converted to a molecule injurious to an entirely different group of organisms. Although such a change in spectrum of toxicity is not quite what is meant by activation in a strict sense, it still is an activation for the second group of organisms because for them the substance that was originally benign is converted to a form that causes harm.

One of the most dramatic instances of a change in spectrum of toxicity occurred in Japanese agriculture. Pentachlorobenzyl alcohol, a compound tolerated by many plant species even at greater than 2000 mg/kg, was introduced for the control of a fungal disease of the rice plant known as rice blast disease. However, 2 years after beginning its use, toxicity to tomato, cucumber, and melon plants was noted, especially in fields where rice straw had been added to the soil the year before. As a result of the marked crop losses, it was found that the pentachlorobenzyl alcohol was converted in the soil to penta-, 2,3,4,6- and/or 2,3,5,6-tetra-, and 2,4,6-trichlorobenzoic acid, all of which, even at extremely low concentrations, were toxic to a variety of plants. As a result of this transformation, the production and marketing of pentachlorobenzyl alcohol immediately ceased (Ishida, 1972).

Many other examples have been described in which the spectrum of toxicity has been changed, presumably as a result of microbial activity (Table 5.1). However, some of the conversions cited could be abiotic. Differing from the case of pentachlorobenzyl alcohol, moreover, the instances cited are not known to have resulted in harm to susceptible species in nature. With other compounds, both the parent molecule and one or more of its metabolic products are toxic to a single species. Thus, atrazine is deethylated in soil, and both parent and daughter are toxic to plants (Sirons *et al.,* 1973), and 2,6-dichlorobenzonitrile (the herbicide known as dichlobenil) is metabolized microbiologically in soil to yield 2,6-dichlorobenzamide, and both precursor and product are phytotoxic (Verloop, 1972). The microbial conversion of avermectin B_{2a} to its 23-keto derivative in soil represents the conversion of one nematicide to another and, as in some of the instances cited, the toxic product is more persistent than the original chemical (Gullo *et al.,* 1983). A variety of fungicides are converted microbiologically to substances that are themselves antifungal; for example, benomyl is converted to the antifungal benzimidazole carbamic acid methyl ester by *Saccharomyces cerevisiae* (Clemons and Sisler, 1969), several microorganisms convert triadimefon to the toxic triadimenol (Deas *et al.,* 1986), and *Neurospora crassa* converts the fungicide captan to the antifungal carbonyl sulfide (Somers *et al.,* 1967). These antifungal-to-antifungal conversions have been described in microbial cultures, but the biodegradation of the fungicide chlorothalonil to the fungicide 1,3-dicarbamoyl-2,4,5,6-tetrachlorobenzene takes place in soil (Rouchaud *et al.,* 1988).

Table 5.1

Changes in Spectrum of Toxicity Arising from Chemical Transformations

Process	Reaction	Reference
Antifungal to carcinogen precursor	Thiram to dimethylamine	Odeyemi and Alexander (1977)
Insecticide to fish toxin	DDT to 1,1-dichloro-*bis*(4-chlorophenyl)ethane	Day (1991)
Insecticide to acaricide	DDT to 1,1-dichloro-*bis*(4-chlorophenyl)ethane	Matsumura *et al.* (1971)
Herbicide to genotoxic product	2,4,5-T to 2,4,5-trichlorophenol	George *et al.* (1992)
	Propanil to 3,3′,4,4′-tetrachloroazobenzene	Prasad (1970)

RISKS FROM BIODEGRADATION

Clearly, many harmful metabolites are generated microbiologically in a variety of environments. These products may represent substantive threats to the health, growth, or vigor of humans and a variety of animals and plants. In view of such conversions, studies of breakdown pathways in natural environments are of particular importance. It is not sufficient merely to measure the persistence and disappearance of the parent substance. What microorganisms do to that chemical may be of as great or even greater importance to human health, agricultural productivity, or populations in natural ecosystems.

The view is often expressed, particularly by the purveyors of a technology, that the technology is not only useful but is also without risk. Whether this view is actually believed or is espoused solely to convince the unwary to use the technology is often difficult to assess.

Every new technology has a risk. That risk may be large or small, but it does exist. Recognizing the issues or factors coupled to the risk is a first step in reducing or avoiding the risk. In the examples of activation in this chapter, such risks have been discussed. These issues are not insubstantial, and by learning more about the dangers associated with microbial metabolites, it should be possible to establish approaches to avoid their occurrence or reduce their concentrations.

The biologically active metabolite formed from a toxicant may not always be toxic. Sometimes, it may be stimulatory. The conversion of toxic compounds to stimulators has been shown particularly for phytotoxic parents. For example, the herbicides DNOC (Bruinsma, 1960), dinoseb (Crafts, 1949), and 2,4-D (Newman *et al.,* 1952) are each converted to breakdown products that actually stimulate plant growth, but the identities of the stimulatory metabolities have not been established.

REFERENCES

Abrahamsson, K., Ekdahl, A., Collén, J., and Pedersén, M., *Limnol. Oceanogr.* **40,** 1321–1326 (1995).

Aldridge, W. N., and Brown, A. W., *in* "The Biological Alkylation of Heavy Elements" (P. J. Craig and F. Glockling, eds.), pp. 147–163. Royal Society of Chemistry, London, 1988.

Allard, A. S., Remberger, M., and Neilson, A. H., *Appl. Environ. Microbiol.* **49,** 279–288 (1985).

Audus, L. J., *Nature (London)* **170,** 886–887 (1952).

Ayanaba, A., and Alexander, M., *Appl. Microbiol.* **25,** 862–868 (1973).

Ayanaba, A., and Alexander, M., *J. Environ. Qual.* **3,** 83–89 (1974).

Ayanaba, A., Verstraete, W., and Alexander, M., *J. Natl. Cancer Inst. (U.S.)* 50, 811–813 (1973).

Belkin, S., Stieber, M., Tiehm, A., Frimmel, F. H., Abeliovich, A., Werner, P., and Ulitzur, S., *Environ. Toxicol. Water Qual.* **9,** 303–309 (1994).

Beynon, K. I., Roberts, T. R., and Wright, A. N., *Pestic. Sci.* **5,** 451–463 (1974).

Bisogni, J. J., Jr., and Lawrence, A. W., *J. Water Pollut. Control Fed.* **47,** 135–152 (1975).

Braman, R. S., *in* "Arsenical Pesticides" (E. A. Woolson, ed.), pp. 108–123. American Chemical Society, Washington, DC, 1975.

Brewer, W. S., Draper, A. C., III, and Wey, S. S., *Environ. Pollut. Ser. B* **1,** 37–83 (1980).

Bruinsma, J., *Plant Soil* **12,** 249–258 (1960).

Byrd, J. T., and Andreae, M. O., *Science* **218,** 565–569 (1982).

Cerniglia, C. E., Kelly, D. W., Freeman, J. P., and Miller, D. W., *Chem.-Biol. Interact.* **57,** 203–216 (1986).

Cheng, C.-N., and Focht, D. D., *Appl. Environ. Microbiol.* **38,** 494–498 (1979).

Clapp, D. W., Naylor, D. V., and Lewis, G. C., *J. Environ. Qual.* **5,** 207–208 (1976).

Clemons, G. P., and Sisler, H. D., *Phytopathology* **59,** 705–706 (1969).

Collins-Thompson, D. L., Sen, N. P., Aris, B., and Schwinghamer, L., *Can. J. Microbiol.* **18,** 1968–1971 (1972).

Cox, D. P., and Alexander, M., *Bull. Environ. Contam. Toxicol.* **9,** 84–88 (1973).

Crafts, A. S., *Hilgardia* **19,** 159–169 (1949).

Day, K. E., *in* "Pesticide Transformation Products" (L. Somasundaram and J. R. Coats, eds.), pp. 217–241. American Chemical Society, Washington, DC, 1991.

Deas, A. H. B., Carter, G. A., Clark, T., Clifford, D. R., and James, C. S., *Pestic. Biochem. Physiol.* **26,** 10–21 (1986).

de Jong, E., Field, J. A., Spinnler, H. E., Wijnberg, J. B. P. A., and De Bont, J. A. M., *Appl. Environ. Microbiol.* **60,** 264–270 (1994).

Duff, W. G., and Menzer, R. E., *Environ. Entomol.* **2,** 309–318 (1973).

Duffy, J. R., and Wong, N., *J. Agric. Food Chem.* **15,** 457–464 (1967).

El Beit, I. O. D., Wheelock, J. V., and Cotton, D. E., *Int. J. Environ. Stud.* **12,** 215–225 (1978).

Elsner, E., Bieniek, D., Klein, W., and Korte, F., *Chemosphere* **1,** 247–250 (1972).

Felsot, A. S., and Pedersen, W. L., *in* "Pesticide Transformation Products" (L. Somasundaram and J. R. Coats, eds.), pp. 172–187. American Chemical Society, Washington, DC, 1991.

Field, J. A., Verhagen, F. J. M., and de Jong, E., *Trends Biotechnol.* **13,** 451–456 (1995).

Flodin, C., Johansson, E., Borén, H., Grimvall, A., Dahlman, O., and Mörck, R., *Environ. Sci. Technol.* **31,** 2464–2468 (1997).

Fujii, K., Kobayashi, M., and Takahashi, E., *J. Sci. Soil Manure Jpn.* **43**(5), 160–164 (1972).

Gaynor, J. D., *Can. J. Soil Sci.* **64,** 283–291 (1984).

George, S. E., Whitehouse, D. A., and Claxton, L. D., *Environ. Toxicol. Chem.* **11,** 733–740 (1992).

Getzin, L. W., and Shanks, C. H., Jr., *J. Econ. Entomol.* **63,** 52–58 (1970).

Giger, W., Brunner, P. H., and Schaffner, C., *Science* **225,** 623–625 (1984).

Gilmour, C. C., Tuttle, J. H., and Means, J. C., *Microb. Ecol.* **14,** 233–242 (1987).

Golab, T., Gramlich, J. V., and Probst, G. W., *Abstr. Pap., 155th Meet., Am. Chem. Soc., San Francisco,* Abstr. No. A-50 (1968).

Greene, S., Alexander, M., and Leggett, D., *J. Environ. Qual.* **10,** 415–420 (1981).

Gullo, V. P., Kempf, A. J., MacConnell, J. G., Mrozik, H., Arison, B., and Putter, I., *Pestic. Sci.* **14,** 153–157 (1983).

Gutenmann, W. H., Loos, M. A., Alexander, M., and Lisk, D. J., *Soil Sci. Soc. Am. Proc.* **28,** 205–207 (1964).

Haiber, G., Jacob, G., Niedan, V., Nkusi, G., and Schöler, H. F., *Chemosphere* **33,** 839–849 (1996).

Hallas, L. E., Means, J. C., and Cooney, J. J., *Science* **215,** 1505–1507 (1982).

Hatzois, K. K., and Penner, D., "Metabolism of Herbicides in Higher Plants." Burgess Publ. Co., Minneapolis, MN, 1982.

Hawksworth, G., and Hill, M. J., *Br. J. Cancer* **25,** 520–526 (1971).

Hjelm, O., Johansson, M.-B., and Öberg-Asplund, G., *Chemosphere* **30,** 2353–2364 (1995).

Ishida, M., *in* "Environmental Toxicology of Pesticides" (F. M. Matsumura, G. M. Boush, and T. Misato, eds.), pp. 281–306. Academic Press, New York, 1972.

Itoh, S.-I., Naito, S., and Unemoto, T., *Water Res.* **19,** 1305–1309 (1985).

Jones, A. S., *J. Agric. Food Chem.* **24,** 115–117 (1976).

Kerr, M. W., *in* "Herbicides and Plant Metabolism" (A. D. Dodge, ed.), pp. 199–210. Cambridge Univ. Press, Cambridge, UK, 1989

Kesner, C. D., and Ries, S. K., *Science* **155,** 210–211 (1967).

Konrad, J. G., Armstrong, D. E., and Chesters, G., *Agron. J.* **59,** 591–594 (1967).

Korte, F., Ludwig, G., and Vogel, J., *Justus Liebig's Ann. Chem.* **656,** 135–140 (1962).

Kozlowski, T. T., *Nature (London)* **205,** 104–105 (1965).

Kunisaki, N., and Hayashi, M., *Appl. Environ. Microbiol.* **37,** 279–282 (1979).

Laturnus, F., *Chemosphere* **31,** 3387–3395 (1995).

Le Patourel, G. N. J., and Wright, D. J., *Comp. Biochem. Physiol.* **53C,** 73–74 (1976).

Lichtenstein, E. P., and Schulz, K. R., *J. Econ. Entomol.* **53,** 192–197 (1960).

MacRae, I. C., Alexander, M., and Rovira, A. D., *J. Gen. Microbiol.* **32,** 69–76 (1963).

Mallik, M. A. B., Tesfai, K., and Pancholy, S. K., *Proc. Okla. Acad. Sci.* **61,** 31–35 (1981).

Matsumura, F., Patil, K. C., and Boush, G. M., *Nature (London)* **230,** 325–336 (1971).

McCormick, N. G., Cornell, J. H., and Kaplan, A. M., *Appl. Environ. Microbiol.* **42,** 817–823 (1981).

Milbarrow, B. V., *Biochem. J.* **87,** 255–258 (1963).

Mills, A. L., and Alexander, M., *J. Environ. Qual.* **5,** 437–440 (1976).

Morimoto, K., and Tatsumi, K., *Chemosphere* **34,** 1277–1283 (1997).

Munnecke, D. M., and Hsieh, D. P. H., *Appl. Environ. Microbiol.* **31,** 63–69 (1976).

Naylor, C. G., in "Pesticide Formulations and Application Systems: 16th Volume" (M. J. Hopkimson, H. M. Collins, and G. R. Gross, eds.), pp. 3–20. American Society for Testing and Materials, West Conshohocken, PA, 1996.

Neurath, G. B., Dünger, M., Pein, F. G., Ambrosius, D., and Schreiber, O., *Food Cosmet. Toxicol.* **15,** 275–282 (1977).

Newman, A. S., Thomas, J. R., and Walker, R. L., *Soil Sci. Soc. Am. Proc.* **16,** 21–24 (1952).

Newman, L. M., and Wackett, L. P., *Appl. Environ. Microbiol.* **57,** 2399–2402 (1991).

Nightingale, P. D., Malin, G., and Liss, P. S., *Limnol. Oceanogr.* **40,** 680–689 (1995).

Öberg, G., Brunberg, H., and Hjelm, O., *Soil Biol. Biochem.* **29,** 191–197 (1997).

Öberg, L. G., Glas, B., Swanson, S. E., Rappe, C., and Paul, C. G., *Arch. Environ. Toxicol. Chem.* **19,** 930–938 (1990).

O'Brien, R. D., "Toxic Phosphorus Esters." Academic Press, New York, 1960.

Odeyemi, O., and Alexander, M., *Appl. Environ. Microbiol.* **33,** 784–790 (1977).

Pieper, D. H., Winkler, R., and Sandermann, H., Jr., *Angew. Chem., Int. Ed. Engl.* **31,** 68–70 (1992).

Prasad, I., *Can. J. Microbiol.* **16,** 369–372 (1970).

Read, D. C., *J. Econ. Entomol.* **62,** 1328–1334 (1969).

Read, D. C., *J. Econ. Entomol.* **64,** 796–800 (1971a).

Read, D. C., *J. Econ. Entomol.* **64,** 800–804 (1971b).

Richardson, M. L., Webb, K. S., and Gough, T. A., *Ecotoxicol. Environ. Saf.* **4,** 207–212 (1980).

Richey, F. A., Jr., Bartley, W. J., and Sheets, K. P., *J. Agric. Food Chem.* **25,** 47–51 (1977).

Roberts, T. R., and Standen, M. E., *Pestic. Biochem. Physiol.* **9,** 322–333 (1978).

Rosenberg, A., and Alexander, M., *Appl. Environ. Microbiol.* **37,** 886–891 (1979).

Rouchaud, J., Roucourt, P., Vanachter, A., Benoit, F., and Ceustermans, N., *Rev. Agric. (Brussels)* **41,** 889–899 (1988).

Routledge, E. J., and Sumpter, J. P., *Environ. Toxicol. Chem.* **15,** 241–248 (1996).

Sander, J., Schweinsberg, E., Ladenstein, M., and Schweinsberg, F., *Zentralbl. Bakteriol., Mikrobiol. Hyg., Abt. I, Orig. A* **227,** 71–80 (1974).

Sethunathan, N., and Yoshida, T., *J. Agric. Food Chem.* **21,** 504–506 (1973a).

Sethunathan, N., and Yoshida, T., *Can. J. Microbiol.* **19,** 873–875 (1973b).

Singh, G., Kathpal, T. S., Spencer, W. F., and Dhankar, J. S., *Environ. Pollut.* **70,** 219–239 (1991).

Sirons, G. J., Frank, R., and Sawyer, T., *J. Agric. Food Chem.* **21,** 1016–1020 (1973).
Somers, E., Richmond, D. V., and Pickard, J. A., *Nature (London)* **215,** 214 (1967).
Svenson, A., Kjeller, L.-O., and Rappe, C., *Environ. Sci. Technol.* **23,** 900–902 (1989).
Tanaka, Y., Iwasaki, H., and Kitamori, S., *Water Sci. Technol.* **34,** 15–20 (1996).
Tate, R. L., III, and Alexander, M., *Soil Sci.* **118,** 317–321 (1974).
Tate, R. L., III, and Alexander, M., *Appl. Environ. Microbiol.* **31,** 399–403 (1976).
Taylor, H. F., and Wain, R. L., *Proc. R. Soc. London, Ser. B.* **156,** 172–186 (1962).
Verloop, A., *Residue Rev.* **43,** 55–103 (1972).
Vlasenko, N. L., Zhuravleva, I. L., Terenina, M. B., Golovnya, R. V., and Ilnitskii, A. P., *Gig. Sanit.* **11,** 15–17 (1981).
Vlitos, A. J., and King, L. J., *Nature (London)* **171,** 523 (1953).
Wannstedt, C., Rotella, D., and Siuda, J. F., *Bull. Environ. Contam. Toxicol.* **44,** 282–287 (1990).
Wheatley, R. E., Millar, S. E., and Griffiths, D. W., *Plant Soil* **181,** 163–167 (1996).
Whitten, C. J., and Bull, D. L., *J. Agric. Food Chem.* **22,** 234–238 (1974).
Wilson, B. H., Smith, G. B., and Rees, J. F., *Environ. Sci. Technol.* **20,** 997–1002 (1986).
Yordy, J. R., and Alexander, M., *J. Environ. Qual.* **10,** 266–270 (1981).
Zwickenpflug, W., and Richter, E., *J. Agric. Food Chem.* **42,** 2333–2337 (1994).

CHAPTER 6

Kinetics*

Knowledge of the kinetics of biodegradation is essential to the evaluation of the persistence of organic pollutants and to assessing exposure of humans, animals, and plants. Once degradation of a chemical commences, the amount disappearing with time and the shape of the disappearance curve will be a function of the compound in question, its concentration, the organisms responsible, and a variety of environmental factors. Information on kinetics is extremely important because it characterizes the concentration of the chemical remaining at any time, permits prediction of the levels likely to be present at some future time, and allows assessment of whether the chemical will be eliminated before it is transported to a site at which susceptible humans, animals, or plants may be exposed. Such knowledge is thus essential for the assessment of the potential risk associated with exposure of susceptible individuals and species to the chemical.

Research on kinetics has focused on two topics. The first is assessing factors that affect the amounts of the compound transformed per unit time. In this regard, much information is available on the influence of temperature, pH, soil moisture, and other C sources on the rates of transformation. The second topic is determining the shapes of the curves that depict the transformation and evaluating which of the patterns of decomposition best fit the metabolism of given chemicals in a microbial culture, in laboratory microcosms, or, occasionally, in the field. This second topic is the subject of this chapter.

In many instances, the information on kinetics comes only from evaluations of the loss of the parent molecule. This is warranted for toxicants whose inhibitory

*Significant portions of this chapter were published earlier. (Reprinted with permission from Alexander and Scow, 1989.)

effects are totally destroyed as a result of the first enzymatic reaction or in metabolic sequences in which intermediates do not accumulate. It is not warranted if intermediates accumulate, especially if they are toxic or their hazard has yet to be evaluated. In other instances, the knowledge comes from tests of mineralization of the compounds; although mineralization reflects detoxication, the pattern of disappearance of toxicity from the natural environment could be quite different from the patterns of conversion of the organic molecule to inorganic products. Unfortunately, essentially no attention has been given to the kinetics of the various steps in transformations in nature that result in product accumulation.

Many models have been proposed to represent the kinetics of biodegradation. An understanding of when to use these models and why they may fail to describe data requires knowledge of the theoretical bases for these models. The models commonly used to fit data from evaluations of biodegradation are, in essence, either empirical or theoretical.

The study of kinetics of biodegradation in natural environments is often empirical, reflecting the rudimentary level of knowledge about microbial populations and activity in these environments. An example of an empirical approach is the power rate model (Hamaker, 1972)

$$-dC/dt = kC^n, \tag{1}$$

where C is substrate concentration, t is time, k is the rate constant for chemical disappearance, and n is a fitting parameter. This model can be fit to substrate–disappearance curves by varying n and k until a good fit is achieved. From this equation, it is evident that the rate is proportional to a power of the substrate concentration. The power-rate law provides a basis for comparison of different curves, but it gives no insight into the reasons for the shapes. Therefore, often it may have no predictive ability. Moreover, investigators interested in kinetics do not always state whether the model they are using has a theoretical basis or is simply empirical, and whether constants in an equation have physical meaning or are only fitting parameters (Bazin *et al.*, 1976).

An appropriate introduction to the kinetics of biodegradation is to consider a pure culture of a single bacterial population that is growing on and degrading a single, soluble organic chemical, and to assume that no barriers exist between the substrate and the cells.

PROCESSES LINKED TO GROWTH

The biodegradation of a particular organic substrate may be carried out by microorganisms that are (a) growing at the expense of that substrate and using it as a source of C, energy, or possibly another nutrient element needed for proliferation; (b) growing at the expense of another organic nutrient that is used as a source of C, energy, or both but metabolizing the substrate of interest (although not using it to supply building blocks for cell synthesis); or (c) not growing as they metabolize the chemical of concern.

Consider the case of a population of cells of one bacterial species using a particular organic compound as its source of C and energy. To simplify the illustration, assume that the organic molecule is water soluble and nontoxic, that the organism is growing in well-aerated liquid media, and that the inorganic nutrients and growth factors needed by the bacterium are present in excess of the organism's need. At low concentrations of the C and energy source, the growth rate of the organism is slow because it is limited by the low level of the substrate. If the organism is inoculated into flasks containing increasing concentrations of the C compound, its growth rate will increase in proportion to the increase in concentration. Above some moderately high level of substrate, the growth rate does not rise as markedly with increasing concentrations. Ultimately, at a quite high level, the growth rate does not increase with further rises in concentration. Monod (1949) mathematically formulated this relationship. It is now commonly written as

$$\mu = \frac{\mu_{\max} S}{K_s + S}, \tag{2}$$

where μ is the specific growth rate of the bacterium, $\mu_{\max}$ is its maximum specific growth rate (which occurs at the higher range of substrate concentrations), S is the substrate concentration, and K_s is a constant that represents the substrate concentration at which the rate of growth is half the maximum rate. The expression for the Monod equation is presented graphically as the hyperbolic curve shown in Fig. 6.1. The value for K_s represents the affinity of the organism for the growth-supporting organic nutrient; the lower the value, the greater is the bacterium's affinity for that molecule.

Values for K_s extend over a considerable range. For a single bacterium, the K_s value differs with different substrates. For a single substrate, the value will depend on the bacterium. Moreover, a single microorganism may have one K_s value at low substrate concentrations and another at high concentrations. Some values are given in Table 6.1. From the data presented, it is immediately evident that the values vary enormously, and no generalizations are evident at first glance. Nevertheless, bacteria characteristic of nutrient-rich environments often have higher K_s values than those obtained from habitats with low levels of organic constituents. Also given in Table 6.1 are the K_s values for metabolism in natural waters, with the data probably representing the species in the water sample that most rapidly metabolizes the added molecule. It is worth pointing out that the concentrations of the C source commonly included in culture media are far higher than the values recorded for K_s.

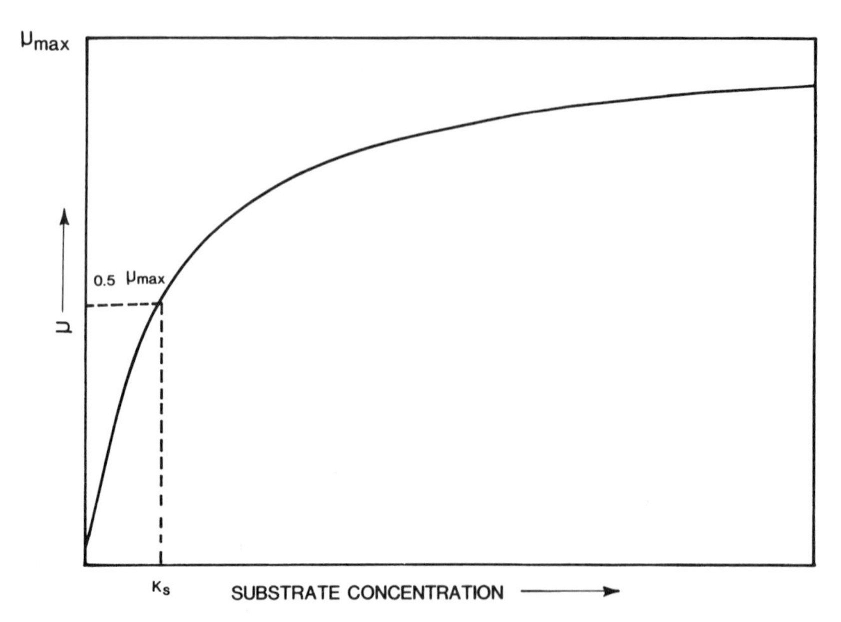

Figure 6.1 Relationship between the growth rate of a bacterium and the concentration of the C source supporting its growth.

Table 6.1

Values for K_s for Bacteria and for Samples of Water

Substrate	Organism or sample	K_s value (mg/liter)	Reference
Glucose	*Flavobacterium* 1	0.0071	van der Kooij and Hijnen (1981)
	Flavobacterium 2	29, 1314	Ishida *et al.* (1982)
	River water	26	Larson (1980)
Glutamate	*Aeromonas* sp.	0.163, 1.3	Ishida *et al.* (1982)
Maltose	*Butyrivibrio fibrisolvens*	2.1	Russell and Bladwin (1979)
Xylose	*B. fibrisolvens*	55	Russell and Baldwin (1979)
m-Cresol	Natural water	0.0006–0.0018	Bartholomew and Pfaender (1983)
Chlorobenzene	Natural water	0.0010–0.0051	Bartholomew and Pfaender (1983)
NTA	Natural water	0.060–0.170	Bartholomew and Pfaender (1983)
Phenol	Wastewater	1.3–270	Rozich *et al.* (1985)

Two K_s values or affinity constants may be evident for a single substrate, both in pure cultures and in microbial communities. One value for pure cultures may be quite low (i.e., a high affinity), in the range of less than 20 μg/liter; the other may be quite high (i.e., a low affinity) and in the range of 1 to more than 10 mg/liter (Ishida *et al.,* 1982; Lewis *et al.,* 1983). Two affinity constants have also been noted for microorganisms in lake water that degrade phenol, the concentrations at which the rates are half the maximum (0.5 μ_{max}) being ca. 5 and 400 μg/liter (Jones and Alexander, 1986). In samples of natural environments, however, the existence of more than one affinity constant may reflect the activity of a single population or, alternatively, different species with dissimilar affinities for the same organic molecule. Similar studies with suspensions of biofilms from river water also show that more than one K_s may characterize the degradation of 4-chlorophenol and glucose in a single environmental sample (Lewis *et al.,* 1988).

Under certain circumstances, Monod kinetics may not be an adequate description of bacterial growth on soluble substrates in pure culture (Koch and Wong, 1982). Nevertheless, such anomalies are not widely known and may be assumed to be uncommon. However, as pointed out in the following, such kinetics may be inappropriate for substrates that are sorbed, insoluble, or toxic, even for pure cultures.

When a pure culture of bacteria is growing in media containing a C source at concentrations far in excess of the K_s, most of the period during the growth cycle occurs without the declining substrate level greatly affecting the growth rate. Therefore, the rate of degradation of the substrate is not markedly influenced by its concentration until nearly all is gone. During this period when the C source (and presumably other nutrients) is in excess, the time for one cell to divide to give two, two to divide to give four, four to give eight, and so on is constant. Thus, for cells that multiply by binary fission, the population density increases in a geometric progression with, in the case of nonlimiting nutrients, a constant time interval between each doubling. This relationship can be expressed mathematically as

$$N_0\,(2^n) = N, \tag{3}$$

where N_0 is the initial cell number and N is the number of cells after n divisions. This is a simple exponential (logarithmic) progression that can be expressed as

$$\ln N - \ln N_0 = kt \tag{4}$$

or in $\log_{10}$,

$$\log N - \log N_0 = kt/2.303, \tag{5}$$

where t is time and k is the specific growth rate constant. On rearrangement, the relationship is

$$\log N = kt/2.303 + \log N_0, \tag{6}$$

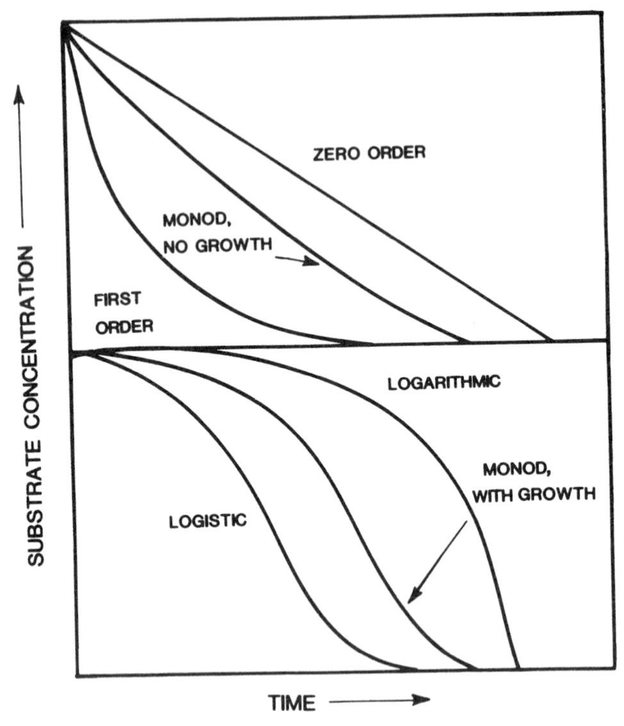

Figure 6.2 Disappearance curves for chemicals that are metabolized by different kinetics. (From Alexander, 1985. Reprinted with permission from the American Chemical Society.)

and a plot of log N versus t yields a straight line with a slope of $k/2.303$ and an intercept of N_0. Thus, during this exponential phase of bacterial growth, a plot of cell number versus time gives a straight line (Neidhardt *et al.*, 1990).

If it is assumed that each cell metabolizes the same amount of organic substrate during this phase, a plot of the logarithm of the amount of *substrate metabolized* (or the amount of C recovered in cells plus all products) versus time should also give a straight line. However, this does not mean that a plot of the logarithm of *substrate remaining* versus time gives a straight line. The shape of the curve that reflects logarithmic growth is depicted in Fig. 6.2. Initially, little loss of the chemical is evident because the cell density is low, but then as the mass of cells becomes large and doubles exponentially, a rapid loss of chemical is evident. Indeed, many doublings in population size occur with little or no loss of C source being evident, and much of the disappearance can be attributed to the last few doublings in population size. Logarithmic kinetics of substrate disappearance can be expressed in differential form as

$$dS/dt = \mu_{\max}(S_0 + X_0 - S) \tag{7}$$

and in integral form as

$$S = S_0 + X_0[1 - \exp(\mu_{\max}t)], \tag{8}$$

where S_0 is initial substrate concentration, S is substrate concentration, and X_0 is the amount of the substrate required to produce the initial population (Simkins and Alexander, 1984). Conditions for logarithmic kinetics are satisfied when a single bacterial population is multiplying and S_0 is much above K_s ($S_0 \gg K_s$). Logarithmic kinetics have been verified for the disappearance of benzoate in cultures of a *Pseudomonas* sp. (Simkins and Alexander, 1984) and apparently for the formation of $^{14}CO_2$ from ^{14}C-labeled 2,4-D added at high concentrations to soil (Kunc and Rybarova, 1983).

The kinetics of growth at initial substrate concentrations much below K_s ($S_0 \ll K_s$) are quite different. From Fig. 6.1, it is evident that the growth rate declines at progressively lower substrate concentrations. Thus, when *Escherichia coli* is multiplying in media with low glucose levels, its growth rate is proportional to glucose concentration (Shehata and Marr, 1971). When the culture multiplies in a medium with such low glucose levels, the cells continue to increase in number, but because the progressively falling substrate concentration is causing a progressively slower growth rate, the period between each doubling becomes progressively longer. Thus, in contrast with the logarithmic phase, in which the doubling time is constant, growth at substrate concentrations below K_s is characterized by increasingly long doubling times even as the cell number is rising. This is known as logistic growth, and the kinetics of this pattern of multiplication will be mirrored in the kinetics of substrate disappearance if each cell metabolizes essentially the same amount of its C source.

The logistic growth curve has interested population ecologists for many years (Odum, 1971). The logistic curve has an S shape and is symmetrical about the point of inflection. This is evident in the plot depicting logistic kinetics of substrate disappearance (Fig. 6.2). Because of the low concentration of substrate, one can only have appreciable increases in cell numbers when the initial population size is small, so that both little substrate and few cells are necessary for this pattern of biodegradation. Logistic kinetics may be written in differential form as

$$-\frac{dS}{dt} = dS\,(S_0 + X_0 - S) \tag{9}$$

or in an integral form as

$$S = \frac{S_0 + X_0}{1 + (X_0/S_0)\exp\,[k\,(S_0 + X_0)t]}, \tag{10}$$

where $k = \mu_{\max}/K_s$ (Simkins and Alexander, 1984). Logistic kinetics have been noted in the mineralization of 2.0 μg of phenol and 2.0 and 7.0 μg of 4-nitrophenol per liter in lake water (Jones and Alexander, 1986).

When bacterial growth follows logistic kinetics, the curve for the time in which the density rises from 10 to 90% of the maximum population density can be closely approximated by a straight line (Schmidt *et al.*, 1985). Thus, when the initial substrate concentration is below K_s ($S_0 \ll K_s$) and few cells are present that are able to use that chemical as a C source of growth, the finding of what appears to be linear kinetics may, in fact, be a reflection of a logistic transformation.

When the initial substrate concentration is approximately the same as K_s ($S_0 \sim K_s$), neither logarithmic nor logistic kinetics apply. The situation is somewhat more complex because μ is not directly dependent on substrate concentration (as when $S_0 \ll K_s$) or largely independent of it (as when $S_0 \gg K_s$) (Fig. 6.1). The pattern of substrate disappearance in this range of concentrations is termed Monod-with-growth kinetics, and the shape of the curve of chemical loss is depicted in Fig. 6.2. Monod-with-growth kinetics can be expressed mathematically in differential form as

$$-\frac{dS}{dt} = \frac{\mu_{max} S (S_0 + X_0 - S)}{K_s + S} \tag{11}$$

and in integral form as

$$K_s \ln (S/S_0) = (S_0 + X_0 + K_s) \ln (X/X_0) - (S_0 + X_0)\mu_{max} t, \tag{12}$$

where X is the amount of substrate to produce the population density. These kinetics describe the metabolism of benzoate by *Pseudomonas* sp. at benzoate levels near K_s (Simkins and Alexander, 1984) and the mineralization of 4-nitrophenol in lake water (Jones and Alexander, 1986).

At high concentrations, many pollutants are toxic to the very microorganisms that use them as C sources. For these chemicals, the typical relationship between the bacterial growth rate and the concentration of its C source, as presented in Fig. 6.1, is not found. Although the growth rate increases with increasing but low substrate concentration, a concentration is reached above which the growth rate falls as the substrate level rises further. This decline is a result of the antimicrobial action of the chemical. Such a relationship between growth rate and concentration of a potentially toxic organic nutrient can be characterized by the Haldane modification of the Monod equation

$$\mu = \frac{\mu_{max} S}{K_s + S + (S^2/K_I)}, \tag{13}$$

where K_I is an inhibition constant that reflects the suppression of the growth rate by the toxic substrate. This equation has been used for describing kinetics of phenol and pentachlorophenol metabolism by microorganisms (Klecka and Maier, 1986; Rozich *et al.*, 1985).

BIODEGRADATION BY NONGROWING ORGANISMS

For bacteria to grow appreciably, the amount of substrate must be sufficiently high relative to the number of cells active on that compound to permit several or many doublings. If the bacterial cell density is high relative to the substrate concentration, little or no increase in cell numbers is possible. Under these conditions, one can again consider three cases: $S_0 \gg K_s$, $S_0 \sim K_s$, and $S_0 \ll K_s$. Such kinetics resemble those of enzyme reactions because multiplication is not involved. The relationship between the rate of an enzyme reaction and the concentration of the substrate for that enzyme is often best expressed by the Michaelis-Menten equation

$$\nu = \frac{V_{max}\, S}{K_m + S}, \tag{14}$$

where ν is reaction rate, V_{max} is the maximum reaction rate, and K_m is a constant called the Michaelis constant. Equation (14) represents the same relationship as the Monod equation [Eq. (2)], differing only in the replacement of ν, V_{max}, and K_m for μ, μ_{max}, and K_s, respectively. Moreover, a graphic representation of the equation would be the same as that shown in Fig. 6.1, with the substitution of ν, V_{max}, and K_m for μ, μ_{max}, and K_s. The essential difference is that in Michaelis-Menten kinetics, the quantity of reactive material (enzyme) is constant, whereas in Monod kinetics, the amount of reactive material (cells) is increasing because of microbial proliferation. It is because Michaelis-Menten kinetics are formulated on the basis of constant reactive material that such kinetics are useful for nongrowing cells.

Referring to Fig. 6.1 and using ν, V_{max}, and K_m in place of μ, μ_{max}, and K_s, the kinetics of biodegradation by nongrowing cells become immediately evident. At initial substrate concentrations much higher than K_m ($S_0 \gg K_m$), the rate does not fall appreciably as the cells transform the organic substrate; that is, moving from the extreme right of the curve to a point somewhat to the left does not greatly alter ν. In other words, the rate is essentially constant as the concentration falls from high levels to concentrations that are lower but still above K_m. The rate is thus linear or, to use the term from chemical kinetics, it follows zero-order kinetics. Conversely, when the initial substrate concentration is much below K_m ($S_0 \ll K_m$) and metabolism of the compound further lowers the concentration, the rate falls in proportion to the decline in substrate concentration because the rate is a direct function of concentration. In this case, the rate is continuously falling as the substrate level falls due to microbial metabolism. This relationship is known as first-order kinetics.

For nongrowing cells, the kinetics when $S_0 \gg K_s$, $S_0 \sim K_s$, and $S_0 \ll K_s$ (referring back to the terms used for growing organisms) are thus termed zero-order. Monod-no-growth, and first-order kinetics. The kinetics are expressed mathematically as follows.

For first-order kinetics, the differential form is

$$-\frac{dS}{dt} = k_1 S \tag{15}$$

and the integral form is

$$S = S_0 \exp(-k_1 t), \tag{16}$$

where t is time, S is substrate concentration at time t, and $k_1 = \mu_{max} (X_0/K_s)$. The term k_1 is the first-order rate constant and is expressed in units of $(\text{time})^{-1}$, that is, h^{-1}, $days^{-1}$, etc. If $k_1 = 0.01\ h^{-1}$, the rate is 1% per hour. This is simply another way of saying a constant percentage is lost per unit time.

For Monod-no-growth kinetics, the differential form is

$$-\frac{dS}{dt} = \frac{k_2 S}{K_s + S} \tag{17}$$

and the integral form is

$$K_s \ln(S/S_0) + S - S_0 = -k_2 t, \tag{18}$$

where $k_2 = \mu_{max} X_0$.

For zero-order kinetics, the differential form is

$$-\frac{dS}{dt} = k_2 \tag{19}$$

and the integral form is

$$S = S_0 - k_2 t \tag{20}$$

(Simkins and Alexander, 1984). In zero-order kinetics, a *constant amount* is lost per unit time.

The patterns of chemical biodegradation that occur by zero-order, Monod-no-growth, and first-order kinetics are presented in Fig. 6.2.

To express Monod-no-growth as Michaelis-Menten kinetics, K_s is replaced by K_m and k_2 is equal to $V_{max}\ B_0$ in both differential and integral equations. The term B_0 is the initial population density. Such Michaelis-Menten or nongrowing-cell kinetics have been reported to describe the kinetics of biodegradation of picloram in soil (Hamaker *et al.*, 1968; Meikle *et al.*, 1973) and the initial metabolism of 3,5-dichlorobenzoate and 4-amino-3,5-dichlorobenzoate by an anaerobic enrichment culture converting these molecules to methane (Suflita *et al.*, 1983), and first-order, Monod-no-growth, and zero-order kinetics sometimes best describe the mineralization of phenol in lake water (Jones and Alexander, 1986).

The terms first and zero order come from chemical kinetics. In a first-order process, the rate is proportional to the concentration of a single reactant; for the purpose of this discussion, that reactant is the substrate. In a zero-order process, the rate is independent of the concentration of the reactants, that is, independent of the substrate concentration. When the concentration is plotted against time, as in Fig. 6.2, the concentration decreases at a constant rate in zero-order processes, but it falls quickly initially and then more slowly in first-order processes. As stated earlier, a constant amount is lost per unit time in zero-order reactions, and a constant percentage disappears per unit time in first-order reactions.

ZERO-ORDER KINETICS

Zero-order kinetics or linear biodegradation of organic substrates (or formation of organic products) has been observed frequently. According to the theory presented in the preceding section, such rates should be evident in processes effected by nongrowing cells when $S_0 \gg K_s$ (or $S_0 \gg K_m$), and they may seem to be evident when bacteria are growing logistically because $S_0 < K_s$. Linear transformation may also occur under the following conditions.

a. The nutrient that limits the growth of the active population becomes available at a constant rate, but the rate does not fully meet the demand of the organisms. For example, several bacteria grow linearly in liquid media when O_2 enters the solution at a rate that limits their further multiplication (Volk and Myrvik, 1953; Brown *et al.*, 1988). The O_2 limitation probably explains why some fungi grown in media with supplemental aeration enter a linear growth phase (Gillie, 1968). Such O_2 limitation to biodegradation is likely to occur at high substrate concentrations. Zero-order kinetics also describe the biodegradation by a mixed culture of anthracene in soil made into a slurry, presumably as a result of the linear rate of dissolution of the hydrocarbon into the aqueous phase (Gray *et al.*, 1994).

b. The organisms use up the supply of some essential nutrient element or growth factor. For example, methane production by *Methanosarcina* switches from a logarithmic to a linear rate when phosphate in the medium becomes depleted (Archer, 1985). A possible reason is that the concentration of some enzyme or enzyme system essential for further multiplication is constant (Monod, 1949).

c. The population size of organisms active on the organic compound has become large as a result of previous addition of the chemical, and a second increment of the compound is introduced. Under these conditions, the biomass of the already large population may not increase as it decomposes the second increment, presumably when the concentration of the second addition is above K_s. This has been observed when an anaerobic enrichment culture that had been acclimated to metabolize 3-chlorobenzoate received a second increment of the chemical (Suflita *et al.*, 1983) and when samples of subsoils are exposed repeatedly to high concentrations of toluene (Allen-King *et al.*, 1994).

d. The population is growing on certain C compounds that have low water solubilities, and the amount in aqueous solution has been totally consumed. The reasons for this are not yet clear. Linear growth has been reported for pure cultures of bacteria, yeasts, and sometimes fungi growing or tri-, tetra-, penta-, hexa-, and octadecane (Lindley and Heydeman, 1986; Yoshida and Yamane, 1971; Thomas *et al.*, 1986), a material known as slack wax, which contains 70 to 90% of straight-chain solid paraffin (Lonsane *et al.*, 1979), cholesterol, β-sitosterol (Goswami *et al.*, 1983), phenanthrene (Stucki and Alexander, 1987), and crystalline or adsorbed polycyclic aromatic hydrocarbons (Volkering *et al.*, 1992). Such linear growth in pure culture frequently follows a period of logarithmic growth (Goswami *et al.*, 1983; Stucki and Alexander, 1987; Lonsane *et al.*, 1979) because the initially small population of microorganisms probably first grows unrestrictedly on the soluble chemical or other dissolved organic nutrients and then, when the supply of those is depleted, uses the chemical that initially is not in the aqueous phase.

Zero-order kinetics have been reported frequently for biodegradation (Table 6.2). Moreover, linear rates have been found at extremely low concentrations (102 pg/liter), which are undoubtedly below the K_s, to such high concentrations (10 g/liter) that they are undoubtedly far above K_s. In a few instances, the metabolic conversion is zero order at high concentrations and first order at low concentrations, for example, for the mineralization of maleic hydrazide in soil (Helweg, 1975) or the mineralization of glucose and linear alcohol ethoxylate in bay water (Vashon and Schwab, 1982).

FIRST-ORDER KINETICS

First-order biodegradation is to be expected when the chemical concentration is below K_s (or K_m) and the organisms are not increasing in abundance, possibly because there is not sufficient available C to support a doubling or because some other limiting nutrient is lacking. A common way of presenting first-order kinetics is to plot the logarithm of the concentration of chemical remaining (or logarithm of S/S_0) as a function of time; if the reaction is first order, a straight line is obtained (Fig. 6.3). First-order kinetics are sometimes termed half-life kinetics because if half of the chemical is gone in time t, half of what is then left will remain at time $2t$, and half again at time $3t$. In other words, if the half-life is 20 days, the amount remaining at 20, 40, 60, and 80 days is ½, ¼, ⅛, and 1/16 of the initial quantity.

First-order kinetics have been observed for the metabolism of glucose by *Salmonella typhimurium* at a concentration (0.4 μg/liter) below that supporting

Table 6.2

Organic Compounds Metabolized by Zero-Order Kinetics in Samples from Natural Environments

Chemical	Concentration (per liter of water or kg of soil)	Environmental sample	Reference
1,1,1-Trichloroethane	100 mg	Biofilm	Wren and Rittman (1996)
Phenol	102 pg–10 g	Lake water	Subba-Rao *et al.* (1982)
2,4-D	1.5 ng	Lake water	Subba-Rao *et al.* (1982)
Aniline	5.7 ng–500 μg	Lake water	Subba-Rao *et al.* (1982)
Diethanolamine	21 ng	Stream water	Boethling and Alexander (1979)
Toluene	380, 3900 ng	Seawater	Button *et al.* (1981)
4,6-Dinitro-2-methylphenol	5–2500 μg	Soil	Hurle and Pfefferkorn (1972)
NTA	10, 200 μg	Estuarine water	Pfaender *et al.* (1985)
Benzylamine	20, 200 μg	Lake water	Subba-Rao *et al.* (1982)
Di(2-ethylhexyl) phthalate	21–200 μg	Lake water	Subba-Rao *et al.* (1982)
N-Nitrosodiethanolamine	54, 940 μg	Lake water	Yordy and Alexander (1980)
	940 μg	Sewage	Yordy and Alexander (1980)
Glucose	1, 10 mg	Bay water	Vashon and Schwab (1982)
2,4-D	25–100 mg	Aquatic	Hemmett and Faust (1969)
Glyphosate	90 mg	Soil	Torstensson and Aamisepp (1977)
Maleic hydrazide	120 mg	Soil	Helweg (1975)
Glucose	400 mg	Activated sludge	Gaudy *et al.* (1963)
Butyrate	1.15 g	Activated sludge	Mateles and Chian (1969)

growth (Schmidt *et al.*, 1985) and for the biodegradation of 200 μg of methyl parathion per liter in seawater (Badawy and El-Dib, 1984), 5.3 μg of methyl parathion per liter in anaerobic sediments (Wolfe *et al.*, 1986), 4.0 mg of hexazinone per kilogram of soil (Bouchard *et al.*, 1985), 0.5 μg of phenol per liter in lake water (Jones and Alexander, 1986), 1,1,1-trichloroethane at concentrations up to 1 mg/liter in a biofilm reactor inoculated with methane-oxidizing bacteria (Arvin,

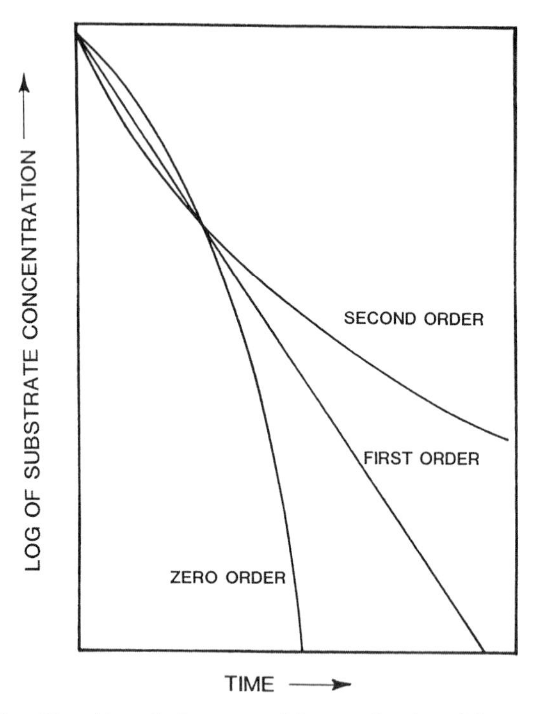

Figure 6.3 A plot of logarithm of substrate remaining as a function of time, as is typical of first-, zero-, and second-order reactions. (From Hamaker, 1966. Reprinted with permission from the American Chemical Society.)

1991), the anaerobic mineralization of hexadecanol in digestor sludge (Nuck and Federle, 1996), 0.04 and 4.0 mg of chlorosulfuron per kilogram of soil (Walker and Brown, 1983), 0.85 to 140 μg of linear alcohol ethoxylate per liter in bay water (Vashon and Schwab, 1982), and a number of other chemicals. Many other claims of first-order transformations cannot be accepted, although they may be correct, because of too few data points to determine the appropriate kinetics and many must be rejected because the data do not fit first-order plots well enough.

First-order kinetics are commonly used to describe biodegradation in environmental fate models because mathematically the expression can be incorporated easily into the models. Unfortunately, this compatibility with models often takes precedence over other evaluation criteria that are more important, and blind acceptance of this type of kinetics can lead to incorrect conclusions on the persistence of toxic chemicals. Many investigators grasp at first-order kinetics because of the ease of presenting and analyzing the data, the simplicity of plotting the logarithm of the chemical remaining versus time as a straight line regardless of the poorness of fit of the line to the points, and the ease of predicting future concentrations once the time is determined for loss of half of the chemical.

It is important to stress that a pollutant whose destruction follows first-order kinetics persists long after the first half-life is over because the level is falling at diminishing rates. This is in contrast with logarithmic or zero-order transformations, the former resulting in more and more being lost per unit time period, the latter resulting in a constant rate until all the chemical is gone.

In developing predictive kinetic models, use has been made of the fact that at substrate concentrations below K_s, the rates of substrate destruction are first order and the cells responsible for the destruction are not growing to any significant degree. In different environments, first-order constants and the number of cells able to metabolize the substrate will differ. However, it has been proposed that for predictive purposes, special use can be made of the value obtained by dividing the first-order rate constant by the number of cells present in natural environments.

One may write the Monod equation as

$$-\frac{dS}{dt} = \frac{\mu_{max} BS}{Y(K_s + S)}, \tag{21}$$

where B is the bacterial cell density and Y is the yield coefficient or the number of bacteria per milligram of substrate. When the substrate concentration is much lower than K_s, this expression can be approximated by

$$-\frac{dS}{dt} = k_b BS, \tag{22}$$

where $k_b = \mu_{max}/YK_s$ and is termed the "second-order" rate constant (Paris and Rogers, 1986; Paris *et al.*, 1981). The use of the term "second-order" is unfortunate because it confuses such expressions with second-order kinetics of chemistry, and it will not be used here for that reason. An important value needed to determine this relationship is the number of cells actually degrading the substrate, but this value is difficult to obtain for populations in nature. Thus, the numbers used are the total bacterial counts, which makes the approach less valuable, especially because the percentage of the total cell number able to degrade a different chemical often will vary greatly in different ecosystems. Such kinetics have been used to characterize the metabolism of malathion in microbial cultures (Falco *et al.*, 1977), methyl parathion and diethyl phthalate by attached microbial growths (Lewis and Holm, 1981), the hydrolysis of an ester of 2,4-D by microorganisms growing on surfaces submerged in fresh water (Lewis *et al.*, 1983), and the transformation of several chemicals in lake water (Paris *et al.*, 1981).

The various orders of reaction, two of which were considered in the foregoing, are commonly summarized by

$$\text{rate} = -\frac{dC}{dt} = kC^n, \tag{23}$$

where k is the rate constant, C is the concentration, and n is the order of the reaction [this was given earlier as Eq. (1)]. From this equation, it is evident that the rate is proportional to a power of the chemical (or substrate, S, in the present context) concentration (Hamaker, 1972). In first- and zero-order reactions, the equation is

$$-\frac{dC}{dt} = k$$

for a zero-order reaction and

$$-\frac{dC}{dt} = kC$$

for a first-order reaction. In a second-order reaction, the rate is proportional to the second power (i.e., the square) of the concentration of a single reactant molecule (rate $= kC^2$) or in other circumstances to the concentration of two reactants (rate $= kC_1C_2$). In both instances, the concentrations of both compounds involved in the reaction change with time. It is in the sense of having two reactants that the term "second order" has been applied to biodegradation: one reactant being the substrate and the other being the microbial biomass. However, both reactants in abiotic processes typically decline in concentration in such transformations, whereas in biodegradation, the cells either multiply or the numbers remain constant. Hence, such kinetics really represent simply first-order rates divided by cell number. In more classical chemical terms, a plot of the logarithm of the chemical concentration remaining, in the case of only a single reacting chemical, against time would give a line that is straight, concave down, or concave up for first, zero, or second (or higher) orders of reaction, respectively (Fig. 6.3).

Frequently, only two or three samples are taken prior to making predictions of the amount of a chemical that will remain at a site in the future. Given the same initial analytical values, the various kinetic models will predict vastly different amounts of chemical remaining at later times. Consider the case of a polluted site with 10 mg/liter initially and 9 mg/liter after 30 days. If predictions were made of the time for the concentration to fall to a regulatory standard of, say, 10 μg/liter, the predictions would be 33 days for the logarithmic model, 300 days for the zero-order model, more than 5 years for the first-order model, and possibly centuries for a model that appears to be initially first order and is followed by a second, slower phase of degradation. In this regard, it is important to note that the similarities in the x axes in Fig. 6.2 could be misleading; the time periods for some models are short and those for other models are very long.

The preceding discussion considered kinetics in relation to the density of bacteria active on the substrate and substrate concentration. The relationship can be depicted in a simple fashion (Fig. 6.4). For the purpose of this presentation, it is considered that either (a) the cells are not growing (the three sectors above the

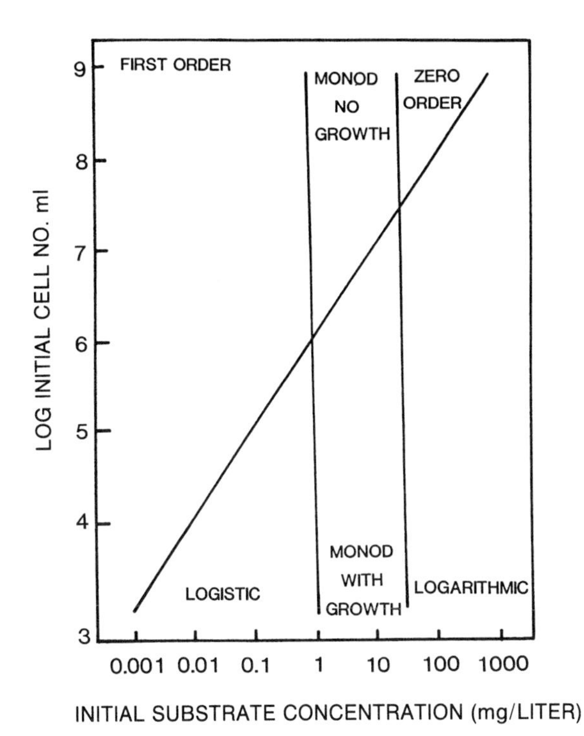

Figure 6.4 Kinetics of biodegradation as a function of substrate concentration and number of bacteria able to metabolize that substrate. Patterns below the diagonal line represent kinetics of biodegradation by bacteria using the substrate as a source of C and energy for growth. (From Simkins and Alexander, 1984. Reprinted with permission from the American Society for Microbiology.)

diagonal line) because the substrate concentration is not sufficiently high to support even a single doubling or (b) they are growing because enough C is present to permit the population to increase in size (the three sectors below the line). A key feature of the illustration is K_s; it can be any value but, for purposes of the example, it is arbitrarily chosen as 1.0 mg/liter and is represented by a vertical line. Values for bacterial density are given per milliliter because that is the convention among microbiologists. For the purposes of this illustration, it is assumed that 1.0 pg of substrate is required to form one bacterial cell, a figure that is often approximated in nature. The precise positions of the vertical lines other than K_s are not certain, but they are placed at positions to indicate that (a) the growth rate, μ, is maintained at about μ_{max} until nearly all the substrate is degraded (the vertical line to the right of K_s) and (b) the rate of degradation per cell varies greatly with substrate concentration (the vertical line to the left of K_s). Thus, the two sectors each at the left, middle, and right denote $S_0 \ll K_s$, S_0, $\sim K_s$, and $S_0 \gg K_s$, and the three sectors above and the three below the diagonal denote nongrowing and growing cells, respectively. From this diagram, the ranges of relative cell densities and substrate concentrations corresponding to each of the six models are evident.

Scow *et al.* (1986) found that the models of the Monod family did not provide good fits to curves depicting the mineralization of low concentrations of phenol, 4-nitrophenol, aniline, 2,4-dichlorophenol, benzylamine, NTA, and cyclohexylamine. The models of the Monod family also did not fit satisfactorily to curves of atrazine, linuron, and picloram disappearance (Hance and McKone, 1971). Other studies have shown that first-order kinetics fit substrate-disappearance curves; however, when biodegradation of more than one concentration of the organic compound was tested, the estimated rate constant often varied with initial concentration, which is not consistent with first-order kinetics.

Natural environments are highly complex, both physically and chemically, the composition of their microbial communities is quite heterogeneous, and the abiotic constituents are commonly reactive. Hence, the application of existing models to biodegradation kinetics in many natural ecosystems often is questionable.

The following factors often confound the facile extrapolation of the kinetics described here to natural circumstances.

a. Diffusional barriers may limit or prevent contact between microbial cells and their organic substrates.

b. Many organic molecules sorb to clay or humus constituents of soils and sediments, and the kinetics of decomposition of sorbed substrates may be quite different from the same compounds free in solution.

c. The presence of other organic molecules that can be metabolized by the biodegrading species may repress or enhance use of the test chemicals.

d. The supply of inorganic nutrients, O_2, or growth factors may govern the rate of transformation, and the process will then be regulated by the diffusion of those nutrients or the rate of their formation or regeneration by other inhabitants of the community.

e. Many species may be matabolizing the same organic compounds simultaneously, and these organisms may have different K_s and K_m values for the substrate.

f. Protozoa or possibly species parasitizing the biodegrading populations may govern the growth, size, or activity of the populations responsible for the biodegradation.

g. Many synthetic chemicals or pollutants have exceedingly low solubilities in water, and the kinetics of their transformation may be wholly dissimilar from compounds that are in the aqueous phase.

h. Cells of the active population may be sorbed or develop in microcolonies, and the kinetics of processes effected by sorbed bacteria or microcolonies are as yet unresolved.

i. Many organic compounds disappear only after an acclimation period, and methods do not now exist that can predict the length of this period or anticipate the percentage of the time between introduction of the chemical and its total destruction. Hence, nearly all the available kinetics models ignore the acclimation period.

DIFFUSION AND SORPTION

Because an important variable in models of biodegradation kinetics is the concentration of the substrate, a process that significantly lowers the concentration should affect the rate of biodegradation. Both physical and chemical processes, for example, diffusion to unavailable sites and sorption, may be involved. A lack of consideration of the kinetics of diffusion and sorption may contribute to the common failure of environmental-fate models in simulating laboratory measurements (van Genuchten *et al.*, 1974; Davidson and Chang, 1972). Sorption is usually treated as a rapid-equilibrium, reversible process, but kinetic studies have shown that sorption is better represented as a two-phase process with an initial fast stage (<1 h) followed by a longer slow phase (days), and that diffusion of the solute to internal sorption sites controls the second rate (Cameron and Klute, 1977; Karickhoff, 1980; McCall and Agin, 1985). The rates of sorption and diffusion to inaccessible sites may be similar to many rates of biodegradation, and so these abiotic processes may be effectively competing with microorganisms for the substrate.

Sorption of a chemical has a major impact on the biodegradation of that compound. Nevertheless, surprisingly little attention has been given to the kinetics of biodegradation of sorbed molecules. A model was proposed by Mihelcic and Luthy (1988), however, that considered diffusion as a controlling factor, and it was based on the assumption that only the compound in the aqueous phase was acted on by microorganisms. A sorption-retarded radial diffusion model that took into account the effect of the size of soil aggregates was developed to describe the degradation of hexachlorocyclohexane, and this model also considered the diffusion of the substrates within the soil particles (Rijnaarts *et al.*, 1990). In the latter study, moreover, the initial rates of biodegradation were greater than the initial rates of desorption, so that the rate of spontaneous desorption may not be an appropriate parameter for the kinetics of biodegradation. More recently, a different model was proposed that considered sorption and biodegradation, as well as diffusion to internal sites in the soil matrix, with biodegradation being described by Monod kinetics and sorption and diffusion by first-order kinetics (Shelton and Doherty, 1997). A different model was proposed by Miller and Alexander (1991) for sorbed organic substrates that are readily desorbed from the surface of solids, and this model gave a good fit to biodegradation of benzylamine that was initially sorbed to montmorillonite clay.

It is likely that diffusion also controls the availability of many organic substrates to microorganisms and influences the rate of degradation of these chemicals. Rovira and Greacen (1957) suggested that much of the native organic matter of soil was protected from microbial attack by its being sequestered within small pores. Other studies have provided evidence that the persistence of 1,2-dibromoethane may result from its entrapment in soil nanopores, which makes it unavailable for microbial degradation (Steinberg *et al.*, 1987; Pignatello *et al.*, 1987). Soil consists of pores of different sizes, and a significant portion of the total pore volume consists of pores with radii <1 μm and even <0.1 μm. Casida (1971) found that most soil bacteria range in size from 0.5 and 0.8 μm, and studies indicate that the mean diameter of soil pores occupied by bacteria is even larger, approximately 2 μm (Kilbertus, 1980). These findings suggest that organic compounds retained within these nanopores will be inaccessible to microorganisms, even to the smallest bacteria. Hence, diffusion of organic compounds into and out of these pores may be an important factor in controlling the rate of biodegradation of the compounds.

The availability of many hydrophobic pollutants is markedly affected if the molecule is in a nonaqueous-phase liquid (NAPL) at the site of pollution. That nonaqueous-phase liquid may be oil from a marine spill, petroleum from a gasoline storage tank, a solvent from a leaking storage tank that is placed within the soil, or a mixture of solvents at a hazardous-waste site. Models for the kinetics of biodegradation of substrates within NAPLs have not been devised. The rate of degradation is undoubtedly affected by the interfacial area, that is, the area of the surface between the NAPL and the aqueous phase, and consideration must be given to the kinetics of microbial growth at the interface, in the aqueous phase only, or at both the interface and in aqueous solution.

An approach to more complex kinetics is exemplified by two-compartment models. In such kinetics models, it is assumed that the substrate exists in two compartments, the identities of which usually are not known. The chemical in one compartment may be unavailable for microbial use, and that in the other compartment may be the form in which the chemical is transformed. In an environment containing particulate matter, a solution phase, and possible air-filled pores such as characterizes soils, one compartment could represent the substrate freely available to microorganisms and subject to rapid mineralization. The second compartment might then be a substrate that is not readily available because it is sorbed to colloidal surfaces or deposited in inaccessible nanopores. After the supply of substrate in the first compartment is depleted, the subsequent rate of biodegradation would be governed by the rate of desorption or diffusion of the substrate from the inaccessible nanopores to sites containing active microorganisms. The rates of mass transfer of the substrate between the two compartments may be designated k_1 and k_2, and the rate of microbial transformation of the substrate in the labile compartment to product may be designated

$$S_1 \underset{k_2}{\overset{k_1}{\rightleftharpoons}} S_2 \xrightarrow{k_3} \text{products.}$$

S_1 and S_2 are the quantities of substrate in the unavailable and available compartment, respectively (Hamaker and Goring, 1976). The substrate in both compartments may be available to some extent, in which case the two-compartment kinetics model may be written as

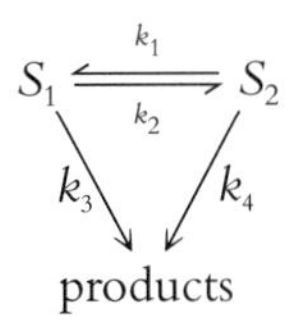

As an example, it is assumed that the substrate is at a concentration too low to support growth and the mass transfer of substrate between compartments follows first-order kinetics; therefore, k_1, k_2, k_3, and k_4 are first-order rate constants. In many cases, for example, when growth occurs, the process will be more complex. The simplified form of the model may be described mathematically by the two differential equations (Scow *et al.*, 1986)

$$\frac{dS_1}{dt} = -(k_1 + k_3)S_1 + k_2 S_2 \tag{24}$$

$$\frac{dS_2}{dt} = k_1 S_1 - (k_2 + k_4)S_2. \tag{25}$$

Of several models tested, the two-compartment model provides the best fit for the mineralization in soil of low concentrations of NTA and phenol (Scow *et al.*, 1986). It also fits the mineralization of aniline at concentrations ranging from 0.3 μg to 500 mg per kilogram. Such kinetics also characterize the metabolism of chloransulam-methyl in soil (Wolt *et al.*, 1996). The two-compartment model also gives good fits for the decomposition of monocrotophos in soil; in this instance, the unavailable compartment is assumed to contain herbicide that is sorbed by soil constituents; and the microflora is assumed to degrade the chemical that is free in soil solution as the available compartment (Furmidge and Osgerby, 1967).

Two-compartment models have been used to describe the biodegradation of organic compounds added to soil. Hamaker and Goring (1976) fit a two-compartment model to curves of degradation of triclopyr in soil. Two first-order curves provided the best fit to the patterns of disappearance in soil of three dinitroaniline herbicides and of metribuzin at 30°C (Hyzak and Zimdahl, 1974; Zimdahl and Gwynn, 1977). At low temperatures, however, first-order kinetics provided the best fit to data. Parker and Doxtader (1983) fit two first-order functions to curves depicting the metabolism of 2,4-D in soil, but the second rate was faster than the first, possibly a result of the growth of the 2,4-D-metabolizing population. The metabolism of dodecane by the fungus *Cladosporium resinae* appears to occur in two linear phases (Lindley and Heydeman, 1986).

Diffusion may also control the rate of biodegradation at high concentrations of the organic chemical. The rate of diffusion of O_2 or inorganic nutrients may be limiting, and such limitations may be especially prominent for bacteria growing in microcolonies (Brunner and Focht, 1984). The rate of diffusion of toxic products away from active organisms may also control their growth and metabolism.

It is difficult to study the effect of diffusion on biodegradation in natural soil given the difficulties in eliminating other potential variables in such a complex system. However, in a defined system in which *Pseudomonas* sp. metabolized glutamate in a gel exclusion-bead matrix, evidence of a role for diffusion in controlling biodegradation was obtained. In this instance, increasing the volume of solution retained inside the beads, which excludes bacteria but not substrate, results in increasingly slower initial rates of mineralization, lower final extents of mineralization, and greater rates in the second, or tail, phase (Scow and Alexander, 1992). Such a defined system enables explicit definition mathematically of the physical and biological processes involved and independent determination, experimentally, of the rates of transfer and degradation.

Two-compartment kinetics also apply to certain circumstances in which products accumulate for some time before they are converted to CO_2. In tests for biodegradation involving determinations of the formation of $^{14}CO_2$ from ^{14}C-labeled substrates, the labeled substrate is essentially one compartment, and the labeled product that temporarily accumulates is the second. Such kinetics also might be evident when two different populations carry out a process that has two separate steps in conversion of the parent chemical to CO_2 (Scow *et al.*, 1986; Simkins *et al.*, 1986).

METABOLISM OF ONE SUBSTRATE DURING GROWTH ON ANOTHER

In the three growth models presented in the preceding section, the cells are multiplying at the expense of the chemical whose biodegradation is being determined. However, the bacteria may be growing at the expense of a different organic compound. The compound whose biodegradation is being measured may still be a substrate, but it is not contributing substantially to the C supply that the cell is using to make more biomass. Such metabolism without providing C to the metabolizing organisms may occur because the concentration of the substance of interest is below the threshold needed to sustain growth or because it is only acted upon by cometabolism.

Mathematical formulations have been developed for the kinetics of biodegradation of one organic chemical when the transformations reflect both the metabolism of that substrate and the simultaneous growth of bacteria on a second compound. The formulations are based on coupling of Monod growth kinetics and

Michaelis–Menten kinetics, which were presented earlier. Nine models have thus been advanced, the nine reflecting linear, logistic, and exponential growth on one substrate and concentrations of the second substrate (whose biodegradation is of interest) that are below, at about, or much above K_m (Table 6.3). The models are arbitrarily numbered I to IX, and differential and integral forms of the equations for each have been published (Schmidt *et al.*, 1985). The shapes of the curves depicting the kinetics of disappearance are presented in Fig. 6.5.

To illustrate the applicability of such models, consider the case of the transformation of a low concentration of a chemical that does not support growth but is acted on by a reasonably large number of cells; its disappearance will be first order. However, if a second organic compound is present at a level that does support growth, the metabolism of the first molecule will reflect both the kinetics of growth on this second compound as well as the kinetics that would normally apply to the enzyme system catalyzing the metabolism of the first.

Tests have been conducted to determine the applicability of the models shown in Table 6.3. Thus, the breakdown of low concentrations of phenol or glucose by two bacteria growing on other C sources is best fit by Models I and IV (Schmidt *et al.*, 1985), and degradation of 4-nitrophenol by a strain of *Pseudomonas* in the presence of glucose is best fit by Model V (Schmidt *et al.*, 1987). The kinetics of mineralization in soil of 5 μg of 4-nitrophenol per kilogram in the presence of increasing concentrations of phenol change from nongrowth kinetics to kinetics that reflect growth of the 4-nitrophenol-mineralizing population on phenol (Scow *et al.*, 1989). The biodegradation of anionic surfactants by biofilm bacteria from polluted rivers is best described by Model I (Lee *et al.*, 1995). Model IV best fits the curve of 4-nitrophenol mineralization in the presence of 5 μg of phenol per kilogram, and Model I provides the best fit in the presence of 10 and 250 mg of phenol per kilogram of soil. Bacterial counts confirm that the population of microorganisms capable of using 4-nitrophenol does indeed multiply. Model IV

Table 6.3

Models for the Kinetics of Biodegradation of Substrates That Do Not Support Growth But Are Metabolized by Populations Growing on Other Organic Substrates

	Concentration of test compound		
Type of growth	$S \ll K_m$	$S \sim K_m$	$S \gg K_m$
Logistic	I	II	III
Logarithmic	IV	V	VI
Linear	VII	VIII	IX

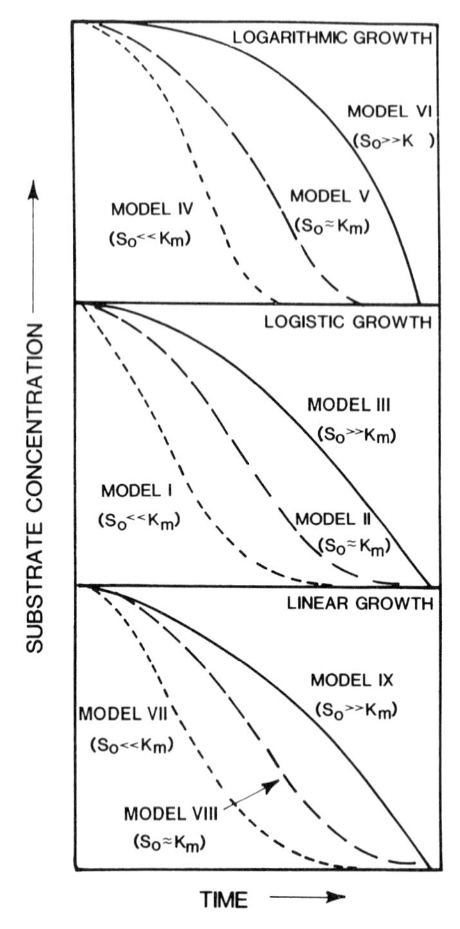

Figure 6.5 Shapes of substrate disappearance curves for nine kinetics models based on growth on one organic substrate and simultaneous metabolism of a second. (From Schmidt *et al.*, 1985. Reprinted with permission from the American Society for Microbiology.)

also gives good fits to the biodegradation of 0.2–20 μg of phenol per liter and comparable levels of 4-nitrophenol in lake water, and Model I also gives good fits to the mineralization of carbonyl-labeled carbofuran (Scow *et al.*, 1990b) and 1.0 and 7.0 μg of phenol per liter of lake water (Jones and Alexander, 1986). The biodegradation of sodium dodecyl sulfate in 16 of 19 samples of river water is also best fit by Model IV (Anderson *et al.*, 1990), whereas the mineralization by *Acinetobacter johnsonii* of phenol at certain N concentrations is best described by Model VII (Hoyle *et al.*, 1995).

The models in Table 6.3 are probably appropriate for compounds acted on by cometabolism, although the possibility has not yet been evaluated experimentally. However, a different model was used to describe the kinetics of cometabolism of trichloroethylene by methane-oxidizing bacteria in the absence of methane (Alvarez-Cohen and McCarty, 1991a), and a somewhat different model has been developed by Criddle (1993). Competitive inhibition was considered in a model for the cometabolism of chlorinated compounds by methane-oxidizing bacteria (Alvarez-Cohen and McCarty, 1991b), and concern with the toxicity to microorganisms of the products of cometabolism has led to the formulation of additional models (Chang and Alvarez-Cohen, 1995; Ely *et al.*, 1995). In some instances, even first- or zero-order kinetics may give good approximations of the kinetics, as in the transformation of propachlor in fresh water and sewage (Novick and Alexander, 1985). It is unlikely that the three models in the right-hand column of Table 6.3 (Models III, VI, and IX) would be realistic for test compounds that could support growth because (a) if S_0 is sufficiently high (the models consider $S_0 \gg K_m$), the organisms would then multiply on the test chemical, or (b) one of two compounds that both support growth would probably repress metabolism of the second; in both cases, these formulations of kinetics would not apply.

THREE-HALF-ORDER KINETICS

Brunner and Focht (1984) proposed the three-half-order model for the mineralization of organic compounds in soil. There are two forms of the model, one assuming linear growth and the other exponential growth. The linear growth form of the three-half-order model can be used when little or no growth occurs. Those kinetics can be represented as

$$P = S_0\{1 - \exp[-k_1t - (k_2t^2)/2]\} + k_0t, \qquad (26)$$

where P is concentration of product, k_1 is a proportionality constant per unit time, k_2 is a constant in units of reciprocal time squared, and k_0 is a zero-order rate constant. The exponential growth model has the form

$$P = S_0\{1 - \exp[k_1t - \frac{E_0}{\mu}(\exp(\mu t) - 1)]\} + k_0t. \qquad (27)$$

Some of the advantages of the three-half-order model are that it fits a set of data containing an acclimation phase, it can be applied to various microbial growth conditions, and it can fit the slow phase of mineralization often observed in the latter portion of mineralization curves. The linear growth form of the three-half-order model has been fit to curves of biphenyl mineralization in soil (Focht and Brunner, 1985) and to curves of mineralization of low concentrations of 4-nitrophenol and benzylamine (Scow *et al.*, 1986). The exponential form of the model has been fit to curves of CO_2 evolution from glucose that was added to soil that was gamma-irradiated to reduce the population density (Brunner and Focht, 1984). The three-half-order model also adequately describes the kinetics of mineralization of di(2-ethylhexyl) phthalate and several surfactants added to soil (Knaebel *et al.*, 1996; Dörfler *et al.*, 1996) and the conversion of two surfactants, glucose, and palmitic acid in anaerobic digestor sludge (Nuck and Federle, 1996).

KINETICS OF FUNGAL PROCESSES

A virtually unexplored area in biodegradation kinetics is the contribution of fungi to the rate of metabolism of organic compounds. In many soils, fungi appear to be more important than bacteria in community respiration (Anderson and Domsch, 1973). However, the theory underlying growth-linked kinetics of biodegradation is derived from the expectations for the multiplication of bacteria. To the extent that the biomass of fungi may increase logarithmically, the kinetics of their metabolism may be assumed to resemble those of bacteria. However, the biomass of fungi increases through hyphal lengthening and branching (not by binary fission) and the organisms undergo morphological changes during the life cycle, so there is no reason to expect that Monod kinetics would be applicable to fungi. Logarithmatic kinetics have been reported to describe the unrestricted growth of fungi in liquid culture (Righelato, 1975). Cubic kinetics, in which a plot of the cube root of the dry weight or respiratory activity of fungi versus time yields a straight line, have been observed in shake cultures for a number of species of fungi and actinomycetes (Marshall and Alexander, 1960). This pattern of biomass increase probably reflects the fact that the growth of many species is largely restricted to the end of the filaments and those filaments develop linearly at a constant rate. If this linear extension then takes place in three dimensions, the hyphal mass of fungi developing as pellets may be viewed as spheres that have radii that increase at a constant rate. For the many fungi that do not form such pellets in mixed, liquid culture, cube-root kinetics presumably do not apply (Mandels, 1965), and they would not apply once unrestricted growth comes to an end. Cubic kinetics may also characterize increases in the biomass of other filamentous microorganisms, such as some actinomycetes (Marshall and Alexander, 1960). The growth rate on the surface of agar media, however, is linear (Trinci, 1970), and it is likely that in a three-dimensional porous matrix such as soil, the kinetics would be different yet.

Direct tests of the kinetics of mineralization in sand or liquid culture of low concentrations of phenol by a strain of *Penicillum* and glucose by species of *Fusarium* and *Rhizoctonia* showed that, of the models of the Monod family, the best fit was obtained with the logistic model (Scow *et al.*, 1990a).

PERSISTENT COMPOUNDS

Compounds that are slowly degraded in soil, and probably in subsoils and aquatic sediments, frequently—and possibly usually—exhibit a different pattern of kinetics. As yet, a mathematical formulation has not been applied to these kinetics. Because the pattern of decline in chemical concentration initially shows a more rapid decline followed by a phase with little or no fall in concentration, somewhat in the shape of a hockey stick, it has been called "hockey stick-shaped" kinetics.

Because these kinetics apply to the behavior of compounds that persist in soil for years and the duration of few laboratory-based studies extend for the requisite time periods, much of the information comes from long-term field evaluations. Typically, the compounds investigated were pesticides and usually the early chlorinated hydrocarbon insecticides that have low volatility, are not subject to abiotic modification, and are not readily metabolized; hence, they persisted. Although several of these insecticides have now been banned in many countries, to a significant degree because of their very persistence, investigations of these pesticides have provided not only an insight into an unexpected pattern of kinetics but also concepts applicable to the behavior of other classes of chemicals.

In a number of the field-monitoring activities with DDT, aldrin and its epoxide (dieldrin), heptachlor, chlordane, lindane, and kepone, a sufficient number of data points were obtained to define the kinetics. In contrast with the assumption that the kinetics are first order, an assumption not based on data with persistent compounds but convenient mathematically, the data in nearly all instances revealed two phases. In the first phase, the compound was being degraded so that its concentration progressively declined with time. The responsible microorganisms were present in the soil, and conditions were suitable for their metabolism, even if the transformation was not overly rapid. In the second phase, the concentration did not fall detectably or it declined at a very slow rate. The phase with little or no activity on the compound was not the result of a change in the soil moisture, temperature, or vegetation because the locations were usually agricultural fields that were being cropped and the studies extended for sufficiently long to minimize the impact of seasonal changes, occasional droughts, and fallow periods (Alexander, 1997).

Differing from the other types of kinetics, hockey stick-shaped kinetics rely on a time-dependent change in bioavailability. The chemicals are intrinsically slowly transformed and hence are liable to physical or chemical modifications that result in alterations in the availability to microorganisms. Processes leading to declining bioavailability, which are known as aging or sequestration, are discussed in Chapter 10.

A typical plot showing hockey stick-shaped kinetics is given in Fig. 6.6. An overall view of the existing information shows that plots differ from the representa-

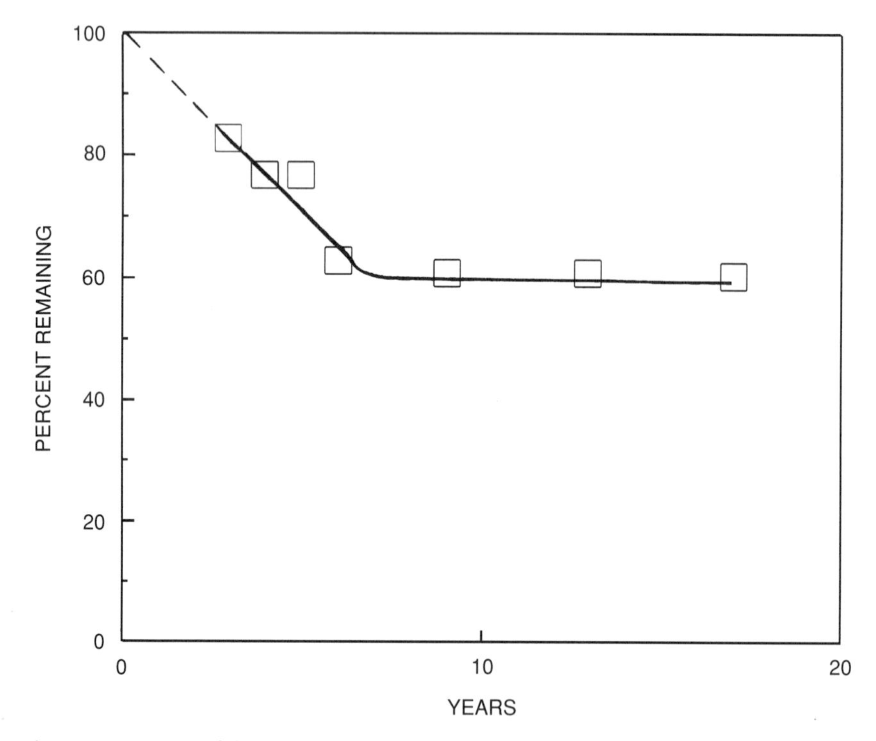

Figure 6.6 Pattern of disappearance of DDT from a field soil (replotted from data of Nash and Woolson, 1967).

tive figures shown in only three significant ways: (a) The time to reach the end of phase one varies appreciably. It may be as short as a year (or somewhat less) or as long as 10 years. (b) The percentage remaining of the originally added compound also varies appreciably. It may extend from approximately 5 to 60%. (c) The second phase may exhibit a small negative slope or the slope may not be statistically distinguishable from zero. At the present time, generalizations for specific compounds or individual soil types with regard to the length of time to reach the end of phase one, the percentage of the original concentration that remains, or the slope of the second phase are not possible.

OVERVIEW

A problem in kinetics analysis is to distinguish among models when the theoretical curves depicting the pattern of decomposition are quite similar. It is not always possible to distinguish among models, even with nonlinear regression techniques, so that the final choices among models are arbitrary. This is especially true for investigations yielding data that are not highly precise or do not result in many points. In instances in which models cannot be distinguished, it is critical to evaluate the models by comparing estimated parameters to independent experimental measurements (Schmidt *et al.,* 1987).

The models formulated to date often are based on sound microbiological and biochemical principles, but they rely on concepts derived from studies of single populations or single enzymes. The fact that the models often adequately characterize data may be a reflection that a single species dominates the process or that the rate is governed by the kinetics of a single enzymatic step. Should several species be involved, predators or parasites act on the biodegrading species, or the species carrying out the transformation require the activities of neighboring populations, more complex models may be necessary. Also, if the substrate is insoluble or retained by abiotic components of the environment, physical and chemical processes may have to be considered in the models. Given the array of chemicals, the complexity of some environments, and the variety of microorganisms that may bring about biodegradation, it is unlikely that a single model or equation would be useful for the description of rates of biodegradation of all organic substrates in all environments.

REFERENCES

Alexander, M., *Environ. Sci. Technol.* **19,** 106–111 (1985).

Alexander, M., *in* "Environmentally Acceptable Endpoints in Soil" (D. G. Linz and D. V. Nakles, eds.), pp. 43–136. American Academy of Environmental Engineers, Annapolis, MD, 1997.

Alexander, M., and Scow, K. M., *in* "Reactions and Movement of Organic Chemicals in Soils" (B. L. Sawhney and K. Brown, eds.), pp. 243–269. Soil Science Society of America, Madison, WI, 1989.

Allen-King, R. M., Barker, J. F., Gillham, R. W., and Jensen, B. K., *Environ. Toxicol. Chem.* **13,** 693–705 (1994).

Alvarez-Cohen, L., and McCarty, P. L., *Environ. Sci. Technol.* **25,** 1381–1387 (1991a).

Alvarez-Cohen, L., and McCarty, P. L., *Appl. Environ. Microbiol.* **57,** 1031–1037 (1991b).

Anderson, D. J., Day, M. J., Russell, N. J., and White, G. F., *Appl. Environ. Microbiol.* **56,** 758–763 (1990).

Anderson, J. P. E., and Domsch, K. H., *Arch Microbiol.* **93,** 113–127 (1973).

Archer, D. B., *Appl. Environ. Microbiol.* **50,** 1233–1237 (1985).

Arvin, E., *Water Res.* **25,** 873–881 (1991).

Badawy, M. I., and El-Dib, M. A., *Bull. Environ. Contam. Toxicol.* **33,** 40–49 (1984).

Bartholomew, G. W., and Pfaender, F. K., *Appl. Environ. Microbiol.* **45,** 103–109 (1983).

Bazin, M. J., Saunders, P. T., and Prosser, J. I., *CRC Crit. Rev. Microbiol.* **4,** 463–498 (1976).

Boethling, R. S., and Alexander, M., *Environ. Sci. Technol.* **13,** 989–991 (1979).

Bouchard, D. C., Lavy, T. L., and Lawson, E. R., *J. Environ. Qual.* **14,** 229–233 (1985).

Brown, D. E., Gaddum, R. N., and McEvoy, A., *Biotechnol. Lett.* **10,** 525–530 (1988).

Brunner, W., and Focht, D. D., *Appl. Environ. Microbiol.* **47,** 167–172 (1984).

Button, D. K., Schell, D. M., and Robertson, B. R., *Appl. Environ. Microbiol.* **41,** 936–941 (1981).
Cameron, D. R., and Klute, A., *Water Resour. Res.* **13,** 183–188 (1977).
Casida, L. E., Jr., *Appl. Environ. Microbiol.* **21,** 1040–1045 (1971).
Chang, H.-L., and Alvarez-Cohen, L., *Environ. Sci. Technol.* **29,** 2357–2367 (1995).
Criddle, C. S., *Biotechnol. Bioeng.* **41,** 1048–1056 (1993).
Davidson, J. M., and Chang, R. K., *Soil Sci. Soc. Am. Proc.* **36,** 257–261 (1972).
Dörfler, U., Haala, R., Matthies, M., and Scheunert, I., *Ecotoxicol. Environ. Saf.* **34,** 216–222 (1996).
Ely, R. L., Hyman, M. R., Arp, D. J., Guenther, R. B., and Williamson, K. J., *Biotechnol. Bioeng.* **46,** 232–245 (1995).
Falco, J. W., Sampson, K. T., and Carsel, R. F., *Dev. Ind. Microbiol.* **18,** 193–202 (1977).
Focht, D. D., and Brunner, W., *Appl. Environ. Microbiol.* **50,** 1058–1063 (1985).
Furmidge, C. G. L., and Osgerby, J. M., *J. Sci. Food Agric.* **18,** 269–273 (1967).
Gaudy, A. F., Jr., Komolrit, K., and Bhatla, M. N., *J. Water Pollut. Control Fed.* **35,** 903–922 (1963).
Gillie, O. J., *J. Gen. Microbiol.* **51,** 179–184 (1968).
Goswami, P. C., Singh, H. D., Bhagat, S. D., and Baruah, J. N., *Biotechnol. Bioeng.* **25,** 2929–2943 (1983).
Gray, M. R., Banerjee, D. K., Fedorak, P. M., Hashimoto, A., Masliyah, J. H., and Pickard, M. A., *Appl. Microbiol. Biotechnol.* **40,** 933–940 (1994).
Hamaker, J. W., *in* "Organic Pesticides in the Environment" (A. A. Rosen and H. F. Kraybill, eds.), pp. 122–131. American Chemical Society, Washington, DC, 1966.
Hamaker, J. W., *in* "Organic Chemicals in the Soil Environment" (C. A. I. Goring and J. W. Hamaker, eds.), pp. 253–340. Dekker, New York, 1972.
Hamaker, J. W., and Goring, C. A. I., *in* "Bound and Conjugated Pesticide Residues" (D. D. Kaufman, G. G. Still, D. D. Paulson, and S. K. Bandal, eds.), pp. 219–243. American Chemical Society, Washington, DC, 1976.
Hamaker, J. W., Youngson, C. R., and Goring, C. A. I., *Weed Res.* **8,** 46–57 (1968).
Hance, R. J., and McKone, C. E., *Pestic. Sci.* **2,** 31–34 (1971).
Helweg, A., *Weed Res.* **15,** 53–58 (1975).
Hemmett, R. B., Jr., and Faust, S. D., *Residue Rev.* **29,** 191–207 (1969).
Hoyle, B. L., Scow, K. M., Fogg, G. E., and Darby, J. L., *Biodegradation* **6,** 283–293 (1995).
Hurle, K., and Pfefferkorn, V., *Proc. Br. Weed Control. Conf., 11th, 1972,* Vol. 2, pp. 806–810 (1972).
Hyzack, D. L., and Zimdahl, R. L., *Weed Sci.* **22,** 75–79 (1974).
Ishida, Y., Imai, I., Miyagaki, T., and Kadota, H., *Microb. Ecol.* **8,** 23–32 (1982).
Jones, S. H., and Alexander, M., *Appl. Environ. Microbiol.* **51,** 891–897 (1986).
Karickhoff, S. W., *in* "Contaminants and Sediments" (R. A. Baker, ed.), Vol. 2, pp. 193–205. Ann Arbor Sci. Publ., Ann Arbor, MI, 1980.
Kilbertus, G., *Rev. Ecol. Biol. Sol* **17,** 543–557 (1980).
Klecka, G. M., and Maier, W. J., *Appl. Environ. Microbiol.* **49,** 46–53 (1985).
Knaebel, D. B., Federle, T. W., McAvoy, D. C., and Vestal, J. R., *Environ. Toxicol. Chem.* **15,** 1865–1875 (1996).
Koch, A. L., and Wong, C. H., *Arch. Microbiol.* **131,** 36–42 (1982).
Kunc, F., and Rybarova, J., *Soil Biol. Biochem.* **15,** 141–144 (1983).
Larson, R. J., *in* "Biotransformation and Fate of Chemicals in the Aquatic Environment" (A. W. Maki, K. L. Dickson, and J. Cairns, Jr., eds.), pp. 67–86. American Society for Microbiology, Washington, DC, 1980.
Lee, C., Russell, N. J., and White, G. F., *Water Res.* **29,** 2491–2497 (1995).
Lewis, D. L., and Holm, H. W., *Appl. Environ. Microbiol.* **42,** 698–703 (1981).
Lewis, D. L., Kollig, H. P., and Hall, T. L., *Appl. Environ. Microbiol.* **46,** 146–151 (1983).
Lewis, D. L., Hodson, R. E., and Hwang, H.-M., *Appl. Environ. Microbiol.* **54,** 2054–2057 (1988).
Lindley, N. D., and Heydeman, M. T., *Appl. Microbiol. Biotechnol.* **23,** 384–388 (1986).
Lonsane, B. K., Singh, H. D., Nigam, J. N., and Baruah, J. N., *Indian J. Exp. Biol.* **17,** 1263–1264 (1979).

Mandels, G. R., *in* "The Fungi" (G. C. Ainsworth and A. S. Sussman, eds.), Vol. 1, pp. 599–612. Academic Press, New York, 1965.

Marshall, K. C., and Alexander, M., *J. Bacteriol.* **80,** 412–416 (1960).

Mateles, R. I., and Chian, S. K., *Environ. Sci. Technol.* **3,** 569–574 (1969).

McCall, P. J., and Agin, G. L., *Environ. Toxicol. Chem.* **4,** 37–44 (1985).

Meikle, R. W., Youngson, C. R., Hedlund, R. T., Goring, C. A. I., Hamaker, J. W., and Addington, W. W., *Weed Sci.* **21,** 549–555 (1973).

Mihelcic, J. M., and Luthy, R. G., *Pap., Int. Conf. Physiochem. Biol. Detox. Hazard. Wastes,* Vol. 2. pp. 708–721. Technomic Publishing Co., Lancaster, PA, 1988.

Miller, M. E., and Alexander, M., *Environ. Sci. Technol.* **25,** 250–245 (1991).

Monod, J., *Annu. Rev. Microbiol.* **3,** 371–394 (1949).

Nash, R. G., and Woolson, E. A., *Science* **157,** 924–927 (1967).

Neidhardt, F. C., Ingraham, J. L., and Schaechter, M., "Physiology of the Bacterial Cell." Sinauer Associates, Sunderland, MA, 1990.

Novick, N. J., and Alexander, M., *Appl. Environ. Microbiol.* **49,** 737–743 (1985).

Nuck, B. A., and Federle, T. W., *Environ. Sci. Technol.* **30,** 3597–3603 (1996).

Odum, E. P., "Fundamentals of Ecology." Saunders, Philadelphia, 1971.

Paris, D. F., and Rogers, J. E., *Appl. Environ. Microbiol.* **51,** 221–225 (1986).

Paris, D. F., Steen, W. C., Baughman, G. L., and Barnett, J. T., Jr., *Appl. Environ. Microbiol.* **41,** 603–609 (1981).

Parker, L. W., and Doxtader, K. G., *J. Environ. Qual.* **12,** 553–558 (1983).

Pfaender, F. K., Shimp, R. J., and Larson, R. J., *Environ. Toxicol. Chem.* **4,** 587–593 (1985).

Pignatello, J. J., Sawhney, B. L., and Frink, C. R., *Science* **236,** 898 (1987).

Righelato, R. C., *in* "The Filamentous Fungi" (J. E. Smith and D. R. Berry, eds.), Vol. 1, pp. 79–103. Edward Arnold, London, 1975.

Rijnaarts, H. H. M., Bachmann, A., Jumelet, J. C., and Zehnder, A. J. B., *Environ. Sci. Technol.* **24,** 1349–1354 (1990).

Rovira, A. D., and Greacen, E. L., *Aust. J. Agric. Res.* **8,** 659–673 (1957).

Rozich, A. F., Gaudy, A. F., Jr., and D'Adamo, P. C., *Water Res.* **19,** 481–490 (1985).

Russell, J. A., and Baldwin, R. L., *Appl. Environ. Microbiol.* **37,** 531–536 (1979).

Schmidt, S. K., Simkins, S., and Alexander, M., *Appl. Environ. Microbiol.* **50,** 323–331 (1985).

Schmidt, S. K., Scow, K. M., and Alexander, M., *Appl. Environ. Microbiol.* **53,** 2617–2623 (1987).

Scow, K. M., and Alexander, M., *Soil Sci. Soc. Am. J.* **56,** 128–134 (1992).

Scow, K. M., Simkins, S., and Alexander, M., *Appl. Environ. Microbiol.* **51,** 1028–1035 (1986).

Scow, K. M., Schmidt, S. K., and Alexander, M., *Soil Biol. Biochem.* **21,** 703–708 (1989).

Scow, K. M., Li, D., Manilal, V., and Alexander, M., *Mycol. Res.* **94,** 793–798 (1990a).

Scow, K. M., Merica, R. R., and Alexander, M., *J. Agric. Food Chem.* **38,** 908–912 (1990b).

Shehata, T. E., and Marr, A. G., *J. Bacteriol.* **107,** 210–216 (1971).

Simkins, S., and Alexander, M., *Appl. Environ. Microbiol.* **47,** 1299–1306 (1984).

Simkins, S., Mukherjee, R., and Alexander, M., *Appl. Environ. Microbiol.* **51,** 1153–1160 (1986).

Steinberg, S. M., Pignatello, J. J., and Sawhney, B. L., *Environ. Sci. Technol.* **21,** 1201–1208 (1987).

Stucki, G., and Alexander, M., *Appl. Environ. Microbiol.* **53,** 292–297 (1987).

Subba-Rao, R. V., Rubin, H. E., and Alexander, M., *Appl. Environ. Microbiol.* **43,** 1139–1150 (1982).

Suflita, J. M., Robinson, J. A., and Tiedje, J. M., *Appl. Environ. Microbiol.* **45,** 1466–1473 (1983).

Thomas, J. M., Yordy, J. R., Amador, J. A., and Alexander, M., *Appl. Environ. Microbiol.* **52,** 290–296 (1986).

Torstensson, N. T. L., and Aamisepp, A., *Weed Res.* **17,** 209–212 (1977).

Trinci, A. P. J., *Arch. Microbiol.* **73,** 353–367 (1970).

van der Kooij, D., and Hijnen, W. A. M., *Appl. Environ. Microbiol.* **41,** 216–221 (1981).

van Genuchten, M. T., Davidson, J. M., and Wierenga, P. J., *Soil Sci. Soc. Am. Proc.* **38,** 29–35 (1974).

Vashon, R. D., and Schwab, B. S., *Environ. Sci. Technol.* **16,** 433–436 (1982).
Volk, W. A., and Myrvik, Q. N., *J. Bacteriol.* **66,** 386–388 (1953).
Volkering, F., Breure, A. M., Sterkenburg, A., and van Endel, J. G., *Appl. Microbiol. Biotechnol.* **36,** 548–552 (1992).
Walker, A., and Brown, P. A., *Bull. Environ. Contam. Toxicol.* **30,** 365–372 (1983).
Wolfe, N. L., Kitchens, B. E., Macalady, D. L., and Grundl, T. J., *Environ. Toxicol. Chem.* **5,** 1019–1026 (1986).
Wolt, J. D., Smith, J. K., Sims, J. K., and Duebelbeis, D. O., *J. Agric. Food Chem.* **44,** 324–332 (1996).
Wrenn, B. A., and Rittmann, B. E., *Biodegradation* **7,** 49–64 (1996).
Yordy, J. R., and Alexander, M., *Appl. Environ. Microbiol.* **39,** 559–565 (1980).
Yoshida, F., and Yamane, T., *Biotechnol. Bioeng.* **13,** 691–695 (1971).
Zimdahl, R. L., and Gwynn, S. M., *Weed Sci.* **25,** 247–251 (1977).

CHAPTER 7

Threshold

Organic pollutants in many surface and groundwaters, soils, and sediments are present at low concentrations. Even at these trace levels, they may be of concern. Among the reasons that they are of practical importance are the following: (a) Risk analyses suggest that many of the chronic toxicants will be injurious to a small portion of the human population consuming waters or foods containing them. Chronic toxicants include a diversity of carcinogens, mutagens, and teratogens. (b) Some of the chemicals at these low concentrations (e.g., micrograms-per-liter levels) are acutely toxic to aquatic organisms. (c) Some are subject to bio-concentration within tissues of organisms in natural food chains and ultimately reach levels that are injurious to species at higher trophic levels in these food chains. (d) Regulatory agencies of national or local governments have established levels of many chemicals that are deemed to be safe, especially for public health, and the concentrations given by these regulatory guidelines or standards are often quite low. The standards for drinking water are often set based on the risk analyses.

Public health and ecological concerns with low chemical concentrations have fostered interest in the biodegradative processes affecting trace concentrations of organic chemicals. In the past, microbiologists have not paid attention to the problem because it was deemed far easier to grow organisms at the high substrate concentrations that would give large cell yields. However, as the interest grew, previously unanticipated phenomena became apparent. One such phenomenon is the existence of a threshold or, more specifically, a concentration of a nutrient source below which microorganisms cannot grow.

To maintain its viability, every organism must expend energy. In animals and humans, the energy used is reflected in basal metabolism. In microorganisms, the amount of energy to permit the organism to remain alive is designated *maintenance energy*. For heterotrophs, this energy is derived from the oxidation of organic compounds. When the concentration of the carbon source for growth is high, diffusion of the substrate from solution to the cell surface and the subsequent transfer of the molecule across the surface into the cell provide enough of the substrate to satisfy the needs for maintenance energy and for processes that lead to increases in cell size, growth, and multiplication. The same is not the case at low substrate levels. Considering only diffusion of the molecule from the liquid to the cell surface (and ignoring transfer across the membrane, which cannot exceed the rate that molecules reach the surface of the organism), as a low substrate concentration is reduced to a still lower level, the energy for maintenance represents an ever higher percentage of substrate-C that reaches the organism by diffusion, and an ever smaller percentage is used for growth and multiplication. At some lower value, all the energy in the form of C that reaches (and enters) the cell is used simply to keep the cell alive, and none is used for growth. At this concentration, therefore, although the substrate is being metabolized, the cells are not growing, and the population size and biomass are not increasing. This concentration represents the threshold.

Moreover, if the population size initially is so small that biodegradation is inconsequential, undetectable, or both, that absence of multiplication will be reflected in the absence of significant or detectable biodegradation, even though the organisms are metabolizing part of the substrate pool to maintain themselves. The threshold is the lowest concentration that sustains growth and represents the level below which a species (that needs to proliferate to cause a detectable change) brings about little or no chemical destruction.

The possible existence of a threshold was first postulated because of the presence of relatively constant levels of dissolved organic C in the oceans. This C, presumably because of its low concentration, was not available to support microbial proliferation and hence mineralization of the C (Jannasch, 1967). The level of such dissolved organic C is approximately 1 mg/liter in marine waters and is commonly less than 5 mg/liter in oligotrophic fresh waters. Moreover, if significant decomposition of this organic matter were occurring, the concentration should fall at increasing distances away from the water's surface, where the organic matter is being generated photosynthetically by the phytoplankton. Because no such marked decline is evident with depth, it was hypothesized that mineralization must be slow. However, this line of evidence in support of the existence of a threshold for growth is weak because (a) much of the organic matter, when concentrated, is intrinsically resistant to microbial degradation (Barber, 1968) and (b) the concentration of some aquatic constituents may represent a steady state, that is, a balance between the continuous formation and continuous mineralization. For example, the concentration of individual amino acids at a site in the Pacific Ocean ranges from less than 0.05 to 3 μg/liter; however, these amino acids are continuously being destroyed, so that they also must be constantly formed to maintain the quantities that are found (Williams *et al.,* 1976).

More convincing evidence has come from studies of biodegradable synthetic compounds in waters and soils. Because these compounds are not formed biologically, their presence at reasonably constant levels or their persistence at low levels indicates that the biodegradation one might expect is not occurring. These studies indicate that no biodegradation occurs in the test period below a certain concentration or that the rate is less than that which might be expected from the rates observed at higher levels (if it is assumed that the rate is proportional to concentration) (Fig. 7.1). Typical data for fresh waters and soils are shown in Table 7.1. It is evident that the threshold is sometimes very low and sometimes reasonably high. Nevertheless, data suggest that the threshold is often at about 0.1 to 5 μg per liter of water or per kilogram of soil. In instances in which water containing the chemical is passing through a solid (e.g., glass beads or soil) on which the microorganisms

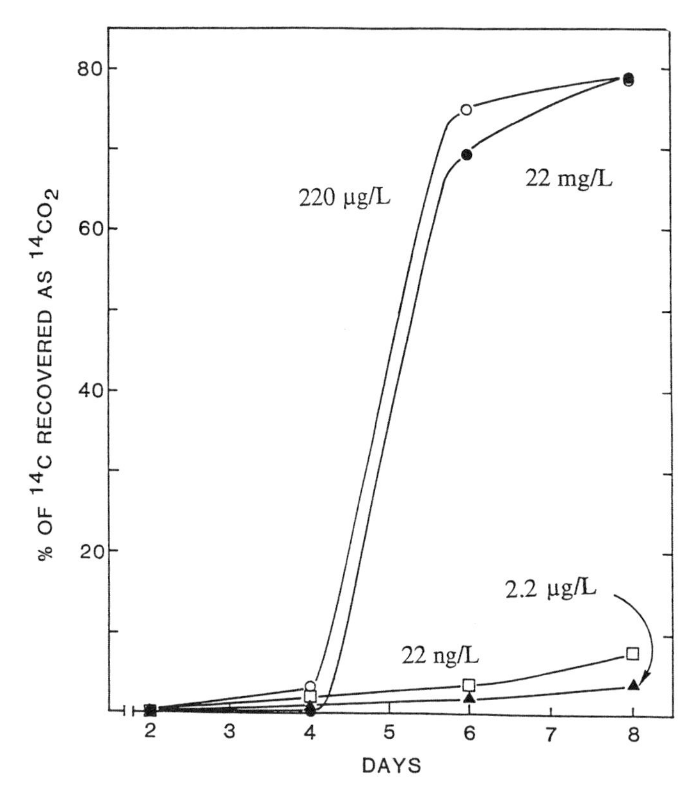

Figure 7.1 Mineralization of 2,4-D added at several concentrations to river water. (From Boethling and Alexander, 1979a. Reprinted with permission from the American Society of Microbiology.)

Table 7.1

Chemical Concentrations at Which Biodegradation Does Not Occur or is Slower Than Predicted

Chemical	Environmental source	Concentration (μg per liter of water or kg of soil)	Reference
2,4-D	Stream	2.2	Boethling and Alexander (1979a)
Sevin	Stream	3.0	Boethling and Alexander (1979a)
Aniline	Lake	0.1	Hoover *et al.* (1986)
4-Nitrophenol	Lake	1.0	Hoover *et al.* (1986)
2,4-Dichlorophenol	Lake	2.0	Hoover *et al.* (1986)
Styrene	Lake	2.5	Fu and Alexander (1992)
Phenol	Lake	0.0015	Rubin and Alexander (1983)
Carbofuran	Soil	10, 100[a]	Chapman *et al.* (1986)
2,4,5-T	Soil	100	McCall *et al.* (1981)
1,2-, 1,3-, and 1,4-Dichlorobenzenes	Biofilm on glass	0.2–7.1[b]	Bouwer and McCarty (1982)

[a] Concentration that did not result in a population increase to rapidly destroy a second addition of carbofuran.

[b] Concentration in effluent from a column containing glass beads supporting microorganisms degrading the chemicals.

reside, the concentration for the apparent threshold may be anomalous because the chemical may escape in the liquid emerging from the bottom of the column of particles before all the chemical, which is being slowly destroyed at these trace levels, is degraded.

A threshold may also exist for the biodegradation of gases. Thus, the threshold concentration for CH_4 degradation in soil is reported to be approximately 0.03 ppmv (Bender and Conrad, 1993).

The experimental values recorded in Table 7.1 are not surprising in view of monitoring data obtained from analyses of samples taken from natural ecosystems. For example, in natural waters of Canada receiving NTA, the average level of this chelating agent was 5 μg/liter (International Joint Commission, 1978). Certain nonylphenoxycarboxylic acids persist in river waters at concentrations of 2 to 116 μg/liter (Ahel *et al.,* 1987). Similarly, a great variety of synthetic chemicals are present in surface waters at low concentrations (Meijers and van der Leer, 1976), and groundwater accidentally contaminated with 2,4-D and 2,4-dichlorophenol still showed their presence at low levels years after an inadvertent release (Faust and Aly, 1964).

Analogous observations have been made when wastewaters are passed through soil as a means of destroying a harmful chemical by microbial action. In experimental trials, the concentrations of many compounds fell to undetectable levels as solutions containing them passed through soil columns, but a reasonable percentage of the 1,2-, 1,3-, and 1,4-dichlorobenzenes and diisobutyl phthalate in the influent water was still present in the effluent, and a readily biodegradable molecule like di(2-ethylhexyl) phthalate at 70 ng/liter did not disappear at all as a result of passage through soil (Bouwer *et al.*, 1981). Benzophenone and diethyl and dibutyl phthalates also have been reported to not disappear when passed at low concentration through soil columns set up to simulate the rapid infiltration of contaminated waters through soil, and waters moving out and away from land-infiltration sites in the field have been found to contain 0.02 μg of toluene, 0.05 to 1.14 μg of xylenes, 0.07 to 0.50 μg of naphthalene, 0.05 to 2.1 μg of benzophenone, and 0.01 to 2.4 μg of individual phthalate esters per liter (Hutchins *et al.*, 1983). Each of these chemicals is mineralized at higher concentrations. Similarly, passing a solution of 1,2-dichlorobenzene through a column of sand reduced the concentration from 25 μg/liter but only to a concentration of 0.1 μg/liter (van der Meer *et al.*, 1987). In a plume of contaminated water derived from secondary sewage effluent subjected to rapid infiltration, a number of compounds were found to have persisted at low concentration in the aquifer for more than 30 years; the average concentrations in the groundwater were 20–70 ng of 2,3-dimethyl-2-butanol, 2-methyl-2-hexanol, ethylbenzene, and propylbenzene isomers (Barber *et al.*, 1988), compounds that are probably all metabolized at higher concentrations.

Thresholds may be evident, although not initially expected, if the compound of concern is present in an oil, organic solvent, or other nonaqueous liquid. If the partitioning of the compound between the nonaqueous liquid and water initially is such that the concentration in the water is very low and most is in the nonaqueous phase, the level in the water—in which the microorganisms function—may be lower than the minimum for biodegradation (Efroymson and Alexander, 1995).

In some instances in which microbial colonization on glass beads is promoted to give biofilms, the minimum concentration below which there is no growth of the biofilm is quite high; for example, 100 to 1000 μg/liter (Rittmann, 1985).

Determining the existence of a threshold concentration for growth of bacteria in pure culture is complicated by the ability of many species to grow in media to which no C source is deliberately added. The liquid or inorganic salts used to formulate the medium, the air in the gas phase above the liquid medium in the flask, or both typically contain sufficient organic matter to support the multiplication of these species, which may reach densities of 10^4 to 10^5 cells per milliliter in such allegedly C-free media. Inasmuch as a population of 10^5 cells per milliliter would probably consume 100 ng of a substrate per milliliter (or 100 μg per liter) and the threshold is usually below 100 μg/liter, it is difficult with such species to show that they cannot grow in solutions with little added C; that is, the growth is nearly entirely at the expense of the uncharacterized, contaminating substrates rather than the test compound. This experimental difficulty is a result of the artificial conditions in pure cultures because most species in nature that actively destroy synthetic organic molecules are probably not effective competitors with their neighboring species for naturally occurring chemicals. Nevertheless, this procedural obstacle in pure culture can be overcome by using species that grow little, if at all, on the contaminating C.

Investigations of pure cultures of bacteria clearly show the existence of a threshold concentration of the C source below which multiplication does not occur. This value is about 18 μg/liter for *Escherichia coli* and *Pseudomonas* sp. growing on glucose (Shehata and Marr, 1971; Boethling and Alexander, 1979b), 180 μg/liter for *Aeromonas hydrophila* growing on starch (van der Kooij *et al.*, 1980), 210 μg/liter for a coryneform bacterium using glucose (Law and Button, 1977), approximately 300 μg/liter for a strain of *Pseudomonas* growing at the expense of 2,4-dichlorophenol (Goldstein *et al.*, 1985), about 5 μg/liter for *Salmonella typhimurium* provided with glucose (Schmidt and Alexander, 1985), and 2 μg/liter for a bacterium mineralizing quinoline (Brockman *et al.*, 1989). A mixture containing an anaerobic benzoate-degrading bacterium and a H_2-metabolizing bacterium transformed benzoate, but it was unable to metabolize the compound below levels of 26 to 790 μg/liter (Hopkins *et al.*, 1995). Such information as well as individual studies of a variety of marine bacteria, for which threshold concentrations of 0.15 to greater than 100 mg/liter were found (Jannasch, 1967), demonstrate that the threshold concentrations below which individual bacterial species are unable to multiply vary enormously. Some species have surprisingly high thresholds, but others are able to grow down to about 2 μg/liter but no lower. These values are of special significance for biodegradation if the population is initially small so that multiplication is essential for appreciable destruction of the substrate. Indeed, it has been noted that the indigenous population of 2,4-dinitrophenol-metabolizing bacteria in soil could not be maintained and would not multiply if the concentration was 0.1 mg/kg, although the bacteria multiplied at higher concentrations (Schmidt and Gier, 1989).

A model has been developed for estimating, on theoretical grounds, the threshold concentration of an organic compound required to support the multiplication of a bacterium. Below the value so calculated, the size of the population should not increase. The model is formulated on the basis of (a) the maximum rate that an organism can acquire energy at a particular concentration of substrate and (b) the rate it uses energy just to maintain its viability. It predicts that a threshold exists when the organism's need for C to supply the energy for maintenance is just equal to the rate of diffusion of the chemical to the cell surface. Below this concentration, too little energy is available to the cell to allow it to be maintained, and thus it will die. The equation for the relationship is

$$\tau = \frac{1/Y_{max}(R_d^2 - R_b^2)/2}{D_{AB}C_b/\rho - (m/\ln 2)(R_d^2 - R_b^2)/2},$$

where τ is the maximum doubling time for the cells, Y_{max} is the yield coefficient, R_b and R_d are the radii of the cell at its first appearance and at the time of its cell division, respectively, D_{AB} is the diffusivity of the chemical, C_b is the chemical concentration, ρ is the dry weight density of the cell (i.e., dry weight divided by the volume of the cell), and m is the maintenance coefficient. Diffusion constants for most organic pollutants are about 10^{-5} cm^2/sec. Common values found for bacteria are 0.55 g dry wt/g of substrate for Y_{max}, 0.31 g dry wt/cm^3 for ρ, and 15 mg of substrate/g dry wt/h for m. Using such common values and assuming that the radius of the cell as it first appears after cell division is 0.50 μm (R_b) and that the radius before the next cell division is 0.63 μm (R_d), the maximum doubling times for the cells at 10, 1.0, 0.5, and 0.20 μg of substrate per liter are 1.71, 21, 57, and infinite hours, respectively, that is, the threshold for such cells would be 0.20 μg/liter (Schmidt *et al.*, 1985). Obviously, however, the actual threshold must be somewhat higher than those suggested by the model because the cell needs energy to grow, not every molecule that reaches the cell surface passes through the membrane, and not every penetrating molecule is utilized. Nevertheless, the model does provide a minimum value. Moreover, because the requirements for maintenance energy differ appreciably among species, so too will the thresholds. Similar assumptions are used to calculate the threshold concentration below which a biofilm of bacteria would not be maintained (Rittmann and McCarty, 1980; Rittmann *et al.*, 1980).

The threshold is lowered if bacteria carrying out the transformation have certain alternative C sources available to them. In continuous culture of a marine bacterium, for example, the threshold concentration for glucose utilization was reported to be 0.48 mg/liter if the sugar was the sole C source, but it was lowered to 8 μg/liter in the presence of arginine and reduced even further in the presence of a mixture of amino acids (Law and Button, 1977). Analogous findings were made with a strain of *Pseudomonas* metabolizing 3-chlorobenzoate, namely that the threshold was lowered if acetate was present (Tros *et al.*, 1996). Similarly, the lower-than-predicted rate of mineralization of 0.39 to 1.5 ng of phenol per liter by a mixture of lake water bacteria was increased to the expected rate if much higher concentrations of arginine were added (Rubin and Alexander, 1983). However, sometimes the threshold is raised in the presence of a second substrate. For example, the threshold for benzoate degradation by a two-membered culture of anaerobes was increased by acetate (Hopkins *et al.*, 1995), and a similar effect was evident for the metabolism of methane by a mixed culture when di- or trichloroethylene was present (Anderson and McCarty, 1997). Too few observations have been made to determine the frequency that the threshold can be changed by second C sources, and the effect of alternative organic substrates is likely to be expressed only when the specific population carrying out the biodegradation is able to compete effectively for the second nutrient in communities containing many other species.

A threshold may also exist for the acclimation of microbial communities. Thus, a freshwater microbial community became acclimated to the mineralization of 4-nitrophenol at levels above but not below 10 μg/liter (Spain and Van Veld, 1983); because such acclimation probably is merely an indication of the time for the cells to become sufficiently numerous to cause a detectable loss of the chemical, the threshold may merely reflect that which characterizes growth per se. However, the induction of metabolic activity in bacterial cells may have a threshold even in the absence of growth, witness the reported induction of 3- and 4-chlorobenzoate degradation by *Acinetobacter calcoaceticus* at concentrations above 160 μg/liter but not below (Reber, 1982).

The threshold phenomenon may not be restricted to C sources, and no growth may take place at concentrations of other nutrients below some threshold value, for example, P (Button, 1985). At this time, however, the occurrence of thresholds for other nutrients and their significance for biodegradation have scarcely been explored.

The fact that the biodegradation of some compounds, both in pure culture and in nature, does not occur below some measurable concentration does not mean that thresholds always exist—or at least at concentrations measurable by currently available methods. Quite the contrary: many organic chemicals are mineralized in natural environments (or in samples collected from these environments and tested in the laboratory) at levels below which organic substrates fail to support growth. In stream water, for example, glucose is mineralized at 1.8 ng/liter and dimethylamine and diethanolamine at less than 10 ng/liter (Boethling and Alexander, 1979b), and a linear alcohol ethoxylate is mineralized in estuarine water at 33 ng/liter (Larson *et al.*, 1983). In lake water, mineralization of benzylamine is evident at less than 1 ng/liter, of phenol at 0.10 ng/liter, of aniline at 5.7 ng/liter, of 2,4-D at 1.5 ng/liter, and of di(2-ethylhexyl) phthalate at 21 ng/liter (Rubin *et al.*, 1982; Subba-Rao *et al.*, 1982) (Fig. 7.2). In many of these instances, the rate, as expected, is directly correlated with concentration, so that the mineralization rate is 10-fold less at 10-fold lower levels. Similarly, mineralization of 0.32 μg of phenol, 0.30 μg of aniline, and 1.0 μg of 4-nitrophenol per kilogram occurs in soil (Scow *et al.*, 1986), and several compounds are mineralized in sediment at 0.5 μg/kg (Ursin, 1985). A threshold also could not be detected in the metabolism of DDT by individual bacteria growing in a rich medium, and the insecticide was degraded even at 10 ng/liter (Katayama *et al.*, 1993). Mineralization at substrate

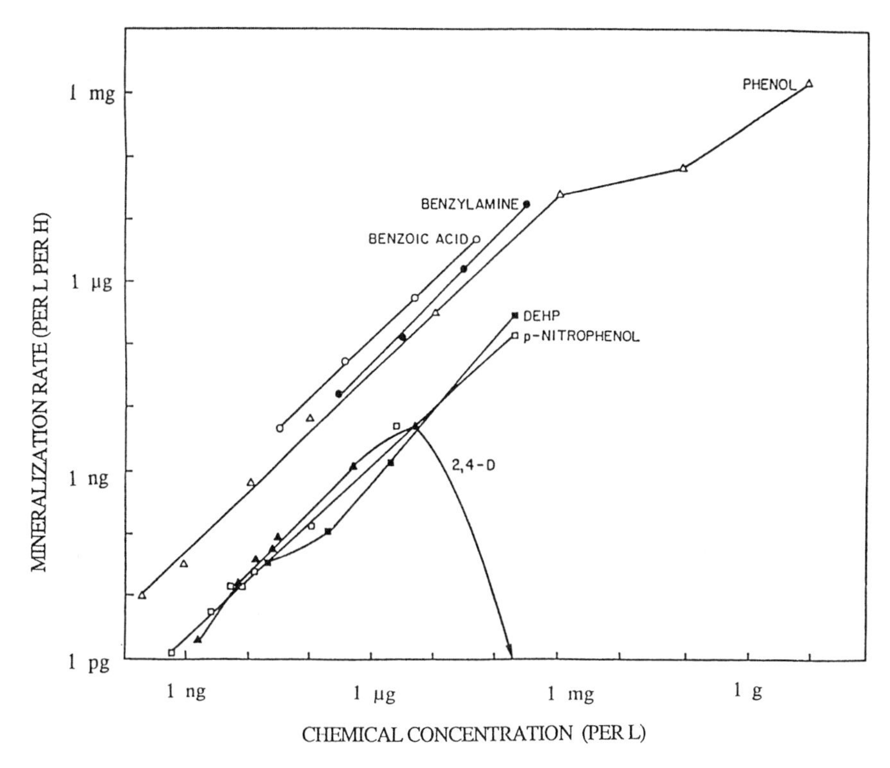

Figure 7.2 Rates of biodegradation in lake water of several organic compounds added at a wide range of concentrations. (From Rubin *et al.*, 1982. Reprinted with permission from the American Society for Microbiology.)

levels below the threshold for replication may be carried out either by nongrowing cells or by cells growing at the expense of other organic compounds present at levels above the threshold. For nongrowing cells to cause a significant change in nature, however, their biomass must be large.

Many environments contain levels of organic C in excess of that needed to support growth, or the levels may be regenerated constantly by excretions of other organisms (e.g., phytoplankton) or by new additions. Under these conditions, the energy needs for maintenance and growth of the populations degrading the compounds of interest may be met by use of these other organic molecules. As will be discussed elsewhere, microorganisms may metabolize two, or sometimes more, organic substrates simultaneously provided that their concentrations are not excessively high. The compound sustaining growth and that is present at levels above the threshold has been called the *primary substrate,* and the compound that is below the threshold but is still catabolized has been designated the *secondary substrate* (Rittmann, 1985). For example, *Salmonella typhimurium,* which has a threshold for growth on glucose of slightly below 5 μg/liter, is able to degrade that sugar at 0.5 μg/liter if the bacterium is multiplying at the expense of arabinose that is initially present at 5.0 mg/liter (Fig. 7.3). Similarly, *Aeromonas hydrophila* destroys starch at concentrations too low to allow for proliferation of the bacterium if it is provided with glucose (van der Kooij *et al.*, 1980). A biofilm composed of a microbial mixture that colonized the surfaces of glass beads, in like fashion, is capable of destroying alanine at 30 μg/liter if the amino acid is a secondary substrate,

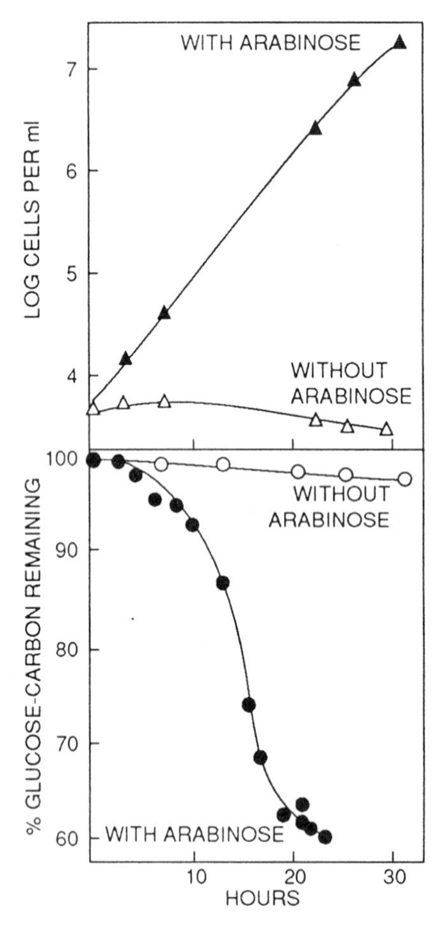

Figure 7.3 Numbers of *Salmonella typhimurium* and mineralization of ^{14}C-labeled glucose (0.5 μg/liter) in media with and without arabinose at 5 mg/liter. Only glucose mineralization was determined. (From Schmidt and Alexander, 1984. Reprinted with permission from the American Society for Microbiology.)

although the threshold for maintenance of the biofilm is 200 μg/liter when it is the sole substrate (Namkung *et al.,* 1983). The microbial destruction of organic molecules present in trace amounts may, in fact, frequently occur because of the presence of primary substrates for the active species. Furthermore, if microorganisms are actively growing on a compound of concern, they sometimes may be able to lower the concentration to undectable levels (Roch and Alexander, 1997).

The apparent existence in natural waters and wastewaters, and possibly in other environments, of traces of potentially degradable organic pollutants may thus be attributable to the thresholds below which growth does not occur. A microorganism whose sole selective advantage in these environments is its ability to grow by using particular novel substrates therefore may not increase in abundance, and the substrate may then not disappear. Moreover, the fact that thresholds exist points to the danger of drawing conclusions about what will happen at low chemical concentrations in nature based on laboratory tests with solutions containing much higher concentrations of the substrate. Nevertheless, it is not presently possible to predict which biodegradable chemicals in what environments will persist because of the threshold and which will be destroyed because of the ability of the responsible populations to function at still lower levels of the substrate.

A significant factor contributing to apparent threshold concentrations in soils, aquifer materials, and sediments may be the mass transfer of the compound to the organisms. This will be true if the compound is not immediately available because it is sorbed or is in inaccessible pores, so that the effective concentration is determined by the rate of diffusion of the molecules from the inaccessible site to the microorganisms (Bosma *et al.,* 1997).

REFERENCES

Ahel, M., Conrad, T., and Giger, W., *Environ. Sci. Technol.* **21,** 697–703 (1987).

Anderson, J. E., and McCarty, P. L., *Environ. Sci. Technol.* **31,** 2204–2210 (1997).

Barber, L. B., II, Thurman, E. M., Schroeder, M. P., and Le Blanc, D. R., *Environ. Sci. Technol.* **22,** 205–211 (1988).

Barber, R. T., *Nature (London)* **220,** 274–275 (1968).

Bender, M., and Conrad, R., *Chemosphere* **26,** 687–696 (1993).

Boethling, R. S., and Alexander, M., *Appl. Environ. Microbiol.* **37,** 1211–1216 (1979a).

Boethling, R. S., and Alexander, M., *Environ. Sci. Technol.* **13,** 989–991 (1979b).

Bosma, T. N. P., Middeldorp, P. J. M., Schraa, G., and Zehnder, A. J. B., *Environ. Sci. Technol.* **31,** 248–252 (1997).

Bouwer, E. J., and McCarty, P. L., *Environ. Sci. Technol.* **16,** 836–843 (1982).

Bouwer, E. J., McCarty, P. L., and Lance, J. C., *Water Res.* **15,** 151–159 (1981).

Brockman, F. J., Denovan, B. A., Hicks, R. J., and Fredrickson, J. F., *Appl. Environ. Microbiol.* **55,** 1029–1032 (1989).

Button, D. K., *Microbiol. Rev.* **49,** 270–297 (1985).

Chapman, R. A., Harris, C. R., and Harris, C., *J. Environ. Sci. Health, Part B* **B21,** 125–141 (1986).

Efroymson, R. A., and Alexander, M., *Environ. Sci. Technol.* **29,** 515–521 (1995).

Faust, S. D., and Aly, O. M., *J. Am. Water Works Assoc.* **56,** 267–279 (1964).
Fu, M. H., and Alexander, M., *Environ. Sci. Technol.* **26,** 1540–1544 (1992).
Goldstein, R. M., Mallory, L. M., and Alexander, M., *Appl. Environ. Microbiol.* **50,** 977–983 (1985).
Hoover, D. G., Borgonovi, G. E., Jones, S. H., and Alexander, M., *Appl. Environ. Microbiol.* **51,** 226–232 (1986).
Hopkins, B. T., McInerney, M. J., and Warikoo, V., *Appl. Environ. Microbiol.* **61,** 526–530 (1995).
Hutchins, S. R., Tomson, M. B., and Ward, C. H., *Environ. Toxicol. Chem.* **2,** 195–216 (1983).
International Joint Commission, 1978. Cited by J. M. Tiedje, *in* "Biotransformation and Fate of Chemicals in the Aquatic Environment" (A. W. Maki, K. L., Dickson, and J. Cairns, Jr., eds.), pp. 114–119. American Society for Microbiology, Washington, DC, 1980.
Jannasch, H. W., *Limnol. Oceanogr.* **12,** 264–271 (1967).
Katayama, A., Fujimura, Y., and Kuwatsuka, S., *J. Pestic. Sci.* **18,** 353–359 (1993).
Larson, R. J., Vashon, R. D., and Games, L. M., *in* "Biodeterioration 5" (T. A. Oxley and S. Barry, eds.), pp. 235–245. Wiley, Chichester, 1983.
Law, A. T., and Button, D. K., *J. Bacteriol.* **129,** 115–123 (1977).
McCall, P. J., Vrona, S. A., and Kelley, S. S., *J. Agric. Food Chem.* **29,** 100–107 (1981).
Meijers, A. P., and van der Leer, R. C., *Water Res.* **10,** 597–604 (1976).
Namkung, E., Stratton, R. G., and Rittmann, B. E., *J. Water Pollut. Control Fed.* **55,** 1366–1372 (1983).
Reber, H. H., *Eur. J. Appl. Microbiol. Biotechnol.* **15,** 138–140 (1982).
Rittmann, B. E., *Sci. Total Environ.* **47,** 99–113 (1985).
Rittmann, B. E., and McCarty, P. L., *Biotechnol. Bioeng.* **22,** 2343–2357 (1980).
Rittmann, B. E., McCarty, P. L., and Roberts, P. V., *Ground Water* **18,** 236–243 (1980).
Roch, F., and Alexander, M., *Environ. Toxicol. Chem.* **16,** 1377–1383 (1997).
Rubin, H. E., and Alexander, M., *Environ. Sci. Technol.* **17,** 104–107 (1983).
Rubin, H. E., Subba-Rao, R. V., and Alexander, M., *Appl. Environ. Microbiol.* **43,** 1133–1138 (1982).
Schmidt, S. K., and Alexander, M., *Appl. Environ. Microbiol.* **49,** 822–827 (1985).
Schmidt, S. K., Alexander, M., and Schuler, M. L., *J. Theor. Biol.* **114,** 1–8 (1985).
Schmidt, S. K., and Gier, M. K., *Microb. Ecol.* **18,** 285–296 (1989).
Scow, K. M., Simkins, S., and Alexander, M., *Appl. Environ. Microbiol.* **51,** 1028–1035 (1986).
Shehata, T. E., and Marr, A. G., *J. Bacteriol.* **107,** 210–216, (1971).
Spain, J. C., and Van Veld, P. A., *Appl. Environ. Microbiol.* **45,** 428–435 (1983).
Subba-Rao, R. V., Rubin, H. E., and Alexander, M., *Appl. Environ. Microbiol.* **43,** 1139–1150 (1982).
Tros, M. E., Bosma, T. N. P., Schraa, G., and Zehnder, A. J. B., *Appl. Environ. Microbiol.* **62,** 3655–3661 (1996).
Ursin, C., *Chemosphere* **14,** 1539–1550 (1985).
van der Kooij, D., Visser, A., and Hijnen, W. A. M., *Appl. Environ. Microbiol.* **39,** 1198–1204 (1980).
van der Meer, I. R., Roelofsen, W., Schraa, G., and Zehnder, A. J. B., *FEMS Microbiol. Ecol.* **45,** 333–341 (1987).
Williams, P. J., Le B., Berman, T., and Holm-Hansen, O., *Mar. Biol. (Berlin)* **35,** 41–47 (1976).

CHAPTER 8

Sorption

Some substances appear to be nonbiodegradable under all circumstances, for example, various synthetic polymers. Many compounds that are potentially subject to microbial attack, however, are not destroyed. It is thus essential to distinguish between a molecule that is biodegradABLE and one that, under particular circumstances, is not biodegradED. The former term indicates a susceptibility to destruction, the latter describes what actually occurs in a particular set of conditions.

Several reasons can be suggested for the lack of biodegradation of a molecule that is biodegradable: (a) The concentration of toxic substances may be so high at the site that microbial proliferation and metabolism are precluded. (b) One or more nutrients needed for microbial growth are at levels too low to permit appreciable growth. (c) The substrate itself may be at a concentration too low to allow for replication of the organisms containing the catabolic enzymes. (d) The substrate may not be in a form that is readily available for the microorganisms. *Bioavailability* is of extreme importance because it frequently accounts for the persistence of compounds that are biodegradable and that might otherwise be assumed to be readily decomposed. It may also limit attempts to bioremediate polluted sites.

The unavailability of an organic molecule could result from its sorption to solids in the environment, its presence in nonaqueous-phase liquids (NAPLs), or its entrapment within the physical matrix of the soil, sediment, or aquifer. These topics will be considered in detail.

The solid surfaces in many environments may dramatically affect the activity of indigenous microorganisms. These surfaces may alter the availability of organic chemicals, change the levels of various organic and inorganic nutrients, modify the pH or O_2 relationships, render inhibitors less toxic, retain microorganisms, or depress the activity of extracellular enzymes. The active surfaces may be clay minerals, the organic fraction (or humic substances) of soils or sediments, other complex carbonaceous matter, or sometimes amorphous Fe or Al oxides or hydroxides. The solid surfaces often act by *adsorption*, which refers to the retention of solutes originally present in solution by the surfaces of a solid material. *Absorption* may also be prominent in certain circumstances, this term referring to the retention of the solute within the mass of the solid rather than on its surfaces. The term *sorption* is used to include both adsorption and absorption. The zone in which sorption occurs represents a microenvironment immediately adjacent to the solid material, but this microenvironment is so different from the surrounding solution and is so important that much attention has been given to its understanding.

A wide array of organic compounds are sorbed by constituents of soils, sediments, wastewaters, subsoil materials, and other natural ecosystems. Included in this array are amino compounds, organic phosphates and phosphonates, nitrogen heterocycles, alkylbenzenesulfonates, cationic surfactants, and certain high-molecular-weight materials, to mention only a few classes of compounds. The properties of the compound will determine whether sorption occurs, the strength of the binding to the particles or other solids, and the specific environmental constituents responsible for the sorption. Certain organic molecules are sorbed more to the clay minerals, and others are bound largely or entirely to the organic matter. Not only are many of the organic substrates sorbed but so too are many of the inorganic nutrients needed by microorganisms.

A number of factors influence sorption of organic compounds. These include the type and concentration of solutes in the surrounding solution, the type and quantity of clay minerals, the amount of organic matter in the soil or sediment, pH, temperature, and the specific compound involved. The type of cation that is saturating the clay (e.g., whether the clay is saturated with Fe, Ca, or H ions) and the exchange capacity and specific surface area of clays also are of importance. Many of the major processes of concern to sorption occur at the surfaces of the clay minerals and humic materials, and these may have large areas per unit of mass; for example, a gram of clay may have a surface area of 20 to 80 m^2.

CHEMISTRY OF SORPTION

Much attention has been given to the sorption of organic compounds both to the clay and to the organic matter of soils and sediments, and it is important to consider the chemistry of sorption by these major constituents of soils and aquatic sediments.

The major clay minerals are of two main types, one in which the Si and Al layers are assembled in a 1 : 1 ratio of Si and Al (−Si·Al·Si·Al·Si·Al−), the second in which the layers are in a 2 : 1 ratio (−Si·Al·Si·Si·Al·Si−). In a 1 : 1 clay, such as kaolinite, the layers are tightly held together. Such nonexpanding clays have smaller surface areas and lesser capacities for sorption than 2 : 1 clays. Molecules are adsorbed on the outer surfaces of the 1 : 1 clays. In a 2 : 1 clay such as montmorillonite, the lattice structure of the clay can expand, and such clays may sorb organic compounds on both external and internal surfaces. With these expanding clays, organic molecules, inorganic nutrients, and water may penetrate between the layers of the mineral crystal. The availability of both outer and internal surfaces of some clay minerals for sorption is often of great importance.

Adsorption may involve physical or van der Waals forces, hydrogen bonding, ion exchange, or chemisorption. Large molecules may be retained on clay surfaces by hydrogen bonding, but for some low-molecular-weight organic compounds of importance as pollutants, of particular significance is ion exchange, in which an ion in solution of one type is exchanged for an ion of another type that is on the solid sorbing material. Clay minerals and colloidal organic materials have a net negative charge, and therefore they attract cations. A clay particle may have H, Ca, K, or Mg ions on its surface, but a positively charged organic molecule may displace another cation already on the surface of the clay and thus become retained by the mineral. Positively charged compounds may also be adsorbed to the organic fraction of soils and sediments since the humic substances also bear negative charges. The surfaces of both clay and humus colloids may retain organic cations by such means. Anionic organic molecules, in contrast, are generally repelled because of the negative charge on the surface (Morrill *et al.*, 1982). As a consequence, it is the molecules that are positively charged at pH values prevailing in nature that are chiefly retained by the negatively charged surfaces.

The capacity of clays to affect biodegradation differs according to clay type. This may be related in many instances to the *cation-exchange capacity* of the clays because the effect is related to the cationic properties of the substrate (or, for large molecules as substrates, the cationic properties of an extracellular enzyme). Montmorillonite, for example, frequently sorbs potential microbial substrates because of its high cation-exchange capacity and its expanding lattice structure. Many organic substrates can enter between the silicate sheets that make up this clay, and they thereby become protected. Typically, the effects of clay on sorption, if effects occur, are marked with montmorillonite and are less prominent with kaolinite and illite.

The organic fraction of soils and sediments is responsible for the sorption of many compounds, particularly those that are hydrophobic. Many polycyclic aromatic hydrocarbons and other nonpolar pollutants are sorbed chiefly by the native organic matter rather than the clay constituents of soils and sediments. The extent of this retention is directly correlated with the octanol–water partition coefficients, which is expressed as the K_{ow} value (a measure of hydrophobicity of chemicals), and the percentage of organic C in the soil or sediment; the more organic matter present in the solid phase, the more the hydrophobic molecule is sorbed. Two views exist on how hydrophobic molecules are retained by the organic matter. One view maintains that the process is physical sorption by the organic matter, in which physical binding of the solute to the organic solids occurs (Calvet, 1989). The other view holds that the hydrophobic molecule exists in the organic matter because it diffuses and partitions into the solid organic matter, much as a hydrophobic compound will partition from aqueous solution into an organic solvent in which it is highly soluble, that is, the compound is within the physical matrix of the solid organic phase that is the sorbent (Chiou, 1989). One view is thus that the molecule is concentrated on the outer surface or within the pores of a solid, where it is sorbed by physical or chemical forces; the other is that the molecule is distributed in the entire volume of the organic matter. These two concepts have markedly different implications for the potential availability of organic molecules to microorganisms.

Ion exchange associated with the native organic matter of soil may account for adsorption of such cationic compounds as the herbicide paraquat, and also of compounds that, following protonation, acquire a positive charge (e.g., the triazine herbicides), and these positively charged molecules may be retained by negatively charged groups (e.g., $R{-}COO^-$) of the complex organic material. The retention of cationic organic compounds because of ion exchange associated with the native organic matter of soils is especially important at neutral to slightly alkaline pH.

Hence, the properties of the sorbent and the chemical of concern determine whether sorption to the inorganic or to the organic surfaces, if both are present, is more important. Cationic organic molecules may be sorbed to the cation-exchange sites of clay minerals, humic surfaces, or both. Anionic compounds are poorly sorbed by clay minerals, but they may be moderately retained by organic surfaces. The complex organic matter of soil and other environments, however, is often the chief sorbent for nonionic organic compounds.

To appreciate the role of sorption in bioavailability and biodegradation, it is essential to realize the importance of the strength of binding. For example, a molecule retained within the expanding lattice of a clay with a 2 : 1 ratio, such as montmorillonite, may be totally unavailable, but the bioavailability would be far different if sorption involves solely the weak retention associated with hydrogen bonds or van der Waals forces. Therefore, the availability of sorbed organic molecules for microbial utilization will depend on the specific sorption mechanism (or mechanisms).

In some instances in which organic compounds interact with the organic matter of soils or sediments, or even the soluble humic materials of natural waters, the interaction is not really sorption in the sense used for clays or even for the hydrophobic binding discussed earlier for nonionic organic compounds. Instead, the change involves the formation of stable linkages. These linkages may sometimes be covalent linkages between the low-molecular-weight chemical and the complex natural humic substances. The result is a new chemical species. These types of interactions are discussed in Chapter 10.

DIMINISHED AVAILABILITY OF SORBED SUBSTRATES

Sorption of the organic substrates of microorganisms has a major impact on their growth and activity. The effect varies with the specific compound, the mechanism by which that compound is bound, the strength of retention if organic cations are the substrates, and the capacity of the microorganisms at the site to use the sorbed compound. The last factor is not well understood, but it is clear that some species are able to use sorbed compounds whereas others can metabolize the same molecules only when they are in aqueous solution.

In some instances, a biodegradable compound that is sorbed becomes completely resistant to microbial attack. For example, when all of the herbicide diquat is sorbed by montmorillonite clay, a mixture of soil microorganisms able to metabolize the organic cation free in solution no longer can mineralize the compound (Fig. 8.1). A complete inhibition as a result of sorption may also occur with compounds sorbed hydrophobically, as shown by the inability of microorganisms in some soils to metabolize certain polycyclic aromatic hydrocarbons (PAHs) that are sorbed (Weissenfels *et al.*, 1992). Similarly, EDB that is freshly added to soil and that can still be easily desorbed is readily metabolized, whereas the same

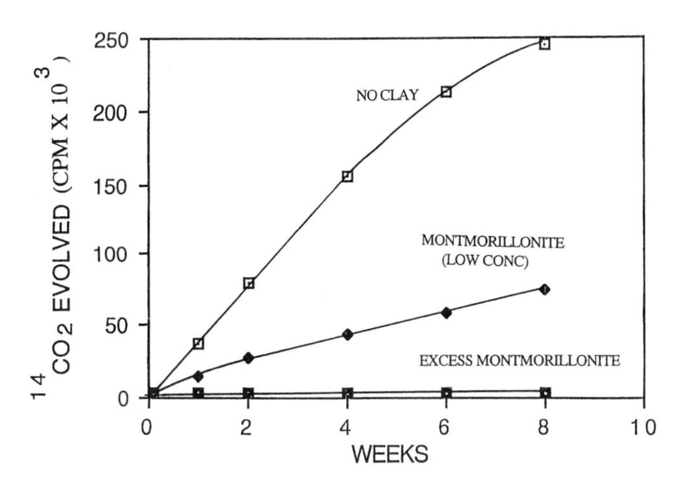

Figure 8.1 Mineralization of ^{14}C-labeled diquat in nutrient solutions with or without clay. (From Weber and Coble, 1968. Reprinted with permission.)

compound that has been in the soil for many years and is scarcely desorbed is likewise almost entirely resistant to microbial degradation (Steinberg *et al.*, 1987).

Reducing the availability of a molecule by sorption sometimes diminishes the rate or extent of transformation but does not wholly abolish the conversion. For example, dodecylbenzenesulfonates with linear or branched side chains are readily adsorbed by colloidal components of soil, and such binding reduces their degradation by a mixture of microorganisms (Inoue *et al.*, 1978), and other surfactants bound to clay or fractions of soil organic matter are metabolized slowly in soil (Knaebel *et al.*, 1994). Similarly, the rate of mineralization of glyphosate in several soils declines as the amount of the herbicide bound by the soil colloids increases (Nomura and Hilton, 1977). Tests with diallate added to mixtures of activated charcoal and soil to give varying percentages of binding of the herbicide demonstrated again a direct relationship between the amount of chemical sorbed and the rate of degradation (Anderson, 1981).

The effect of clays in suppressing decomposition has been verified with many organic molecules. Thus, the extent of mineralization of low concentrations (20 ng to 200 μg per liter) of benzylamine in lake water is reduced by montmorillonite, apparently because the clay reduces the amount of substrate available to the lake water microflora by binding the aromatic amine (Subba-Rao and Alexander, 1982). Alkylamines such as *n*-heptyl-, *n*-octyl-, and *n*-decylamines are also protected from attack by amine-utilizing bacteria if the compounds are bound to montmorillonite (Wszolek and Alexander, 1979).

The role of sorption to the organic fraction, rather than to the clay constituents of soil, in reducing biodegradation is suggested by findings that herbicides known to be sorbed by soil organic matter are more persistent in soils rich in organic matter than those having low humus levels, for example, pyrazon (Smith and Meggitt, 1967), simazine, and atrazine (Briška *et al.*, 1974). The decrease with time in the rate of breakdown of *N*′-(4-chlorophenoxy)phenyl-*N*,*N*-dimethylurea in a humus-rich soil was attributed to the slow rate of desorption of this herbicide from the natural organic complexes (Geissbühler *et al.*, 1963). Humic acid, a major component of humic materials, may increase the acclimation period and also retard the mineralization of benzylamine, and the extent of benzylamine mineralization declines as the percentage of the amine initially bound increases (Amador and Alexander, 1988). In the latter instance, however, the binding of an amine to humic acid probably is not simple sorption, but rather involves the formation of a complex with stable chemical bonds.

Soils, clay, and organic matter also protect substrates against hydrolysis by individual enzymes; thus, although alkaline phosphatase hydrolyzes Guthion and parathion and acid phosphatase hydrolyzes parathion and pirimiphos-methyl, none is cleaved if these organophosphate insecticides are sorbed to soil (Heuer *et al.*, 1976). In addition, proteins that complex with soil organic matter are resistant to hydrolysis by a proteolytic enzyme of *Streptomyces griseus* (Burns *et al.*, 1972), and complexing of proteins with lignin makes them resistant to hydrolysis by chymotrypsin or by several bacteria (Estermann *et al.*, 1959). Proteins adsorbed on kaolinite or montmorillonite clays are protected to some degree from hydrolysis by chymotrypsin, although the sorbed proteins are still slowly attacked by the proteolytic enzymes (Ensminger and Gieseking, 1942; McLaren, 1954). Complexes of dextrans or hydroxyethylcellulose with montmorillonite are less readily attacked by microorganisms than are the free polysaccharides (Lynch and Cotnoir, 1956; Olness and Clapp, 1972); in these instances, microorganisms probably act on the polysaccharides by extracellular enzymes. Cationic compounds, such as paraquat, that enter into the clay lattice are resistant to biodegradation (Burns and Audus, 1970). Similarly, purines, amino acids, and peptides that enter the interlayer region of expanding lattice clays such as montmorillonite may become protected from microbial attack (Greaves and Wilson, 1973), although extracellular enzymes excreted by microorganisms may slowly metabolize chemicals sorbed between lattices of such clays (Estermann *et al.*, 1959). It should be kept in mind, however, that low-molecular-weight compounds are generally believed to be degraded by intra- and not extracellular enzymes. Nevertheless, these studies have relevancy to approaches to bioremediation involving immobilized enzymes. It is highly likely that compounds degraded solely by intracellular enzymes would be resistant to biodegradation if located within the lattices of expanding-lattice clay minerals because of the inaccessibility of such chemicals to the microbial cell.

Several reasons can be advanced to explain the diminished rate of biodegradation because of the presence of surfaces. The diminished transformation often results from the fact that the substrate becomes less available or wholly unavailable in the sorbed state because biodegradation requires the compound to enter the cell to be acted on by intracellular enzymes. The chemical bound to a solid surface is not free to be transported across the outer surface of the cell to be transformed by the catalysts within the confines of the cell. Although most, and often all, of the enzymatic steps in the metabolism of low-molecular-weight compounds are intracellular, some steps in the microbial metabolism of low-molecular-weight substrates may be extracellular, and the initial phases in the decomposition of high-molecular-weight molecules are also extracellular. Reactions catalyzed by such extracellular enzymes are markedly affected by reactive surfaces because the enzymes are subject to sorption, and they may then lose catalytic activity; for example, pronase in the presence of montmorillonite (Griffith and Thomas, 1979), uricase sorbed on montmorillonite (Durand, 1964), acid phosphatase on montmorillonite, illite, and kaolinite (Makboul and Ottow, 1979), a protease, amylases, and cellulase on allophane, montmorillonite, and halloysite (Aomine and Kobayashi, 1964), and arylsulfatase on montmorillonite and kaolinite (Hughes and Simpson, 1978). In addition, availability of a substrate acted on by extracellular enzymes may be diminished or become negligible because it is less accessible to the enzyme, possibly because the chemical is shielded from the catalyst or because of steric effects preventing the formation of the substrate-enzyme complex necessary for the enzyme to catalyze the alteration of the substrate. As indicated earlier, however, studies of sorbed substrates acted on by individual enzymes have little relevancy for most environmental pollutants, for which it is generally believed that metabolism is entirely intracellular.

Sorption may affect biodegradation in a number of ways in addition to removing the organic substrate from solution or binding extracellular enzymes. (a) Inorganic nutrients and growth factors are also sorbed, and the removal from solution of such essentials for microbial replication may reduce the rate or extent of growth. (b) The microenvironment around the surface may be less favorable for the transformation than the surrounding solution because of the frequently lower pH immediately around negatively charged surfaces (because these surfaces attract and concentrate H^+ from the solution). (c) Conversely, sorption concentrates the nutrients at the surface of the adsorbent, so that growth of organisms near the surface may be enhanced and the biodegradation may be stimulated, especially if the surrounding solution has a low concentration of nutrients. (d) Moreover, the microorganisms themselves are sorbed, and frequently most bacterial cells in ecosystems with much clay and particulate organic matter are associated with the solids rather than the free liquid. It is likely, indeed, that most bacteria active in degradation in soils, sediments, and aquifers are retained by the solids, a view supported by the finding that, as naphthalene is being metabolized in soil, the population of naphthalene degraders that are adsorbed is two orders of magnitude greater than those present in the water phase (Di Grazio *et al.*, 1990). Some bacteria that become attached may adhere reversibly, but some adhere irreversibly and are not released to the ambient solution (Kefford *et al.*, 1982). At low cell densities, nearly all bacteria in the liquid may become adsorbed, but a small percentage may be retained at high cell densities (Gordon *et al.*, 1983). Once the cells become attached to the surface, their physiological activity may alter and their metabolic activity may be greater than, less than, or sometimes not different from the cells free in solution. Furthermore, attached cells may promote desorption, as suggested by a study of a model system containing a bacterium of the genus *Sphingomonas* and 3-chlorodibenzofuran sorbed to porous Teflon granules (Harms and Zehnder, 1995).

UTILIZATION OF SORBED COMPOUNDS

As indicated earlier, although sorption often reduces the rate and extent of biodegradation, it does not necessarily prevent it. Many sorbed molecules can be used by microorganisms as sources of C, energy, N, and probably other elements, and the compounds are thereby transformed, frequently slowly, but sometimes at reasonable rates. Biodegradation is evident even when all the chemical is sorbed

(Fig. 8.2). Such utilization occurs with compounds sorbed by clays as well as those retained by hydrophobic mechanisms. For example, certain amino acids or peptides that are bound to clays and are not spontaneously desorbed may serve as sources of C, N, or both for individual species of bacteria (Dashman and Stotzky, 1986). Many chemicals apparently adsorbed to the external surfaces of clay minerals can be metabolized, as is adenine present at the exposed edges of montmorillonite (Greaves and Wilson, 1973) and protein adsorbed on the outer surface of the same clay, especially if present on the clay in more than one layer (Pinck *et al.*, 1954). Some proteins adsorbed in monolayers on montmorillonite or kaolinite can be hydrolyzed by pure cultures of bacteria or by a purified proteolytic enzyme (Estermann *et al.*, 1959). The susceptibility to microbial attack of proteins complexed with clay varies with the type of protein, the site on the clay to which the protein is adsorbed, and the cation saturating the clay (Stotzky, 1986).

The availability of sorbed hydrophobic chemicals is evident in a study of phenanthrene sorbed to polyacrylic beads or sediment. On the one hand, a bacterium (strain P3) isolated in the usual way—with the test compound not being sorbed—could not mineralize the hydrocarbon that was sorbed to the beads or lake-bottom sediment. On the other hand, a second bacterium (strain P5-2), which was obtained by enrichment culture with the C source being phenanthrene bound to the beads, mineralized the compound sorbed to either beads or the aquatic sediment (Fig. 8.3). The second bacterium was not using the compound as it was being desorbed by abiotic mechanisms because, if that were true, the first organism

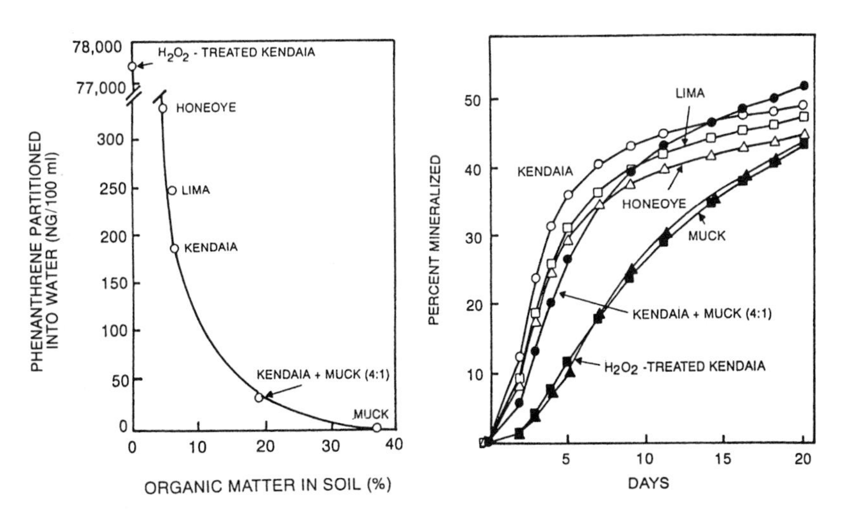

Figure 8.2 Sorption (left) and mineralization (right) of phenanthrene in three mineral soils (Honeoye, Lima, and Kendaia), a muck soil, a soil-muck mixture, and soil treated with H_2O_2 to remove organic matter. (From Manilal and Alexander, 1991. Reprinted with permission from Springer-Verlag.)

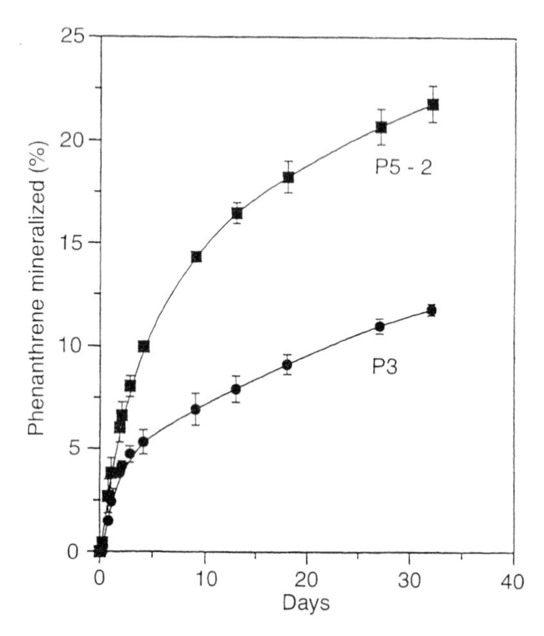

Figure 8.3 Mineralization by two bacteria of phenanthrene sorbed to polyacrylic beads. Error bars represent one standard deviation. (From Tang *et al.*, 1998. Reprinted with permission from Springer-Verlag.)

also should have been able to carry out the transformation. These observations suggest that bacteria acting on a sorbed molecule may need to have two physiological traits—the necessary catabolic enzymes and the capacity to make the sorbed molecule available. The first trait alone is not sufficient. This view is supported by a report that only one of two naphthalene-utilizing bacteria was able to degrade naphthalene sorbed by soil (Guerin and Boyd, 1992) and by investigations showing that certain bacteria capable of mineralizing biphenyl in solution were unable to use the molecule when sorbed to polyacrylic beads but a mixed culture could metabolize the sorbed compound (Calvillo and Alexander, 1996) and that proteins adsorbed to glass surfaces are readily available to marine bacteria (Taylor, 1995). It is also evident that part of the total quantity of hydrophobic compounds that become sorbed to soil or aquifer solids, and which are not the desorption-resistant fraction (i..e, not appearing in solution in reasonable periods of time), can be utilized by microorganisms (White and Alexander, 1996). This desorption-resistant fraction is of particular importance because the fraction that is readily desorbed would be removed from soil with rainfall events or would disappear by biodegradation, leaving the sorbed molecules with the solid phase.

These findings point to a shortcoming of many microbiological studies of the biodegradation of sorbed molecules: the organisms are nearly always isolated because of their ability to use the compound in solution or in a nonsorbed form, even if the issue ultimately to be addressed is biodegradation of the bound molecule. Because some bacteria appear to have physiological mechanisms that allow them to make use of sorbed compounds, investigations with species not having those mechanisms may provide results that have little environmental relevance or have little bearing on means to bring about bioremediation.

A methodological problem exists in verifying that an organism is actually using a sorbed compound: at equilibrium, even though most of the substrate may be retained by the solids, some is in aqueous solution. Then, as microorganisms grow and use up the substrate in the aqueous phase, more of the compound desorbs, and it may be the molecules initially present in solution and those subsequently desorbing and entering the liquid that sustain growth and that are being metabolized rather than those molecules that are sorbed (Fig. 8.4). However, even when little or no desorbed phenanthrene is detectable, it is readily mineralized in organic soils (Aronstein *et al.*, 1991; Manilal and Alexander, 1991), suggesting that the sorbed molecule is being degraded. Phenol and dialkyl quaternary ammonium compounds sorbed to sediments (Shimp and Young, 1988) and 4-nitrophenol and phenol sorbed to granular activated carbon (Speitel *et al.*, 1989a,b) also appear to be biodegraded, presumably by microorganisms acting on the molecules that are actually sorbed rather than those that are in solution. Organic molecules localized on other types of surfaces may be utilized, as shown by the ability of bacteria to use stearic acid or palmitic acid coated on glass (Hermansson and Marshall, 1985;

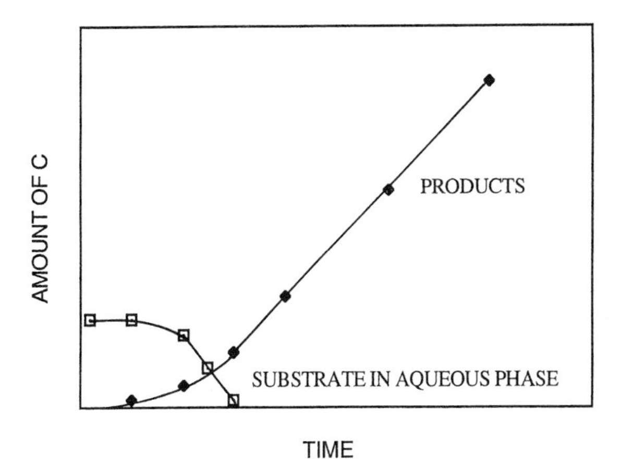

Figure 8.4 Microbial utilization of a compound in solution prior to metabolism of a substrate that was initially sorbed.

Thomas and Alexander, 1987). Nevertheless, many sorbed substrates are not readily used or are not degraded at all, and any biodegradation of these compounds that occurs is at the expense of chemical in solution. Should the substrate be irreversibly sorbed so that there is no desorption and should there be no organism able to make use of the sorbed substrate, that molecule would be rendered resistant to degradation.

It is still not clear how sorbed molecules become available to microorganisms. To be used for growth and to be degraded, the chemical must be assimilated into the cell. Three hypotheses can be advanced to explain the mechanism of utilization. (a) The organism uses the chemical that is initially in solution, and it also metabolizes the compound that enters the aqueous phase as a result of spontaneous desorption from the solid. At the outset, an equilibrium exists between the chemical that is retained on the surface and that which is in the ambient liquid:

$$\text{sorbed chemical} \underset{k_2}{\overset{k_1}{\rightleftharpoons}} \text{chemical in solution.}$$

Once the chemical in solution is consumed, which will occur if the degrading population is initially large or when the cell density becomes high as the organisms grow at the expense of the substrate, the subsequent rate of metabolism will be governed by the rate at which additional amounts of the substrate enter the soluble phase, that is, the desorption rate (k_1). Thus, when the concentration of the compound in solution is effectively zero because all that was initially present and that which subsequently enters solution by desorption is consumed, the subsequent rate of degradation is governed by the rate of desorption. Figure 8.5 depicts the

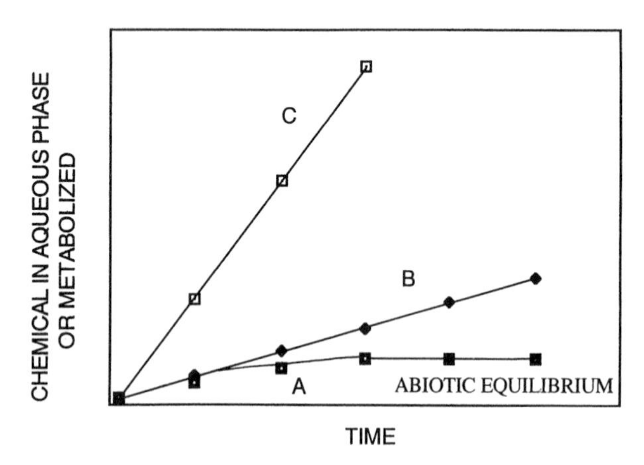

Figure 8.5 (A) Spontaneous desorption and abiotic equilibrium in the aqueous phase of a reversibly sorbed compound, (B) biodegradation of a compound limited by the rate of abiotic desorption, and (C) biodegradation of a compound that is used more rapidly than the rate of abiotic desorption.

rate of spontaneous desorption and the attainment of equilibrium in the absence of microorganisms (A); desorption-limited biodegradation is also shown (B). In other words, desorption is the rate-limiting step. Studies of the mineralization in lake water of benzylamine sorbed to montmorillonite suggest that once the amine in solution is degraded, the subsequent mineralization is limited by desorption (Subba-Rao and Alexander, 1982). At very low cell densities, the rate of transformation is low, and the organisms may be subsisting on the chemical that is originally in solution. However, as the cell density increases, the microbial demand for organic nutrients increases concomitantly until none is left in solution; at this point, the controlling role of desorption becomes evident.

(b) According to the second mechanism of utilization, microorganisms excrete metabolites that facilitate desorption so that the rate of biodegradation is actually greater than the rate of spontaneous desorption as determined in the absence of microorganisms. This is depicted in Fig. 8.5 by the plot labeled C. A study of decylamine-montmorillonite complexes showed that the rate of mineralization of the amine at high bacterial densities exceeds the spontaneous desorption rate and suggested that bacteria facilitate removal of the chemical from the clay (Wszolek and Alexander, 1979). Similarly, the initial rate of biodegradation of α-hexachlorocyclohexane in soil suspensions is often faster than the initial rate of desorption, also suggesting that the microflora enhances the desorption (Rijnaarts *et al.*, 1990). In some instances, spontaneous desorption is slow, yet biodegradation is rapid, as with the mineralization of biphenyl sorbed to polyvinylstyrene beads (Calvillo and Alexander, 1996) or of PAHs in a soil rich in organic matter (Carmichael *et al.*, 1997). Under such circumstances, the rate of biodegradation is not limited by the spontaneous desorption rate. Although the mechanism of facilitated desorption is unresolved, it could involve the elaboration of surfactants that release the chemical from the surface, cations that displace charged compounds, or extracellular enzymes, or it may be associated with a microbiologically induced change in the pH at the surface of the charged material. Surfactants desorb a number of hydrophobic compounds sorbed to soil, for example, anthracene, phenanthrene, pyrene (Liu *et al.*, 1991), and PCBs (McDermott *et al.*, 1989). However, high concentrations of most synthetic surfactants are needed both to desorb the hydrocarbons and to bring into aqueous solution these compounds of low water solubility. On the other hand, low concentrations of two nonionic alcohol ethoxylate surfactants markedly enhance the rate and/or extent of mineralization of phenanthrene and biphenyl sorbed to soils or aquifer solids, even though little of the compounds is desorbed (Aronstein *et al.*, 1991; Aronstein and Alexander, 1992, and several other surfactants stimulated pyrene degradation in samples of soil (Thibault *et al.*, 1996). Hence, it is plausible to believe that surfactants produced by microorganisms, which are typically excreted in low concentrations, may facilitate the use of bound compounds by these organisms.

(c) According to the third hypothesis, the sorbed compound is directly utilized by microorganisms that adhere to the same surface. The organism may come into direct contact with the compound, which may then penetrate into the cell without entering the surrounding liquid. In the case of hydrophobic molecules, the transfer might involve passage of the sorbed compound from the solid to the cell through the hydrophobic surface of the microorganism. Consistent with this hypothesis are findings that microbial adherence to solids facilitates the biodegradation of compounds that are sorbed (Calvillo and Alexander 1996; Harms and Zehnder, 1995). This would be comparable to the utilization of water-insoluble organic compounds by bacteria that adhere to the substance undergoing attack (Thomas and Alexander, 1987).

KINETICS

A number of models have been proposed for the biodegradation of compounds that are initially sorbed. Typically, these models assume that the substrate must be in aqueous solution for it to be metabolized. In an early model, for example, use was made of rate constants for desorption (k_1), adsorption (k_2), and biodegradation (k_3) (Furmidge and Osgerby, 1967):

$$\text{sorbed chemical} \underset{k_2}{\overset{k_1}{\rightleftharpoons}} \text{chemical in solution} \xrightarrow{k_3} \text{products of biodegradation}$$

The desorption term in such a formulation is usually considered to reflect spontaneous desorption, and facilitated desorption, direct use of the sorbed compound, or changes in the rate constants as the degrading population multiplies are not considered. If one makes some simplifying assumptions, assumes first-order kinetics of biodegradation, and introduces a partition coefficient (k_p) to reflect the affinity of the chemical for the sorbent, one might expect that the half-life of the chemical would be equal to k_p/k_3 (Briggs, 1976). The formulation in the preceding equation has been extended to consider the special cases in which the products of biodegradation themselves are sorbed by introducing rate constants for the sorption (k_4) and desorption (k_5) of these products (Holm *et al.*, 1980). The model of Dao and Lavy (1987) also uses rate constants for adsorption and desorption, as well as a rate constant for biodegradation, but their model assumes that only the substrate that is in aqueous solution is metabolized and that adsorption and desorption are essentially instantaneous. The model of Miller and Alexander (1991) was designed for sorbed molecules that are readily desorbed and that are used by nongrowing populations, and their model gives good fits to the biodegradation by a pseudomonad of benzylamine sorbed to montmorillonite. If the compound is desorbed rapidly relative to the rate of microbial utilization, the latter rate possibly being slow because of a small biomass or an intrinsically slow transformation, then the kinetics of biodegradation will resemble those occurring in the absence of a sorbent (Scow and Johnson, 1997). Several approaches are based on radial diffusion models for kinetics of sorption and desorption, in which the compounds are assumed to diffuse into and out of small pores within soil aggregates (Rijnaarts *et al.*, 1990), and again it is commonly assumed that only the compound in solution is used (Mihelcic and Luthy, 1991). Several other models have been proposed for describing the rate of biodegradation of sorbed compounds (Guerin and Boyd, 1992; Scow and Hutson, 1992; Scow and Johnson, 1997). However, the applicability of these models to a variety of substrate–sorbent systems and how well they describe the kinetics have yet to be determined.

STIMULATORY EFFECTS

The presence of particulate material sometimes is stimulatory to microorganisms, in contrast to the frequent findings that surfaces only reduce activity. This enhancement may be evident as an increase in the growth rate of marine oligotrophs by glass beads (Carlucci *et al.*, 1986) or the stimulation of specific activities, such as starch decomposition in the presence of montmorillonite (Filip, 1973), the degradation of uric acid by *Pseudomonas* sp. in the presence of bentonite (Durand, 1964), the degradation of hexamethylenediamine by *Bacillus subtilis* in the presence of several clay minerals (Garbara and Rotmistrov, 1982), the mineralization of several aldehydes in soils to which montmorillonite is added (Kunc and Stotzky, 1977), the mineralization of glucose and growth of several actinomycetes upon the addition of montmorillonite to liquid media (Martin *et al.*, 1976), or the hydrolysis of proteins by *Pseudomonas* sp. in the presence of kaolinite (Estermann and McLaren, 1959). The extent of mineralization may also be increased in the presence of clay minerals, and the enhancement in the rate or extent of mineralization may occur at one but not another substrate concentration (Subba-Rao and Alexander, 1982).

Definitive evidence to support one or another explanation for the stimulation is not at hand. Because the enhancement is frequently evident at low concentrations of organic chemicals, it has been proposed that the stimulation by charged surfaces, such as those of clays, results from the surface acting to concentrate the substrate, thereby promoting the metabolism of compounds present in solution at concentrations below K_m or K_s. The stimulation may result from the surface acting to reduce the concentration of a toxin in solution or by moderating the pH changes that may occur in poorly buffered environments.

From scientific, engineering, and predictive viewpoints, one can only view current knowledge of the biodegradation of sorbed compounds with considerable unease. Some molecules are totally resistant, some are partially degraded, and some are fully available to microorganisms. At times, sorption is stimulatory but often it retards biodegradation. It is unclear how microorganisms make use of sorbed substrates. Models for the kinetics of biodegradation of such molecules have been applied only to an occasional compound. It is hoped that future research will provide meaningful information to fill this vast void of knowledge.

REFERENCES

Amador, J. A., and Alexander, M., *Soil Biol. Biochem.* **20,** 185–191 (1988).

Anderson, J. P. E., *Soil Biol. Biochem.* **13,** 155–161 (1981).

Aomine, S., and Kobayashi, Y., *Trans. Int. Congr. Soil Sci., 8th, 1964,* Vol. 3, pp. 697–703 (1964).

Aronstein, B. N., and Alexander, M., *Environ. Toxicol. Chem.* **11,** 1227–1233 (1992).

Aronstein, B. N., Calvillo, Y. M., and Alexander, M., *Environ. Sci. Technol.* **25,** 1728–1731 (1991).

Briggs, G. G., *in* "The Persistence of Insecticides and Herbicides." *Monogr.—Br. Crop Prot. Counc.* **17,** 41–54 (1976).

Briška, A., Cencelj, J., Hočevar, J., Maček, J., and Sišakovič, V., *Agrohemija* **1/2,** 37–43 (1973); cited in *Weed Abstr.* **23,** 1391 (1974).

Burns, R. G., and Audus, L. J., *Weed Res.* **10,** 49–58 (1970).

Burns, R. G., Pukite, A. H., and McLaren, A. D., *Soil Sci. Soc. Am. Proc.* **36,** 308–311 (1972).

Calvet, R., *Environ. Health Perspect.* **83,** 145–177 (1989).

Calvillo, Y. M., and Alexander, M., *Appl. Microbiol. Biotechnol.* **45,** 383–390 (1996).

Carlucci, A. F., Shimp, S. L., and Craven, D. B., *FEMS Microb. Ecol.* **38,** 1–10 (1986).

Carmichael, L. M., Christman, R. F., and Pfaender, F. K., *Environ. Sci. Technol.* **31,** 126–132 (1997).

Chiou, C. T., *in* "Reactions and Movement of Organic Chemicals in Soils" (B. L. Sawhney and K. Brown, eds.), pp 1–29. Soil Science Society of America, Madison, WI, 1989.

Dao, T. H., and Lavy, T. L., *Soil Sci.* **143,** 66–72 (1987).

Dashman, T., and Stotzky, G., *Soil Biol. Biochem.* **18,** 5–14 (1986).

Di Grazio, P. M., Blackburn, J. W., Bienkowski, P. R., Hilton, B., Reed, G. D., King, J. M. H., and Sayler, G. S., *Appl. Biochem. Biotechnol.* **24/25,** 237–252 (1990).

Durand, G., *Ann. Inst. Pasteur, Paris* **107,** Suppl. 3, 136–147 (1964).

Ensminger, L. E., and Gieseking, J. E., *Soil Sci.* **53,** 205–209 (1942).

Estermann, E. F., and McLaren, A. D., *J. Soil Sci.* **10,** 64–78 (1959).

Estermann, E. F., Peterson, G. H., and McLaren, A. D., *Soil Sci. Soc. Am. Proc.* **23,** 31–36 (1959).

Filip, Z., *Folia Microbiol. (Prague)* **18,** 56–74 (1973).

Furmidge, C. G. L., and Osgerby, J. M., *J. Sci. Food Agric.* **18,** 269–273 (1967).

Garbara, S. V., and Rotmistrov, M. N., *Mikrobiologiya* **51,** 332–335 (1982).

Geissbühler, H., Haselbach, C., Aebi, H., and Ebner, L., *Weed Res.* **3,** 277–297 (1963).

Gordon, A. S., Gerchakov, S. M., and Millero, F. J., *Appl. Environ. Microbiol.* **45,** 411–417 (1983).

Greaves, M. P., and Wilson, M. J., *Soil Biol. Biochem.* **5,** 275–276 (1973).

Griffith, S. M., and Thomas, R. L., *Soil Sci. Soc. Am. J.* **43,** 1138–1140 (1979).

Guerin, W. F., and Boyd, S. A., *Appl. Environ. Microbiol.* **58,** 1142–1152 (1992).

Harms, H., and Zehnder, A. J. B., *Appl. Environ. Microbiol.* **61,** 27–33 (1995).

Hermansson, M., and Marshall, K. C., *Microb. Ecol.* **11,** 91–105 (1985).

Heuer, B., Birk, Y., and Yaron, B., *J. Agric. Food Chem.* **24,** 611–614 (1976).

Holm, T. R., Anderson, M. A., Stanforth, R. R., and Iverson, D. G., *Limnol. Oceanogr.* **25,** 23–30 (1980).
Hughes, J. D., and Simpson, G. H., *Aust. J. Soil Res.* **16,** 35–40 (1978).
Inoue, K., Kaneko, K., and Yoshida, M., *Soil Sci. Plant Nutr.* **24,** 91–102 (1978).
Kefford, B., Kjelleberg, S., and Marshall, K. C., *Arch. Microbiol.* **133,** 257–260 (1982).
Knaebel, D. B., Federle, T. W., McAvoy, D. C., and Vestal, J. R., *Appl. Environ. Microbiol.* **60,** 4500–4508 (1994).
Kunc, F., and Stotzky, G., *Soil Sci.* **124,** 167–172 (1977).
Liu, Z., Laha, S., and Luthy, R. G., *Water Sci. Technol.* **24,** 475–485 (1991).
Lynch, D. L., and Cotnoir, L. J., Jr., *Soil Sci. Soc. Am. Proc.* **20,** 367–370 (1956).
Makboul, H. E., and Ottow, J. C. G., *Microb. Ecol.* **5,** 207–213 (1979).
Manilal, V. B., and Alexander, M., *Appl. Microbiol. Biotechnol.* **35,** 401–405 (1991).
Martin, J. P., Filip, Z., and Haider, K., *Soil Biol. Biochem.* **8,** 409–413 (1976).
McDermott, J. B., Unterman, R., Brennan, M. J., Brooks, R. E., Mobley, D. P., Schwartz, C. C., and Dietrich, D. K., *Environ. Prog.* **8,** 46–55 (1989).
McLaren, A. D., *Soil Sci. Soc. Am. Proc.* **18,** 170–174 (1954).
Mihelcic, J. R., and Luthy, R. G., *Environ. Sci. Technol.* **25,** 169–177 (1991).
Miller, M. E., and Alexander, M., *Environ. Sci. Technol.* **25,** 240–245 (1991).
Morrill, L. G., Mahilum, B., and Mohiuddin, S. H., "Organic Compounds in Soils: Sorption, Degradation and Persistence." Ann Arbor Sci. Publ., Ann Arbor, MI, 1982.
Nomura, N. S., and Hilton, H. W., *Weed Res.* **17,** 113–121 (1977).
Olness, A., and Clapp, C. E., *Soil Sci. Soc. Am. Proc.* **36,** 179–181 (1972).
Pinck, L. A., Dyal, R. S., and Allison, F. E., *Soil Sci.* **78,** 109–118 (1954).
Rijnaarts, H. H. M., Bachmann, A., Jumelet, J. C., and Zehnder, A. J. B., *Environ. Sci. Technol.* **24,** 1349–1354 (1990).
Scow, K. M., and Hutson, J., *Soil Sci. Soc. Am. J.* **56,** 119–127 (1992).
Scow, K. M., and Johnson, C. R., *Adv. Agron.* **58,** 1–56 (1997).
Shimp, R. J., and Young, R. L., *Ecotoxicol. Environ. Saf.* **15,** 31–45 (1988).
Smith, D. T., and Meggitt, W. F., *Weed Sci.* **18,** 260–264 (1967).
Speitel, G. E., Jr., Lu, C.-J., and Zhu, X.-J., *Environ. Sci. Technol.* **23,** 68–74 (1989a).
Speitel, G. E., Jr., Turahia, M. H., and Lu, C.-J., *J. Am. Water Works Assoc.* **81**(4), 168–176 (1989b).
Steinberg, S. M., Pignatello, J. J., and Sawhney, B. L., *Environ. Sci. Technol.* **21,** 1201–1208 (1987).
Stotzky, G., *in* "Interactions of Soil Minerals with Natural Organics and Microbes" (P. M. Huang and M. Schnitzer, eds.), pp. 305–428. Soil Science Society of America, Madison, WI, 1986.
Subba-Rao, R. V., and Alexander, M., *Appl. Environ. Microbiol.* **44,** 659–668 (1982).
Tang, W.-C., White, J. C., and Alexander, M., *Appl. Microbiol. Biotechnol.* **49,** 117–121 (1998).
Taylor, G. T., *Limnol. Oceanogr.* **40,** 875–885 (1995).
Thibault, S. L., Anderson, M., and Frankenberger, W. T., Jr., *Appl. Environ. Microbiol.* **62,** 283–287 (1996).
Thomas, J. M., and Alexander, M., *Microb. Ecol.* **14,** 75–80 (1987).
Weber, J. B., and Coble, H. D., *J. Agric. Food Chem.* **16,** 475–478 (1968).
Weissenfels, W. D., Klewer, H.-J., and Langhoff, J., *Appl. Microbiol. Biotechnol.* **36,** 689–696 (1992).
White, J. C., and Alexander, M., *Environ. Toxicol. Chem.* **15,** 1973–1978 (1996).
Wszolek, P. C., and Alexander, M., *J. Agric. Food Chem.* **27,** 410–414 (1979).

CHAPTER 9

Nonaqueous-Phase Liquids and Compounds with Low Water Solubility

Many pollutants exist at contaminated sites not in the aqueous phase or sorbed to solids but rather in liquids that are immiscible with water. As such, the availability of the contaminants to biodegradation and bioremediation may be drastically reduced. These nonaqueous-phase liquids (NAPLs) containing the environmental contaminants are present in aquifers, subsoils, sediments, soils, and at the top of the water column of marine, estuarine, and fresh waters. NAPLs are most widely known because of spills or leakages from oil tankers, and these crude-oil NAPLs have contaminated surface waters, marine sediments, and coastal beaches of the Atlantic Ocean, Caribbean Sea, and Prince William Sound, Alaska. Much oil that is inadvertently discharged in marine waters eventually sinks to the bottom and persists in the sediments. Comparable spills of oil or petroleum products have occurred on land as a result of tank car spills or oil pipeline leaks. Gasoline, petroleum products, or industrial solvents have contaminated aquifers and groundwaters at uncounted numbers of sites because of underground storage tanks that, after many years of burial, corrode and begin to leak their contents. In addition, many hazardous-waste sites contain industrial solvents, and not uncommonly, these organic solvents move from the site and enter adjacent groundwaters, frequently making nonpotable what was previously a safe water supply. If the spilled material in subsoils is dense, it will move downward and come to rest and remain as a pool at the bottom of the aquifer; these are called dense NAPLs (DNAPLs).

NAPLs usually contain an array of organic molecules, although a spill of solvent from an industrial source or a leak from an underground storage tank may contain a single chemical. Typically, NAPLs are composed of molecules that have low water solubilities and high solubilities in organic solvents, and the concentration in the water phase is thus quite low. However, the NAPL represents a long-term source of water pollution because the contaminants will continue to enter the water phase to replace that which is transported away from the site, is biodegraded, or is removed by some remediation technology.

Because nearly all the research on microbial physiology and the metabolism of organic substrates has been centered on molecules that are in aqueous solution, it is often assumed that the portion of a compound that is not dissolved in water is not readily accessible for utilization. The chemical must presumably make contact with the cell surface, so that it can enter the organism to be acted on by intracellular enzymes. If a pollutant is present initially in a NAPL, rapid uptake of that compound by the cell from the aqueous phase—and hence rapid growth on that compound as a C source—cannot occur since little will be present in the water, unless a special mechanism exists for uptake and assimilation of the substrate by the cell containing the degradative enzymes.

If a NAPL is a pure solvent (i.e., a single chemical), the water solubility of that solvent is of special importance. In this regard, it is worth considering not only liquids that are composed of a single type of molecule but also solids. Solids composed of a single compound are not environmental pollutants, but they are considered here because research on them has helped lay a foundation for our understanding of how chemicals of low water solubilities are metabolized.

The solubility of several chemicals in water is given in Table 9.1. Different values are reported by different investigators, but the data are typical of those that have been published. The values may be somewhat different in seawater or waters containing humus constituents than in distilled water (Boehm and Quinn, 1973; Sutton and Calder, 1974), but the differences are not of sufficient magnitude to convert a compound that otherwise has a very low solubility into one of appreciably greater solubility.

Biodegradation of the aliphatic hydrocarbons, whose solubilities decline as the number of carbons in the molecule increases (Table 9.1), has been studied extensively. Much of this research has been prompted by their presence in oil spills, but some of the interest has been a result of curiosity about how such poorly soluble substrates are assimilated. Alkanes up to at least a molecular weight of 618 ($C_{44}H_{90}$) can be mineralized by microorganisms (Haines and Alexander, 1974), even in instances in which the solubility apparently is less than 1 ng/liter. The quantity of such aliphatic hydrocarbons in water would allow for the growth of fewer than one bacterial cell per milliliter, so that clearly the organisms must be able to use the insoluble phase when extensive utilization occurs.

Aromatic hydrocarbons with such low solubilities in water are also decomposed microbiologically. For example, anthracene, phenanthrene, pyrene, 1,2-benzpyrene, and chrysene are destroyed microbiologically in soil (Bossert and Bartha, 1986), and anthracene and naphthalene are mineralized by the microorgan-

Table 9.1

Solubility of Several Organic Compounds in Water

Group	Chemical	mg/liter
Aliphatic hydrocarbons[a]	Heptane (C_7H_{16})	2.9
	Octane (C_8H_{18})	0.66
	Nonane (C_9H_{20})	0.22
	Decane ($C_{10}H_{22}$)	0.052
	Hexadecane ($C_{16}H_{34}$)	0.000020
	Eicosane ($C_{20}H_{42}$)	0.00000011
Aromatic hydrocarbons[b]	Naphthalene	31
	Biphenyl	7.2
	Acenaphthene	4.3
	Anthracene	0.050
	Phenanthrene	1.1
	Pyrene	0.13
	Chrysene	0.0020
	1,2-Benzpyrene	0.0053
Others[c]	4-Chlorobiphenyl	0.96
	Palmitic acid	0.0035
	DEHP	0.29

[a] Coates *et al.* (1985).
[b] Yalkowsky *et al.* (1983).
[c] Thomas *et al.* (1986).

isms of sediments (Bauer and Capone, 1985). In pure culture, PAHs such as naphthalene, anthracene, and phenanthrene serve as carbon sources for bacterial growth (Stucki and Alexander, 1987; Wodzinski and Bertolini, 1972). The concentration of several of these aromatic compounds in water is so low that the finding of extensive bacterial growth, chemical loss, or mineralization shows that at least some, if not all, of the chemical that is not in the water phase is being utilized.

If a NAPL is not a pure solvent but instead contains two or—as is common in oil spills, leakages of underground storage tanks containing gasoline, or hazardous-waste sites—a multitude of hydrophobic compounds, a critical factor is the amount of the compound present in the aqueous phase that is in equilibrium with the NAPL phase. The chemical partitions in the NAPL and in the water on the basis of its relative solubilities in these two phases, the more hydrophobic compounds being at higher concentrations in hydrophobic NAPLs and the more hydrophilic ones existing at higher concentrations in the water. This relative partitioning is usually expressed by determining the amounts present in an arbitrarily chosen organic solvent (*n*-octanol) and in water at equilibrium and is designated as the octanol-water partition coefficient K_{ow}, where K_{ow} is the ratio of the concentration of the test chemical (or solute) in octanol to the concentration present in water. Because the values for K_{ow} often are very high, the values are usually expressed logarithmically, as log K_{ow}. A compound with a high value is hydrophobic, and little would exist in water that is in equilibrium with octanol or with NAPLs. One with a low value would exist at higher concentrations in water. Values of log K_{ow} for a number of compounds are given in Table 9.2. Because the extent of sorption to the native organic matter of soils and sediments is correlated with the hydrophobicity of the molecule being retained, such log K_{ow} values are good predictors of the extent of binding to the humic fraction of these environments.

Field observations show that NAPLs are often extremely persistent. This is quite evident in sediments and in subterranean sites. Soils on which factories were built to convert coal to flammable gases that were used in cities for heat and lighting still contain the tarry materials that were deposited more than 100 years ago. The compounds that are present within these NAPLs obviously are also persistent. Under more defined conditions in the laboratory, the resistance of compounds present in NAPLs can be demonstrated, even otherwise readily metabolized compounds. For example, in a soil in which ^{14}C-labeled naphthalene, hexadec-

Table 9.2

Octanol–Water Partition Coefficients of Organic Compounds[a]

Chemical	Log K_{ow}
Dioxane	−1.1
Acetone	0.23
Pyridine	0.71
Benzene	2.0
Toluene	2.4
Xylene	3.1
Diethyl phthalate	3.3
Diphenyl ether	4.3
Decane	5.6
Tetradecane	7.6
Dioctyl phthalate	8.8

[a] From Laane *et al.* (1987a).

ane, phenanthrene, or di(2-ethylhexyl) phthalate (DEHP) is added with 2,2,4,4,6,8,8-heptamethylnonane as NAPL, two of the compounds are readily mineralized, phenanthrene is mineralized only after an acclimation, and little degradation of DEHP occurs (Fig. 9.1). Heptamethylnonane was chosen for these experiments because of its very low toxicity to microorganisms and its persistence. Similar tests confirm that some NAPLs reduce the availability of their hydrophobic constituents in samples of soils and subsoils (Efroymson and Alexander, 1994a). The degree of protection is rarely known, but the finding of individual compounds in NAPLs introduced many years ago into subsoils or aquifers is ample proof of a high degree of unavailability of the constituents that remain.

The same compound in different NAPLs will be metabolized at considerably different rates. This is well known from measurements of the disappearance of individual alkanes in oils of different composition following spills of crude oil or under experimental conditions (Atlas, 1981). Similar observations have been made with soil into which DEHP in several NAPLs was introduced; the phthalate is metabolized quickly in the absence of a NAPL and slowly with one of the NAPLs, but almost no destruction is observed with three of the NAPLs (Fig. 9.2). The data show not only a reduced availability of substrates within NAPLs but also that biodegradation does take place. This is also evident in studies of individual bacteria.

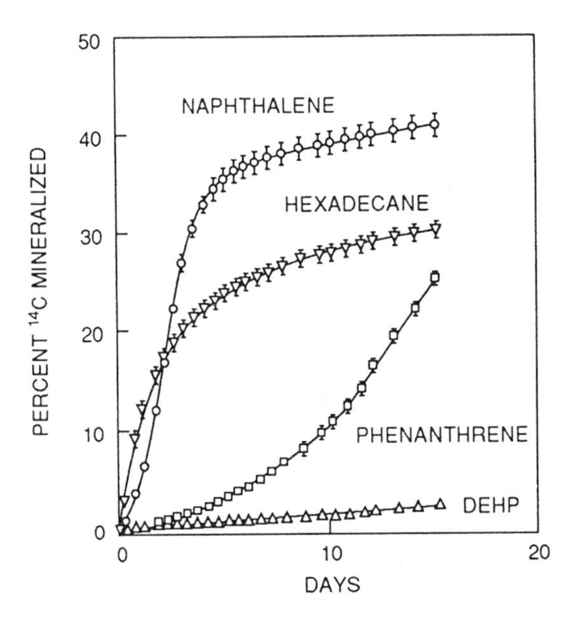

Figure 9.1 Mineralization in soil of four hydrophobic compounds present in 2,2,4,4,6,8,8-heptamethylnonane. (From Efroymson and Alexander, 1994. Reprinted with permission from Pergamon Press Ltd., Headington Hill Hall, Oxford OX3 OBW, U.K.)

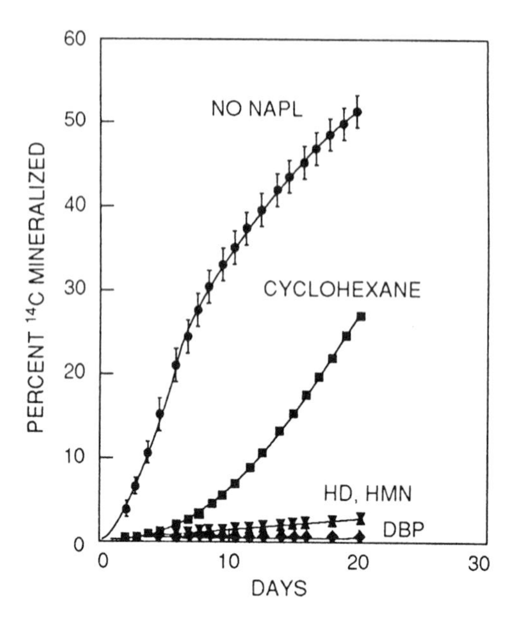

Figure 9.2 Mineralization in soil of DEHP initially dissolved in heptamethylnonane (HMN), cyclohexane, hexadecane (HD), or dibutyl phthalate (DBP) or added with no NAPL. (From Efroymson and Alexander, 1994. Reprinted with permission from Pergamon Press Ltd., Headington Hill Hall, Oxford OX3 OBW, U.K.)

Thus, *Corynebacterium equi* oxidizes 1- and 2-tetradecanol present initially in isooctane (Takazawa *et al.,* 1984), a strain of *Arthrobacter* mineralizes naphthalene and hexadecane dissolved in heptamethylnonane (Efroymson and Alexander, 1991), and several bacterial species utilize individual hydrocarbons present in crude oil (Foght *et al.,* 1990).

Many of the compounds present in a NAPL may be degraded simultaneously, or some may be metabolized initially and others may be destroyed only after the more susceptible molecules are transformed. Simultaneous metabolism is readily evident during the degradation of crude oils, which are composed of a highly heterogeneous mixture of aliphatic and aromatic hydrocarbons, heterocyclic compounds, and other chemical classes. Gas chromatograms depicting oil before and after incubation with a mixture of marine organisms are shown in Fig. 9.3. It is evident that many of the alkanes are destroyed even in this short period under laboratory conditions, the bacteria using the aliphatic hydrocarbons essentially at the same time. One must not extrapolate from such data to nutrient-poor or cool aquatic environments, in which the conversions may occur quite slowly, and especially not to NAPLs not spread out in the thin films common to surface

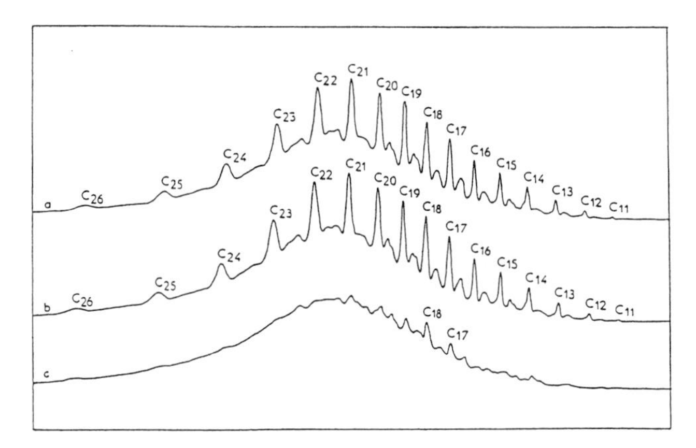

Figure 9.3 Gas chromatograms of alkanes in oil in a medium at (a) 0 h and after (b) 2 and (c) 5 days incubation with a mixture of marine microorganisms. The solution was supplemented with N and P. (From LePetit and Barthelemy, 1968. Reprinted with permission from Elsevier Science Publishers.)

waters. Because the exposed surface area is a key factor determining the rate of biodegradation, components of a NAPL existing as a thin film at the surface of the water column are far more likely to be biodegraded quickly than constituents of a NAPL that is thick and therefore has a lower surface to volume ratio. In contrast with the nearly simultaneous utilization of some components of heterogeneous NAPLs, other components are used only after some constituents have largely been destroyed (Oberbremer and Müller-Hurtig, 1989).

A number of factors are known to influence the biodegradation of individual constituents of NAPLs. These include (a) the intrinsic ability of the microflora at the site to transform the constituents, (b) the rate of partitioning (i.e., mass transfer) of the constituent from the NAPL to the water phase, (c) viscosity of the NAPL, (d) toxicity of the nonaqueous liquid, (e) the area of the interface between the NAPL and the aqueous phase, (f) the presence of other constituents that are degraded more readily than the specific compound of interest, and (g) an equilibrium concentration in the aqueous phase too low to allow for the responsible species to grow, i.e., a threshold. In addition, should the chemical of interest be transformed only by cometabolism (discussed in Chapter 13), that compound might be transformed more readily if the NAPL has constituents that support growth of the cometabolizing species.

The rates of partitioning, or mass transfer, from NAPL to water vary with the identity of the NAPL (Carroquino and Alexander, 1998) and the compound of concern. The rates may be slow or they may be rapid. If the microbial transformation is slow because of a small biomass, nutrient limitation, or toxicity, the compound may enter the aqueous phase faster than it is utilized so that partitioning does not limit the degradation. However, the rate of biodegradation is sometimes correlated with or controlled (or limited) by the spontaneous transfer of the microbial substrate from the nonaqueous liquid to the water phase (Efroymson and Alexander, 1994b; Ghoshal and Luthy, 1996; Ghoshal *et al.*, 1996).

The viscosities of NAPLs extend over a wide range. Some of the tars, for example, are so viscous that they almost resemble solids and flow very slowly. Others are nonviscous and can be mixed readily and flow freely. Studies with phenanthrene added to NAPLs having dissimilar viscosities have shown that the rate and extent of mineralization in soil slurries decreased with increasing viscosities of petroleum hydrocarbons in which the PAH was initially dissolved (Table 9.3). Inasmuch as viscosity influences partitioning of PAHs from a NAPL, such as oil, to water (Chen *et al.*, 1994), it is likely that viscosity affects the biodegradation rate because of the slower diffusion of phenanthrene within the more viscous NAPLs and the consequently slower partitioning from the NAPL to the water.

A major factor determining the biodegradation of compounds present in a NAPL is its toxicity. This toxicity results from the single solvent itself if the NAPL is a single compound, or it may be the result of the major solvent or of one or more minor components of a heterogeneous NAPL. Many organic solvents suppress microbial proliferation and metabolism. As a rule, organic solvents with high values for log K_{ow} (4.0 or greater) do not suppress microbial activity, whereas those with

Table 9.3

Biodegradation of Phenanthrene Added to Soil Slurries in Different NAPLs[a]

NAPL	Mineralization rate (ng/ml/h)[b]	Extent of mineralization at 136 h (%)	NAPL viscosity (cSt)[c]
Jet fuel	29.0A	51.6A	8.45
Diesel oil	21.7B	44.5B	12.2
Sentry 19 oil	19.6C	46.9B	33.8
Sentry 39 oil	16.2D	33.2C,D	67.6
Solvent 600 oil	15.5D	35.1C	179
Sentry 69 oil	15.2D	30.1D	118
Sentry 129 oil	10.9E	32.5C,D	218
150 Bright stock oil	8.89F	15.8F	256
Crude oil	6.91G	14.1E	260

[a] From Birman and Alexander (1996b).
[b] Values in a column followed by the same letter are not significantly different ($p = 0.05$).
[c] cSt represents a centistoke, or 0.01 stoke, a unit of viscosity [g/(s × cm × density)].

low values for log K_{ow} (often 2.0 or lower) are highly toxic (Inoue and Horikoshi, 1991; Laane *et al.,* 1987b). Nevertheless, some bacteria are tolerant of solvents with low log K_{ow} values, and a NAPL with a high log K_{ow} value may still prevent biodegradation because it contains a highly toxic component. For some bacteria, even a compound with a high log K_{ow} may be toxic (Jimenez and Bartha, 1996).

As indicated earlier, the exposed area of the NAPL is of great significance; the greater the interfacial area between the NAPL and the water, the more rapid will be the degradation. Presumably those molecules that are at or near the surface of the NAPL will be initially utilized, so that the more surface of NAPL that is exposed to the water or the microorganisms, the faster it or its components will be metabolized. The best evidence for the key role of interfacial area comes from studies of individual compounds that exist as liquids and are sparingly soluble in water. For example, for aliphatic hydrocarbons present as droplets in water, the smaller the droplet, the larger is the interfacial area; as the surface area exposed is thus increased, so too does the growth rate of the organisms and hence the rate of their biodegradation (Moo-Young *et al.,* 1971). Studies with a strain of *Candida* growing on ethyl butyrate, which was chiefly in the silicone oil (the NAPL) phase of a silicone oil–water mixture, suggest that the rate of biodegradation may also be related to the interfacial area because a large part of the biomass, which characteristically is hydrophobic, adheres to the NAPL–water interface as a biofilm (Ascón-Cabrera and Lebault, 1995), and the larger the interfacial area, the larger the biomass. A role of the NAPL–water interfacial area was also indicated in the biodegradation by *Pseudomonas aeruginosa* of phenanthrene in a mixture of 2,2,4,4,6,8,8-heptamethylnonane and water (Kohler *et al.,* 1994). However, increasing the surface area may not always cause such a stimulatory response (Fogel *et al.,* 1985).

In a NAPL that is chiefly composed of a compound (or compounds) that is used more rapidly than the chemical of interest, utilization of the former may suppress degradation of the latter. Microorganisms metabolizing the more readily available substrate compete with those species utilizing the less readily available compounds. In this competition, the likely nutrients being competed for are O_2, N, or P. The faster growth, larger biomass, or both of the microflora active on the preferred substrate results in a depletion of the limiting nutrient(s) and the consequent depressed degradation of the less quickly utilized constituent (Morrison and Alexander, 1997).

The concentration of the NAPL constituent that is present in the water phase may also affect the biodegradation because that concentration, although possibly high in the nonaqueous liquid, may be so low in the aqueous phase that it is below the threshold concentration for growth. This would be true for compounds that have a high NAPL–water partition coefficient. As with the octanol–water partition coefficient, K_{ow}, the concentration of a hydrophobic compound is much higher in the organic solvent, be it octanol or some other NAPL, than in water. When that coefficient is high, the concentration in water may be below that which will permit microbial replication. For example, a phenanthrene-degrading mixed culture failed to mineralize that compound when it was initially dissolved in di(2-ethylhexyl) phthalate (the NAPL) at concentrations of 0.6–20 mg/liter. The reason: the concentration in water was <1 μg/liter. Other organisms, presumably those able to use other carbon sources in the water, were able to metabolize the phenanthrene, but the rate was unexpectedly slow (Efroymson and Alexander, 1995).

Scant attention has been paid to the kinetics of biodegradation of constituents of NAPLs. In the case of a strain of *Pseudomonas* grown on phenanthrene initially dissolved in silicone oil, the bacterium grew exponentially at first; however, when the rate of phenanthrene utilization had increased and reached the rate of its partitioning from NAPL to water, a new growth phase apparently equivalent to that rate of mass transfer was initiated (Bouchez *et al.,* 1995). In contrast, a strain of *Arthrobacter* mineralized naphthalene dissolved in di(2-ethylhexyl) phthalate initially slowly, but its subsequent activity was rapid. The initial, slow phase apparently resulted from bacterial cells growing in the water, but the bacterium then extensively colonized the NAPL–water interface, where the large surface-adhering biomass brought about a transformation more rapid than that predicted from the rate of partitioning determined in the absence of microorganisms (Ortega-Calvo and Alexander, 1994). The kinetics of biodegradation of NAPLs in nature likewise have received little attention, although it has been noted that the mineralization in soil samples of several hydrophobic compounds dissolved in NAPLs is linear (Efroymson and Alexander, 1994a). In addition, measurements have been made of the kinetics of growth and sometimes mineralization by pure cultures of microorganisms acting on a number of pure hydrocarbons, both liquids and solids, that have low solubilities in water. Sometimes, the process is logarithmic, as in the mineralization of biphenyl by species of *Moraxella* and *Pseudomonas* (Stucki and Alexander, 1987), of palmitic acid by *Pseudomonas pseudoflava* (Thomas and Alexander, 1987), or the mineralization of palmitic acid, DEHP, and Sevin by a mixture of bacteria (Thomas *et al.,* 1986). Often linear kinetics are evident, as in the mineralization of octadecane by a mixture of bacteria (Thomas *et al.,* 1986), the decomposition of tri-, tetra-, penta-, hexa-, and octadecane by the fungus *Cladosporium resinae* (Lindley and Heydeman, 1986), the growth of *Torulopsis* sp. on aliphatic hydrocarbons (McLee and Davies, 1972), and the multiplication of three pseudomonads on a mixture of solid hydrocarbons known as slack wax (Amin *et al.,* 1973). Such linear growth or biodegradation might reflect a linear rate of partitioning of the organic substrate from the insoluble to the soluble phase. However, the growth of other species on poorly soluble chemicals is initially logarithmic, but the logarithmic phase is followed by a period of linear growth, as in the growth of *Flavobacterium* sp. and *Beijerinckia* sp. on phenanthrene (Stucki and Alexander, 1987). *Arthrobacter* sp. on a sterol (Goswami *et al.*, 1983), and *Candida tropicalis* on hexadecane (Blanch and Einsele, 1973).

Studies of the utilization of individual alkanes by pure cultures have led to the formulation of a series of models to describe the kinetics of microbial growth on alkanes. These models are based on several assumptions, including the assumptions that the hydrocarbon is solubilized by the organisms before it is assimilated (Goma and Ribot, 1978), that growth takes place initially on the soluble fraction but subsequently by a microbial dissolution of the chemical (Chakravarty *et al.*, 1975), and that the cells obtain the C they need by attachment to or direct contact with the insoluble fraction (Moo-Young *et al.*, 1971; Mallee and Blanch, 1977). The validity of models of these sorts has yet to be tested for individual solvents or heterogeneous NAPLs in natural or polluted environments.

MECHANISMS OF UTILIZATION

Many scientists have been intrigued by the mechanism by which microorganisms make use of organic substrates that are not in the aqueous phase, through which cells get much of their nutrient supply. Because the literature on microbial metabolism is largely derived from studies of water-soluble substrates, a molecule not in solution appears to be an anomaly: how does the organism assimilate it into the cell, wherein the enzymes bring about its transformation? In instances in which the substrate is acted on by a hydrolytic enzyme excreted by the microorganism, the mechanism is straightforward and involves the cleavage by the enzyme of small, soluble fragments from the insoluble molecule. However, such hydrolytic enzymes, although involved in initiation of the transformation of such polymers as cellulose and chitin, are apparently not responsible for the initial phase of utilization of the low-molecular-weight compounds that are important pollutants. For these low-molecular-weight, water-insoluble substrates, the mechanism of initial uptake by the cell is quite different.

Three mechanisms appear to explain how microorganisms utilize compounds in NAPLs or metabolize organic solvents having low solubilities in water. These mechanisms focus on how the chemical is transferred from the environment surrounding the organism to the cell surface, from which point it is transported through the membrane and to intracellular sites of enzymatic activity. (a) Only the chemical in the water phase is used. Once the supply is assimilated into the cell, the organism can only use molecules that enter the aqueous phase by spontaneous partitioning, and hence further degradation would depend on the rate of spontaneous partitioning into the water phase. (b) The microorganism excretes products that convert the substrate into droplets with sizes less than 1 μm, and these are then assimilated by the organism. Because of the small sizes of the droplets or particles, this process is sometimes called pseudosolubilization. The process commonly involves the excretion by the microorganism of surfactants that facilitate the pseudosolubilization. (c) The cells come into direct contact with the NAPL, on the surfaces of which the population develops, and the chemical at or near the point of contact with the organism passes through the cell surface into the cytoplasm. Microorganisms have been described that use NAPLs or constituents of NAPLs by each of these three mechanisms.

Microbial utilization of only that portion of the compound in a NAPL (or of a sparingly soluble chemical) that is in the aqueous phase is likely with substrates that have low log K_{ow} values or compounds that dissolve rapidly in water as the supply of organic solutes in the aqueous phase is depleted by degradation. Such an organism would multiply as it uses the chemical in solution, but as its population size or biomass increases, its subsequent growth rate—and hence biodegradation—would often be limited by the rate of dissolution. A microorganism of this type would grow in the aqueous phase in the presence of a NAPL and would presumably not excrete a surfactant. In Fig. 9.4, which is identical to Fig. 8.5, A designates the rate of partitioning from the NAPL to the water phase, and the abiotic equilibrium reflects the partition coefficient, that is, how much of the compound would be in the water when equilibrium is reached between chemical in the NAPL and in the water. The rate of degradation of a substrate used by this mechanism is then shown by the line B.

One indication that some bacteria use only the substrate in the aqueous phase comes from studies showing that the growth rates of certain pure cultures of

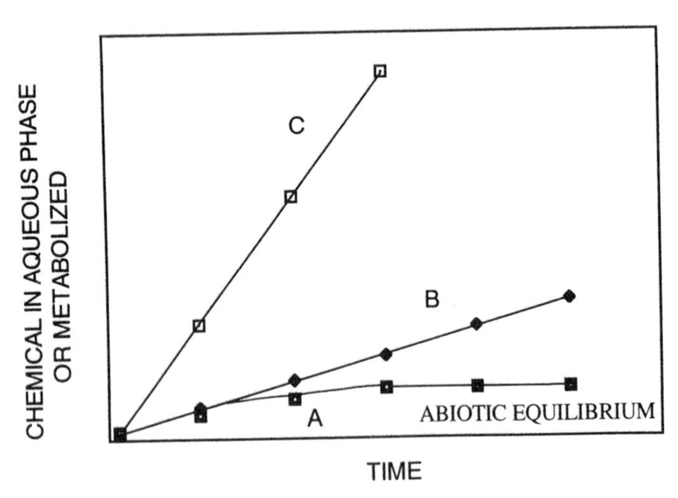

Figure 9.4 (A) Rate of partitioning of a chemical from a NAPL to the surrounding water and the equilibrium reached in the absence of microorganisms, (B) rate of biodegradation by microorganisms using only substrate entering the aqueous phase, and (C) biodegradation using the substrates by some other mechanism.

hydrocarbon-utilizing bacteria increase on aromatic hydrocarbons with increasing water solubilities—growing at a more rapid rate on naphthalene than on phenanthrene and more rapidly on phenanthrene than on anthracene (Wodzinski and Johnson, 1968). Since the concentrations in solution are probably far below the organism's presumed K_s for those substrates, the growth rate should increase with more chemical in solution. Further evidence comes from studies demonstrating that the rate of growth of certain bacteria on phenanthrene (or, with other bacteria, on naphthalene, bibenzyl, or diphenylmethane) is the same if the medium contains only the dissolved hydrocarbon or if it contains both dissolved and the insoluble chemical (Wodzinski and Bertolini, 1972; Wodzinski and Coyle, 1974; Wodzinski and Larocca, 1977).

Another indication of the importance of the water-soluble phase comes from a study in which bacteria were found to grow readily on the portion of naphthalene and 4-chlorobiphenyl that was in solution in culture media, the rate of dissolution of the chemicals exceeding the rate of microbial degradation as long as the biomass was small. However, when the concentration of organic substrate in solution became undetectable, the bacterial growth rate fell dramatically (Thomas *et al.*, 1986). In considering the exclusive use of the chemical present in the water phase, both the concentration initially in the water and the spontaneous dissolution rate must be considered. If the organisms grow slowly, the rate of dissolution may be sufficiently rapid that soluble substrate is always available; the dissolution rate would then exceed the degradation rate. However, as the biomass increases, the microbial demand for the C source may ultimately exceed the dissolution rate, and subsequent activity will be limited by the rate of partitioning of the chemical from the NAPL (or the solid) to the water. In the case of phenanthrene biodegradation by *Flavobacterium* sp. and *Beijerinckia* sp., for example, the degradation appears to be limited by the rate of spontaneous dissolution of the PAH (Stucki and Alexander, 1987). Similarly, findings that the degradation by *Pseudomonas* sp. of phenanthrene initially present in silicone oil (Bouchez *et al.*, 1995) and the metabolism in slurries of naphthalene initially in coal tar (Ghoshal and Luthy, 1996) were governed by the rates of partitioning support the validity of this mechanism.

Under certain circumstances, however, biodegradation sometimes is faster than the rate of spontaneous partitioning, i.e., the rate of mass transfer from NAPL to water determined in the absence of microbial activity. This was found to be true for the rates of biodegradation of phenanthrene (Efroymson and Alexander, 1994b, 1995), pyrene (Bouchez *et al.*, 1997), and styrene (Osswald *et al.*, 1996) initially in several NAPLs. Therefore, the rate of biodegradation sometimes cannot be predicted from the rate of the spontaneous mass transfer to water, and other mechanisms of utilization must be invoked. These mechanisms appear to involve the excretion of emulsifiers or surfactants and adherence of the cells to the NAPL–water interface.

The production of surfactants that facilitate the partitioning of the chemical from the NAPL to the water phase, thus resulting in enhanced biodegradation, has received considerable attention in the case of NAPLs that are composed of single compounds (specifically, the liquid alkanes), oils, or oil products. Similar studies have been conducted with some compounds or materials that are solids at ambient temperatures. The substance excreted by the microorganisms, commonly termed an emulsifier or a surfactant, brings about the conversion of the NAPL to small droplets. In contrast with a true solution, in which two or more substances are mixed homogeneously at the molecular or ionic level, an emulsion is an immiscible liquid that is dispersed intimately in another immiscible liquid as droplets that have diameters usually greater than 0.1 μm. In the present context, the first immiscible liquid is the NAPL and the second is water. If the droplets in suspension in the water are very small, that suspension has many of the properties of a solution. A typical emulsion is homogenized milk, in which water-insoluble fat is suspended as droplets so small that they are stable for long periods in the water phase. The smaller the droplets, the more stable the emulsion. The emulsifying agent, that is, the substance that confers stability on these emulsions, dramatically increases the surface of the interface between the NAPL and water; witness the million-fold increase in area of this interfacial surface if 10 ml of oil is converted to droplets with a diameter of 0.2 μm. A surface-active agent, usually called a surfactant, may improve the stability of the emulsion and thus acts as an emulsifier, the surfactant reducing the surface or interfacial tension of the emulsion. Most emulsifying agents are surfactants, but not all surfactants are emulsifying agents (Becher, 1965). The surfactant, by definition, lowers the surface tension of a liquid, but the emulsifier may or may not do so; that is, the emulsifier may act by reducing interfacial tension or by some dissimilar mechanism. Microbial emulsifying agents are typically surfactants.

A surfactant molecule has a hydrophobic and a hydrophilic portion. At low concentrations, surfactants are fully soluble in water. If the concentration is increased, the molecules of the surfactant associate to form extremely small aggregates that are called micelles. The lowest concentration at which micelles begin to form is known as the critical micelle concentration (CMC). In water, the hydrophobic ends of the surfactant molecules are clustered in the center of the micelle, and the hydrophilic ends are on the outside toward the water phase (Fig. 9.5). A hydrophobic substrate derived from a NAPL would then be incorporated, presumably with some of the NAPL, in the inner region of the micelle and appear to be dissolved in the water, although it is really in a quasi-soluble form because it is entrapped within these small micelles. As a rule, the pseudosolubilization or apparent solubility becomes evident only at a surfactant concentration above but not below the CMC, but some surfactants may increase the water pseudosolubility of hydrophobic molecules below the CMC (Kile and Chiou, 1989).

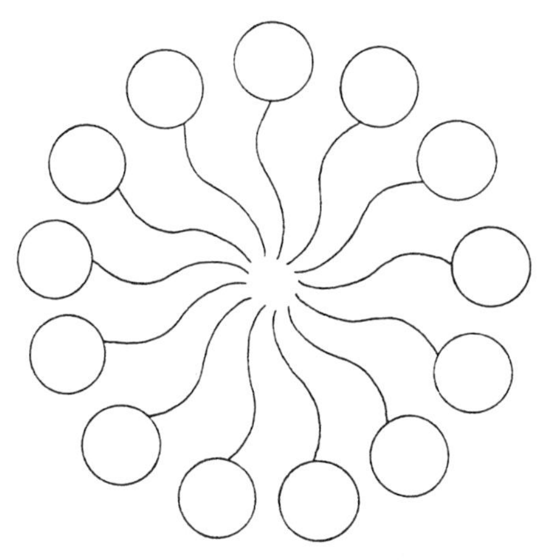

Figure 9.5 Schematic representation of a micelle. (From Robotham and Gill, 1989. Reprinted with permission from Elsevier Science Publishers.)

Many species growing on and degrading NAPLs that are pure alkanes or oils excrete surface-active or emulsifying agents, and these convert the NAPL to droplets or particles with diameters of 0.1 to 1.0 μm (Einsele *et al.*, 1975) that lead to the apparent solubility or "pseudosolubility" of molecules originally present in the NAPL. The rate of apparent solubilization caused by the excretions is often great enough to account for utilization of the substance by the microorganisms and their consequent growth (Cameotra *et al.*, 1983; Goswami and Singh, 1991). These excretions often are capable of emulsifying various types of oil and mixtures of hydrocarbons (Rosenberg *et al.*, 1979). In general, the smaller the droplet produced by the emulsifier, the faster the microorganism grows and degrades its organic substrate (Singer and Finnerty, 1984). The linear growth and degradation so often observed in pure cultures may result from the fact that the exposed surface area of these droplets is limiting the rate of decomposition. The microbial surfactants that have been characterized chemically are polysaccharides, polysaccharide-protein complexes, or glycolipids (Rosenberg, 1986).

The role of surfactants in the degradation of NAPLs other than pure aliphatic hydrocarbons remains uncertain. Nevertheless, evidence exists for their importance in the degradation of sterols by *Arthrobacter* sp. (Goswami *et al.*, 1983) and phthalate esters by strains of *Mycobacterium* and *Nocardia* (Gibbons and Alexander, 1989).

According to the third mechanism, cells carrying out the biodegradation become attached directly to the NAPL and there, at the surface, metabolize the NAPL constituents. Bacteria that grow on aliphatic hydrocarbons not in aqueous solution often adhere to their substrates, and if these are droplets, the cells retained by the droplets may form agglomerates containing clumps of cells together with some of the hydrocarbon (Miura, 1978). Bacteria may also attach to such solid substrates as palmitic acid and sterol particles, and they then multiply over the surfaces to which they attach. Colonization of the surface may begin or become prominent only after the compound in solution is destroyed (Goswami *et al.*, 1983; Thomas and Alexander, 1987). Some microorganisms have a strong affinity for the NAPL, but others have a lesser degree of binding (Rosenberg and Rosenberg, 1985).

For certain microorganisms at least, this attachment of the cells is of great importance and may be a prerequisite for the degradation. For example, a strain of *Arthrobacter* has been described that biodegrades hexadecane dissolved in a NAPL (2,2,4,4,6,8,8-heptamethylnonane) without excreting products that increase the water solubility of hexadecane. In this instance, spontaneous partitioning of hexadecane into the water phase can be ruled out because no such partitioning is detectable. Instead, the bacterium becomes attached to the NAPL–water interface and there is able to obtain its substrate, hexadecane, from the NAPL (Efroymson and Alexander, 1991). This need for direct contact of the cells with the NAPL–water interface was further confirmed by showing that the addition of Triton X-100, a surfactant that suppresses cell adherence but yet is not toxic to the bacterium at the concentration used, prevented mineralization of hexadecane dissolved in the heptamethylnonane (Fig. 9.6). Analogous results were obtained for the utilization of naphthalene dissolved in di(2-ethylhexyl) phthalate (Ortega-Calvo and Alexander, 1994). A similar conclusion on the need for adherence comes from a study of *Acinetobacter calcoaceticus* in which comparisons were made of the parent culture with a mutant that did not adhere to a NAPL composed of a pure alkane. Cells of the adherent parent grow on hexane, but the mutant does not grow for a long period. When

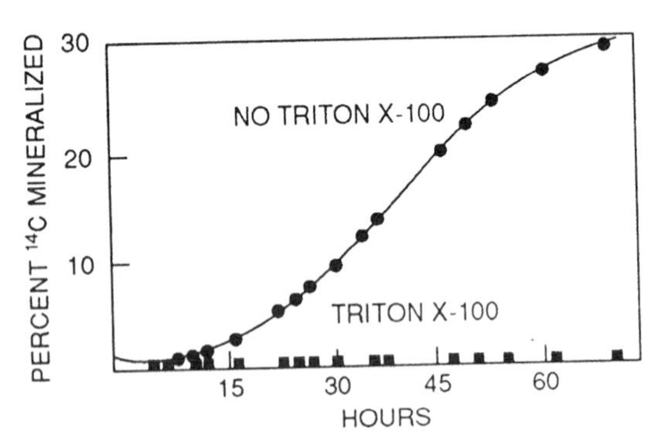

Figure 9.6 Effect of Triton X-100 on the mineralization by *Arthrobacter* sp. of hexadecane dissolved in heptamethylnonane. (From Efroymson and Alexander, 1991. Reprinted with permission from the American Society of Microbiology.)

an emulsifying agent is added, however, the nonadhering mutant is able to utilize the insoluble C source (Rosenberg and Rosenberg, 1981). An identical conclusion was reached from a study of a strain of *Pseudomonas* that does not produce an extracellular surfactant or emulsifier but does adhere to the surface of hexadecane (Goswami and Singh, 1991). The substrate appears to be transferred from the NAPL directly to the bacterium. The utilization of pyrene in a NAPL phase by *Rhodococcus* sp. also seems to entail uptake of the substrate directly at the NAPL–water interface (Bouchez *et al.*, 1997). Indeed, much of the biomass of certain microorganisms adheres as a biofilm to the NAPL, as with a strain of *Candida* growing at the surface of silicone oil and obtaining its carbon (ethyl butyrate) from that NAPL (Ascon-Cabrera and Lebeault, 1995). The cell surface of the organism may be critical in the adherence, possibly a lipophilic layer that facilitates binding of cell and substrate (Kaeppeli and Fiechter, 1976), or the process may involve an emulsifier that is present as a thin layer over the surface of the cell (Pines and Gutnick, 1986). In some species, hair-like structures or thin fimbriae appear to be formed by the microorganisms, and these thin appendages may permit the organism to transport the insoluble compound into the cell, so that the organism can make use of it and bring about its metabolism (Käppeli *et al.*, 1984; Rosenberg *et al.*, 1982). Because cells adhering to the NAPL exist at a site where the concentration of the compound is greater than at a distance away in the aqueous phase, it is also possible that the concentration gradient of that compound is greater than in the free water phase, so that the partitioning rate in the biofilm is greater than that determined by the usual measurement procedures (Ortega-Calvo and Alexander, 1994).

Although most of the reactions involved in the degradation of constituents of NAPLs and the metabolism of sparingly soluble organic compounds probably take place within the cell, this may not always be the case. If an extracellular enzyme is involved in an initial step in the biodegradative sequence, that enzyme would probably function at the interface between the NAPL and the water (Mattson and Volpenhein, 1966).

ENHANCEMENT OF BIODEGRADATION

Several methods are known by which the biodegradation of chemicals in NAPLs may be enhanced. Five ways have been explored, although chiefly in the laboratory rather than in the field: the addition of surfactants, intense mixing/aeration, supplementation with inorganic nutrients, additions of bacteria, and means of obviating toxicity of the NAPLs. Most concern has been with oil. A commerical preparation known as Sugee 2, which disperses crude oil, promotes the degradation of those *n*-alkanes in crude oil with 17 to 28 C atoms in the molecule (Mulkins-Phillips and Stewart, 1974); a commercial dispersant known as Corexit together with a microbial enrichment enhances the degradation of lubricating oil in soil slurries (Rittmann and Johnson, 1989); nonionic surfactants stimulate the destruction by a strain of *Acinetobacter* of components of oil (Lupton and Marshall, 1979); and surfactants also stimulate the degradation of a defined mixture of aliphatic and aromatic hydrocarbons in a soil suspension (Oberbremer *et al.*, 1990). Not all surfactants are stimulatory, but a few enhance the rate, extent, or both of the mineralization of hydrophobic constituents of NAPLs in soil samples; surfactants with hydrophile–lipophile balance number of 11 and above are beneficial with several NAPLs. Some of the surfactants are beneficial even below their apparent CMC values (Fu and Alexander, 1995). Many surfactants are reasonably toxic or, if not harmful to microorganisms, may dislodge cells adhering to the NAPL and thus preclude their acquiring their substrate; hence, choices have to be made judiciously. There is general agreement that surfactants that enhance the transformation do so by increasing the partitioning rate or providing a larger interfacial area that not only favors mass transfer from the NAPL but also allows for greater microbial colonization.

Procedures that result in intensive mixing and aeration often promote the rate of biodegradation of NAPLs and individual constituents. This may be accomplished by slurrying soil or aquifer solids or mixing suspensions containing the NAPLs, a procedure that enhances the mineralization of phenanthrene, biphenyl, and di(2-ethylhexyl) phthalate present in several NAPLs (Kohler *et al.*, 1994; Labare and Alexander, 1995; Fu and Alexander, 1995). Because the biodegradation of hydrocarbons and some other pollutants frequently requires O_2 and because O_2 is rapidly metabolized at the NAPL–water interface, the mixing or slurrying may be beneficial because it increases O_2 diffusion through the liquid. The same type of benefit is accomplished by injecting air, pure O_2, or H_2O_2-containing water into NAPL-contaminated aquifers. At the surface of marine or fresh waters polluted with oil, the supply of dissolved O_2 is usually sufficient and the diffusion of O_2 from the overlying air is usually rapid enough so that this requisite for aerobic bacteria does not limit the rate of biodegradation. The same is probably not true in soils, and an insufficient supply of O_2 probably often limits the rate of degradation of components of NAPLs. The mixing or slurrying also increases the rate of partitioning because it breaks up the NAPL into small droplets that have a larger NAPL–water interfacial area and thus more exposed surface for mass transfer to the water. The larger surface area would also permit greater microbial colonization and result in a greater biomass.

The stimulation by inorganic nutrients is not unexpected. The NAPL is made up of one or a variety of organic compounds, and at the NAPL—water interface, the amount of C available to the potential degrading species is large. This results in a large microbial demand for inorganic nutrients, especially N and P. However, the concentration of these nutrients is frequently low or the rate of diffusion from the surrounding water or soil to the vicinity of the NAPL, where the degradation is occurring, is too slow to meet the large demand. The concentration in the immediate vicinity of crude oil that provides a marked effect may range from 1 to 11 mg of N and about 0.07 mg of P per liter of seawater (Floodgate, 1984). An inadequate supply of inorganic nutrients in the immediate vicinity of the NAPL is common in subsoils and soils. For example, this is shown by a laboratory study in which phenanthrene was introduced into the subsoil either in a NAPL (DEHP or hexane) or without a NAPL. The addition of N and P markedly increases both the rate and the extent of mineralization of phenanthrene dissolved in the first NAPL and had an effect, albeit less pronounced, on phenanthrene metabolism in the second NAPL. In contrast, the addition of N and P has little or no influence on biodegradation of phenanthrene in subsoil having no NAPL (Efroymson and Alexander, 1994a), apparently because the nutrient supply is adequate when the low concentration of phenanthrene is mixed throughout the subsoil. Because the Fe concentration in the ocean is low and much of the Fe is probably not readily assimilated at the pH values of marine water, this element may also sometimes limit the rate of degradation of oil or other NAPLs in seawater (Dibble and Bartha, 1976).

Several laboratory investigations have suggested that the addition of specialized bacteria may be beneficial. For example, the rate and extent of mineralization of phenanthrene, biphenyl, and di(2-ethylhexyl) phthalate in NAPLs in soil slurries and biphenyl in slurries of aquifer solids were enhanced by the addition of acclimated microorganisms, even though the uninoculated samples had some activity. In these cases, consideration was given to whether the added microorganisms should have been acclimated either solely to the target compounds or to those chemicals in the NAPL phase, the latter species possibly being able to acquire the substrate by mechanisms other than relying on abiotic partitioning (i.e., by producing surfactants or by adhering to the NAPL) (Labare and Alexander, 1995; Birman and Alexander, 1996a). Inasmuch as the added organisms could be sensitive to surfactants, that toxicity might be avoided by adding the inoculum some time after the surfactant, whose toxicity might be reduced because some of it entered the NAPL (Birman and Alexander, 1996a). The practical usefulness of inoculation, sometimes termed *bioaugmentation,* needs to be evaluated in the field. Inoculation and bioaugmentation are discussed further in Chapter 15.

A major problem in bioremediation of some NAPLs is their toxicity, and unless this toxicity can be minimized or overcome, a biological approach to remediation is not a practical option. One possible means to obviate the problem is to add species that are able to proliferate in the presence of these toxic NAPLs, and bacteria and yeasts tolerant to a variety of solvents have been isolated. Some of these are able to degrade hydrocarbons dissolved in several organic solvents (Abe *et al.,* 1995; Fukumaki *et al.,* 1994; Moriya and Horikoshi, 1993). Such organisms may be fresh isolates from nature or mutants from laboratory cultures that have become resistant to the inhibitory solvents. A very different approach to mitigate the toxicity involves the use of a three-phase system containing a toxic NAPL, water, and a nontoxic NAPL. The second NAPL is kept apart from the toxic liquid. In a two-phase system, the toxic constituents partition into the water to a concentration that inhibits the microorganisms that otherwise would carry out the biodegradation. However, the nontoxic organic solvent serves as a trap for much of the toxicant that had entered the aqueous phase, lowering the concentration and thereby permitting the biodegradation of compounds entering the water from the toxic NAPL and reentering from the solvent trap. In this way, 1,2-dichlorobenzene- and toluene-sensitive bacteria were able to metabolize phenanthrene initially in those otherwise toxic NAPLs (Robertson and Alexander, 1996).

REFERENCES

Abe, A., Inoue, A., Usami, R., Moriya, K., and Horikoshi, K., *Biosci. Biotechnol. Biochem.* **59,** 1154–1156 (1995).

Amin, P. M., Nigam, J. N., Lonsane, B. K., Baruah, B., Singh, H. D., Baruah, J. N., and Iyengar, M. S., *Folia Microbiol. (Prague)* **18,** 49–55 (1973).

Ascón-Cabrera, M. A., and Lebeault, J.-M., *Appl. Microbiol. Biotechnol.* **43,** 1136–1141 (1995).

Atlas, R. M., *Microbiol. Rev.* **45,** 180–209 (1981).

Bauer, J. E., and Capone, D. G., *Appl. Environ. Microbiol.* **50,** 81–90 (1985).

Becher, P., "Emulsions: Theory and Practice." Reinhold, New York, 1965.

Birman, I., and Alexander, M., *Appl. Microbiol. Biotechnol.* **45,** 267–272 (1996a).

Birman, I., and Alexander, M., *Environ. Toxicol. Chem.* **15,** 1683–1686 (1996b).

Blanch, H. W., and Einsele, A., *Biotechnol. Bioeng.* **15,** 861–877 (1973).

Boehm, P. D., and Quinn, J. G., *Geochim. Cosmochim. Acta* **37,** 2459–2477 (1973).

Bossert, I. D., and Bartha, R., *Bull. Environ. Contam. Toxicol.* **37,** 490–495 (1986).

Bouchez, M., Blanchet, D., and Vandecasteele, J.-P., *Appl. Microbiol. Biotechnol.* **43,** 952–960 (1995).

Bouchez, M., Blanchet, D., and Vandecasteele, J.-P., *Microbiology* **143,** 1087–1093 (1997).

Cameotra, S. S., Singh, H. D., Hazarika, A. K., and Baruah, J. N. *Biotechnol. Bioeng.* **25,** 2945–2956 (1983).

Carroquino, M. J., and Alexander, M., *Environ. Toxicol. Chem.* **17,** 265–270 (1998).

Chakravarty, M., Singh, H. D., and Baruah, J. N., *Biotechnol. Bioeng.* **17,** 399–412 (1975).

Chen, C. S.-H., Delfino, J. J., and Rao, P. S. C., *Chemosphere* **28,** 1385–1400 (1994).

Coates, M., Connell, D. W., and Barron, D., *Environ. Sci. Technol.* **19,** 628–632 (1985).

Dibble, J. T., and Bartha, R., *Appl. Environ. Microbiol.* **31,** 544–550 (1976).

Efroymson, R. A., and Alexander, M., *Appl. Environ. Microbiol.* **57,** 1441–1447 (1991).

Efroymson, R. A., and Alexander, M., *Environ. Toxicol. Chem.* **13,** 405–411 (1994a).

Efroymson, R. A., and Alexander, M., *Environ. Sci. Technol.* **28,** 1172–1179 (1994b).

Efroymson, R. A., and Alexander, M., *Environ. Sci. Technol.* **29,** 515–521 (1995).

Einsele, A., Schneider, H., and Fiechter, A., *J. Ferment. Technol.* **53,** 241–243 (1975).

Floodgate, G. D., *in* "Petroleum Microbiology" (R. M. Atlas, ed.), pp. 354–397. Macmillan, New York, 1984.

Fogel, S., Lancione, R., Sewall, A., and Boethling, R. S., *Chemosphere* **14,** 375–382 (1985).

Foght, J. M., Fedorak, P. M., and Westlake, D. W. S., *Can. J. Microbiol.* **36,** 169–175 (1990).

Fu, M. H., and Alexander, M., *Appl. Microbiol. Biotechnol.* **43,** 551–558 (1995).

Fukumaki, T., Inoue, A., Moriya, K., and Horikoshi, K., *Biosci. Biotechnol. Biochem.* **58,** 1784–1788 (1994).

Ghoshal, S., and Luthy, R. G., *Environ. Toxicol. Chem.* **15,** 1894–1900 (1996).

Ghoshal, S., Ramaswami, A., and Luthy, R. G., *Environ. Sci. Technol.* **30,** 1282–1291 (1996).

Gibbons, J. A., and Alexander, M., *Environ. Toxicol. Chem.* **8,** 283–291 (1989).

Goma, G., and Ribot, D., *Biotechnol. Bioeng.* **20,** 1723–1734 (1978).

Goswami, P. C., and Singh, H. D., *Biotechnol. Bioeng.* **37,** 1–11 (1991).

Goswami, P. C., Singh, H. D., Bhagat, S. D., and Baruah, J. N., *Biotechnol. Bioeng.* **25,** 2929–2943 (1983).

Haines, J. R., and Alexander, M., *Appl. Microbiol.* **28,** 1084–1085 (1974).

Inoue, A., and Horikoshi, K., *J. Ferment. Bioeng.* **71,** 194–196 (1991).

Jimenez, I. Y., and Bartha, R., *Appl. Environ. Microbiol.* **62,** 2311–2316 (1996).

Kaeppeli, O., and Fiechter, A., *Biotechnol. Bioeng.* **18,** 967–974 (1976).

Käppeli, O., Walther, P., Mueller, M., and Fiechter, A., *Arch. Microbiol.* **138,** 279–282 (1984).

Kile, D. E., and Chiou, C. T., *Environ. Sci. Technol.* **23,** 832–838 (1989).

Köhler, A., Schüttoff, M., Bryniók, D., and Knackmuss, H.-J., *Biodegradation* **5,** 93–103 (1994).

Laane, C., Boeren, S., Hilhorst, R., and Veeger, C., *in* "Biocatalysis in Organic Media" (C. Laane, J. Tramper, and M. D. Lilly, eds.), pp. 65–84. Elsevier, Amsterdam, 1987a.

Laane, C., Boeren, S., Vos, K., and Veeger, C., *Biotechnol. Bioeng.* **30,** 81–87 (1987b).

Labare, M. P., and Alexander, M., *Environ. Toxicol. Chem.* **14,** 257–265 (1995).

LePetit, J., and Barthelemy, M. H., *Ann. Inst. Pasteur, Paris* **114,** 149–158 (1968).

Lindley, N. D., and Heydeman, M. T., *Appl. Microbiol. Biotechnol.* **23,** 384–388 (1986).

Lupton, F. S., and Marshall, K. C., *Geomicrobiol. J.* **1,** 235–247 (1979).

Mallee, F. M., and Blanch, H. W., *Biotechnol. Bioeng.* **19,** 1793–1816 (1977).

Mattson, F. H., and Volpenhein, R. A., *J. Am. Oil Chem. Soc.* **43,** 286–289 (1966).

McLee, A. G., and Davies, S. L., *Can. J. Microbiol.* **18,** 315–319 (1972).

Miura, Y. *Adv. Biochem. Eng.* **9,** 31–56 (1978).

Moo-Young, M., Shimuzu, T., and Whitworth, D. A., *Biotechnol. Bioeng.* **13,** 741–760 (1971).

Moriya, K., and Horikoshi, K., *J. Ferment. Bioeng.* **76,** 168–173 (1993).

Morrison, D. E., and Alexander, M., *Environ. Toxicol. Chem.* **16,** 1561–1567 (1997).

Mulkins-Phillips, G. J., and Stewart, J. E., *Appl. Microbiol.* **28,** 547–552 (1974).

Oberbremer, A., and Müller-Hurtig, R., *Appl. Microbiol Biotechnol.* **31,** 582–586 (1989).

Oberbremer. A., Müller-Hurtig, R., and Wagner, F., *Appl. Microbiol. Biotechnol.* **32,** 485–489 (1990).

Ortega-Calvo, J.-J., and Alexander, M., *Appl. Environ. Microbiol.* **60,** 2643–2646 (1994).

Osswald, P., Baveye, P., and Block, J. C., *Biodegradation* **7,** 297–302 (1996).

Pines, O., and Gutnick, D., *Appl. Environ. Microbiol.* **51,** 661–663 (1986).

Rittmann, B. E., and Johnson, N. M., *Water Sci. Technol.* **21**(4/5), 209–219 (1989).

Robertson, B. K., and Alexander, M., *Environ. Sci. Technol.* **30,** 2066–2070 (1996).

Robotham, P. W. J., and Gill, R. A., *in* "The Fate and Effects of Oil in Freshwater" (J. Green and M. W. Trett, eds.), pp. 41–79. Elsevier Applied Science, London, 1989.

Rosenberg, E., *CRC Crit. Rev. Biotechnol.* **3,** 109–132 (1986).

Rosenberg, E., Perry, A., Gibson, D. T., and Gutnick, D. L., *Appl. Environ. Microbiol.* **37,** 409–413 (1979).

Rosenberg, M., and Rosenberg, E., *J. Bacteriol.* **148,** 51–57 (1981).

Rosenberg, M., and Rosenberg, E., *Oil Petrochem. Pollut.* **2,** 155–162 (1985).

Rosenberg, M., Bayer, E. A., Delarea, J., and Rosenberg, E., *Appl. Environ. Microbiol.* **44,** 929–937 (1982).

Singer, M., and Finnerty, W. R., *in* "Petroleum Microbiology" (R. M. Atlas, ed.), pp. 1–59. Macmillan, New York, 1984.

Stucki, G., and Alexander, M., *Appl. Environ. Microbiol.* **53,** 292–297 (1987).

Sutton, C., and Calder, J. A., *Environ. Sci. Technol.* **8,** 654–657 (1974).

Takazawa, Y., Sato, S., and Takahashi, J., *Agric. Biol. Chem.* **48,** 2489–2495 (1984).
Thomas, J. M., and Alexander, M., *Microb. Ecol.* **14,** 75–80 (1987).
Thomas, J. M., Yordy, J. R., Amador, J. A., and Alexander, M., *Appl. Environ. Microbiol.* **52,** 290–296 (1986).
Wodzinski, R. S., and Bertolini, D., *Appl. Microbiol.* **23,** 1077–1081 (1972).
Wodzinski, R. S., and Coyle, J. E., *Appl. Microbiol.* **27,** 1081–1084 (1974).
Wodzinski, R. S., and Johnson, M. J., *Appl. Microbiol.* **16,** 1886–1891 (1968).
Wodzinski, R. S., and Larocca, D., *Appl. Environ. Microbiol.* **33,** 660–665 (1977).
Yalkowsky, S. H., Valvani, S. C., and Mackay, D., *Residue Rev.* **85,** 43–55 (1983).

CHAPTER 10

Bioavailability: Aging, Sequestering, and Complexing

The availability of many chemicals is affected by a series of ill-defined, often uncharacterized processes. In some of these processes, the compound is readily evident and can be easily removed from the soil, sediment, or aquifer by conventional extraction procedures; the evidence for reduced bioavailability of these compounds is the marked decline in the rate of biodegradation. In other processes, the compound is still present, but it can only be removed from the environmental sample by highly vigorous extraction techniques. The evidence for reduced bioavailability of such a compound is the marked decline in the rate of biodegradation with time or the almost complete resistance of the molecule to microbial destruction. This is evident by the data for an insecticide in Fig. 10.1. In the case of the illustration, the compound added to soil was aldrin, but the aldrin is converted by a simple epoxidation to a closely related molecule, dieldrin; data shown are the percentage of aldrin plus dieldrin present in the soil.

The compound that remains is often termed an *aged* (or sometimes *weathered*) chemical or *aged residue*. Residue refers to the residual nature of the compound, that is, that it persists. A substrate that is only slowly metabolized, as shown in Fig. 10.1, has considerable time to interact with physical or chemical components of soils, subsoils, or sediments and thus have its physical, chemical, or biological behavior altered. The term "aged" is unfortunate because it implies a change in the identity rather than the behavior of the molecule. Because the compound is still intact and unchanged but has become hidden and inaccessible, such aged molecules are often termed *sequestered,* and the process is considered to be a *sequestration*.

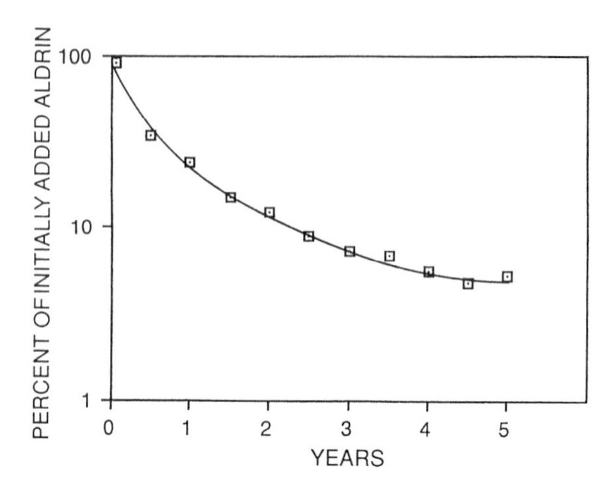

Figure 10.1 Logarithmic plot of the disappearance of aldrin (plus the dieldrin formed from it) in 12 field sites in various countries. The data represent the means of the percentages remaining at all sites in areas receiving 2.24 or 4.48 kg of aldrin/ha. (From Elgar, 1975. Reprinted with permission from Georg Thieme Verlag, Stuttgart.)

With other synthetic compounds, the molecule is converted to a form that cannot be removed by even the most vigorous extraction with nonpolar or polar organic solvents, yet the molecule remains in a form that is recognizable as the parent compound, or sometimes as a nonextractable metabolite derived from it. These highly persistent materials are commonly termed *bound residues.* Many insecticides, herbicides, fungicides, and undoubtedly other classes of chemicals undergo changes, especially in soil, that result in the formation of bound residues. In this instance, the word "bound" is also unfortunate and misleading because it implies a binding mechanism of rendering the compound far less available for biodegradation. Many of the bound residues appear to result from complexing with humic materials in soils, and they may then be new molecular species and not the parent molecules in a strict sense.

At this stage in the development of knowledge of a subject that is still characterized by confusion, it might be prudent to envision three separate categories of molecules of reduced bioavailability. These are in addition to compounds that are poorly available because they are sorbed to solid surfaces or present within NAPLs *in the immediate vicinity of microbial cells* having the requisite catabolic enzymes.

(a) Nonsorbed compounds in micropores at some distance from cells having the requisite enzymes. Distance in the present context refers to micrometer distances, which are those that are important to cells in a porous or nonfluid environment.

(b) Compounds that either enter into, and are retained within, nanopores or that partition into the solids themselves. These nanopores are far smaller than micropores and have dimensions too small for even the tiniest bacterium to penetrate. Molecules present within environmental solids are also inaccessible to all organisms.

(c) Chemicals that complex with humic materials or other environmental constituents to form molecular species that, although containing the parent molecules or metabolites generated from them, are in fact new molecular species.

One or more of these models may be inappropriate representations of what transpires in nature. Nevertheless, the confusing state of knowledge of these poorly available substrates, for the present at least, can best be understood by considering these three separate possibilities.

REMOTE COMPOUNDS

A chemical may become less available or essentially wholly unavailable for biodegradation if it enters or is deposited in a micropore that is inaccessible to microorganisms. Soils, subsoils, and sediments are characterized by particles of various sizes, and between these particles are pores that obviously are also of various dimensions, both large and small. The various sizes of solids in a soil are depicted in Fig. 10.2. The pores may be filled entirely with water, as in sediments or below the water table in land areas, or they may be filled with air and water, as in soils above the water table. A chemical would move out of a micropore by diffusion to a site containing a bacterium with the capacity to bring about its destruction, but the path for diffusion of the molecule in an environment with small particles

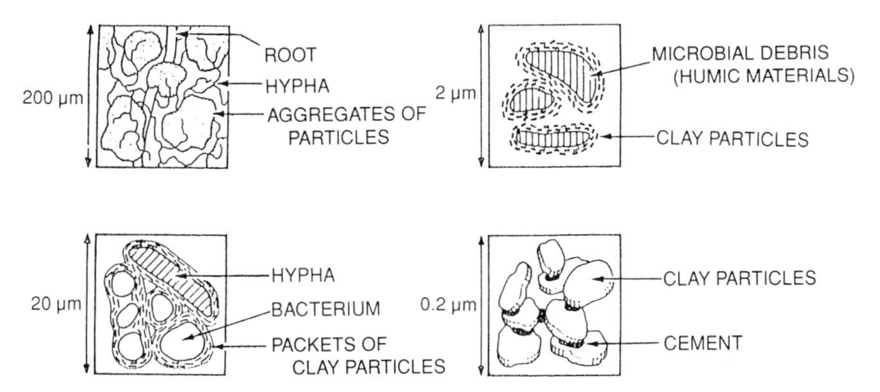

Figure 10.2 Models of the various particle sizes in soil. (From Tisdall and Oades, 1982. Reprinted with permission from Blackwell Scientific Publications.)

is far from straight. Instead, it is highly tortuous. This *tortuosity* may increase enormously the path the molecule must traverse before it reaches the appropriate cells (Fig. 10.3). The longer the path, the greater the influence on biodegradation. Models have been developed to describe the effect of diffusion on biodegradation (Priesack, 1991; Scow and Alexander, 1992; Scow and Hutson, 1992).

In a habitat such as soil, most of the bacterial cells exist on the surfaces of particulate matter, and these cells are frequently present as small colonies or aggregates of cells. From the microscale vantage point of the bacterium, large portions of the clay and humus surfaces are free of living organisms. Most of the cells thus sorbed probably do not move freely, and a chemical occluded at a distant site, distant on a microscopical scale, is inaccessible to the cell fixed to the surface. Moreover, the pores between small particles have narrow necks that would impede movement of even mobile cells that otherwise might be transported to a point in a larger pore at the other side of the thin neck. An appreciation of the small dimensions of these pores can be gained from data indicating that pores with diameters less than 0.2 μm occupy 30% of the total volume of some soils (Hassink *et al.,* 1993).

Some direct evidence exists that organic compounds may be physically sequestered and thereby protected from microbial attack. Thus, a thin layer of glass microbeads placed between a population of a *Pseudomonas* strain and chitin results in a long delay before the onset of mineralization of this polysaccharide by the organism. This delay in activity is not a result of sorption of the substrate but rather results from physical separation of the bacterium from its C source (Ou and Alexander, 1974). Similar evidence suggesting the importance of accessibility of a substrate comes from a study in which starch deposited in micropores of artificially created soil aggregates was found to be protected from microbial attack (Adu and Oades, 1978). In both of these investigations, the substrates did not diffuse through the aqueous phase, but low-molecular-weight, nonsorbed compounds would diffuse to the physical sites occupied by bacteria so that a slow, diffusion-limited degradation would take place.

Bioavailability would be dramatically affected if a chemical is not only physically remote from potentially active microorganisms but is sorbed to solid surfaces

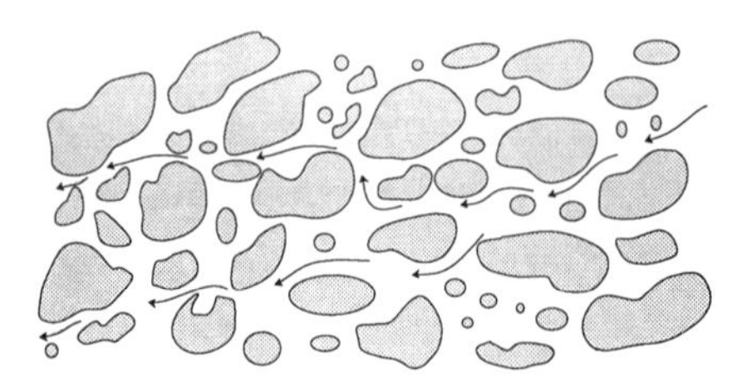

Figure 10.3 Pathways of a molecule diffusing through an environment containing small particulates.

associated with that remote micropore. The substrate would then need to be desorbed and diffuse through a tortuous pathway, all the while being subject to resorption. This combination of sorption–desorption, diffusion, and tortuosity imposes major limitations on the metabolism of an otherwise biodegradable molecule.

SEQUESTRATION AND AGING

Organic compounds that persist in soil often undergo a time-dependent decline in bioavailability. The same process probably occurs in sediments, but data for sediments are sparse. Because the process is slow and time dependent, and thus is appropriately designated aging, it is—at least given the present state of knowledge—somewhat different from that involved in rendering the "remote" substrates less available. The modification in availability to microorganisms as a result of sequestration occurring during aging, which has been invoked as the explanation for hockey stick-shaped kinetics (discussed in Chapter 6), is illustrated by the three plots in Fig. 10.4. In the initial period, the compound disappears as a result of biodegradation, and possibly by other loss mechanisms, but little or none of the compound is destroyed after it has resided in the soil for some time. It is noteworthy that the times to reach the period after which little or no biodegradation of DDT occurs and the percentage of the compound that remains differ markedly among the three soils and locations shown in Fig. 10.4. A similar phenomenon appears to occur in sediments, witness the finding that although 80% of the hexachlorobenzene deposited in the early 1970s in a lake bottom sediment was dechlorinated in the succeeding 20 years, all sediment cores still contained 40 μg/kg in a form that was dechlorinated extremely slowly, if at all (Beurskens *et al.*, 1993). Because most polluted soils, subsoils, and sediments were contaminated many years ago and often one or more decades ago, sequestration is particularly important, especially as it relates to bioremediation and the toxicological significance of poorly available pollutants.

Data that serve as the basis for hockey-stick shaped kinetics, which largely come from the long-term monitoring of treated field sites, show that much of the chemical can be removed from the soil by vigorous extractants but that the soil-aged compound is apparently not metabolized, or not at an appreciable rate, by microorganisms. In contrast, the unaged compound in soil is transformed at reasonable rates. This time-dependent change, which has been observed with a number of insecticides, was the first line of evidence for sequestration (Alexander, 1995, 1997). These aged compounds are obviously not bound residues because their recovery does not entail the use of vigorous hydrolysis by acids or bases. Because the aged molecules are solvent extractable, albeit by vigorous treatment, they are

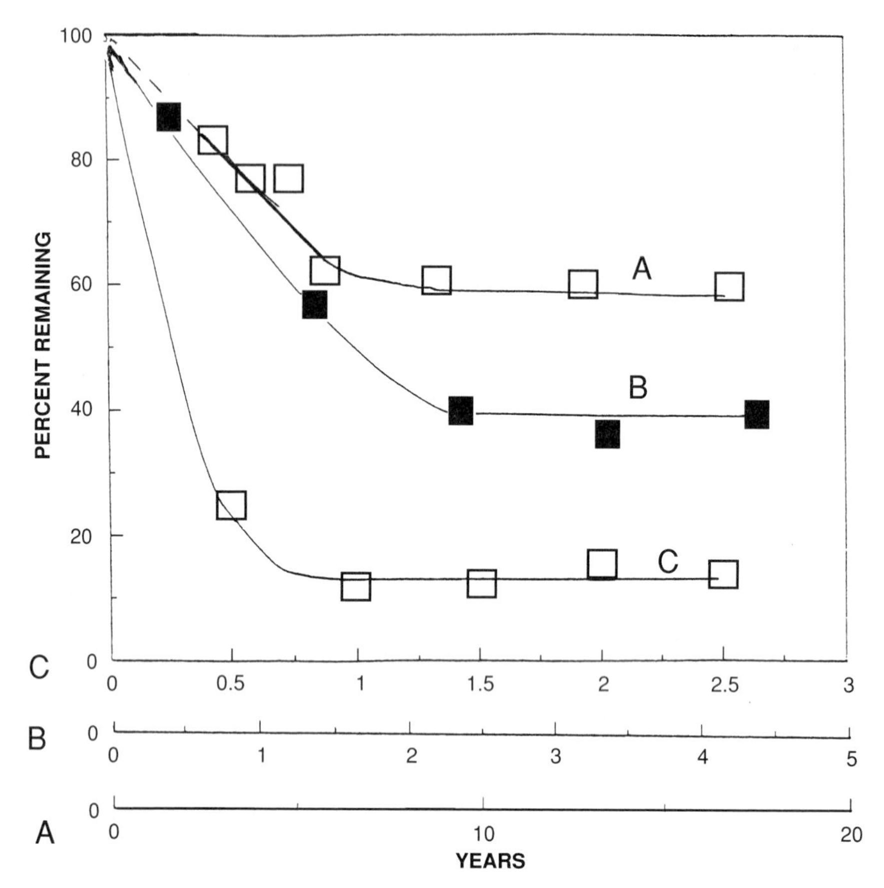

Figure 10.4 Concentrations of DDT in soils at three different field sites. The amounts initially added were (A) 200 mg/kg (Nash and Woolson, 1967), (B) 4.42 mg/kg (Lichtenstein *et al.*, 1960), and (C) 11.2 kg/ha (Onsager *et al.*, 1970). Note the differences in periods of time shown by the *x* axes.

assumed to be present in an uncomplexed form and thus are considered to be pollutants and are subject to regulation by governmental agencies.

A second line of evidence for sequestration comes from laboratory studies of samples of soil treated many months or many years earlier with pesticides that persisted and to which a freshly added chemical was applied. To distinguish between the aged, field-applied and the unaged, laboratory-applied pesticide, the latter was radiolabeled with ^{14}C. For example, Steinberg *et al.* (1987) compared the availability of 1,2-dibromoethane (EDB) present in soils treated 0.9 and 3 years earlier with a freshly added compound and found that the aged pesticide was resistant to biodegradation but that the freshly added chemical was readily and extensively mineralized (Fig. 10.5). Thus, although the compounds are identical, the one that has persisted in the soil is somehow protected from attack. It is hidden or sequestered in some fashion. Similarly, when ^{14}C-labeled simazine was added to a soil that contained the herbicide as a result of application for 20 consecutive years, the concentration of the former declined appreciably in 7 weeks, whereas the levels of the latter were unaltered (Scribner *et al.*, 1992). The same type of apparent sequestration is evident from an investigation in which it was observed that phenanthrene, anthracene, fluorene, pyrene, chrysene, and other PAHs remaining in the soil from a closed coking plant were not susceptible to biodegradation, but were metabolized when extracted from and added back to the soil (Weissenfels *et al.*, 1992). The marked differences between unaged and aged compounds in the rate and extent of biodegradation suggest that misinterpretations will arise when using

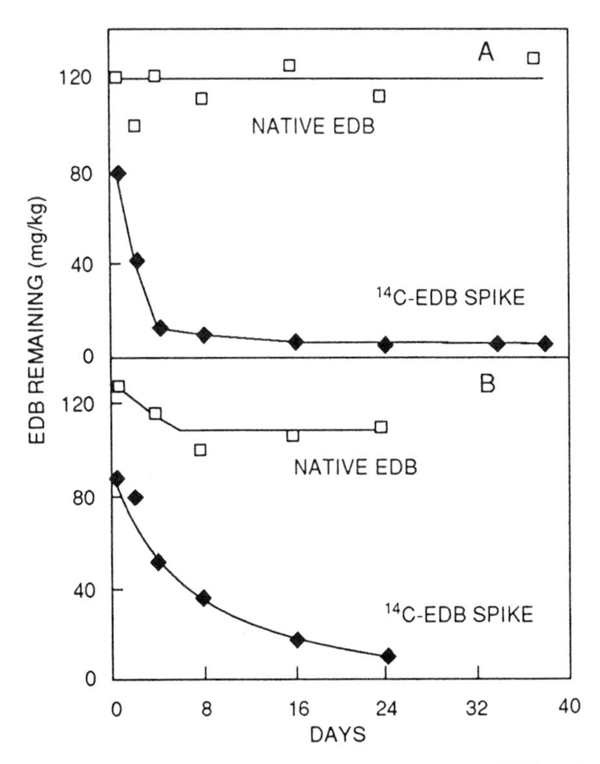

Figure 10.5 Degradation of EDB in suspensions of field soil containing EDB applied earlier (native EDB) or containing newly added ^{14}C-labeled EDB (^{14}C-EDB spike). The soils used in this particular study had received unlabeled EDB 0.9 (A) and 3 (B) years earlier, and they are not the soils in which EDB was found to persist for 19 years. (From Steinberg *et al.*, 1987. Reprinted with permission from the American Chemical Society.)

freshly added compounds to understand the behavior, bioavailability, or susceptibility to the bioremediation of aged molecules.

A third line of evidence comes from studies of individual compounds that were allowed to remain in soil in the laboratory for varying periods to simulate natural aging. To prevent biodegradation of the chemicals during the aging period, the soils were initially sterilized. Moreover, the test compounds were selected to have low volatility and were not subject to significant abiotic breakdown. Then, after varying residence periods in the soil, samples of sterile soil containing the same amount of a test chemical were inoculated with a bacterium capable of mineralizing the compound. In tests with several PAHs, atrazine, and 4-nitrophenol, for example, the extent of mineralization declined with increasing periods of aging (Hatzinger and Alexander, 1995; Kelsey *et al.*, 1997; Tang *et al.*, 1998). When mass balances were performed on the uninoculated soils, all or nearly all of the compounds could be recovered. Moreover, analysis of the PAH-amended soil after biodegradation had largely ceased, using a vigorous extraction procedure, confirmed that aging had resulted in an increased amount of the PAH resisting decomposition (Tang *et al.*, 1998).

Changes in extractability provide a fourth line of evidence for the occurrence of sequestration. It has long been known that it is increasingly difficult to remove hydrophobic compounds from soil with mild extractants as the residence time of those compounds in the soil increases. For example, dieldrin and heptachlor epoxide were progressively more difficult to remove from soil by several solvent systems as the residence time increased up to 2 months (Johnson and Starr, 1967), and the recovery of dieldrin was similarly found to be less in another study in which soils with freshly added and the aged insecticide were compared (Chiba and Morley, 1968). These studies have been extended more recently to other classes of compounds added to soil; for example, phenanthrene and atrazine (Chung and Alexander, 1998). This phenomenon is not restricted to soils, witness the decline in extractability of the benzene or anthracene added to stream sediments (Haddock *et al.*, 1983). Thus, nonbiological approaches also indicate a time-dependent diminution in availability.

The amount of a compound that is sequestered increases with time. Expressed in another way, the percentage of the chemical that is biologically available diminishes with increasing persistence. This presumably occurs because more of the molecules are diffusing to inaccessible sites. However, after a period of time that varies with the soil and the compound, sequestration of additional quantities slows and possibly stops. The reason for this rate decline is not presently known.

The decline in availability affects other species in addition to microorganisms. This was recognized long ago as a result of measurements of the toxicity of DDT to the fruit fly, *Drosophila melanogaster* (Peterson *et al.*, 1971). More recent studies showed that even when DDT and dieldrin scarcely disappear from sterile samples of soil into which these insecticides were added, their lethal effects on three insect species declined and often disappeared (Robertson and Alexander, 1998). Similarly, the amount of phenanthrene, anthracene, fluoranthene, and pyrene assimilated by earthworms from samples of sterile soil diminished with increasing times of aging (Tang *et al.*, 1998). These declines in the availability of aged chemicals to species of higher organisms are relevant to bioremediation since only a portion of the toxicant present in sites that are being considered for, or are undergoing, treatment is present in a form that is toxicologically important. That sites in which aging has occurred still pose risks despite some sequestration is evident from data indicating that dieldrin and heptachlor at recommended rates of application still killed termites 28 years after their addition to soil in the field (Grace *et al.*, 1993) and that part of the 2,3,7,8-TCDD that had persisted for several years in soil was still available to rats (Shu *et al.*, 1988).

The rate and extent of sequestration vary among different soils, and universal values cannot be assigned to individual compounds. This is suggested by the plots in Fig. 10.4, although a conclusion from those curves about the effects of soil type can only be tentative because temperature, rainfall, and other conditions at the field sites were not the same. In a more definitive study under laboratory conditions, however, it has been clearly demonstrated that the decline in the availability of phenanthrene and atrazine for bacterial mineralization differs appreciably among soils. In some, little sequestration is evident. In others, the rate is fast and much of the compound becomes inaccessible to bacteria. Some relationships to individual soil properties are also evident (Chung and Alexander, 1998).

The relevance of sequestration to bioremediation is manifestly clear from analyses of the results of biological treatment to destroy petroleum hydrocarbons in soil. Characteristically, the hydrocarbons disappear rapidly at first, but then the rate of degradation slows and possibly stops. Often, both in field and laboratory tests, the concentration does not fall to zero, but a reasonable amount of the compounds frequently remains. Should that amount be above the level permitted by regulatory agencies, the bioremediation might be deemed to have not met its objective (Loehr and Webster, 1996, 1997). Although these numerous observations suggest that the residual fraction represents sequestered hydrocarbons, that conclusion might be questioned because the remaining compounds might merely be those hydrocarbons that are intrinsically slowly metabolized. However, some of the PAHs that remain, if present in culture medium, would be transformed rapidly. Such objections do not pertain to bioremediations of soils in which the fate of individual compounds is being traced. Thus, a remediation by use of an anaerobic slurry reactor showed that 25% of the DDT and 65% of the DDE were not available to the anaerobes (White and Herndon, 1995). Similarly, part of the PCP remained in soil despite 2 years of bioremediation in the field, and further microbial treatment in the laboratory failed to bring about an appreciable reduction in the PCP concentration (Salkinoja-Salonen *et al.*, 1990).

Despite the decline in bioavailability of aged compounds, a part remains in a form that can be metabolized by microorganisms and, if the substance is toxic, a part is still present in a form that can do harm. Hence, remediation is still warranted by some means, and bioremediation is feasible. That the microbiological treatment is effective in reducing the total concentration and the bioavailability of the aged compounds is evident from a laboratory investigation in which several PAHs were aged in sterile soil, and the uptake of the aged PAHs by earthworms was determined before aging, after aging but before inoculation, and after the addition of PAH-utilizing bacteria to soil containing the aged or freshly added hydrocarbon. Data for one of the hydrocarbons, which are given in Table 10.1, show the decline in availability of the compound to the bacteria and the earthworms as a result of aging as well as the combined effect of the processes of sequestration and biodegradation in reducing the amount of fluoranthene assimilated by the animals. It is also noteworthy that a small amount of the hydrocarbon was still found in the worm tissue even after aging and bioremediation. Similar effects have been observed with plants (Tang *et al.*, 1998).

The decline in bioavailability as a result of the time-dependent sequestration is not reflected by the analytical methods for pollutants that are currently the basis for risk analyses and governmental regulation. This is true both for the bioavailability of toxicants for microbial metabolism and for the uptake and toxicity to higher organisms. Those methods typically rely on a vigorous extraction followed by gas chromatography or high-performance liquid chromatography. Vigorous extraction, however, removes both the bioavailable and the sequestered fractions of the targeted substance and thus overestimates the concentrations accessible to living organisms

Table 10.1

Effect of Laboratory-Scale Bioremediation on Availability of Unaged and Aged Fluoranthene to *Eisenia foetida*[a]

		% of fluoranthene assimilated by worms	
Time of aging (days)	Concentration in soil after bioremediation[b] (mg/kg)	Before remediation	After remediation
0	1.65	38.2	7.81
75	4.56	26.0	1.89
107	5.85	25.7	1.85
140	7.72	27.1	1.21

[a] From Tang *et al.* (1998).
[b] Initial concentration: 100 mg/kg.

(Kelsey and Alexander, 1997). In the study from which the data in Table 10.1 were obtained, for example, essentially the same amount of fluoranthrene was present initially and after 140 days, as determined by vigorous extraction, yet the amount available to the bacterium had declined appreciably. Even more striking in these data are the values from vigorous extractions, which show greater than fourfold higher levels of fluoranthrene remaining after biomediation of the 140-day-aged than the unaged compound, yet a more than sixfold lower percentage uptake by the worms. From the viewpoint of actual field bioremediations, the frequent observations that bioremediations do not reduce the levels of target pollutants below regulatory levels may be misleading because the amount that remains, which is determined by vigorous extraction of the soil, may be largely unavailable and thus pose little risk. The remediation, therefore, may have met its actual objectives even if the analytical procedure suggests a failure.

Attempts are being made to establish chemical procedures that provide results that reflect actual bioavailability. Although bioassays would seem to be a more appropriate way of assessing bioavailability, such assays typically have low precision, they are usually slow and expensive to perform, and they are not as familiar to regulatory authorities as chemical methodologies. One approach to predict bioavailability of sequestered pollutants is to use a mild and selective extractant that gives data that parallel the availability of organic compounds to living organisms. Preliminary data with phenanthrene and atrazine suggest that such an approach is feasible (Kelsey *et al.*, 1997). However, before a chemical assay to predict bioavailability can be accepted, more studies will be required with different soils, a variety of chemicals, various remediation methodologies, and with correlations of data so obtained with bioassays involving a range of dissimilar species.

Evidence from laboratory studies indicate that a sequestered compound may be made more accessible for biodegradation. Some greater accessibility may be brought about by slurrying the soil, although this effect may be more related to increasing the accessibility of compounds tentatively designated as "remote" in the context discussed earlier. In addition, the introduction of a PAH not utilized by a strain of *Pseudomonas* into a soil containing aged phenanthrene made more of that phenanthene available for biodegradation by the bacterium, possibly as a result of the added PAH displacing part of the sequestered chemical (J. C. White, M. Alexander, and J. J. Pignatello, unpublished data). However, repeated wetting and drying of soil sometimes decreases the amount of a PAH that is subject to microbial attack (White *et al.*, 1997).

An idea of what is transpiring in the process of sequestration can be obtained from studies of desorption. Although desorption is often assumed to be rapid, this is clearly not true for many chemicals. What is observed instead is an initially rapid desorption but then a progressively slower rate. The observations are commonly taken to reflect the localization of the molecule at sites very near to and progressively further from the particle–water interface, those molecules near the surface being desorbed rapidly, and the more remote molecules entering the solution more slowly. In essence, sequestration might be viewed as a reflection of the remoteness of molecules from the outer surfaces of soil particles, from which the substrate for biodegradation would be obtained.

Views on the mechanisms of sequestration have been further developed. In a simple sense, the molecule of concern has been sorbed in some fashion so that it is no longer accessible to microorganisms. However, that sorption is not a two-dimensional process in which the molecule is retained on the outer, exposed surface of a soil particle, rather it is a three-dimensional process in which the molecule of concern has moved in that third dimension (Brusseau and Rao, 1991). The compound is somehow within soil particles and is therefore not directly accessible even to the smallest organisms. Appreciable penetration of the molecules in that third dimension is apparently slow; therefore, whatever the mechanism be, it requires the passage of considerable time. Hence, the word *aging* is appropriate. The longer the period of aging, the more of the chemical entering that third dimension and presumably the further is much of the compound from the outer, available surface of the soil particle.

Two hypotheses have been invoked to explain where the chemicals are located in a biologically inaccessible locale within the soil matrix. One assumes that the molecules are retained within nanopores that are so small as well as so remote that they are not penetrable by any organism, even the smallest bacterium. The second assumes that the molecules enter into solid organic matter in soil and that solid-state diffusion moves the molecule to the interior of the solid.

Nanopores, which are the basis for the first hypothesis, are not only common but are surprisingly abundant in soils, sediments, and aquifer solids. Pores with millimeter diameters are obvious to the unaided eye, and those in the micrometer-diameter range have been characterized as well. However, using appropriate porosimeters, nanopores with diameters as small as 5 nm can be shown to represent a large portion of the total soil volume. Such nanopores have also been found at depths of at least 1 m (Newman and Thomasson, 1979), in the solid phase of aquifers (Ball *et al.*, 1990), and in coastal sediments (Mayer, 1994a,b). Moreover, the total surface area of these minute pores is enormous, and it has been found that most of the surface area of coastal sediments is in pores that are smaller than 10 nm (Mayer, 1994a,b); thus they have the capacity to sorb a large amount of a compound that reaches their remote location within the solid matrix. The distribution of nanopores in three soils is depicted in Fig. 10.6. Microscopic examination of the crystals of geothite, an iron oxide mineral common in soils, has shown the presence of pores 20–30 nm wide that become still narrower toward the interior of the crystals and narrow to diameters of 2 nm or less (Fischer *et al.*, 1996).

The term nanopores has been used in several ways. The scheme of the International Union of Pure and Applied Chemistry classifies macropores, meso-

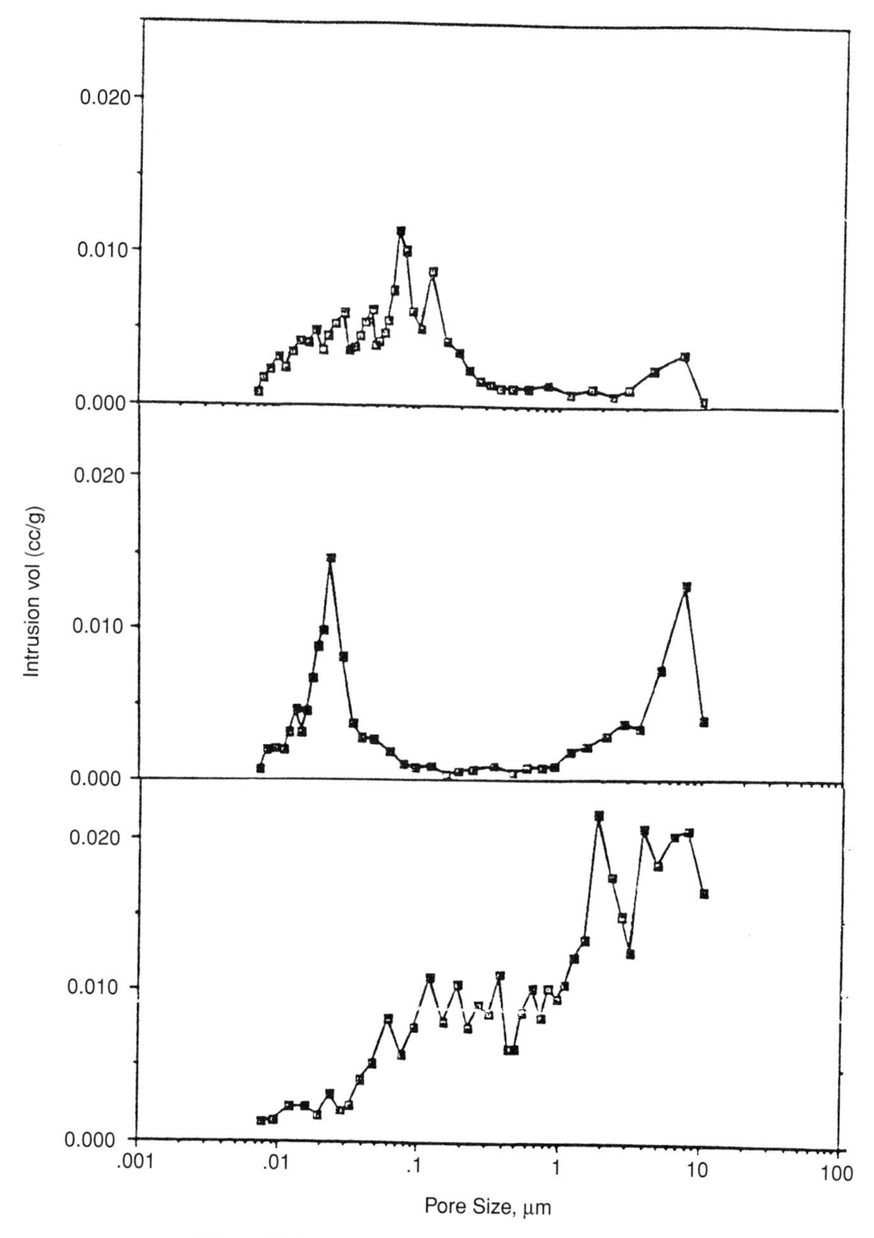

Figure 10.6 Distribution of nanopores in three soils.

pores, and micropores as those with diameters >50, 2–50, and <2 nm, whereas soil scientists, who have been measuring porosity for many years, often consider micropores to have diameters of 5–30 μm. The term *nanopore,* the prefix *nano* referring to sizes smaller than those designated by the prefix *micro,* has been adopted for environmental science because it is consistent with the terminology in soil science.

Because fungi and many bacteria have diameters greater than 1000 nm and no free-living organism has a dimension less than 100 nm, a molecule within these smaller nanopores is not available for metabolism, so long as it is retained within the pores and does not diffuse into the larger pores inhabited by microorganisms. Nanopores that do not sorb the compounds of interest do not sequester molecules of concern because those molecules would readily diffuse out and to microsites where they can be metabolized (Nam and Alexander, 1998). However, diffusion is retarded appreciably not only by sorption within micropores but also by tortuosity. Tortuosity refers to the long, twisted path the molecule must traverse as it moves in the channels of liquid between the innumerable physical obstructions created by solids. The path is not straight and short; it is twisted and quite long. In addition, after the molecule is desorbed from a remote nanopore and moves a short distance on its tortuous path, it is repeatedly resorbed and then desorbed and then again resorbed—all along this twisted way to the site of accessibility (Ball and Roberts, 1991a,b; Wu and Gschwind 1986). The entrapment and sorption in remote nanopores coupled with repeated desorption and sorption associated with tortuosity make for an attractive hypothesis.

The hypothesis that sequestration results from molecules that become entrapped within the solid phase of soil organic matter, which has been proposed for hydrophobic compounds, envisions the process as equivalent to the partitioning of a compound from water into an organic solvent during an extraction (Chiou, 1989). In these organic solids, the rate of diffusion is orders of magnitude slower than in water (Brusseau and Rao, 1991), so that a molecule would move only very slowly from the outer surface of the solid to inner sites—consistent with times for aging—and the slow solid-state diffusion from these sites to the outer surface would likewise be very slow, which is consistent with the poor bioavailability of the sequestered substrate. This hypothesis has been suggested as being important for the slow phases of desorption of hydrophobic compounds (Brusseau *et al.,* 1991). Even this hypothesis, however, relies on void spaces because diffusion in solids reflects the movement within voids of molecular dimensions.

The results of studies of model solids are consistent with both hypotheses. For example, artificial beads containing nanopores with hydrophobic surfaces almost completely reduce the bacterial biodegradation of phenanthrene, although beads with pores not hydrophobically coated or beads with hydrophobic coatings but no pores have no effect, indicating a specific need for pores with hydrophobic coatings to prevent metabolism of a hydrophobic substrate (Nam and Alexander,

1998). Alternatively, phenanthrene present in a number of nonporous organic solids is, as expected, unavailable to bacteria (Hatzinger and Alexander, 1997). Although the former data are consistent with the intraparticle diffusion hypothesis and the latter observations are consistent with the intraorganic matter diffusion hypothesis, the results provide no direct proof for either hypothesis.

COMPLEXED SUBSTRATES

Many pesticides are reported to be converted to "bound residues" in soil. However, the very definition of bound residues serves to obscure chemical interpretation. Indeed, a common procedure for measuring the quantity of a pesticide converted to bound residue—namely, measuring the amount of ^{14}C remaining in soil amended with ^{14}C-labeled pesticide after incubation and then removing any solvent-extractable ^{14}C—shows the chemical inadequacy of such a concept of bound residues, since the microbial biomass, polymers formed enzymatically from the parent molecule, and strongly sorbed products of microbial metabolism would all be considered as bound residues. The definition is merely a description of the results of an analytical procedure and is of no use in describing chemical behavior or toxicological significance.

Nevertheless, it is evident that a portion—whether large or small is often uncertain—of the fraction termed "bound residues" represents new molecular species formed from the parent compound or from structurally similar products derived from it. These new molecules are often complexes of the parent or a metabolite with organic constituents of soils, and probably of sediments and subsoils as well. That the new molecules are derived from the parent compounds or sometimes structurally related metabolites is evident from the finding that hydrolysis of soil with strong alkali or acid yields the parent molecules or related metabolites (Fuchsbichler *et al.,* 1978; Hsu and Bartha, 1976; Singh and Agarwal, 1992). To be chemically meaningful and also in light of recent information on sequestration, an appropriate definition of bound residue is the residue of a compound that is not extracted with water or by vigorous treatment with organic solvents or aqueous–organic solvent mixtures but which yields the original compound or structurally similar products derived from it on hydrolysis with strong alkali or acid. Much of the complexing is with the organic or humus fraction of soils and probably usually involves the formation of covalent bonds between the compound and the humic material. The organic fraction of soil is frequently subdivided into three complex, heterogeneous mixtures known as humin, fulvic acid, and humic acid. These represent portions that are (a) not soluble in the dilute alkali used for the initial extraction, (b) soluble in dilute base and weak acid, and (c) soluble in base but not acid, respectively. Because they can be thus brought into solution, more chemistry has been performed on the fulvic and humic acid fractions (Kim *et al.*, 1997; Li *et al.*, 1997). Because of their poorly characterized structure, the heterogeneity of these three fractions, and their high molecular weights, the identities and structures of the complexes have been difficult to establish. Most of the complexes that have been studied have these characteristics: they are strongly associated with the humic materials, the stable linkages probably being a result of covalent bonding between the original chemical and the humic polymers; they are not readily extractable with organic solvents; part or sometimes all of the complexes may be cleaved with strong acid or alkali; and they are reasonably resistant to microbial degradation and persist for long periods.

The complexes probably are formed either by an attachment of the compound to reactive sites on the surfaces of the organic colloids or by incorporation of the compound into the structure of humic and fulvic acids that are being formed microbiologically (Stevenson, 1976). Much of the research has been done on amines (or on nitro compounds that presumably are reduced microbiologically to amines before the complex is formed), phenols, or quinones. The amines probably react with polyphenols or quinones in humus, and the phenols or catechols probably complex with amino groups of the humic materials (Bartha *et al.*, 1983; Calderbank, 1989). Recent studies have dealt with the complexing of aniline or a N-containing herbicide, with data indicating that the formation of the bound residue results, at least in part, from a binding of the N in the molecule to quinones or other carbonyl groups (Kim *et al.*, 1997; Thorn *et al.*, 1996). Such reactions do not account for the many complexes formed from molecules with neither amino nor hydroxyl groups. The complexes may have molecular weights ranging from 2100 to 100,000 (Meikle *et al.*, 1976).

Organic pollutants may also interact or complex with soluble humic substances or high-molecular-weight substances in aquatic environments. Such complexing has been reported for TNT (Held *et al.*, 1997), DDT (Singh and Agarwal, 1995), PCBs, and PAHs (Eadie *et al.*, 1990; Jota and Hassett, 1991) as well as other compounds. Complexes formed between test chemicals and such water-soluble humic substances may have molecular weights of 700–5000 (Madhun *et al.*, 1986). It is also likely that these complexes are resistant to biodegradation, although this has not been investigated.

In some and possibly many instances, complex formation involves microbial metabolism to a greater or lesser degree. For example, because the presence of amino or hydroxyl groups is necessary for the formation of some complexes, microbial reduction of the nitro moiety of a molecule to an amino group or the hydroxylation of an aromatic ring may be a necessary first step. A comparison of the formation of complexes generated from parathion in sterile and nonsterile soil provides more direct evidence for a key role of microorganisms, since sterilization markedly reduces the magnitude of complex formation (Katan *et al.*, 1976; Katan and Lichtenstein, 1977). Although the conversion of most organic pollutants to simple products is probably chiefly an intracellular process, it is possible that the biosynthesis of complexes results in part from extracellular enzymes. This possibility is suggested by the observation that phenol oxidase of the fungus *Rhizoctonia solani* catalyzes reactions that lead to the formation of a series of oligomers from mixtures containing chloroanilines or 2,4-dichlorophenol and hydroxylated aromatic acids (Bollag *et al.,* 1980, 1983).

The compounds contained in the complexes are far less readily degraded than the free molecules. Under laboratory or field conditions, their prolonged persistence is immediately and strikingly evident. The bonds associated with creating the complexes therefore are obviously not readily cleaved enzymatically, since such cleavage would yield the parent molecules, which are usually metabolized at reasonable rates. The resistance is evident from data showing that only 1% of the ^{14}C from labeled dinitroaniline complexes is converted to $^{14}CO_2$ in soil in 21 weeks (Helling, 1976) and that only 1.2–10% of the added ^{14}C of a series of chlorophenol complexes is mineralized in 13 weeks (Dec and Bollag, 1988). An atrazine complex is even found in soil 9 years after this herbicide was applied in the field (Capriel *et al.,* 1985). Nevertheless, the complexes are slowly destroyed, and mineralization tests with individual bacteria or fungi show a slow release of $^{14}CO_2$ from ^{14}C-labeled substrates converted to complexes in soil or in liquid culture.

TOXICOLOGICAL SIGNIFICANCE

A pollutant that becomes sequestered as a result of aging is not readily accessible to microorganisms or to sensitive animals and plants. Nevertheless, it is regulated as if it were fully available because governmental agencies assess exposure and risk based on the results of analyses using vigorous extractants. Obviously, data from such analyses, from the toxicological viewpoint, are misleading. Similarly, the amount of a pollutant remaining after bioremediation may be largely in a sequestered state, the microorganisms having destroyed much of the bioavailable form of the pollutant. Analyses based on vigorous extractions, in this instance too, may thus also be misleading, and a bioremediation technology initially deemed to have failed to reduce the concentration to an acceptable level from the viewpoint of risk may, in fact, have fully accomplished its purpose.

Because bound residues are formed from molecules that are acknowledged pollutants, some are pesticides, and others probably affect animals, plants, or microorganisms, considerable attention has been devoted to assessing whether bound residues are cleaved in nature to give detectable levels of the original compound and whether the complexes are assimilated by animals and plants. Are they problems of present or future toxicological importance? From a practical viewpoint, no convincing answer can be given, although governmental agencies typically do not regulate them, viewing the complexes as nontoxic and as yielding the parent molecules in concentrations too low to represent a risk. Whether one can be confident of the absence of an environmental concern is questionable because at least some complexes can be assimilated and some converted from a form not extractable with organic solvents to a solvent-extractable state. For example, plants assimilate and translocate butralin complexes formed in soil (Helling and Krivonak, 1978), and both earthworms and oat plants not only assimilate complexes of parathion formed in soil but modify them to yield products that are soluble in organic solvents and even in water and thus have simpler structures (Racke and Lichtenstein, 1985). To obtain more convincing evidence of the risks, or lack thereof, will require further investigation.

Because bound residues have little or no toxicity, it has been proposed that their formation should be deliberately promoted. This would cause a major reduction in the risk from the toxic parent compound. Such an approach to remediation, although promising, has not yet been exploited from a practical standpoint.

REFERENCES

Adu, J. K., and Oades, J. M., *Soil Biol. Biochem.* **10,** 109–115 (1978).

Alexander, M., *Environ. Sci. Technol.* **29,** 2713–2717 (1995).

Alexander, M., *in* "Environmentally Acceptable Endpoints in Soil" (D. G. Linz and D. V. Nakles, eds.), pp. 43–136. American Academy of Environmental Engineers, Annapolis, MD, 1997.

Ball, W. P., Buehler, C., Harmon, T. C., MacKay, D. M., and Roberts, P. V., *J. Contam. Hydrol.* **5,** 253–295 (1990).

Ball, W. P., and Roberts, P. V., *Environ. Sci. Technol.* **25,** 1223–1237. (1991a).

Ball, W. P., and Roberts, P. V., *Environ. Sci. Technol.* **25,** 1237–1249 (1991b).

Bartha, R., You, I.-S., and Saxena, A., *in* "Pesticide Chemistry: Human Welfare and the Environment" (S. Matsunaka, D. H. Hutson, and S. D. Murphy, eds.), Vol. 3, pp. 345–350. Pergamon, Oxford, 1983.

Beurskens, J. E. M., Dekker, C. G. C., Jonkhoff, J., and Pompstra, L., *Biogeochemistry* **19,** 61–81 (1993).

Bollag, J.-M., Liu, S.-Y., and Minard, R. D., *Soil Sci. Soc. Am. J.* **44,** 52–56 (1980).

Bollag, J.-M., Minard, R. D., and Liu, S.-Y., *Environ. Sci. Technol.* **17,** 72–80 (1983).

Brusseau, M. L., Jessup, R. E., and Rao, P. S. C., *Environ. Sci. Technol.* **25,** 134–142 (1991).

Brusseau, M. L., and Rao, P. S. C., *in* "Rates of Soil Chemical Processes" (D. L. Sparks and D. L. Suarez, eds.), pp. 281–302. Soil Science Society of America, Madison, WI, 1991.

Calderbank, A., *Rev. Environ. Contam. Toxicol.* **108,** 71–103 (1989).

Capriel, P., Haisch, A., and Khan, S. U., *J. Agric. Food Chem.* **33,** 567–569 (1985).

Chiba, M., and Morley, M. V., *J. Agric. Food Chem.* **16,** 916–922 (1968).

Chiou, C. T., *in* "Reactions and Movement of Organic Chemicals in Soils" (B. L. Sawhney and K. Brown, eds.), pp. 1–29. Soil Science Society of America, Madison, WI, 1989.

Chung, N., and Alexander, M., *Environ. Sci. Technol.* **32,** 855–860 (1998).

Dec, J., and Bollag, J.-M., *Soil Sci. Soc. Am. J.* **52,** 1366–1371 (1988).

Eadie, B. J., Morehead, N. R., and Landrum, P. F., *Chemosphere* **20,** 161–178 (1990).

Elgar, K. E., *Environ. Qual. Saf.* **3** Suppl., 250–257 (1975).

Fisher, L., Zur Mühlen, E., Brümmer, G. W., and Niehus, H., *Eur. J. Soil Sci.* **47,** 329–334 (1996).

Fuchsbichler, G., Süss, A., and Wallnöfer, P., *Z. Pflanzenkr. Pflanzenschutz* **85,** 724–734 (1978).

Grace, J. K., Yates, J. R., Tamashiro, M., and Yamamoto, R. T., *J. Econ. Entomol.* **86,** 761–766 (1993).

Haddock, J. D., Landrum, P. F., and Glesy, J. P., *Anal. Chem.* **55,** 1197–1200 (1983).

Hassink, J., Bouwman, L. A., Zwart, K. B., and Brussard, L., *Soil Biol. Biochem.* **25,** 47–55 (1993).
Hatzinger, P. B., and Alexander, M., *Environ. Sci. Technol.* **29,** 537–545 (1995).
Hatzinger, P. B., and Alexander, M., *Environ. Toxicol. Chem.* **16,** 2215–2221 (1997).
Held, T., Draude, G., Schmidt, F. R. J., Brokamp, A., and Reis, K. H., *Environ. Technol.* **18,** 479–487 (1997).
Helling, C. S., *in* "Bound and Conjugated Pesticide Residues" (D. D. Kaufman, G. G. Still, G. D. Paulson, and S. K. Bandal, eds.), pp. 366–367. American Chemical Society, Washington, DC, 1976.
Helling, C. S., and Krivonak, A. E., *J. Agric. Food Chem.* **26,** 1164–1172 (1978).
Hsu, T.-S., and Bartha, R., *J. Agric. Food Chem.* **24,** 118–122 (1976).
Johnson, R. E., and Starr, R. I., *J. Econ. Entomol.* **60,** 1679–1682 (1967).
Jota, M. A. T., and Hassett, J. P., *Environ Toxicol. Chem.* **10,** 483–491 (1991).
Katan, J., and Lichtenstein, E. P., *J. Agric. Food Chem.* **25,** 1404–1408 (1977).
Katan, J., Fuhremann, T. W., and Lichtenstein, E. P., *Science* **193,** 891–894 (1976).
Kelsey, J. W., and Alexander, M., *Environ. Toxicol. Chem.* **16,** 582–585 (1997).
Kelsey, J. W., Kottler, B. D., and Alexander, M., *Environ. Sci. Technol.* **31,** 214–217 (1997).
Kim, J.-E., Fernandes, E., and Bollag, J.-M., *Environ. Sci. Technol.* **31,** 2392–2398 (1997).
Li, A. Z., Marx, K. A., Walker, J., and Kaplan, D. L., *Environ. Sci. Technol.* **31,** 584–589 (1997).
Lichtenstein, E. P., De Pew, L. J., Eshbaugh, E. L., and Sleesman, J. P., *J. Econ. Entomol.* **53,** 136–142 (1960).
Loehr, R. C., and Webster, M. T., *J. Soil Contam.* **5,** 361–383 (1996).
Loehr, R. C., and Webster, M. T., *in* "Environmentally Acceptable Endpoints in Soil" (D. G. Linz and D. V. Nakles, eds.), pp. 137–386. American Academy of Environmental Engineers, Annapolis, MD, 1997.
Madhun, Y. A., Young, J. L., and Freed, V. H., *J. Environ. Qual.* **15,** 64–68 (1986).
Mayer, L. M., *Chem. Geol.* **114,** 347–363 (1994a).
Mayer, L. M., *Geochim. Cosmochim. Acta* **58,** 1271–1284 (1994b)
Meikle, R. W., Regoli, A. J., Kurihara, N. H., and Laskowski, D. A., *in* "Bound and Conjugated Pesticide Residues" (D. D. Kaufman, G. G. Still, G. D. Paulson, and S. K. Bandal, eds.), pp. 272–284. American Chemical Society, Washington, DC, 1976.
Nam, K., and Alexander, M., *Environ. Sci. Technol.* **32,** 71–74 (1998).
Nash, R. G., and Woolson, E. A., *Science* **157,** 924–927 (1967).
Newman, A. C. D., and Thomasson, A. J., *J. Soil Sci.* **30,** 415–439 (1979).
Onsager, J. A., Rusk, H. W., and Butler, L. I., *J. Econ. Entomol.* **63,** 1143–1146 (1970).
Ou, L.-T., and Alexander, M., *Soil Sci.* **118,** 164–167 (1974).
Priesack, E., *Soil Sci. Soc. Am. J.* **55,** 1227–1230 (1991).
Racke, K. D., and Lichtenstein, E. P., *J. Agric. Food. Chem.* **33,** 938–943 (1985).
Robertson, B. K., and Alexander, M., *Environ. Toxicol. Chem.* **17,** 1034–1038 (1998).
Salkinoja-Salonen, M., Middledorp, P., Briglia, M., Valo, R., Häggblom, M., and McBain, A., *in* "Biotechnology and Biodegradation" (D. Kamely, A. Chakrabarty, and G. S. Omenn, eds.), pp. 347–365. Portfolio Publishing Co., The Woodlands, TX, 1990.
Scow, K. M., and Alexander, M., *Soil Sci. Soc. Am. J.* **56,** 128–134 (1992).
Scow, K. M., and Hutson, J., *Soil Sci. Soc. Am. J.* **56,** 119–127 (1992).
Scribner, S. L., Benzing, T. R., Sua, S., and Boyd, S. A., *J. Environ. Qual.* **21,** 115–120 (1992).
Shu, H. D., Pautenbach, D., Murray, F. J., Maple, L., Brunk, B., Dei Rossi, D., Webb, A. S., and Teitelbaum, P. *Fundam. Appl. Toxicol.* **10,** 648–654 (1988).
Singh, D. K., and Agarwal, H. C., *J. Agric. Food Chem.* **40,** 1713–1716 (1992).
Singh, D. K., and Agarwal, H. C., *Environ. Sci. Technol.* **29,** 2301–2304 (1995).
Steinberg, S. M., Pignatello, J. J., and Sawhney, B. L., *Environ. Sci. Technol.* **21,** 1201–1208 (1987).
Stevenson, F. J., *in* "Bound and Conjugated Pesticide Residues" (D. D. Kaufman, G. G. Still, G. D. Paulson, and S. K. Bandal, eds.), pp. 180–207. American Chemical Society, Washington, DC, 1976.

Tang, J., Carroguino, M. J., Robertson, B. K., and Alexander, M., *Environ. Sci. Technol.* **32,** 3586–3590 (1998).

Thorn, K. A., Pettigrew, P. J., and Goldenberg, W. S., *Environ. Sci. Technol.* **30,** 2764–2775 (1996).

Tisdall, J. M., and Oades, J. M., *J. Soil Sci.* **33,** 141–163 (1982).

Weissenfels, W. D., Klewer, H. J., and Langhoff, J., *Appl. Microbiol. Biotechnol.* **36,** 689–696 (1992).

White, J. C., Kelsey, J. W., Hatzinger, P. B., and Alexander, M., *Environ. Toxicol. Chem.* **16,** 2040–2045 (1997).

White, T. E., and Herndon, F. G., *in* "Proc. 50th Ind. Waste Conf. Purdue Univ.," pp. 41–48, 1995.

Wu, S.-C., and Gschwend, P. M. *Environ. Sci. Technol.* **20,** 717–725 (1986).

CHAPTER 11

Effect of Chemical Structure on Biodegradation

Many practical reasons exist for the need to predict whether a particular compound will be biodegradable and, if biodegraded, what products will be formed. First, in developing new compounds for industrial use, it is important to know whether a compound or class of chemicals will persist or be converted to toxic products in nature and whether there may be potential health or ecological problems. The synthesis, evaluation of efficacy of chemicals for their targeted purpose, and toxicological assessment are expensive, and the industrial researcher would like to avoid the expense of these various activities if it is likely that the substance to be developed will persist or be converted to undesirable intermediates. Second, in seeking the least costly but yet effective means of ridding the area of the problem chemicals, the manager of a polluted site would like to be reasonably certain that bioremediation will work and that the cleanup by this relatively inexpensive technology would generate only nontoxic products; if not, more costly technologies would have to be employed. Third, in countries in which regulatory agencies must give their approval before the commercial production of new chemicals begins, particularly those chemicals that ultimately will enter soils, waters, sewage, or aquifers, decisions are difficult if it is not possible to anticipate whether the chemical of interest will persist and whether it will be transformed to toxicants that may pose a hazard to humans, animals, or plants. Such predictions are made difficult because of the millions of organic compounds that are known, the many different classes of chemicals, and the lack of attention paid in microbiology, biochemistry, and toxicology to all but a few of these classes.

Predicting whether a compound is *biodegradable* is not the same as predicting whether it will be *biodegraded*. For a compound that can be metabolized by microorganisms to actually be transformed, appropriate microorganisms must be present at the site, requisite inorganic nutrients (and possibly growth factors) must be present, the compound must be in a bioavailable form (as sorption, sequestration, or presence in a NAPL may render it largely unavailable), the site must not have toxic substances inimical to microbial growth and activity, and the concentration must be above the threshold level if that molecule is to be acted on by populations using it as a C and energy source.

Shortly following the beginning of the widespread use of pesticides and detergents, it was recognized that members of individual classes of organic compounds had markedly different periods of persistence in soils and waters. In many instances, a slight modification in the structure of the molecule was found to make it considerably more or less susceptible to destruction in these environments. Because it was soon evident that the compounds with short lives were destroyed microbiologically, the conclusion was reached that these modest alterations in chemical structure changed the suitability of the molecules as substrates for growth or metabolism by the resident community of microorganisms. An apparently modest alteration in the molecule, for example, the substitution of one atom or substituent for another, rendered the molecule appreciably more or less susceptible to microbial metabolism. Because of the enormous economic importance of these pesticides and surfactants and the growing concern with their contribution to environmental deterioration if they persisted, considerable effort was directed to establishing the structural features that governed the suitability of these chemicals for microbial degradation. Since that time, studies have been conducted on a wide variety of other chemical classes, including molecules that have different uses, and these have provided a large literature on the relationship between structure and biodegradation.

The choices of compounds for these investigations have rarely been made to establish generalizations or scientific principles relative to the underlying mechanisms, but rather have fit in with the needs of the particular industry involved. Thus, a large base of information exists on certain pesticides and surfactants that are major components of widely used detergents, but only a limited number of compounds of other classes have been investigated. As a result, the types of compounds that might be selected to establish more widely useful generalizations have not been tested, and the generalizations that are possible are still only a few in number.

In the two major industries (those concerned with pesticides and surfactants) initially concerned with the relationship between chemical structure and biodegradation, the compelling reason to conduct research was to replace the more persistent chemicals with new but often structurally similar molecules that were more readily metabolized. The public outcry against surfactants that persisted for long periods in water and the concern with the ecological or health consequences of the highly persistent pesticides motivated industry and, in many instances, regulatory agencies of government to seek replacements. The continued use of these classes of chemicals was threatened because of their longevity in waters and soils, and this longevity was specifically associated with characteristics of the molecules that made them less suitable for microbial metabolism and growth. Their replacement with new compounds is witness to the success in the search for more readily decomposable compounds.

The approach to finding biodegradable replacements for persistent but efficacious compounds largely remains one of trial and error. Such trial-and-error approaches are characteristic of the industrial research focused on pesticides, surfactants, the detergent builders that accompany the surfactants, polymers, and other classes of materials. Such approaches are expensive and time-consuming, and they often fail. A far more suitable approach would rely on basic principles that explain the relationships between structure and biodegradability. Unfortunately, the information base for establishing meaningful relationships is still quite small. The underlying issue is the specificity of microorganisms for their substrates, a specificity that is linked, in large part, with the specificity of enzymes to catalyze only certain types of chemical reactions. Each enzyme is largely restricted to carrying out only a single type of reaction on a narrow and often unpredictable range of substrates of very similar structures.

GENERALIZATIONS

At the present time, only a few generalizations are possible on the influence of structure on biodegradation, and exceptions to these generalizations or to other generalizations that might be made are many. Several reasons may be advanced for the few generalizations and the many exceptions. (a) Different microorganisms are present in dissimilar environments, and the development of one population may result in one group of related chemicals being degraded in the first environment, but because of the proliferation of other organisms in a different habitat, another set of compounds may be destroyed in the second environment. (b) Structural features of organic substrates often alter their availability to microorganisms (e.g., by sorption or by partitioning into a NAPL), and thus a molecule of one structure may be readily degraded in environments in which it is freely available but will persist where its bioavailability is low. (c) In those instances in which a lengthy acclimation period occurs before biodegradation is detected, it is likely that a somewhat unique population appears for the degradation; those populations arising as a result of acclimation probably will not be the same in dissimilar environments. (d) The physical or chemical characteristics of two environments are quite different, for example, because one is aerobic and another anaerobic or one is at low pH and another is at neutral pH, and the populations that assume dominance in the transformation will not only often be different but will rely on different enzymes. These dissimilar enzymes will likely act on different members of groups of closely related chemical structures.

A few of these difficulties are well illustrated in studies of cultures of individual bacterial species. Investigations of several organisms that are able to utilize the same organic substrate as a C source for growth show that they proliferate by using or cometabolizing a somewhat different range of chemicals of an individual class of substrates. Some of the molecules are metabolized by one species but not a second, and the second organism will use a few but not others that support growth or metabolism of the first species. Such investigations establish the catabolic potential of an isolate and provide more definitive answers on the effect of chemical structure on microbial utilization than would be obtained in studies of natural environments that contain a multitude of species with dissimilar degradative potentials. Nevertheless, generalizations derived from studies of individual microorganisms suffer from the fact that they may not apply to an environment where the tested species is not present and where a population with an entirely different range of substrates assumes dominance in a particular biodegradation. Individual organisms have their physiological and catabolic idiosyncrasies, and the idiosyncracies may not be related to the intrinsic resistance of chemicals to biodegradation. Because microbial strains, species, and genera have enzymes with dissimilar substrate specificities and probably different cell permeabilities, it is more difficult to establish generalizations than in chemistry, in which the role of structure on chemical reactivity is also addressed. For the present, it is prudent to assume that susceptibility to biodegradation is an attribute of a chemical class in a particular ecosystem with particularly important environmental variables, including O_2 status, and that the biochemical potential of the entire community, and not just individual species, needs to be assessed.

The difficulty in making generalizations is particularly evident among chemicals that are quite persistent and then, after a long acclimation period, suddenly disappear. In these instances, probably no organism was initially able to grow rapidly on the compound. However, after some time, a rare organism becomes prominent or a genetic change occurs in one of the indigenous species such that it is now able to metabolize the chemical. That newly emerging population then acts on a range of chemicals that otherwise would have appeared to be persistent in tests of the original environment.

REASONS FOR PERSISTENCE

Life appeared on earth hundreds of millions of years ago, and the biochemistry of living organisms has evolved in a limited number of ways. Comparisons of ancient fossils with modern organisms disclose that the chemistry of cells has not changed drastically, suggesting that only some of the countless reactions possible of organic molecules have been exploited in the processes necessary for metabolism. The resulting anabolic, and presumably catabolic, processes are thus few in number. Many possible chemical reactions are thus foreign to macro- and microorganisms. All the catabolic reactions that characterize living cells are consistent with chemical principles, but because of the few reactions that can be catalyzed by the enzymes that have appeared during the course of biochemical evolution as well as the very complexity of enzymes as catalysts and microorganisms as integrated assemblages of catalysts, the reactions that microorganisms effect may not be those that would be expected based on the chemical principles established for the same organic molecules whose change is brought about by nonbiological agencies.

Given the relatively few catabolic pathways that characterize microbial cells, it is not surprising that an organic chemical that is not a product of biosynthesis, sometimes termed a *xenobiotic,* will be degraded to an *appreciable extent* only if an enzyme or an enzyme system exists that is able to catalyze its conversion to a product that is an intermediate or a substrate of one of those pathways. (In contrast, a chemical may be modified to a *slight extent,* i.e., cometabolized, but the one or few enzymes involved convert the substrate to a product having many of the features of the parent molecule.) The greater the difference in structure of the xenobiotic from the constituents of living organisms or the less common the substituent in living matter, the less the likelihood of extensive biodegradation or the slower the transformation.

Furthermore, if nearly all microorganisms able to metabolize a chemical do so by one or a few similar metabolic pathways, then modification of the molecule to render it somewhat different from the intermediates or substrates of those pathways probably will prevent, slow, or delay the initiation of the biodegradation. If the altered molecule can be modified enzymatically to yield a natural intermediate, then the degradation will proceed—although the process may be slow if the organism that converts the xenobiotic to a natural intermediate grows slowly when that molecule is the C source. However, biodegradation will be evident only after a long period if (a) the microorganism that can convert the xenobiotic is present initially at low cell densities and must proliferate before a significant loss of the parent compound is detected, (b) a mutant must appear, or (c) the organism that destroys the chemical does so by ignoring the change that makes the natural product into a xenobiotic, that is, acts on a portion of the molecule not affected by the change.

Organic compounds that are mineralizable may be rendered partially or possibly wholly resistant to mineralization by the addition of a single substituent. These substituents may be termed *xenophores,* that is, substituents that are physiologically uncommon or that are entirely nonphysiological. Thus, because addition of a single Cl, NO_2, SO_3H, Br, CN, or CF_3 to simple aromatic molecules, fatty acids, or other readily utilizable substrates greatly increases their resistance, they are xenophores. These substituents are alien to most organisms, and hence their removal is not effeced by many species, or their presence hinders the functioning of otherwise common pathways. Sometimes CH_3, NH_2, OH, and OCH_3 act as xenophores. Often the degradation of organic compounds is stimulated by the presence of OH, COOH, or an amide, ester, or anhydride functional group.

Stated in other terms, the identity of the substituent added to a molecule affects its biodegradation, or at least its mineralization. This is evident by the findings in Table 11.1 for a group of compounds. The resistance associated with the presence of halogens and NO_2 is widespread, but only a few compounds containing CN or CF_3 have been tested. The impact of the substituent will vary, however, with the structure of the rest of the molecule, witness that some compounds with NO_2 (e.g., 4-nitrophenol), SO_3H (e.g., some surfactants), and CN groups are quickly destroyed.

The presence of several potential xenophores on a molecule makes it less likely for a xenobiotic to be transformed to intermediates in normal metabolic

Table 11.1

Substituents Whose Addition Slows Extensive Aerobic Biodegradation of Organic Compounds

Test chemical	Substituent slowing degradation[a]	Test environment	Reference
Aniline	3-Br, 3-CH_3, 3-NO_2, 3-CN, 3-OCH_3	Water	Paris and Wolfe (1987)
Aniline	4-Cl	Soil	Süss *et al.* (1978)
Benzene	NO_2, SO_3H[b]	Soil suspension	Alexander and Lustigman (1966)
Benzoic acid	2-, 3-, or 4-NO_2; 2-, 3-, or 4-Cl	Wastewater	Haller (1978)
Cyanuric acid	NH_2	Soil	Hauck and Stephenson (1964)
Diphenylmethane	3-CF_3	Water	Saeger and Thompson (1980)
IPC	3-Cl	Soil	Clark and Wright (1970)
Phenylacetic acid	4-NO_2, 4-Cl	Soil suspension	Subba-Rao and Alexander (1977)
Pyridine	2-, 3-, or 4-NH_2	Soil suspension	Sims and Sommers (1986)
Pyridine	2-Cl; 2-, 3-, or 4-OH	Soil suspension	Naik *et al.* (1972)
Valeric acid	3-CH_3	Soil suspension	Hammond and Alexander (1972)

[a] The number designates the position of the substituent.
[b] Compared to phenol.

pathways or results in slower rates of transformation; hence, they are more resistant to biodegradation. Typically, the addition of two xenophores, whether they be identical or different, to a biodegradable molecule makes it even less likely for rapid degradation to occur or results in a still longer acclimation period before a population of sufficient size develops and causes rapid degradation. Addition of a third xenophore, the same or a different one, renders the molecule even more resistant or results in the need for a still longer acclimation period for aerobic biodegradation. Some typical examples in which the added substituents are Cl, NH_2, NO_2, OH, or CH_3 are shown in Table 11.2. Entries in the third column indicate whether the molecule is mono-, di-, or trisubstituted with the xenophore or the position of the substituent on the otherwise readily degradable substrate. Such retarding effects by increasing the number of substituents may not occur under anaerobic conditions, at least not for chlorophenols (Mikesell and Boyd, 1985) and chlorophenoxyacetic acids (Suflita *et al.*, 1984).

Table 11.2

Effect of Several Xenophores on Extensive Aerobic Biodegradation of Organic Compounds

Test chemical	Substituent	Degradation rate	Test environment	Reference
Acetic acid	Cl	Mono > di > tri	Soil	Kaufman (1966)
Aniline	NH_2	4- > 3,4-	Soil	Süss *et al.* (1978)
Benzoic acid	Cl	Mono > 2,4-	Sewage	DiGeronimo *et al.*, (1979)
Benzoic acid	NO_2	3-, 4- > 3,5-	Sewage	Hallas and Alexander (1983)
Cyanuric acid	NH_2[a]	Mono > di > tri	Soil	Hauck and Stephenson (1964)
Diphenylmethane	OH	4- > 4,4′-	Soil suspension	Subba-Rao and Alexander (1977)
Fatty acid	Cl	Mono > di	Sewage	Dias and Alexander (1971)
Fatty acid	CH_3	Mono > di	Soil suspension	Hammond and Alexander (1972)
IPC	Cl[b]	4- > 2,4- > 2,4,5-	Soil	Kaufman (1966)
Phenoxyacetic acid	Cl	2,4- > 2,4,5-	Soil	Burger *et al.* (1962)
Propionic acid	Cl	2- > 2,2- > 2,3,3-	Soil	Kaufman (1966)
Pyridine	OH	Mono > di	Soil suspension	Sims and Sommers (1986)

[a] NH_2 replaces the OH of cyanuric acid.
[b] Cl on the ring.

The position of the xenophore on the molecule has a pronounced influence on the degradation. At some positions, it may have little impact; in others, it may drastically reduce the rate of microbial utilization. Because different environments contain dissimilar populations, the effect on biodegradation of the position of the substituent may not be the same in all localities. This is shown by data in Table 11.3, from which it is evident that a substituent in one position may enhance degradation in one environment and depress it in another. Moreover, one xenophore in a given position on the molecule may have a different effect than another in the same location. Thus, generalizations on the effect of position of substituents do not appear to be applicable to all environments. That the effect is dependent on the idiosyncracies of the particular populations that appear is suggested by the finding that some bacteria readily degrade di-, tri-, tetra-, and pentachlorophenols (Apajalahti and Salkinoja-Salonen, 1986), yet the populations in most environments do not act readily on polychlorinated phenols. Thus, if conditions favor development of a population that behaves atypically, as may occur because of prior additions of specific compounds, the populations in that environment will degrade substituted compounds at different relative rates than in some other environment in which that type of population is not favored.

An early, highly dramatic, and widely recognized effect of structure on biodegradation was evident when synthetic detergents first became widely used. The lakes and rivers into which the washwaters were introduced showed an obvious failing of the microflora: they were covered with froth and foam. The reason: those early surfactants in the detergent preparations were not quickly destroyed. The ensuing public outcry compelled industry to seek the cause for the persistence, and industrial researchers quickly found that extensive methyl branching on the alkyl moiety of the surfactant created obstacles for the mineralizing populations.

These early detergents contained alkylbenzene sulfonates (ABSs) as the surfactant constituents (Fig. 11.1). The sulfonate was at different positions on the benzene ring, but the alkyl portion invariably contained many methyl branches. This large number of methyl groups on the alkyl moiety was responsible for the longevity of the surfactants in surface waters and in sewage-treatment facilities. A single methyl group usually had little or no noticeable influence among the many tested chemicals that had alkyl chains of various lengths, but many methyl branches on the alkyl portion rendered the molecule less readily biodegradable. Moreover, if these ABSs had two methyl groups on the penultimate C, making it into a quaternary C (Fig. 11.1), the compound showed considerable resistance. In the face of widespread public pressure and the threat of government regulation, the soap and detergent industry quickly not only found the structures that were refractory but also provided replacements. These readily degraded alternatives did not have alkyl moieties with many methyl branches but rather were linear [$RCH_2CH_2(CH_2)_n$-CH_2CH_3] and did not contain the quaternary C atoms that were the culprits (Huddleston and Allred, 1967; Swisher, 1987).

Table 11.3

Effect of Position of Substituent on Biodegradation

Compound	Added substituent	Effect of substituent position on biodegradation		Environment	Reference
		Rapid	Slow		
Phenol	Cl	Not *meta*	*meta*	Soil suspension	Alexander and Aleem (1961)
	Cl	Not *meta*	*meta*	Soil	Baker and Mayfield (1980)
	Cl	2–, 3–	4–	Wastewater	Haller (1978)
	Cl	2–	4–	Soil	Boyd *et al.* (1983)
	Cl	2–	4–	Sludge	Mikesell and Boyd (1985)
Benzoic acid	Cl	3–	2–, 4–	Wastewater	Haller (1978)
	Cl	3,4–	2,4–	Sewage	DiGeronimo *et al.* (1979)
Aliphatic acid	Halogen	ω–	α–, β–	Sewage	Dias and Alexander (1971)
Phenol	CH_3	2–	4–	Soil	Boyd *et al.* (1983)
	CH_3	4–	3–, 2–	Soil	Smolenski and Suflita (1987)
Phenoxyalkanoic acid	Cl	4–	3–	Soil	Burger *et al.* (1962)
	Cl	4–	2–	Soil	Audus (1960)
Aliphatic acid	Phenoxy	ω–	α–	Soil	Burger *et al.* (1962)
Benzamide	Cl	3,6–	2,6–	Soil	Fournier (1974)
Diphenylmethane	Cl	2,4–	2,5–	Sludge	Saeger and Thompson (1980)

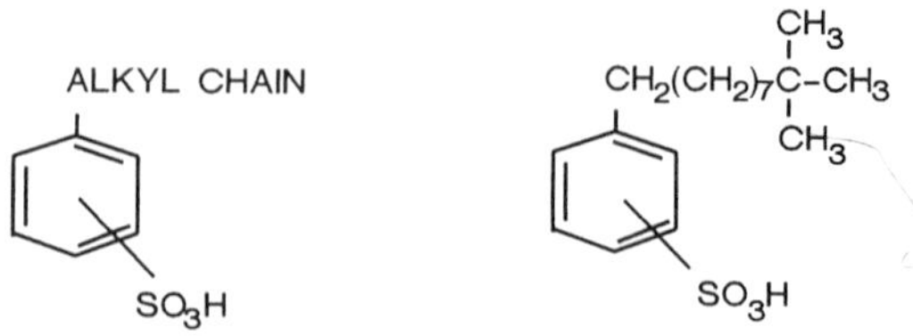

Figure 11.1 ABS and ABS with a quaternary C.

Methyl branching is also associated with persistence of aliphatic hydrocarbons, aliphatic acids, alcohols, and other chemicals. Among the alkanes that are major components of oil, the nonbranched molecules entering soils and waters usually are destroyed more readily than those alkanes having several or many methyl branches on the chain. Among the more resistant of these hydrocarbons are the highly branched phytane and pristane (Bossert and Bartha, 1984; Jobson *et al.*, 1974; Kator, 1973):

$$CH_3CH(CH_3)(CH_2)_3CH(CH_3)(CH_2)_3CH(CH_3)(CH_2)_3CH(CH_3)CH_3 \quad \text{Pristane}$$

$$CH_3CH(CH_3)(CH_2)_3CH(CH_3)(CH_2)_3CH(CH_3)(CH_2)_3CH(CH_3)CH_2CH_3 \quad \text{Phytane}$$

Similarly, mono- and dicarboxylic acids with no methyl branches are quickly degraded by soil microorganisms. These molecules have the structures

$$CH_3(CH_2)_nCOOH$$

$$HOOC(CH_2)_nCOOH$$

However, if the two hydrogens on one of the methylene (CH_2) carbons are replaced with two methyl groups, thereby giving a quaternary C, mineralization is markedly retarded (Hammond and Alexander, 1972). In like fashion, linear aliphatic alcohols with the structure

$$CH_3(CH_2)_nCH_2OH$$

are quickly destroyed microbiologically, but alcohols with quaternary carbons tend to be attacked only slowly (Dias and Alexander, 1971; Fukuda and Brannon, 1971; McKinney and Jeris, 1955).

The reason for this influence of structure on biodegradation has been established. Alkanes or compounds, like ABSs, with alkyl side chains are usually metabolized to give the corresponding carboxylic acids:

$$R(CH_2)_nCH_2CH_2CH_3 \rightarrow R(CH_2)_nCH_2CH_2CH_2OH \rightarrow R(CH_2)_nCH_2CH_2CHO \rightarrow R(CH_2)_nCH_2CH_2COOH$$

Alcohols are intermediates in this sequence, and primary linear alcohols [$R(CH_2)_n$-$CH_2CH_2CH_2OH$] in nature appear to be metabolized as depicted in this sequence. The carboxylic acid [$R(CH_2)_nCH_2CH_2COOH$] is then destroyed by a sequence known as β-oxidation. The sequence was so named because the β-carbon is oxidized as shown in Fig. 11.2. The new carboxylic acid thus generated has two carbons fewer than the first one and, in turn, is metabolized in the same fashion to give CH_3COOH and a third carboxylic acid, which has two carbons fewer than its predecessor. The process is repeated until the molecule is converted to simple carboxylic acids, which are readily oxidized. The reaction inside the cell, in fact, involves coenzyme A, but the coenzyme-complexed acids are not found outside the cell.

Consider the α and β carbons in this sequence. In the pathway of conversion, they bear H, OH, or O but no methyl branches. The enzymatic transformation will not proceed if a methyl is present. However, individual methyl groups may be removed enzymatically. If two methyl groups are present on either of the C atoms, the likelihood is small of their both being removed readily. Hence, such a hindrance to β-oxidation operates if the C containing two methyl groups (the quaternary C) is near the end of the chain—either where the COOH is present or where a COOH will be formed from the CH_3 or CH_2OH. The quaternary C apparently does not initially pose a hindrance if it is some distance from the end of the chain. Under these conditions, the chain is shortened, but only until the shortening brings the COOH close to the quaternary C, at which point further degradation is difficult to achieve. This shorter chain bearing the quaternary C then persists (Catelani *et al.*, 1977; Hammond and Alexander, 1972) (Fig. 11.3). With time, however, these compounds probably are destroyed, possibly because of the growth of microorganisms that possess a different mechanism of degrading such compounds, as by the oxidation of the methyl groups attached to the quaternary C.

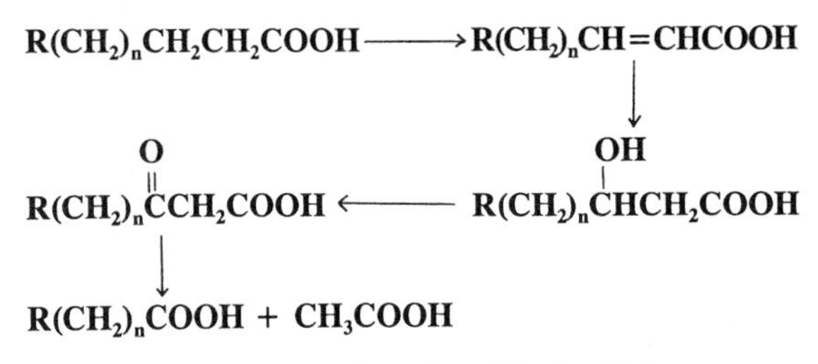

Figure 11.2 Products formed in β-oxidation.

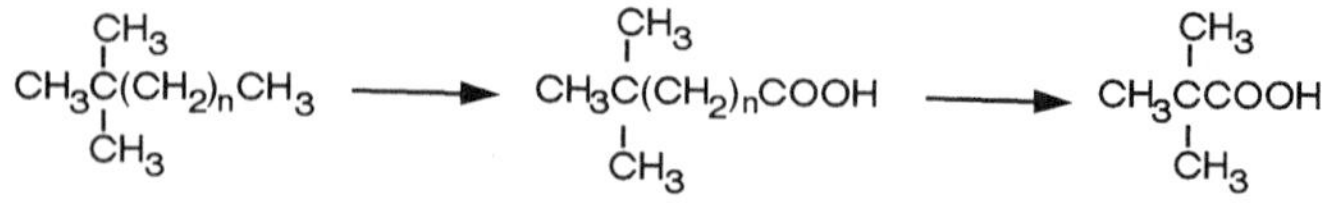

CH3
CH3C(CH2)nCH3
CH3
CH3
CH3C(CH2)nCOOH
CH3
CH3
CH3CCOOH
CH3

Figure 11.3 Conversion of dimethylalkanes and dimethylalkanoic acids to trimethylacetic acid (synonym, pivalic acid).

Another influence of structure on biodegradation is evident among molecules containing both an aromatic and an alkyl or aliphatic acid moiety. This effect is related to the place at which the alkyl or aliphatic acid portion is linked to the benzene ring. Thus, among the phenoxy herbicides in which biodegradation is not delayed because of a *meta* chlorine on the ring, detoxication and loss of the parent molecule are rapid if the phenoxy portion is linked to the last (the ω-position) and not the α C of the aliphatic acid (Burger *et al.*, 1962) (Fig. 11.4). Similarly, 4-phenylbutyric acid is readily destroyed by sewage microorganisms, but 2-phenylbutryic acid is resistant (Dias and Alexander, 1971) (Fig. 11.4). An analogous relationship is evident among some of the ABSs that do not have methyl branches; that is, the isomers with the phenyl linked near the end of the alkyl

RAPIDLY
DEGRADED
SLOWLY
DEGRADED
OCH2(CH2)nCH2COOH
Cl
Cl
CH3(CH2)nCHCOOH
O
Cl
Cl
CH2CH2CH2COOH
CH3CH2CHCOOH
CH2(CH2)10CH3
SO3H
CH3(CH2)3CH(CH2)6CH3
SO3H

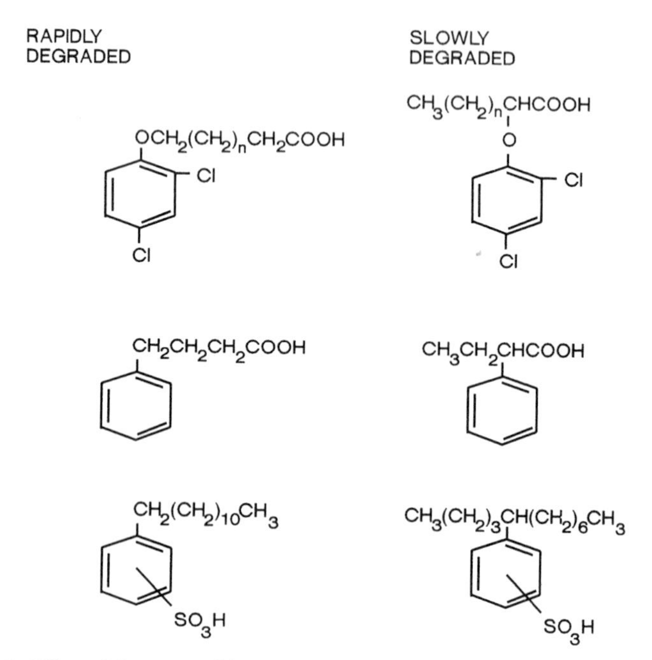

Figure 11.4 Effect of the point of linkage of the aromatic ring to the alkane or aliphatic acid on biodegradation.

moiety are destroyed more quickly than those in which the linkage is with the central portion of the chain (Swisher, 1987). These observations are consistent with the view that the phenyl or phenoxy moiety hinders β-oxidation, the obstruction nearer to the carboxyl or the end of the chain being more likely to hinder this pathway. However, the point of attachment of the alkyl moiety to the aromatic ring sometimes may have no effect (Larson, 1990).

A marked effect on biodegradability is evident among the PAHs in soil. Anthracene, phenanthrene, and acenaphthylene contain three rings (as well as pyrene, a tetracyclic molecule) and are destroyed at reasonable rates when O_2 is present (Fig. 11.5). In contrast, in an intertidal sand flat, although phenanthrene had disappeared in 256 days, anthracene and pyrene were still present (Wilcox *et al.*, 1996). As a rule, PAHs with four and five rings are highly persistent.

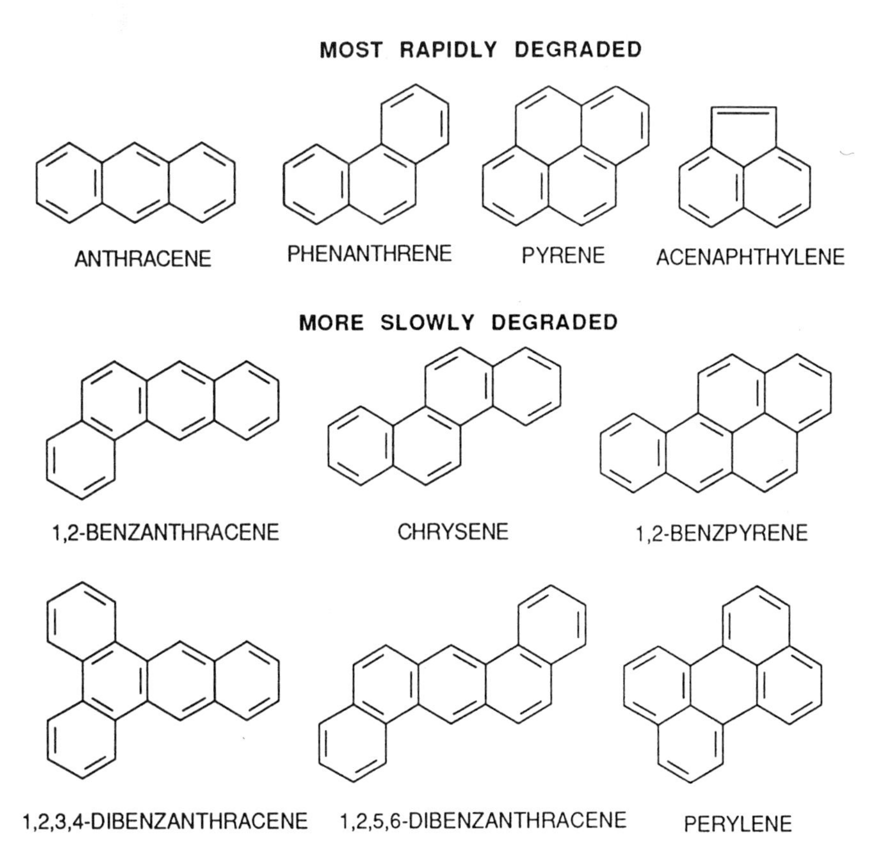

Figure 11.5 Relative rates of biodegradation of polyaromatic hydrocarbons in soil. (Reprinted with permission from Bossert and Bartha, 1986.)

Generalizations about structure–biodegradability relationships, which are sometimes termed SBRs, in aerobic environments do not seem to be applicable to anaerobic environments. For example, the findings that 3-chlorobenzoate is degraded but not 2- and 4-chlorobenzoates and that 2,4,5-trichlorobenzoate is metabolized but not mono- and dichlorobenzoates (DeWeerd *et al.*, 1986) are totally dissimilar from the patterns in aerobic habitats. Similar differences in structure–biodegradability relationships have been noted during the metabolism of phenols, benzoates, and phenoxyalkanoates in anaerobic sludge from municipal sewage (Buisson *et al.*, 1986) and in anaerobic sediments (Genthner *et al.*, 1989).

Other structural characteristics of chemicals are associated with slow mineralization. As a rule, however, observations with a single class of chemicals are few in number so that generalizations are still risky. However, many studies have been performed with pure cultures, but it is not clear which data represent valid generalizations for microbial communities and which merely reflect the idiosyncracies of the particular species being tested. Consider, for example, a group of homologous chemicals (designated A, B, C, D, E, and F) that differ in some slight way from one another, and assume that four species exist in nature with somewhat different abilities to use the group of homologues as sources of C and energy for growth. Based on their dissimilar capacities to utilize these potential nutrients, environments containing one or more of these species would show markedly different patterns of degradation in an evaluation of the effect of chemical structure on the fate of the chemical. Some of these possibilities are illustrated in Table 11.4. In the examples, given in Table 11.4, the first species metabolizes only A for growth, the second uses A, B, or C, the third can grow on A or D, and the fourth can use C, D, or E as nutrients for growth. The outcomes of these activities in nature are given in the footnote of Table 11.4.

Table 11.4

Biodegradability of a Homologous Series of Compounds That Support Growth of Different Species[a]

Species	Growth rate of species	Occurrence of species	Homologous compound supporting growth					
			A	B	C	D	E	F
1	Rapid	Widespread	+	—	—	—	—	—
2	Rapid	Rare	+	+	+	—	—	—
3	Slow	Widespread	+	—	—	+	—	—
4	Slow	Rare	—	—	+	+	+	—

[a] A, always degraded rapidly; B, degraded rapidly but only in some environments; C, degraded rapidly, slowly, or not at all; D, always degraded slowly; E, degraded slowly but only in some environments; F, not degraded.

Generalizations on the effect of structure on biodegradation often do not apply following growth of a single population able to destroy a less readily degradable isomer or member of a homologous series of compounds. The activity then represents the metabolic potential of the previously rare organism, and the spectrum of substrates acted on by this population may be quite different from that of the original microbial community. This is evident in the finding that once samples of sewage become acclimated to pentachlorophenol, the populations that proliferate during the acclimation period are able to metabolize tri- and tetrachlorophenols (Etzel and Kirsch, 1975) that, in an environment with no such acclimation, would be highly persistent.

PREDICTING BIODEGRADABILITY

Given the broad array of organic chemicals that are of potential ecological and public health concern, attempts have been made to develop an overview of why structure affects degradability. This overview is needed for the development of a predictive capacity for previously untested chemicals. Such a predictive ability is important (a) in industry so that chemicals may be synthesized that will not only be useful for the particular purpose but will also have relatively short lives in nature and (b) for regulatory agencies of government that must decide whether a new chemical will be persistent or not before it is introduced commercially. The need by regulatory agencies for a predictive capacity is obvious in light of the fact that probably less than 1% of the more than 2000 new compounds submitted to the U.S. Environmental Protection Agency each year for regulatory review contain data on biodegradability (Boethling and Sabjlić, 1989). For those new commercial products that will be released into natural environments, either deliberately or inadvertently, that information is important to assess potential exposure.

One approach to predicting biodegradability from the properties of a molecule invokes the similarity of the test chemical to substrates and intermediates in known metabolic pathways. Chemicals that do not differ to a marked extent or have only a single xenophore are presumably decomposed reasonably quickly. Those that differ to a significant extent are more resistant, presumably because they are converted to these intermediates slowly due to the need for several enzymes (found only in rare species) to convert the synthetic molecules to natural products. This approach is biochemical because it relies on comparisons with biochemical precedents.

In contrast with the use of biochemical principles or analogies to predict biodegradability, other approaches rely on physical or chemical properties of the compounds, their structural features, descriptors of the molecules, or molecular fragments. If such approaches are to prove useful, these properties or structural features should correlate with existing data on biodegradability or biodegradation rates, and they would then be useful for roughly predicting the rates of disappearance of compounds that have not yet been tested.

Among the properties suggested for use in predicting biodegradability are water solubility, melting point, boiling point, molecular weight, molecular volume, molar refractivity, density, and log K_{ow} or some other descriptor of hydrophobicity. In recent studies, some of the descriptors that have been used are the van der Waals radii of individual substituents, van der Waals volume of the entire molecule, Hammett's sigma constant, ionization constant (pK_a), dipole moment, energy of the highest and lowest molecular orbitals (Damborsky and Schultz, 1997), electronic and hydrophobic constants of substituents (Utkin *et al.*, 1995), electrophilic superdelocalizability, and Sterimol substituents length (Dearden and Cronin, 1996).

Such approaches are designed to predict degradability, qualitatively at least, based on chemical properties that can be determined either experimentally or from considerations of the structure of the compound. For example, the biodegradation rates of some compounds are correlated with the rate constants for their alkaline hydrolysis, but the biodegradability of *N*-methyl arylcarbamates is not correlated with their susceptibility to abiotic hydrolysis, even though the initial metabolic step is a hydrolysis of the carbamate (Pussemier *et al.*, 1989). The rates of anaerobic biodegradation of halogenated aromatic compounds in sediments are correlated with the strength of the C–halogen bond that is cleaved, a not unsurprising finding because this is the bond broken in the rate-determining step of the microbial conversion (Peijnenburg *et al.*, 1992). In the case of PAHs in intertidal sand flats, the relative abundance of several polycyclics after 256 days of biodegradation was correlated with their molecular weights and molecular volumes (Wilcox *et al.*, 1996). Use also is made of the group contribution approach, which entails dividing the molecule into functional groups or molecular fragments (Govind and Lei, 1996; Dessai *et al.*, 1990), and of electronic and steric parameters of the molecule (Degner *et al.*, 1991).

Predictions of biodegradability also are based on molecular topology, which deals with such structural features of molecules as their shapes and sizes, the presence of branching, or the types of atom-to-atom connections. Of particular interest in molecular topology is molecular connectivity. Molecular connectivities can be determined from the structural formula of the chemical, and their use appears to have promise for some classes of compounds (Boethling, 1986; Boethling and Sabljić, 1989; Dearden and Cronin, 1996; Sabljić and Piver, 1992). The rates of decomposition also appear to be correlated with the van der Waals radius (the van der Waals radius being a property of the substituents of the compound) for phenols with several substituents in the *para* position (Paris *et al.*, 1982, 1983) and anilines with several substituents in the *meta* position (Paris and Wolfe, 1987). The rate of decomposition of a few chemicals is also correlated with the Hammett substituent constant (Reineke and Knackmuss, 1978; Pitter, 1985).

In face of the enormous number and variety of organic molecules that have yet to be tested, it must be admitted that predicting biodegradability among classes for which there are few precedents remains difficult. Nevertheless, models have been devised that have good predictive abilities for increasingly larger numbers of compounds, and these correctly classify the relative biodegradability of many compounds under aerobic conditions (Howard *et al.,* 1991; 1992). As the data base grows and more generalizations and models appear, it should be possible to better select structures that not only have efficacy for the purposes they were made but also degrade sufficiently rapidly to prevent untoward consequences.

REFERENCES

Alexander, M., and Aleem, M. I. H., *J. Agric. Food Chem.* **9,** 44–47 (1961).

Alexander, M., and Lustigman, B. K., *J. Agric. Food Chem.* **14,** 410–413 (1966).

Apajalahti, J. H. A., and Salkinoja-Salonen, M. S., *Appl. Microbiol. Biotechnol.* **25,** 62–67 (1986).

Audus, L. J., *in* "Herbicides and the Soil" (E. K. Woodford and G. R. Sagar, eds.), pp. 1–19. Blackwell, Oxford, 1960.

Baker, M. D., and Mayfield, C. I., *Water Air Soil Pollut.* **13,** 411–424 (1980).

Boethling, R. S., *Environ. Toxicol. Chem.* **5,** 797–806 (1986).

Boethling, R. S., and Sabjlić, A., *Environ. Sci. Technol.* **23,** 672–679 (1989).

Bossert, I., and Bartha, R., *in* "Petroleum Microbiology" (R. M. Atlas, ed.), pp. 435–473. Macmillan, New York, 1984.

Bossert, I. D., and Bartha, R., *Bull. Environ. Contam. Toxicol.* **37,** 490–495 (1986).

Boyd, S. A., Shelton, D. R., Berry, D., and Tiedje, J. M., *Appl. Environ. Microbiol.* **46,** 50–54 (1983).

Buisson, R. S. K., Kirk, P. W. W., Lester, J. N., and Campbell, J. A., *Water Pollut. Control* **85,** 387–394 (1986).

Burger, K., MacRae, I. C., and Alexander, M., *Soil Sci. Soc. Am. Proc.* **26,** 243–246 (1962).

Catelani, D., Colombi, A., Sorlini, C., and Treccani, V., *Appl. Environ. Microbiol.* **34,** 351–254 (1977).

Clark, C. G., and Wright, S. J. L., *Soil Biol. Biochem.* **2,** 19–26 (1970).

Damborsky, J., and Schultz, T. W., *Chemosphere* **34,** 429–446 (1997).

Dearden, J. C., and Cronin, M. T. D., *in* "Biodegradability Prediction" (W. J. G. M. Peijnenburg and J. Damborsky, eds.), pp. 93–104. Kluwer, Dordrecht, The Netherlands, 1996.

Degner, P., Nendza, M., and Klein, W., *Sci. Total Environ.* **109/110,** 253–259 (1991).

Dessai, S. M., Govind, R., and Tabak, H. H., *Environ. Toxicol. Chem.* **9,** 473–477 (1990).

DeWeerd, K. A., Suflita, J. M., Linkfield, T., Tiedje, J. M., and Pritchard, P. H., *FEMS Microbiol. Ecol.* **38,** 331–339 (1986).

Dias, F. F., and Alexander, M., *Appl. Microbiol.* **22,** 1114–1118 (1971).

DiGeronimo, M. J., Nikaido, M., and Alexander, M., *Appl. Environ. Microbiol.* **37,** 619–625 (1979).

Etzel, J. E., and Kirsch, E. J., *Dev. Ind. Microbiol.* **16,** 287–295 (1975).

Fournier, J.-C., *Chemosphere* **3,** 77–82 (1974).

Fukuda, D. S., and Brannon, D. R., *Appl. Microbiol.* **21,** 550–551 (1971).

Genthner, B. R. S., Price, W. A., II, and Pritchard, P. H., *Appl. Environ. Microbiol.* **55,** 1466–1471 (1989).

Govind, R., and Lei, L., *in* "Biodegradability Prediction" (W. J. G. M. Peijnenburg and J. Damborsky, eds.), pp. 115–138. Kluwer, Dordrecht, The Netherlands, 1996.

Hallas, L. E., and Alexander, M., *Appl. Environ. Microbiol.* **45,** 1234–1241 (1983).

Haller, H. D., *J. Water Pollut. Control Fed.* **50,** 2771–2777 (1978).

Hammond, M. W., and Alexander, M., *Environ. Sci. Technol.* **6,** 732–735 (1972).

Hauck, R. D., and Stephenson, H. F., *J. Agric. Food Chem.* **12,** 147–151 (1964).
Howard, P. H., Boethling, R. S., Stiteler, W., Meylan, W., and Beauman, J., *Sci. Total Environ.* **109/110,** 635–641 (1991).
Howard, P. H., Boethling, R. S., Stiteler, W. M., Meylan, W. M., Hueber, A. E., Beauman, J. A., and Larosche, M. E., *Environ. Toxicol. Chem.* **11,** 593–603 (1992).
Huddleston, R. L., and Allred, R. C., *in* "Soil Biochemistry" (A. D. McLaren and G. H. Peterson, eds.), pp. 343–370. Dekker, New York, 1967.
Jobson, A., McLaughlin, M., Cook, F. D., and Westlake, D. W. S., *Appl. Microbiol.* **27,** 166–171 (1974).
Kator, H., *in* "The Microbial Degradation of Oil Pollutants" (D. G. Ahearn and S. P. Meyers, eds.), pp. 47–65. Louisiana State University, Center for Wetlands Resources, Baton Rouge, 1973.
Kaufman, D. D., *in* "Pesticides and Their Effects on Soils and Water" (M. E. Bloodworth, ed.), pp. 85–94. Soil Science Society of America, Madison, WI, 1966.
Larson, R. J., *Environ. Sci. Technol.* **24,** 1241–1246 (1990).
McKinney, R. E., and Jeris, J. S., *Sewage Ind. Wastes* **27,** 728–735 (1955).
Mikesell, M. D., and Boyd, S. A., *J. Environ. Qual.* **14,** 337–340 (1985).
Naik, M. N., Jackson, R. B., Stokes, J., and Swaby, R. J., *Soil Biol. Biochem.* **4,** 313–323 (1972).
Paris, D. F., and Wolfe, N. L., *Appl. Environ. Microbiol.* **53,** 911–916 (1987).
Paris, D. F., Wolfe, N. L., and Steen, W. C., *Appl. Environ. Microbiol.* **44,** 153–158 (1982).
Paris, D. F., Wolfe, N. L., Steen, W. C., and Baughman, G. L., *Appl. Environ. Microbiol.* **45,** 1153–1155 (1983).
Peijnenburg, W. J. M., Hart, M. J. 'T., den Hollander, H. A., van de Meent, D., Verboom, H. H., and Wolfe, N. L., *Environ. Toxicol. Chem.* **11,** 301–314 (1992).
Pitter, P., *Acta Hydrochim. Hydrobiol.* **13,** 453–460 (1985).
Pussemier, L., DeBorger, R., Cloos, P., and van Bladel, R., *J. Environ. Sci. Health Part B,* **B24,** 117–129 (1989).
Reineke, W., and Knackmuss, H.-J., *Biochim. Biophys. Acta* **542,** 412–423 (1978).
Sabjlić, A., and Piver, W. T., *Environ. Toxicol. Chem.* **11,** 961–972 (1992).
Saeger, V. W., and Thompson, Q. E., *Environ. Sci. Technol.* **14,** 705–709 (1980).
Sims, G. K., and Sommers, L. E., *Environ. Toxicol. Chem.* **5,** 503–509 (1986).
Smolenski, W. J., and Suflita, J. M., *Appl. Environ. Microbiol.* **53,** 710–716 (1987).
Subba-Rao, R. V., and Alexander, M., *J. Agric. Food Chem.* **25,** 327–329 (1977).
Suflita, J. M., Stout, J., and Tiedje, J. M., *J. Agric. Food Chem.* **32,** 218–221 (1984).
Süss, A., Fuchsbichler, G., and Eben, C., *Z. Pflanzenernaehr. Bodenkd.* **141,** 57–66 (1978).
Swisher, R. D., "Surfactant Biodegradation." Dekker, New York, 1987.
Utkin, I., Dalton, D. D., and Wiegel, J., *Appl. Environ. Microbiol.* **61,** 346–351 (1995).
Wilcock, R. J., Corban, G. A., Northcott, G. L., Wilkins, A. L., and Langdon, A. G., *Environ. Toxicol. Chem.* **15,** 670–676 (1996).

CHAPTER 12

Predicting Products of Biodegradation

The practical reasons for the need for information on the products of biodegradation were presented at the beginning of Chapter 11. This chapter deals specifically with approaches to predicting the identities of those products.

In mineralization, the substrate is ultimately converted to inorganic products, microbial biomass, and polymeric substances that are generally not well characterized but are believed to be toxicologically unimportant. Nevertheless, some of the intermediates in mineralization sequences may be excreted and remain for some time outside of the cells or hyphae of the active species. In cometabolism (see Chapter 13), in contrast, inorganic products are generally not formed, at least not initially, and C in the substrate is not incorporated into biomass; in these instances, low-molecular-weight intermediates and products are excreted and may persist for some time. Of the metabolic products that are generated, environmental relevancy applies to those that (a) are excreted, (b) reach a concentration that may be of environmental or toxicological importance, (c) persist for a reasonable period of time, and (d) are either known to be toxic or whose toxicity has not been determined. Conversely, an intermediate that remains within the confines of the cell, that does not appear in sufficiently high concentrations, does not persist long enough to result in the exposure of potentially sensitive populations, or is known to be nontoxic is not of environmental relevance.

Because of the lack of evidence for their excretion, products of many enzyme reactions and many enzymological or biochemical studies are not considered in this discussion. In terms of environmental importance, moreover, information from investigations of individual microorganisms sometimes may also not be of toxicological or ecological relevance because the products formed in those cultures, even if excreted and reaching relatively high concentrations, may never appear at detectable levels in nature because other organisms residing in the same habitat destroy those products as quickly as they are formed.

Biochemistry, microbial physiology, and pharmacology provide information on products of many metabolic sequences and conversions catalyzed by individual enzymes. The substrates in these instances are sugars, amino acids, purines, pyrimidines and their oligomers and polymers (i.e., polysaccharides, proteins, and nucleic acids), organic acids, phosphate esters, and pharmaceutical agents, but few of these compounds are important from a toxicological viewpoint. The literature relevant to issues of biodegradation and bioremediation deals with different compounds and comes from studies of pesticides, petroleum products, surfactants and detergent constituents, solvents, and materials synthesized by the chemical industries. Of particular relevance are those products and reaction sequences found to occur in soils, surface and groundwaters, sediments, and sewage.

Representatives of some of the categories of compounds that are, or may become, of environmental importance are shown in Fig. 12.1. The diversity in structures is immediately evident and so, too, would the diversity of metabolic products and pathways.

It is sometimes easier to predict what would not occur than what may take place. For example, a direct reductive cleavage of the $-CH_2-CH_2-$ moiety of a molecule to give methyl groups ($-CH_3$) and the aerobic cleavage of the benzene ring without the prior formation of a hydroxylated intermediate are improbable. However, a large number of reaction types have been established, and these can serve as guidelines for predictive purposes.

Although the degradative reactions (i.e., those that ultimately lead to a simplification of the molecule, thus excluding additions, conjugations, oligomerizations, and polymerizations) are many and disparate, their role is essentially the same in bacteria and fungi using the compounds as sources of carbon and energy for growth: the reactions transform the molecules to substrates in those intracellular processes that generate energy or form the precursors for the synthesis of cell constituents. Although these reactions have no role in bacteria and fungi that cometabolize the compounds, many cometabolic reactions mimic steps that ultimately lead to pathways and sequences that provide energy and precursors for biosyntheses; hence, they are presented below in the same way.

The growth-linked biodegradation of substrate A can be depicted by this sequence.

$$A \rightarrow B \rightarrow C \rightarrow D \rightarrow\rightarrow\rightarrow CO_2 + \text{energy} + \text{cell C}$$

For anaerobes, CO_2 and CH_4 may be formed in such a sequence, and low-molecular-weight compounds characteristic of bacterial fermentations will be ex-

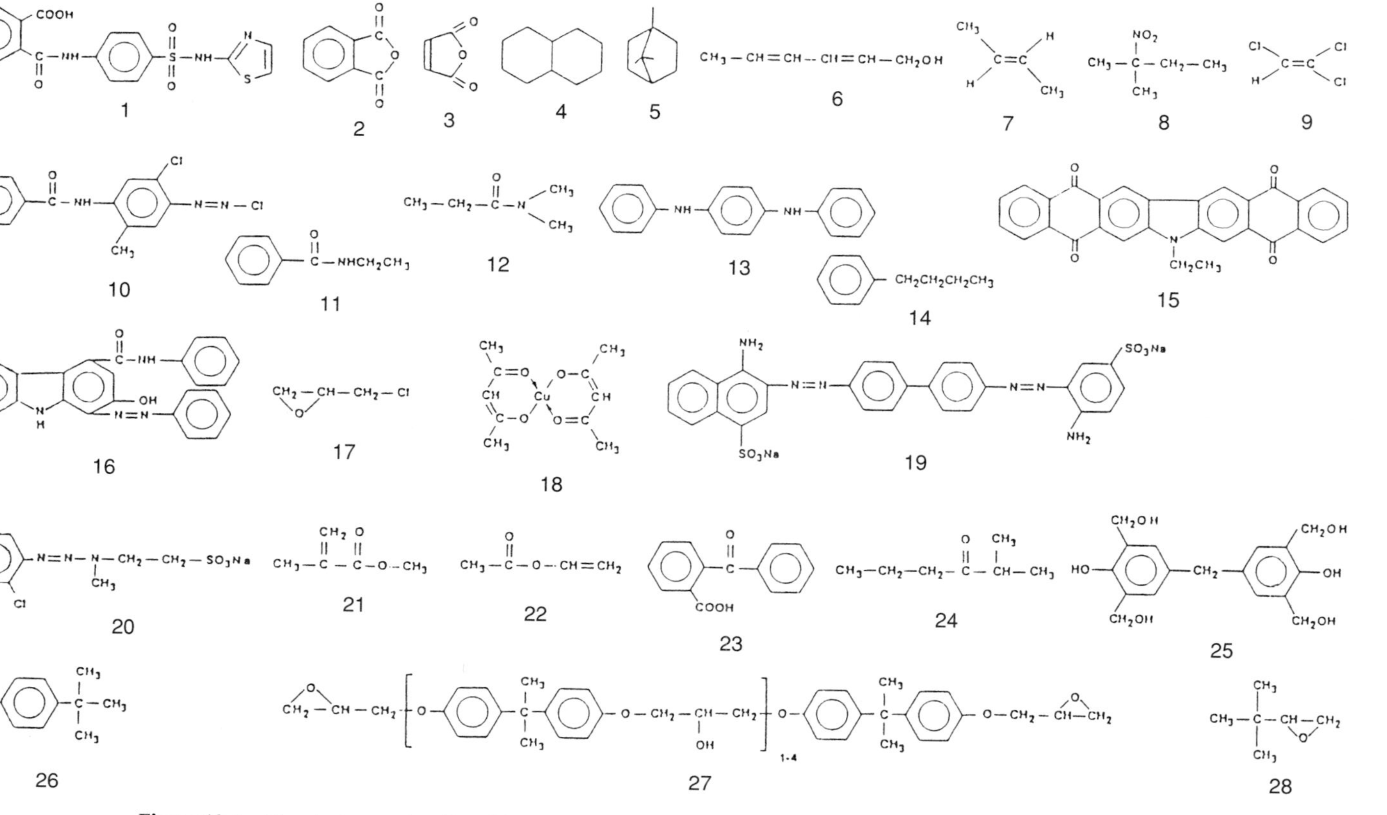

Figure 12.1 Chemical categories of possible environmental importance. (From Boethling *et al.*, 1989.)

creted and detected. If the conversion of substrate A is a result of cometabolism (see Chapter 13), the sequences might be written as one of the following:

$$A \rightarrow B$$

$$A \rightarrow B \rightarrow C$$

$$A \rightarrow B \rightarrow C \rightarrow D$$

The end product of the activity of a cometabolizing population may be utilized rapidly or slowly by other organisms, or it may persist for some time because no other microorganisms are active on it and no abiotic mechanism exists to bring about its disappearance.

In considering the reaction types and products of the first metabolic steps, as presented below, it should be kept in mind that the number of substrates (designated A) is vast and, although the number of enzymes is smaller (albeit unknown), those enzymes will catalyze the formation of a vast array of products (designated B, C, and D). The reaction types represent that small number of enzymes that constitute a metabolic sequence. For natural environments, in contrast with microbial cultures in the laboratory, it is not known whether a particular product is (a) an intermediate that is excreted by species that will then destroy that molecule or (b) an intermediate or final product resulting from the actions of cometabolizing populations. Hence, for natural or polluted environments, it is unclear whether a reaction type may merely be a step in a series of processes in cells of organisms that use the compound for growth or the sole conversion that the organisms can perform on that molecule.

For the sake of simplification, *Ar, Alk,* and *R* are used to designate the aryl (aromatic), alkyl (aliphatic), and some other portion of the substrate molecule. To designate products detected in soil, sediment, sewage (or waste water), groundwater, surface water, or samples thereof in the laboratory or in microbial cultures (pure or mixed cultures), the symbols *so, sd, se, gw, sw,* and *mc* are used. In some instances in which anaerobiosis is important, the symbol *an* is used.

In the degradation of the listed compounds by a given microorganism, only a few or possibly none of the products shown will be excreted; which products, if any, will be found in the culture medium will depend on the physiological properties and idiosyncracies of the particular species, as well as environmental conditions. The presence or absence of O_2 is a major factor in determining the identities of the products. Furthermore, depending on which microorganisms are present and their relative abundance and activity, different products will be formed, accumulate, and persist in different natural and polluted environments, treatment systems, bioremediation sites, or mixed cultures or microbial consortia. The products and pathways given here indicate what *may* occur; the populations in the environment, treatment system, or culture—and the abiotic factors at the site that affect microorganisms, especially O_2—will determine what actually will be present.

Most degradative reactions (i.e., not including addition, conjugation, oligomerization, and polymerization reactions) involve (a) the addition to the substrate of OH and H from H_2O, 2H or 2 protons, O_2, or $\frac{1}{2}O_2$ or (b) the removal of H_2O, 2H, or HX where X represents some substituent. The differences among the many reaction types presented below are usually not differences in the nature of the step catalyzed by the enzyme because those steps typically are the additions or removals given earlier. Rather it is the structure of the substrate to or from which the addition or removal takes place. The OH, H, 2H, or O_2 is added or OH, H, or 2H is removed from the substrate in the process of degradation. The OH, H, 2H, or O_2 may act directly on the molecule or may function by removing a substituent; for example, $-NH_2$, $-NO_2$, -Cl, $-OSO_3H$, or $-OPO_3H_2$. Usually, the site at which the H_2O, H, or OH is inserted or removed is specific and predictable, but sometimes, especially with OH, $\frac{1}{2}O_2$, or O_2 added to aromatic molecules, it is reasonably nonspecific and largely unpredictable. Often, the insertion of H_2O, H, or OH leads to release of an inorganic ion such as ammonium, nitrite, chloride, sulfite, sulfate, or phosphate from the organic substrate. However, sometimes the insertion leads only to organic products.

AROMATICS, MONO- AND POLYCYCLIC: HYDROXYLATION AND KETONE FORMATION

A very common initial or early step, and sometimes the only one, is a conversion of the substrate to a corresponding hydroxylated or ketone product. Indeed, it might even seem that hydroxylating enzymes are promiscuous, acting on many different kinds of molecules and, often, on several positions on a single molecule. However, it appears that two types of reactions occur. One is a prelude to further metabolism of the molecule, frequently leading to opening of the benzene ring or cleavage of one of the PAH rings; in such processes, the enzyme acts on specific substrates and at specific locations on the substrate. The second is apparently a nonspecific hydroxylation (sometimes leading to a ketone or a quinone), the enzyme catalyzing a reaction that adds an OH to one or more of several positions on the substrate.

Representative reactions from recent publications are shown in Table 12.1. Quite commonly, the product is a dihydrodiol, a compound in which two OH groups are introduced, usually on adjacent C atoms. Although products of these types are frequently found in pure and mixed cultures of bacteria and fungi, as well as in enrichment cultures, they may not be found by analyses of natural or polluted environments because they could react abiotically with humic or colloidal materials to give complexes or be further degraded by the producing or other microorganisms.

Table 12.1

Substrates and Products of Hydroxylation and Reactions Leading to Ketones

Substrate	Products	System[a]	Reference
Acenaphthene	1-Acenaphthenol, 1-acenaphthenone	mc	Komatsu *et al.* (1993)
Anthracene	Anthracene *trans*-1,2-dihydrodiol, 9, 10-anthraquinone	mc	Bezalel *et al.* (1996)
Benzene	Phenol, hydroquinone	mc	Burback and Perry (1993)
Carbofuran	3-Hydroxycarbofuran, 3-ketocarbofuran	so	Johnson and Lavy (1995)
Chlorobenzene	4-Chlorophenol	mc	Burback and Perry (1993)
Chrysene	*trans*-1,2-Dihydroxy-1,2-dihydrochrysene	mc	Kiehlmann *et al.* (1996)
Dibenzothiophene	*cis*-1,2-Dihydroxy-1,2-dihydrobenzothiophene	mc	Resnick and Gibson (1996)
Ethylbenzene	4-Ethylphenol	mc	Burback and Perry (1993)
Fluorene	9-Fluorenol, 9-fluorenone	mc	Bezalel *et al.* (1996)
Halogenated phenols	Corresponding catechols	mc	Hofrichter *et al.* (1994)
Naphthalene	1-Naphthalenol	mc	Liu *et al.* (1992)
Phenanthrene	Phenanthrene *cis*-3,4-dihydrodiol	sd	MacGillivray and Shiaris (1994)
Pyrene	Pyrene *cis*- and *trans*-4,5-dihydrodiol and 1,4- and 1,8-dihydroxypyrene	mc	Bezalel *et al.* (1996), Cullen *et al.* (1994), Lambert *et al.* (1994)
1,2,3-Trichlorobenzene	2,3,4- and 2,3,5-Trichlorophenols	mc	Sullivan and Chase (1996)
p-Xylene	3,6-Dimethylcatechol	mc	Arcangeli and Arvin (1995)

[a] Explained on p. 198.

Analogous reactions occur with dibenzofuran (Cerniglia *et al.*, 1979) and dibenzothiophene (Laborde and Gibson, 1977), at least in microbial cultures. These are compounds with two benzene rings linked together by a C–C and O or a C–C and S bridge, respectively. The products are mono- or dihydroxy derivatives.

Other examples, these from the earlier literature, are given in Table 12.2.

MONOCYCLIC AROMATICS: REDUCTION OF DOUBLE BONDS

Tests with microbial cultures have shown the reduction of one or more double bonds of the benzene ring under anaerobic conditions. Typical of such

Table 12.2

Substrates and Products of Hydroxylation and Reactions Leading to Ketones or Quinones

Substrate	Products	System	Reference
Benzo(*a*)pyrene	*cis*-9,10-Dihydroxy-9,10-dihydro-benzo(*a*)pyrene	mc	Gibson *et al.* (1975)
Benzoic acid	3-Hydroxybenzoate	mc	Wheelis *et al.* (1967)
Biphenyl	2,3-Dihydroxybiphenyl	mc	Gibson *et al.* (1973)
3-Chloroaniline	2-Amino-4-chlorophenol	mc	Fletcher and Kaufman (1973)
Chlorobenzene	3-Chlorocatechol	mc	Gibson *et al.* (1978)
Chlorobiphenyl	4-Chloro-4′-hydroxybiphenyl	mc	Neu and Ballschmiter (1977)
Chlorophenol	4-Chlorocatechol	mc	Knackmuss and Hellwig (1978)
o-Cresol	3-Methylcatechol	se	Masunaga *et al.* (1986)
2,4-Dichlorophenoxyacetic acid (2,4-D)	2,4-Dichloro-5-hydroxyphenoxyacetic acid	mc	Faulkner and Woodock (1964)
Dicamba	5-Hydroxydicamba	sw	Yu *et al.* (1975)
1,4-Dichlorobenzene	3,6-Dichlorocatechol	mc	Schraa *et al.* (1986)
2,4-Dichlorophenol	3,5-Dichlorocatechol	mc	Engelhardt *et al.* (1979)
Hydroquinone	1,4-Benzoquinone	mc	Harbison and Belly (1982)
1-Naphthol	1,4-Naphthoquinone	sw	Lamberton and Claeys (1970)
Phenol	Catechol	mc	Evans (1947)
o-Xylene	2,3-Dimethylphenol	mc	Baggi *et al.* (1987)

reactions are the conversion of benzene to cyclohexene (Grbić-Galić and Vogel, 1987), benzoic acid to 1-cyclohexene-1-carboxylic acid and cyclohexane carboxylic acid (Keith *et al.,* 1978), and toluene to 4-methylcyclohexanol (Grbić-Galić and Vogel, 1987) (Fig. 12.2).

MONOCYCLIC AROMATICS: RING CLEAVAGE

A wealth of information exists on the pathways, intermediates, and enzymes involved in the cleavage of monocyclic aromatic compounds. Few of the intermediates have been detected in natural or polluted environments, during bioremediation, or in engineered treatment systems, possibly because of their reactivity or because they are rapidly metabolized. Nevertheless, some of the products of the cleavage

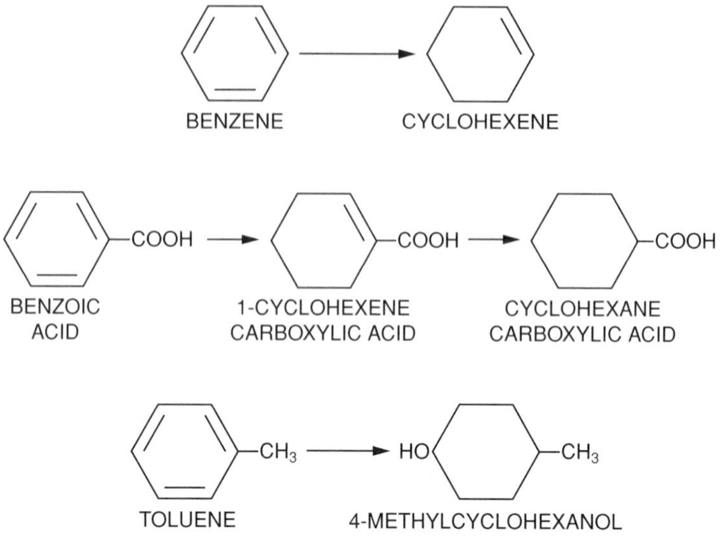

BENZENE → CYCLOHEXENE

BENZOIC ACID —COOH → 1-CYCLOHEXENE CARBOXYLIC ACID —COOH → CYCLOHEXANE CARBOXYLIC ACID —COOH

TOLUENE —CH_3 → HO 4-METHYLCYCLOHEXANOL —CH_3

Figure 12.2 Reduction of double bonds of monocyclic aromatics.

of the aromatic rings of synthetic aromatic compounds differ from natural products, such as chlorinated metabolites generated from 2,4-D or chlorophenols or the nitro-containing hexanoic and pentanoic acid derivatives formed microbiologically from 2,4- and 2,6-dinitrophenols (Lenke *et al.*, 1992; Ecker *et al.*, 1992.

PAHS: RING CLEAVAGE

The extensive degradation of PAHs proceeds by cleavage of one, several, or all of the rings. Certain characteristic types of products are formed, although the precise structure depends on the substrate and the particular microorganism. Several typical products formed under aerobic conditions are listed in Table 12.3. The feature in common for all of these products is that they have one ring fewer than the substrates and that they have, almost as a memory of the lost ring, either two carboxyls or one carboxyl and one hydroxyl group.

A similar residue of the old ring is produced in the degradation of diphenylmethanes; for example, in the conversion of 4,4′-dichlorodiphenylmethane to 4-chlorophenylacetic acid (Focht and Alexander, 1970). This is illustrated by the following, where R is 4-chlorophenyl:

$$RCH_2R \rightarrow RCH_2COOH$$

Table 12.3

Products of Cleavage of PAHs by Microbial Cultures

Substrate	Products	System	Reference
Acenaphthylene	1,8-Naphthylenedicarboxylic acid	mc	Komatsu *et al.* (1993)
Anthracene	3-Hydroxy-2-naphthoic acid	mc	Rogoff and Wender (1957)
1- and 2-Ethylnaphthalenes	Corresponding salicylic acids	mc	Bestetti *et al.* (1994)
Fluorene	Phthalic acid	mc	Grifoll *et al.* (1994)
Naphthalene	2-Hydroxybenzoic acid	mc	Liu *et al.* (1992)
Phenanthrene	1-Hydroxy-2-naphthoic acid	sd	MacGillivray and Shiaris (1994)

MONOCYCLIC AROMATICS: CARBOXYLATION

Sometimes, at least in mixed or pure cultures, the product is a carboxylated derivative of the substrate. Carboxylation reactions occur under both aerobic and anaerobic conditions:

$$\mathrm{ArH} \rightarrow \mathrm{ArCOOH}$$

This is evident in the conversion of benzene to benzoic acid (Chaudhuri and Wiesmann, 1995), phenol to benzoic acid (Li *et al.,* 1996), and 2-chlorophenol to 3-chlorobenzoic acid (Bisaillon *et al.,* 1993), aniline to anthranilic acid (Aoki *et al.,* 1984), and *m*-cresol to 4-hydroxy-2-methylbenzoate (Ramanand and Sulflita, 1991). The carboxylation of phenol may also occur in sewage sludge (Knoll and Winter, 1987).

CYCLOALKANES: OXIDATION

A number of cycloalkanes or compounds containing cycloalkane moieties can be converted to hydroxy or keto derivatives or they may be dehydrogenated. Hydroxylation is evident in the conversion of cyclohexane to cyclohexanol in microbial cultures (deKlerk and van der Linden, 1973) or the cyclohexyl moiety of the pesticide hexazinone to the corresponding hydroxy compound in soil (Rhodes, 1980). The formation of the keto derivative is evident in the conversion of cyclohexane to cyclohexanone (Beam and Perry, 1973) and cyclopentanecarboxylic acid to 2-ketocyclopentanecarboxylic acid in microbial cultures (Hasegawa *et al.,* 1980). The dehydrogenation of 3-cyclohexenecarboxylic acid to yield benzoic acid represents another type of conversion effected in culture (Blakley and Papish, 1982). These transformations may be represented in the following way, in which the -CH_2CH_2- is part of cycloalkane:

$$-CH_2CH_2- \rightarrow -CH(OH)-CH_2-$$

$$-CH_2-CH_2- \rightarrow -C(=O)-CH_2-$$

$$-CH_2-CH_2- \rightarrow -HC{=}CH-$$

METHYL GROUPS (RCH_3): OXIDATION

Many organic pollutants and natural products contain methyl groups. These are present at the terminal end of alkanes in oil products, the hydrophobic alkyl portions of surfactants, and in simple aromatics such as toluene and xylenes, as well as a variety of other groups of chemicals. They are commonly oxidized by a series of reactions that yield the corresponding alcohols, aldehydes, and carboxylic acids:

$$RCH_3 \rightarrow RCH_2OH \rightarrow RCHO \rightarrow RCOOH$$

In any given environment, microbial mixture, or pure culture, not all of the products will be detected. Some examples are presented in Table 12.4. Oxidations

Table 12.4

Substrates and Products of Reactions in Which Methyl Groups Are Oxidized

Substrate	Products	System	Reference
Eicosane ($C_{20}H_{42}$)	$CH_3(CH_2)_{18}COOH$	so	Amblès *et al.* (1994)
Ethylbenzene	1-Phenylethanol, phenylacetaldehyde, phenylacetic acid	mc	Corkery *et al.* (1994)
Ethylbenzene	1-Phenylethanol	mc/an	Ball *et al.* (1996)
4-Nitrotoluene	4-Nitrobenzaldehyde, 4-nitrobenzoic acid	mc	Rhys *et al.* (1993)
TNT	4-Amino-2,6-dinitrobenzoic acid	mc	Vanderberg *et al.* (1995)
Toluene	Benzaldehyde, benzoic acid	mc/an	Seyfried *et al.* (1994)
o-Xylene	2-Methylbenzylalcohol, -benzylaldehyde, and -benzoic acid	mc	Arcangeli and Arvin (1995)

of methyl groups are common among aerobic microorganisms, but some anaerobes also carry out identical types of conversions.

As indicated by the pathway by which methyl groups are oxidized, alcohols and aldehydes are intermediates. These too may be oxidized. Some examples of such conversions as well as early reports of the oxidation of methyl groups are listed in Table 12.5.

ALKANES $[CH_3(CH_2)_nCH_3]$: DEHYDROGENATION

Under certain circumstances, an alkane can be dehydrogenated in a step not related to β-oxidation and sometimes at a position distant from the end of the molecule:

$$RCH_2CH_2R' \rightarrow RCH{=}CHR'$$

This is evident in the conversion of eicosane ($C_{20}H_{42}$) to eicos-9-ene in soil (Amblès *et al.*, 1994), in the formation of 1-dodecene (a 12-C compound) from tetradecane

Table 12.5

Substrates and Products of Reactions in Which Methyl, Hydroxymethyl, or Aldehyde Groups Are Oxidized

Substrate	Products	System	Reference
Bromacil	Hydroxymethyl derivative	so	Gardiner *et al.* (1969)
4-Chlorobenzyl alcohol	4-Chlorobenzoic acid	mc	Omori and Alexander (1978)
p-Cresol	4-Hydroxybenzoic acid	so	Smolenski and Suflita (1987)
Decane	Decyl aldehyde	mc	Iizuka *et al.* (1961)
Denmert	Hydroxymethyl derivative	so	Ohkawa *et al.* (1976)
Dodecyltri-methylammonium	9-Carboxynonyltri-methylammonium	mc	Dean-Raymond and Alexander (1977)
Ethane	Acetic acid	mc	Leadbetter and Foster (1959)
Ethylene glycol	Glyoxylic acid	mc	Child and Willetts (1978)
Linuron	Hydroxymethyl derivative	mc	Tillmanns *et al.* (1978)
2-Methylnaphthalene	2-Naphthoic acid	mc	Raymond *et al.* (1967)
3-Methylpyridine	Nicotinic acid	mc	Skryabin *et al.* (1969)
Paraquat	4-Carboxy-1-methylpyridinium	mc	Funderburk and Bozarth (1967)
Pentachlorobenzyl alcohol	Pentachlorobenzoic acid	so	Ishida (1972)
Vanillin	Vanillic acid	so	Kunc (1971)

($C_{14}H_{30}$) by anaerobic bacteria (Morikawa *et al.*, 1996), and of 1-heptene from *n*-heptane in microbial culture (Chouteau *et al.*, 1962).

ALKYL GROUPS [$R(CH_2)_nCH_3$]: SUBTERMINAL OXIDATION

In some instances, oxidation may occur at other than the terminal C of an alkyl chain to generate a ketone or hydroxy compound. This is evident in the conversion of ethylbenzene to acetophenone (Ball *et al.*, 1996):

$$ArCH_2CH_3 \rightarrow ArC(=O)CH_3$$

Similarly, hexane is converted to 2-hydroxyhexane and 2-ketohexane (Patel *et al.*, 1980a,b). Even if the methylene (-CH_2-) is between two aromatic rings, it may be oxidized, as indicated by the oxidation in soil of 4,4′-methylene-bis(2-chloroaniline) (MBOCA), a cross-linking agent used in the production of polymers and resins (Voorman and Penner, 1986):

$$ArCH_2Ar \longrightarrow ArC(=O)Ar$$

ALKENES AND OTHER COMPOUNDS WITH DOUBLE BONDS: REDUCTION, OXIDATION, AND HYDRATION

Some microorganisms in laboratory media can reduce a substrate at the double bond. This type of conversion occurs in the metabolism of cinnamic acid in culture (Blakley and Simpson, 1964):

$$RCH{=}CHR' \rightarrow RCH_2CH_2R'$$

or in the transformation of products of DDT breakdown in water or microbial culture (Pfaender and Alexander, 1972; Wedemeyer, 1967; Patel *et al.*, 1972):

$$Ar_2C{=}CH_2 \rightarrow Ar_2CHCH_3$$

$$Ar_2C{=}CHCl \rightarrow Ar_2CHCH_2\,Cl$$

An oxidation at the double bond is exemplified by the hydroxylation by a fungus of the herbicide dicryl (Wallnöfer *et al.*, 1973),

$$\underset{\displaystyle R'}{RC}{=}CH_2 \longrightarrow \underset{\displaystyle R'}{RC}OHCH_2OH$$

and a hydration is shown by the microbial conversion of 1,1-diphenylether to 2,2-diphenylethanol (Focht and Joseph, 1974):

$$\underset{\displaystyle R'}{RC}{=}CH_2 \longrightarrow \underset{\displaystyle R'}{RC}HCH_2OH$$

ALKENES AND OTHER COMPOUNDS WITH DOUBLE BONDS: EPOXIDE FORMATION

A number of compounds containing double bonds are converted to epoxides:

$$RCH{=}CHR' \longrightarrow RCH\overset{O}{—}CHR'$$

Widely known examples in soil and microbial cultures are the conversion of aldrin to dieldrin (Lichtenstein and Schulz, 1960; Murado Garcia, 1974) and heptachlor to heptachlor epoxide (Duffy and Wong, 1967; Elsner *et al.*, 1972). Similarly, styrene is transformed by a bacterium to styrene oxide (O'Connor *et al.*, 1995), the latter being a carcinogen, and octene-1 is oxidized to octene-1,2-epoxide (van der Linden, 1963).

ALKYNES AND OTHER COMPOUNDS WITH TRIPLE BONDS: REDUCTION

The reduction of the triple bond may also be accomplished microbiologically:

$$RC{\equiv}CH \rightarrow RCH{=}CH_2$$

For example, such a reaction occurs in the transformation of the herbicide buturon in soil (Haque *et al.*, 1977) and by a fungus (Tillmanns *et al.*, 1978). Nitrogen-fixing bacteria, by virtue of their possession of the enzyme nitrogenase, reduce acetylene to ethylene:

$$HC{\equiv}CH \rightarrow H_2C{=}CH_2$$

CARBOXYLIC ACIDS (RCOOH): DECARBOXYLATION AND REDUCTION

Aromatic, linear, and aromatic compounds with side chains may lose their carboxyl groups:

$$ArCOOH \longrightarrow ArH$$

$$\underset{\displaystyle CH_3}{RCHCOOH} \longrightarrow RCH_2CH_3$$

$$Ar_2CHCOOH \longrightarrow Ar_2CH_2$$

Some examples are given in Table 12.6. The compounds listed in the table include several herbicides (bifenox, dichlorofop-methyl, fluroxypur, and picloram). The last two compounds have a pyridine ring (a N heterocycle) rather than a benzene ring.

Not as well studied are the reductions of carboxylic acids to the corresponding aldehydes and/or alcohols:

$$RCOOH \rightarrow RCHO \rightarrow RCH_2OH$$

The product that accumulates may be the aldehyde, the alcohol, or both. Such a transformation is illustrated by the microbial conversion of the herbicide 2,4,5-T to the corresponding alcohol (Nakajima *et al.*, 1973), the reduction of *p*-aminobenzoate to the aldehyde (Ramano and Shanmugasundaram, 1962) or alcohol (Sloane

Table 12.6

Substrates and Products of Decarboxylation Reactions

Substrate	Products	System	Reference
Bifenox	Nitrofen	so	Leather and Foy (1977)
Caffeic acid	4-Vinylcatechol	mc	Finkle *et al.* (1971)
Dichlorfop-methyl	4-(2,4-Dichlorophenoxy) phenetole	so	Smith (1977)
2,4-Dihydroxybenzoate	Resorcinol	mc	Halvorson (1963)
3-Fluorophthalate	3-Fluorobenzoate	mc	Aftring and Taylor (1981)
Fluroxypur	Decarboxylated derivative	so	Lehman *et al.* (1990)
4-Hydroxybenzoate	Phenol	mc	Patel and Grant (1969)
Phthalate	Benzoate	mc	Taylor and Ribbons (1983)
Picloram	4-Amino-3,5,6-trichloropyridine	mc	Rieck (1970)

et al., 1963), and furfuraldehyde to the corresponding alcohol (Searles and French, 1964), all in microbial cultures.

CARBOXYLIC ACIDS (RCOOH) AND ALCOHOLS (RCH_2OH): ESTER FORMATION

$$(R\overset{O}{\overset{\|}{C}}OCH_2R') \qquad \text{(Structure of an ester)}$$

Occasionally, a carboxylic acid may combine with an alcohol to give an ester, as occurs when an acid and an alcohol are formed in alkane oxidation (Kallio, 1969):

$$RCH_2OH + HO\overset{O}{\overset{\|}{C}}R \longrightarrow RCH_2O\overset{O}{\overset{\|}{C}}R + H_2O$$

The two reactants sometimes may not be derived from the same precursor, however, as in the formation of ethyl dichloroacetate during the decomposition of the insecticide dichlorvos (Lieberman and Alexander, 1983):

$$RCH_2OH + HO\overset{O}{\overset{\|}{C}}R' \longrightarrow RCH_2O\overset{O}{\overset{\|}{C}}R' + H_2O$$

ALKANOIC ACIDS [$R(CH_2)_nCOOH$], ALKANES [$H(CH_2)_nCH_3$], AND ALKYL GROUPS [$R(CH_2)_nCH_3$]: β-OXIDATION

Compounds containing both an aromatic and an alkyl or aliphatic moiety may undergo β-oxidation. The same conversion occurs with alkanes and aliphatic acids. If the initial substrate has a terminal methyl group, as with $Ar(CH_2)_nCH_3$ or $H(CH_2)_nCH_3$, that methyl group is first converted to a carboxyl as shown on p. 204. The resulting acid then undergoes the sequence depicted in Fig. 12.3. Only a few of the intermediates formed in this sequence, which converts the alkyl portion to a series of products with 2 C atoms less, accumulate. However, the sequence has been reported for fatty acids, surfactants with alkyl portions, and phenoxyalkanoic acids. Some examples are given in Table 12.7.

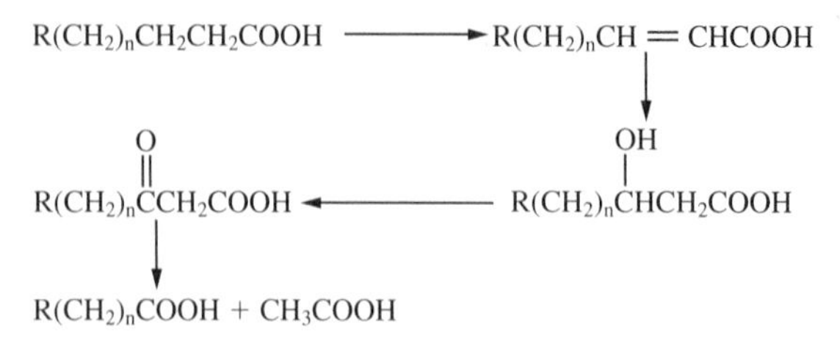

Figure 12.3 β-oxidation of alkanoic acids.

HYDROXYL GROUPS (ROH): METHYLATION AND ETHER (ROR′) FORMATION

Hydroxyl groups can be converted to the methoxy derivatives in a process known as *O*-methylation:

$$ROH \rightarrow ROCH_3$$

This type of reaction is common among chlorinated phenols, and the resulting *O*-methylated product is termed an anisole. Some of these *O*-methylations are given in Table 12.8. Some PAHs, such as phenanthrene and pyrene, can be converted to methoxy derivatives, presumably after one of the PAH rings is hydroxylated:

$$Ar \rightarrow ArOH \rightarrow ArOCH_3$$

Table 12.7

Substrates and Products of β-Oxidation

Substrate	Products	System	Reference
Azelaic acid	Pimelic acid	mc	Janota-Bassalik and Wright (1964)
10-(2,4-Dichlorophenoxy)octanoic acid	2,4-D	so	Gutenmann *et al.* (1964)
Dodecyltrimethylammonium	7-Carboxyheptyltrimethylammonium	mc	Dean-Raymond and Alexander (1977)
Diheptyl ether	2-*n*-Heptoxyacetic acid	mc	Modrzakowski and Finnerty (1980)
Hexane	Butyric acid	mc	Heringa *et al.* (1961)
MCPB	MCPA	so	Smith and Hayden (1981)
1-Phenyldecane	Phenylacetic acid	mc	Webley *et al.* (1956)

Table 12.8

Substrates and Products of Microbial Processes Leading to Ethers

Substrate	Products	System	Reference
Atrazine	Methoxy derivative	so	Tafuri *et al.* (1982)
2,4-D	2,4-Dichloroanisole	so	Smith (1985)
Diphenylmethane	1,1,1′,1′-Tetraphenyldimethyl ether	mc	Subba-Rao and Alexander (1977)
Endosulfan	Endosulfan ether	so	Ranga Rao and Murthy (1981)
Pentachlorophenol	Pentachloroanisole	so, mc	Kuwatsuka and Igarishi (1975), Cserjesi and Johnson (1972)
Phenanthrene	1-Methoxyphenanthrene	mc	Narro *et al.* (1992)
Pyrene	1,6-Dimethoxypyrene	mc	Wunder *et al.* (1997)
2,4,5-T	2,4,5-Trichloroanisole	so	McCall *et al.* (1981)
2,3,4,6-Tetrachlorophenol	2,3,4,6-Tetrachloroanisole	mc	Curtis *et al.* (1972)
2,4,6-Trichlororophenol	2,4,6-Trichlorophenylethyl ether	so	Schmitzer *et al.* (1989)

Phenoxy herbicides, such as 2,4-D and 2,4,5-T, are initially metabolized to the corresponding phenols (2,4-di- and 2,4,5-trichlorophenols), which are then methylated:

$$ArOCH_2COOH \rightarrow ArOH \rightarrow ArOCH_3$$

Even a N heterocycle like atrazine can be methylated, presumably after an initial hydroxylation.

Also shown in Table 12.8 is an *O*-ethylation:

$$ArOH \rightarrow ArOCH_2CH_3$$

More complex ethers are also possible microbial products, as indicated by the conversion of diphenylmethane to 1,1,1′1′-tetraphenyldimethyl ether:

$$Ar_2CH_2 \longrightarrow Ar_2CH{-}O{-}CHAr_2$$

Presumably here too the intermediate is the hydroxy compound.

ESTERS ($RCOCH_2R'$, with C=O): HYDROLYSIS

Esters are usually readily hydrolyzed microbiologically to give rise to the corresponding carboxylic acid and alcohol:

$$R\overset{O}{\overset{\|}{C}}OCH_2R' + H_2O \longrightarrow R\overset{O}{\overset{\|}{C}}OH + HOCH_2R'$$

These esterase-catalyzed reactions have been described for diverse substrates, including phthalate esters (Chauret *et al.*, 1995; Sanborn *et al.*, 1975; Engelhardt and Wallnöfer, 1978) and a variety of pesticides, including malathion (Walker, 1976), bromoxynil (Smith, 1980), and dichlorfop-methyl (Smith, 1977), and the conversion has been noted in microbial cultures, soil, and natural waters. Often, only one of the products is detected, either the acid or the alcohol, the other presumably being metabolized so readily that none is found.

ETHERS (ROR′): CLEAVAGE

Although ethers once were deemed to be resistant to microbial degradation, this generalization was quickly found to be untrue once experiments were performed. Many ethers are indeed cleaved. The substrates, some of which are presented in Table 12.9, include dialkyl and diaryl ethers, phenoxy herbicides, polyethylene glycols, and polyethoxylate surfactants. Such surfactants and polyethylene glycols have many ether linkages [$R\text{-}O\text{-}(CH_2CH_2O)_nCH_2OH$] and are readily cleaved. The conversion of methoxy compounds ($ROCH_3$), such as in chloroneb, dicamba, and anisoles ($ArOCH_3$), to the corresponding hydroxy derivatives represents another class of ether-cleavage reactions, which are also termed *O*-demethylations.

HALOGENATED AROMATICS: REDUCTIVE DEHALOGENATION

Under anaerobic conditions, many aromatic compounds with one to several chlorines can be dehalogenated, the chlorines being replaced by hydrogens:

$$ArCl + [2H] \rightarrow ArH + H^+ + Cl^-$$

Table 12.9

Ethers Cleaved by Microorganisms

Substrate	Products	System	Reference
Chloroneb	2,5-Dichlorohydroquinone	mc	Wiese and Vargas (1973)
2,4-D	2,4-Dichlorophenol	so	Smith (1974)
Dicamba	3,6-Dichlorosalicylate	so	Smith (1985)
2,4-Dichlorophenoxyalkanoic acids	Alkanoic acids	mc	MacRae *et al.* (1963)
2,7-Dichlorodibenzo-*p*-dioxin	1,2,4-Trihydroxybenzene	mc	Valli *et al.* (1992)
Dimethyl ether	Methanol	mc	Hyman *et al.* (1994)
Diphenyl ether	Phenol	mc	Schmidt *et al.* (1992)
Docecyl triethoxy sulfate	Ethylene glycol sulfate	mc	Hales *et al.* (1982)
MTBE	*tert*-Butyl alcohol	mc	Salanitro *et al.* (1994)
Nonylphenol polyethoxylate	Nonylphenol diethoxylate	mc	Kvestak and Ahel (1995)
2,4,5-T	2,4,5-Trichlorophenol	sd/an/mc	Bryant (1992), Rosenberg and Alexander (1980)

If the conversion is extensive, all or nearly all of the halogens can be removed. Thus, hexachlorobenzene and pentachlorobenzene are converted in sequence to one or more tetra-, tri-, and dichlorobenzenes, and finally chlorobenzene is formed anaerobically in a soil slurry (Ramanand *et al.*, 1993). However, different tetra- (e.g., 1,2,4,5- or 1,2,3,4-), tri- (1,2,4- or 1,2,3-), or dichloro- (1,2- or 1,4-) isomers are formed by different microbial assemblages. Likewise, pentachlorophenol (PCP) is sequentially converted by anaerobic cultures of microorganisms to various tetra-, tri-, and dichlorophenols, the identities of the isomers varying with the organisms, incubation conditions, or both (Chang *et al.*, 1996; Nicholson *et al.*, 1992); some anaerobic organisms will convert polychlorinated phenols to a monochlorophenol (Gardin *et al.*, 1994) and some will ultimately form phenol (Basu *et al.*, 1996). Thus, the anaerobic transformation may be envisioned as:

$$ArCl_6 \rightarrow HArCl_5 \rightarrow H_2ArCl_4 \rightarrow H_3ArCl_3 \rightarrow H_4ArCl_2 \rightarrow H_5ArCl \rightarrow H_6Ar$$

for hexachlorobenzene, and a similar sequence occurs with pentachlorophenol, with the isomers that are produced and the extent of dehalogenation depending on the populations and environmental conditions. Other chlorinated compounds that are reductively dehalogenated in this fashion are 2,4-dichlorophenoxyacetic acid (2,4-D) and 2,4,5-trichlorophenoxyacetic acid (2,4,5-T) to yield 4-chlorophenoxyacetic acid and 2,5-dichlorophenoxyacetic acid, respectively, in anerobic sediments (Bryant, 1992); hepta-, hexa-, and pentachlorinated dibenzo-*p*-dioxins to tetrachlorinated dibenzo-*p*-dioxins in anerobic sediments (Adriaens *et al.*, 1995); and the tetrachlorinated to tri-, di-, and monochlorinated dibenzo-*p*-dioxins by anaerobic enrichment cultures (Beurskens *et al.*, 1995). Generalizations cannot now be made about which isomer will be generated.

Additional examples are given in Table 12.10. Although most of these reductive dehalogenations are anaerobic, the process may sometimes be aerobic, as in the example of pentachloronitrobenzene (PCNB).

HALOGENATED ALKANES AND ALKENES: REDUCTIVE DEHALOGENATION

Halogen-containing alkanes and alkenes are also reductively dehalogenated by anaerobes. Examples are listed in Table 12.11. The Cl atoms are replaced with H atoms, and if the chlorinated alkane has two, three, or four chlorine substituents, each is reduced sequentially:

$$CCl_4 \rightarrow HCCl_3 \rightarrow H_2CCl_2$$

$$Cl_2CHCHCl_2 \rightarrow Cl_2CHCH_2Cl \rightarrow ClCH_2CH_2Cl \rightarrow ClCH_2CH_3$$

Table 12.10

Substrates and Products of Reductive Dehalogenation of Aromatic Compounds

Substrate	Products	System	Reference
Bromacil[a]	3-*sec*-Butyl-6-methyluracil	mc	Adrian and Suflita (1990)
Bromobenzoates	Benzoate	mc	Suflita *et al.* (1982)
4-Chlorophenol	Phenol	sd	Gibson and Suflita (1986)
Dicamba	2,4-Dihydroxy-3,6-dichlorosalicylate	mc	Taraban *et al.* (1993)
Dichlorobenzoates	Monochlorobenzoates	sd	Gibson and Suflita (1986)
2,4-Dichlorophenol	Phenol	sd	Gibson and Suflita (1986)
Diuron	3-Chlorophenylurea	mc	Stepp *et al.* (1985)
Pentachlorobenzyl alcohol	Tetrachlorobenzoates	so	Ishida (1972)
Pentachloronitrobenzene	Tetrachloroaniline	so, mc	deVos *et al.* (1974), Mora Torres *et al.* (1996)
2,3,5-Triiodobenzoate	Diiodobenzoates	so	Moy and Ebert (1972)

[a] A pyrimidine (N heterocycle) derivative.

Table 12.11

Substrates and Products of Reductive Dehalogenation of Halogenated Alkanes and Alkenes

Substrate	Products	System	Reference
Carbon tetrachloride	Chloroform, dichloromethane	mc/an	Stromeyer *et al.* (1992)
Chloroform	Dichloromethane	mc/an	Gupta *et al.* (1996)
Trichlorofluoromethane	Dichlorofluoromethane	mc/an	Sonier *et al.* (1994)
1,1,2,2-Tetrachloroethane	1,1,2-Trichloroethane, *cis*- and *trans*-1,2-dichloroethylene	mc/an	Chen *et al.* (1996)
1,1,1-Trichloroethane	1,1-Dichloroethane, vinyl chloride, dichloromethane	mc/an	Ahlert and Ensminger (1992)
1,1,2-Trichloroethane	1,2-Dichloroethane, vinyl chloride	mc/an	Chen *et al.* (1996)
1,2-Dichloroethane	Chloroethane, ethylene	mc/an	Chen *et al.* (1996)
Tetrachloroethylene (PCE)	TCE, *cis*- and *trans*-1,2- and 1,1-dichloroethylene, vinyl chloride, ethylene	mc/an	Fathepure and Tiedje (1994), Holliger *et al.* (1993), Skeen *et al.* (1995)
Trichloroethylene (TCE)	*cis*-1,2-Dichloroethylene, vinyl chloride, ethylene	mc/an	Wild *et al.* (1995)

The same is true of chlorinated alkenes such as tetra- and trichloroethylene (also known as PCE and TCE):

$$Cl_2C{=}CCl_2 \rightarrow Cl_2C{=}CHCl \rightarrow HClC{=}CHCl \text{ or } Cl_2C{=}CH_2 \rightarrow HClC{=}CH_2 \rightarrow H_2C{=}CH_2$$

The penultimate compound in the sequence (chloroethylene) is commonly called vinyl chloride.

Some earlier examples of reductive dehalogenation are given in Table 12.12.

Reductive dehalogenation is not limited to aromatic compounds, alkanes, and alkenes. For example, microorganisms in the rat cecum reductively dechlorinate trichloroacetic acid (Maghaddam *et al.*, 1996):

$$Cl_3CCOOH \rightarrow Cl_2HCCOOH$$

An analogous transformation may occur with substrates, such as DDT, that have a trichloroaklyl moiety:

$$\underset{\displaystyle CCl_3}{\underset{|}{ArCHAr}} \longrightarrow \underset{\displaystyle CHCl_2}{\underset{|}{ArCHAr}}$$

The product formed from DDT is known as DDD and it is formed in soil (Duffy

Table 12.12

Substrates and Products of Reductive Dehalogenation of Alkanes and Alkenes

Substrate	Products	System	Reference
Bromoethane	Ethane	mc	Belay and Daniels (1987)
Bromotrichloromethane	Chloroform	mc	Lam and Vilker (1987)
Chloropicrin	Nitromethane	mc	Castro *et al.* (1983)
1,2-Dichloroethane	Ethane	mc	Holliger *et al.* (1990)
Dichloroethylenes	Chloroethylene	gw	Wilson *et al.* (1982)
Tetrachloroethylene	Chloroethylene	sd, mc	Parsons *et al.* (1985), DiStefano *et al.* (1991)
Tetrachloromethane	Dichloromethane	mc	Egli *et al.* (1987)
1,1,1-Trichloroethane	1,1-Dichloroethane	sd	Parsons *et al.* (1985)
Trichloroethylene	Chloroethylene, 1,2-dichloroethylene, ethylene	gw, so, mc	Wilson *et al.* (1986), Kloepfer *et al.* (1985), Freedman and Gossett (1989)

and Wong, 1967), sea water (Juengst and Alexander, 1975), and microbial cultures (Wedemeyer, 1967); this conversion occurs both anaerobically and aerobically.

HALOGENATED COMPOUNDS: HYDROLYTIC DEHALOGENATION

Some microorganisms remove halogens in a hydrolytic process, with the halogen in the aromatic or other molecule being replaced by OH. The product is often metabolized further, sometimes by conversion of the hydroxyl to a carbonyl or quinone:

$$RCl + H_2O \rightarrow ROH + H^+ + Cl^-$$

Thus, *trans*-1,3-dichloropropene is converted to *trans*-3-chloroallyl alcohol in soil (Ou *et al.*, 1995), hexachlorobenzene is converted to pentachlorophenol in a model ecosystem (Lu *et al.*, 1978), and fungi are able to hydrolytically remove one Cl from 2,4-5-trichlorophenol and oxidize the product further (Joshi and Gold, 1993). The chlorine attached to the triazine ring of atrazine can be replaced with OH by a fungus (Couch *et al.*, 1965), and the chloromethyl of propachlor is converted to a hydroxymethyl in soil (Kaufman *et al.*, 1971).

HALOGENATED COMPOUNDS: DEHYDRODEHALOGENATION

Another means of transforming halogenated substrates involves the simultaneous removal of both halogen and H. Such reactions may occur with C atoms having three, two, or one chlorines:

$$R_2CHCCl_3 \rightarrow R_2C{=}CCl_2$$
$$R_2CHCHCl_2 \rightarrow R_2C{=}CHCl$$
$$R_2CHCH_2Cl \rightarrow R_2C{=}CH_2$$

These types of reactions occur in the metabolism of DDT. The first reaction is exemplified by the conversion of DDT to a product known as DDE, and the second reaction occurs later in the degradation when a subsequent product (DDMU) is dehydrodehalogenated. These transformations occur in culture (Wedemeyer, 1967), soil (Duffy and Wong, 1967), sea water (Juengst and Alexander, 1975), and sewage (Albone *et al.*, 1978). Lindane (an insecticide; α-1,2,3,4,5,6-hexachlorocyclohexane) also undergoes a dehydrodehalogenation in soil (Mathur and Saha, 1975) and culture (Ohisa *et al.*, 1980) in which the cycloalkane loses Cl and H to give an unsaturated ring. The metabolism in culture of 1,2-dibromoethane (EDB) appears to entail a similar dehydrodehalogenation (Belay and Daniels, 1987).

HALOGENATED COMPOUNDS: HALOGEN MIGRATION

In certain instances, a reaction is brought about that results in the migration of a chlorine from one C to another. Thus, a fungus converts 2,4-dichlorophenoxyacetic acid (2,4-D) to 2,5-dichloro-4-hydroxyphenoxyacetic acid (Faulkner and Woodcock, 1964), the Cl being displaced from C number 4 and moving to C number 5. In another example, one of the chlorines of 1,1,2-trichloroethylene (TCE) moves so that all three chlorines are on one C atom, as in the bacterial conversion of TCE to 2,2,2-trichloroethanol (Vanderberg *et al.*, 1995) and trichloroacetic acid (Uchiyama *et al.*, 1992).

TRIHALOMETHYL-CONTAINING COMPOUNDS ($RCCl_3$, RCF_3): TRANSFORMATION

Several agricultural chemicals have a trichloromethyl or trifluoromethyl substituent. These can be metabolized to give rise to the corresponding carboxylic acids:

$$RCCl_3 \rightarrow RCOOH$$

A well-documented instance is the degradation of DDT in culture (Wedemeyer, 1967) and fresh water (Pfaender and Alexander, 1972), where R in the equation is 4-chlorophenyl:

$$\underset{\displaystyle CCl_3}{\overset{\displaystyle RCHR}{|}} \longrightarrow \underset{\displaystyle COOH}{\overset{\displaystyle RCHR}{|}}$$

In this process, several intermediates are generated. Processes that effectively resemble the same type of conversion occur in the metabolism of nitrapyrin [2-chloro-6-(trichloromethyl)pyridine] to 6-chloropicolinic acid in culture (Vanelli and Hooper, 1992) and in soil (Redemann *et al.*, 1964) and the replacement of the trifluoromethyl of the insecticide fluvalinate by a carboxyl in soil (Staiger and Quistad, 1983).

HALOGENATED COMPOUNDS: CONVERSION TO METHYLTHIO DERIVATIVES

An interesting process has been reported in which a microorganism in culture replaces one of the chlorines of 2,4-dichloro-1-nitrobenzene or the fungicide chlorothalonil with a methylthio group (Tahara *et al.*, 1981; Katayama *et al.*, 1997):

$$ArCl \rightarrow ArSCH_3$$

The intermediate and the precise compound to which the methylthio is added are unknown. Metribuzin, a nonchlorinated pesticide, is also converted to a methylthio compound (Webster and Reiner, 1976), so if there is a common thread, the halogen must be converted to another intermediate, which in turn is converted to the methylthio product.

AMINES: REDUCTIVE DEAMINATION

The N in primary amines (RNH_2) may be removed by a reductive mechanism, a hydrolysis, or a dehydrodeamination. Amino acids are the most extensively studied group of primary amino compounds, and a model α-amino acid

$$RCH_2\overset{\displaystyle NH_2}{\overset{|}{C}}HCOOH$$

may be initially metabolized to yield ammonium plus the following:

Reductive	RCH_2CH_2COOH
Hydrolytic	$RCH_2CH(OH)COOH$
Dehydrodeamination	$RCH{=}CHCOOH$

A typical reductive process occurs as anaerobic bacteria remove the amino groups of triaminotoluene to yield toluene (Boopathy *et al.,* 1993). For one amino group, the process is:

$$ArNH_2 \rightarrow ArH$$

The amino linked to the N in N heterocycles may also be reductively deaminated, as in the degradation of the herbicide metribuzin in soil (Webster and Reiner, 1976) and the herbicide metamitron in culture (Engelhardt and Wallnöfer, 1978):

$$>N{-}NH_2 \longrightarrow >NH$$

AMINES: HYDROLYTIC DEAMINATION

In addition to amino acids, a number of synthetic compounds can be hydrolytically deaminated. The product is the corresponding hydroxy compound:

$$RNH_2 \rightarrow ROH$$

Several examples are given in Table 12.13.

Some amines may be converted to carbonyl compounds. The reactions may be written as:

$$RCH_2NH_2 \longrightarrow RCHO$$

$$\begin{matrix} R \\ R' \end{matrix}\!>CHNH_2 \longrightarrow \begin{matrix} R \\ R' \end{matrix}\!>C{=}O$$

Such a conversion has been reported in culture for benzylamine (Durham and Perry, 1978), the herbicide glufosinate (Bartsch and Tebbe, 1989), and cyclohexylamine (Tokieda *et al.,* 1979). It is likely that the carbonyl compound is formed following an initial hydrolytic deamination:

$$RCH_2NH_2 \rightarrow RCH_2OH \rightarrow RCHO$$

Table 12.13

Substrates and Products of Hydrolytic Deamination

Substrate	Products	System	Reference
Anthranilic acid	2,3-Dihydroxybenzoate	mc	Staron *et al.* (1966)
Asulam	4-Hydroxysulfonate	mc	Balba *et al.* (1979)
Chlornitrofen	4-(2,4,6-Trichlorophenoxy)phenol	so	Oyamada and Kuwatsuka (1979)
2-Chloro-1,3,5-triazine-4,6-diamine	Hydroxy deivative	mc	Grossenbacher *et al.* (1984)
Cyclohexylamine	Cyclohexanol	mc	Tokieda *et al.* (1979)
Dinoseb	Hydroxy derivative	mc	Kaake *et al.* (1995)
1,4-Diaminobenzene	4-Hydroxyaniline	se	Udod (1972)

AMINES: ACYLATION

A common conjugation carried out by microorganisms is *N*-acylation. In these processes, an aromatic amine is converted to an *N*-acyl derivative. Most common apparently are *N*-acetylations and *N*-formylations:

$$\mathrm{ArNH_2} \longrightarrow \mathrm{ArNH\overset{\overset{\displaystyle O}{\|}}{C}CH_3} \text{ or } \mathrm{ArNH\overset{\overset{\displaystyle O}{\|}}{C}H}$$

Sometimes, the *N*-propionyl derivatives ($\mathrm{ArNH\overset{\overset{O}{\|}}{C}CH_2CH_3}$) are formed. Early examples are listed in Table 12.14. More recently, *N*-acetylations have been found to occur as microorganisms reduce di- and trinitro compounds, including TNT, to corresponding aromatic amines and then acetylate the amines (Bruns-Nagel *et al.,* 1996; Noguera and Friedman, 1996).

AMINES: *N*-METHYLATION

Primary amines may be *N*-methylated, and the product is the monomethyl, dimethyl, or both products: For example, microorganisms in culture can convert aniline to *N*-methyl- and *N,N*-dimethylaniline (Lenfant *et al.,* 1970).

Table 12.14

Substrates and Products of *N*-Acylation of Aromatic Amines

Substrate	Products	System	Reference
4-Aminoazobenzene	Acetanilides	mc	Idaka *et al.* (1982)
Aniline	Acetanilide, formanilide	mc	Cerniglia *et al.* (1981)
Anthranilic aicd	*N*-Acetylanthranilic acid	mc	Lübbe *et al.* (1986)
Benzidine	Mono- and diacetyl derivative	so	Lu *et al.* (1977)
Bifenox	Acetyl and formyl derivatives	so	Ohyama and Kuwatsuka (1978)
Chlomethoxynil	Acetyl and propionyl derivatives	so	Niki and Kuwatsuka (1976)
4-Chloroaniline	Acetyl and propionyl derivatives	mc	Engelhardt *et al.* (1977)
4-Chloroaniline	Acetyl and formyl derivatives	mc	Freitag *et al.* (1984)
Chlornitrofen	Acetyl and formyl derivatives	so	Oyamada and Kuwatsuka (1979)
3,4-Dichloro-4-nitroaniline	Formyl derivative	so	Kearney and Plimmer (1972)
Dinoseb	Acetyl derivative	mc	Wallnöfer *et al.* (1978)
Linuron	3,4-Dichloroacetanilide	mc	Funtikova (1979)
4-Toluidine	Formyl and acetyl derivative	se	Hallas and Alexander (1983)

SECONDARY AMINES, TERTIARY AMINES, AND QUATERNARY N COMPOUNDS: TRANSFORMATIONS

Some secondary amines, tertiary amines, and quaternary N compounds are environmentally significant. When the *N*-linked substituents are alkyl groups, those portions can be removed. This is illustrated by the removal of the *N*-linked ethyl and isopropyl groups from atrazine to give *N*-dealkylated products, a process that occurs in soil (Shelton *et al.*, 1995) and microbial cultures (Masaphy *et al.*, 1996; Behki and Khan, 1994):

$$RNH(CH_2)_nCH_3 \rightarrow RNH_2$$

In these cases, the moiety removed from the N is replaced by H.

The metabolism by bacteria of a tertiary amine may result in the formation of a secondary amine, as when dodecyldimethylamine is converted to dimethylamine (Kroon *et al.*, 1994):

$$CH_3(CH_2)_{11}N(CH_3)_2 \rightarrow HN(CH_3)_2$$

Similarly, trimethylamine remains as a product when bacteria degrade the quaternary N compound, hexadecyltrimethylammonium chloride (Ginkel *et al.*, 1992):

$$CH_3(CH_2)_{15}\overset{+}{N}(CH_3)_3 \rightarrow N(CH_3)_3$$

AMINES: N OXIDATION

The N in primary amines may be oxidized. The products that accumulate may be one or several of the following:

$$RNH_2 \rightarrow RNHOH \rightarrow RNO \rightarrow RNO_2$$

The most common products appear to be the nitro compounds (RNO_2), but the hydroxylamino (RNHOH) or nitroso (RNO) derivatives may also appear.

Secondary amines (RNHR′) or tertiary amines [R(R′)NR″] may be converted to the corresponding *N* oxides.

$$R(R')NH \longrightarrow R(R')N \rightarrow O$$

$$R(R')(R'')N \longrightarrow R(R')(R'')N \rightarrow O$$

Several examples are listed in Table 12.15.

AMINES: CONVERSION TO N HETEROCYCLES

A novel series of products is sometimes formed in the metabolism of aromatic amines in which a second ring compound is produced. If the substrate contains one N, the second ring bears that N. However, if the substrate contains a second nitrogen *ortho* to the amino group, the second ring contains both N atoms. Thus, aniline is converted in sewage to 2-methylquinoline, and 2-nitroaniline is transformed to yield 2-methylbenzimidazole (Hallas and Alexander, 1983). This is illustrated in Fig. 12.4. Some herbicides contain diethylamino [$RN(CH_2CH_3)_2$] or dipropylamino [$RN(CH_2CH_2CH_3)_2$] groups on the aromatic ring, and *ortho* to these groups is a nitro group. These two are converted to benzimidazoles in a reaction similar to that shown in Fig. 12.4. This type of transformation has been reported for dinitramine in soil (Smith *et al.*, 1973) and culture (Laanio *et al.*,

Table 12.15

Substrates and Products of *N*-Oxidation of Amines

Substrate	Products	System	Reference
4-Aminobenzoate	4-Nitrobenzoate	mc	Sloane *et al.* (1963)
2-Amino-4-nitrophenol	2,4-Dinitrophenol	mc	Madhosingh (1961)
Aniline	Phenylhydroxylamine, nitrobenzene	mc	Lyons *et al.* (1984), Russel *et al.* (1979)
4-Chloroaniline	4-Chlorophenylhydroxylamine, 4-chloronitrosobenzene,	so, mc	Freitag *et al.* (1984), Kaufman *et al.* (1973)
3,4-Dichloroaniline	3,4-Dichlorophenyl-hydroxylamine, 3,4-dichloronitrobenzene	mc	Lee and Kim (1978)
2,6-Dimethylpyridine	*N*-oxide derivative	mc	Kost *et al.* (1977)
Tridemorph	Tridemorph *N*-oxide	so	Otto and Drescher (1973)

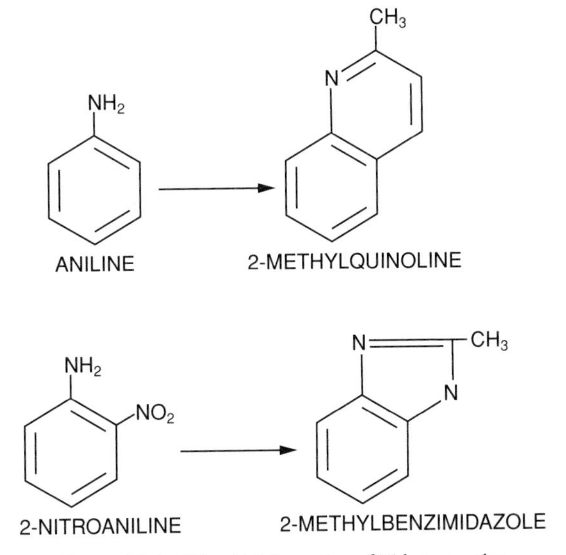

Figure 12.4 Microbial formation of N heterocycles.

1973), fluchloralin in soil (Kearney *et al.*, 1976), and oryzalin in soil (Golab *et al.*, 1975). In such conversions, presumably the nitro group is reduced, a conjugate is formed to add C to one of the two N atoms, and the ring is closed.

AMINES: DIMERIZATON

A number of aromatic amines undergo a dimerization. The products may be azobenzenes:

$$ArNH_2 \rightarrow ArN{=}NAr$$

or azoxybenzenes:

$$ArNH_2 \rightarrow Ar\overset{\overset{O}{\uparrow}}{N}{=}NAr$$

Sometimes, an additional N is somehow introduced to generate a triazene:

$$ArNH_2 \rightarrow ArNHN{=}NAr$$

The process may occur following the introduction of the aromatic amine to cultures or soil or it may require initial steps to generate the amine that then dimerizes; e.g., the reduction of nitro groups of TNT or trifluralin or a series of other steps with the pesticides imugam, karsil, propanil, or swep to yield the appropriate aromatic. Several examples are given in Table 12.16. It is possible that the dimerization step may be nonbiological (Noguera and Freedman, 1996; Haidour and Ramos, 1996).

SECONDARY AMINES (RNHR′): *N*-NITROSATION

Nitroso groups (-NO) may be added to secondary amines. This process, which is termed *N*-nitrosation, is of considerable importance because many of these *N*-nitroso compounds (often called nitrosamines) are chronic toxicants, often at extremely low concentrations. Many, or most, are carcinogenic, mutagenic, and teratogenic. The actual *N*-nitrosation reaction involves a secondary amine and nitrite:

$$RR'NH + NO_2^- \longrightarrow RR'N{-}NO + OH^-$$

Nitrite in concentrations sufficient for *N*-nitrosation is formed from nitrate or ammonium in many natural environments. Secondary amines are ubiquitous and

Table 12.16

Substrates and Products of the Dimerization of Amines

Substrate	Products	System	Reference
4-Chloroaniline	4,4′-Dichloroazobenzene, 4,4′-dichloroazoxybenzene	mc, so	Kaufman *et al.* (1973), Freitag *et al.* (1987)
4-Chloroaniline	1,3-Bis(4-chlorophenyl)triazene	mc	Minard *et al.* (1977)
3,4-Dichloroaniline	3,3′,4,4′-Tetrachloroazobenzene	so, mc	Bartha and Pramer (1967), Kaufman *et al.* (1972)
3,4-Dichloroaniline	1,3-Bis(3,4-dichlorophenyl)triazene	mc	Corke *et al.* (1979)
2,4-Dinitrotoluene	2,2′-Dinitro-4,4′-azoxytoluene	mc	McCormick *et al.* (1978)
Imugam	3,3′,4,4′-Tetrachloroazobenzene	so	Satiriou *et al.*, (1976)
Karsil	3,3′,4,4′-Tetrachloroazobenzene	so	Bartha (1968)
Swep	3,3′,4,4′-Tetrachloroazobenzene	so	Bartha and Pramer (1969)
TNT	4,4′-Azoxy-2,2′,6,6′-tetranitrotoluene	mc	McCormick *et al.* (1976)
Trifluralin	Azoxybenzene derivative	so	Golab *et al.* (1979)

they may be generated in natural or polluted environments from tertiary amines or quaternary N compounds:

$$RR'N^{+}R''R''' \longrightarrow RR'N{-}R'' \longrightarrow RR'NH$$

N-nitrosation has been reported for dimethylamine in soil and microbial cultures (Ayanaba *et al.*, 1973; Mills and Alexander, 1976), diphenylamine (Ayanaba and Alexander, 1973) and pyrrolidine (Hawksworth and Hill, 1971) in culture, diethanolamine in lake water and sewage (Yordy and Alexander, 1982), and trimethylamine in sewage, lake water, soil, and culture (Ayanaba and Alexander, 1974; Ayanaba *et al.*, 1973).

AMINES: S ADDITION

A novel conversion, which has not been observed often, is the addition of S containing a methyl group to an amine. The substrate for the microbial culture was 2,6-dinitro-4-trifluoromethylaniline (Lusby *et al.*, 1980):

$$ArNH_2 \longrightarrow ArNH\overset{\overset{O}{\parallel}}{S}CH_3$$

ALKYLAMINES [RNHALK, RN(ALK)$_2$, R$\overset{+}{N}$(ALK)$_3$]: DEALKYLATION

Various synthetic and naturally occurring amines contain one, two, or three alkyl groups. Many herbicides have these structures. The synthetic and natural products may be secondary (RNHR′) or tertiary [R(R′)NR″] amines, and the alkyl moieties may be methyl, ethyl, or propyl groups. This *N*-dealkylation is a reductive process:

$$RNHAlk \longrightarrow RNH_2$$

$$RNAlk_2 \longrightarrow RNHAlk \longrightarrow RNH_2$$

$$>NAlk \longrightarrow >NH$$

Rarely is the identity established of the portion removed. Some examples are given in Table 12.17.

CARBAMATES (R$\overset{\overset{O}{\parallel}}{C}$—NHR′) AND AMIDES (R$\overset{\overset{O}{\parallel}}{C}NH_2$): CLEAVAGE

A number of herbicides, insecticides, and fungicides are carbamates, and amides are found among industrial chemicals. These may be converted to the corresponding carboxylic acid, amine, or both. Rarely are both products of the cleavage identified:

$$R\overset{\overset{O}{\parallel}}{C}NHR' \longrightarrow RCOOH + H_2NR'$$

$$R\overset{\overset{O}{\parallel}}{C}NH_2 \longrightarrow RCOOH + NH_3$$

Table 12.17

Substrates and Products of *N*-Dealkylation

Substrate	Products	System	Reference
Atrazine	Mono- and didealkyl derivatives	so, mc	Beynon *et al.* (1972a), Behki and Khan (1994)
Chlorotoluron	Mono- and didemethyl derivatives	so	Gross *et al.* (1979)
Dimethylethylamine	Ethylamine	mc	Fahlbusch *et al.* (1983)
Dinitramine	Mono- and dideethyl derivatives	so	Smith *et al.* (1973)
Diuron	Mono- and didemethyl derivatives	so, mc	Dalton *et al.* (1966), Tillmans *et al.* (1978)
Fluchloralin	Dealkyl derivative	so	Kearney *et al.* (1976)
Fluometuron	Mono- and didemethyl derivatives	so, mc	Rickard and Camper (1978), Bozarth and Funderburk (1971)
Linuron	Mono- and didemethyl derivatives	mc	Tillmans *et al.* (1978)
Monuron	Mono- and dideethyl derivatives	mc	Tillmans *et al.* (1978)
Profluralin	Depropyl derivative	mc	Stralka and Camper (1981)
Simazine	Mono- and dideethyl derivatives	so, mc	Beynon *et al.* (1972b), Kearney *et al.* (1965)
Trifluralin	Mono- and didepropyl derivatives	so, mc	Kearney *et al.* (1976)
Trimethylamine	Dimethylamine	se, sw, mc	Ayanaba and Aelxander (1973, 1974)

Some examples are given in Table 12.18.

NITRILES (RC≡N): CONVERSION TO AMIDES AND CARBOXYLIC ACIDS

A number of nitriles are of importance in industry or as pesticides. They may be converted to the corresponding amides, carboxylic acids, or both. The reaction sequence appears to be:

Table 12.18

Substrates and Products of Cleavage of Carbamates and Amides

Substrate	Products	System	Reference
Acrylamide	Acrylic acid	so	Nishikawa *et al.* (1979)
Alachlor	2,4-Diethylaniline	mc	Tiedje and Hagedorn (1975)
Asulam	Sulfanilamide	so, mc	Smith and Milward (1983), Balba *et al.* (1979)
Barban	3-Chloroaniline	so, mc	Quilt *et al.* (1979), Wright and Forey (1972)
Carbendazim	2-Aminobenzimidazole	so	Baude *et al.* (1974)
2-Chlorobenzamide	2-Chlorobenzoic acid	mc	Fournier and Catroux (1972)
Diflubenzuron	4-Chlorophenylurea	so, mc	Verloop *et al.* (1975), Seuferer *et al.* (1979)
Fluometuron	3-Trifluoromethylaniline	so	Bozarth and Funderburk (1971)
Linuron	3,4-Dichloroaniline	mc	Wallnöfer (1969)
Metobromuron	4-Bromoaniline	mc	Tweedy *et al.* (1970)
Propanil	3,4-Dichloroaniline	so	Bartha and Pramer (1967)

$$RC{\equiv}N \longrightarrow R\overset{\displaystyle O}{\overset{\|}{C}}NH_2 \longrightarrow R\overset{\displaystyle O}{\overset{\|}{C}}OH$$

This type of conversion occurs with three dihalogenated benzonitriles used as pesticides, namely dichlobenil in soil (Verloop, 1972; Fournier and Salle, 1974), bromoxynil in soil and microbial cultures (Collins, 1973; Smith and Cullimore, 1974), and ioxynil in culture (Hsu and Camper, 1979). Analogous conversions occur with acrylonitrile (Yamada *et al.*, 1979) and triacrylonitrile (Asano *et al.*, 1981) in culture and the herbicide cyanazine in soil (Beynon *et al.*, 1972b).

N-NITROSO COMPOUNDS (NITROSAMINES): DENITROSATION

As indicated earlier, *N*-nitroso compounds are of importance because many or most are carcinogenic, teratogenic, and mutagenic at very low concentrations. In culture, they may be converted to the corresponding secondary amines, as has been reported for *N*-nitrosodimethylamine (Royland and Grasso, 1975):

$$(CH_3)_2NNO \rightarrow (CH_3)_2NH$$

AZOBENZENES: REDUCTION

Microorganisms are sometimes able to reduce the *N,N*-dimer, azobenzene, to the corresponding amines, as shown in culture for 4-aminoazobenzene (Idaka *et al.*, 1982):

$$ArN{=}NAr' \rightarrow ArNH_2 + H_2NAr'$$

NITRO COMPOUNDS (RNO_2): REDUCTION

In the metabolism of nitroaromatic compounds containing one nitro group ($ArNO_2$), the transformation often entails a sequential reduction to yield the corresponding nitroso, hydroxylamino, and amino compounds:

$$ArNO_2 \rightarrow ArNO \rightarrow ArNHOH \rightarrow ArNH_2$$

However, the two intermediates frequently may not be detected, although occasionally they are excreted and persist. The same type of sequence may occur when the substrate contains two or three nitro groups, but one or another of the nitro groups in these instances may be completely reduced to $-NH_2$, partly reduced to -NO or -NHOH, or remain unaltered. Which of these products is found depends on the species carrying out the transformation. The complete reduction is illustrated by the conversion of nitrophenols, 4-nitrobenzoic acid, and 4-nitroaniline to the corresponding amino compounds by anaerobes (Gorontsy *et al.*, 1993). Such reductions may also be carried out by some aerobic microorganisms (Bumpus and Tatarko, 1994) and in aerobic environments. The formation of 2-nitroso-, 2-hydroxylamino-, and 2-amino-6-nitrotoluene from 2,6-dinitrotoluene by a bacterium illustrates the stepwise conversion but only of one of the nitro substituents (Sayama *et al.*, 1992), whereas another bacterium forms two monohydroxylamino dinitrotoluenes [$(O_2N)_2ArNHOH$], two aminodinitroluenes [$O_2N)_2ArNH_2$], and a diaminomononitrotoluene [$(O_2N)Ar(NH_2)_2$] from 2,4,6-trinitroluene (TNT) (Haidour and Ramos, 1996). Given the possibility of one, two, or three substituents that may be $-NO_2$, -NO, -NHOH, or $-NH_2$, the accumulating compound clearly can be one of many, and many have been found. Moreover, in some instances, as in the bacterial conversion of 2,4-dinitrophenol to 2-hydroxy-5-nitropenta-2,4-dienoic acid (Ecker *et al.*, 1992), a nitro group is retained even after cleavage of the benzene ring.

Additional examples are given in Table 12.19.

NITRO COMPOUNDS: HYDROLYTIC DENITRATION

Some microorganisms are capable of hydrolytically removing a nitro group and leaving a hydroxyl in its place. The N is presumably liberated as nitrite:

Table 12.19

Substrates and Products of Reduction of Nitro Groups

Substrate	Products	System	Reference
4-Chloronitrobenzene	Nitroso, hydroxylamino, and amino derivatives	mc	Corbett and Corbett (1981)
2,6-Dichloro-4-nitroaniline	Amino derivative	so	Van Alfen and Kosuge (1976)
1,2-Dinitrobenzene	Nitroaniline	se	Hallas and Alexander (1983)
2,4-Dinitrophenol	2-Amino-4-nitrophenol	mc	Madhosingh (1961)
Methyl parathion	Amino methyl parathion	so	Adhya *et al.* (1981)
Nitrobenzene	Aniline	se, mc	Hallas and Alexander (1983), Cartwright and Cain (1959)
4-Nitrobenzoic acid	4-Hydroxylamino-benzoic acid	mc	Gingell (1973)
Nitroluenes	Toluidines	se	Hallas and Alexander (1983)
Parathion	Aminoparathion	so	Adhya *et al.* (1981)
RDX	Trinitrosotriazine	se	McCormick *et al.* (1980)
TNT	4-Amino-2,6-dinitrotoluene	gw, so	Pereira *et al.* (1979), Pennington and Patrick (1990)
3-Trifluoromethyl-4-nitrophenol	Amino derivative	sw	Bothwell *et al.* (1973)

$$RNO_2 + H_2O \rightarrow ROH + NO_2^- + H^+$$

This is illustrated by the conversion in microbial culture of 4-nitrophenol to hydroquinone (Spain and Gibson, 1991), the herbicide DNOC (4,6-dinitro-*o*-cresol) to 2,3,5-trihydroxytoluene (Tewfik and Evans, 1966), and 1,3-dinitrobenzene to nitrophenol (Dey and Godbole, 1986).

NITRO COMPOUNDS: REDUCTIVE DENITRATION

$$ArNO_2 \rightarrow ArH$$

Such a conversion is illustrated by a bacterium that converts 1,5-dinitrobenzene to 5-nitrobenzene (Boopathy *et al.*, 1994).

NITRATE ESTERS ($RONO_2$): CLEAVAGE

$$RONO_2 \rightarrow ROH$$

Mono-, di-, and trinitrate esters are subject to biodegradation in which an -OH replaces -ONO_2. The products are the corresponding mono-, di-, or trihydroxy compounds. For example, glyceryl trinitrate (more commonly called nitroglycerin) is stepwise converted by microbial cultures to glycerol dinitrate and then to glycerol mononitrate isomers and finally to glycerol (Wendt *et al.*, 1978; Meng *et al.*, 1995). Likewise, ethylene glycol dinitrate is degraded by bacteria to ethylene glycol (Ramos *et al.*, 1996). In these nitrate esters, the sequence thus can be depicted as follows:

$$R(ONO_2)_3 \rightarrow HOR(ONO_2)_2 \rightarrow (HO)_2RONO_2 \rightarrow (HO)_3R$$

C–S BOND: CLEAVAGE

A number of environmentally important compounds contain C–S bonds. These include thioethers (RSR′) and sulfonic acids (RSO_3H). Frequently, the nature of the cleavage reaction is uncertain because of the few products isolated. With thioethers, a thiol (RSH) is generated, as in the case of methionine and ethionine in soil (Banwart and Bremner, 1975). A similar type of conversion leads to the breakdown of ordram in culture (Golovleva *et al.*, 1978):

$$RSR' \rightarrow RSH$$

In the degradation of the insecticide malathion in soil and culture, both products of the cleavage have been identified, albeit under different test conditions, and the reaction appears to be represented by this equation (Matsumura and Boush, 1966; Rosenberg and Alexander, 1979):

$$RSR' \rightarrow RSH + HR'$$

Malathion is not a thioether but the S instead is linked to one C and one P.

In the metabolism of other compounds, cleavage of the C–S bond results in the formation of a hydroxy derivative, as with the herbicide ametryn in culture (Cook and Hütter, 1982):

$$RSR' \rightarrow ROH$$

or sometimes a carbonyl derivative is generated, as with the fungicide denmert in soil (Ohkawa *et al.*, 1976):

$$\begin{matrix} R \\ R' \end{matrix} \rangle CHSR'' \longrightarrow \begin{matrix} R \\ R' \end{matrix} \rangle C{=}O$$

However, the herbicide orbencarb is degraded to yield a sulfonic acid (Ikeda *et al.*, 1986):

$$RCH_2SCR' \longrightarrow RCH_2SO_3H$$
(with O double-bonded to the C in RCH₂SCR′)

Many surfactants are sulfonic acids or sulfonates. These are used widely in detergents. In culture, microorganisms metabolize simple sulfonates, such as benzene sulfonate and 4-toluene sulfonate to hydroxy compounds. The sulfur is liberated as sulfate or sulfite (Cain and Farr, 1968; Kertesz *et al.*, 1994):

$$RSO_3H \rightarrow ROH$$

In the degradation of alkylbenzene sulfonates, in which the sulfonate is attached to the benzene ring, the cleavage also yields the hydroxy compound and sulfite or sulfate (van Ginkel 1996; Willetts and Cain, 1970):

$$ArSO_3H \rightarrow ArOH$$

SULFATE ESTERS ($ROSO_3H$): CLEAVAGE

The cleavage of these esters gives rise to the corresponding alcohols. Sulfate may be the inorganic product (van Ginkel, 1996):

$$ROSO_3H \rightarrow ROH$$

Two examples are the microbial conversion of sodium dodecyl sulfate (the surfactant SDS) to dodecanol (Marchesi *et al.*, 1994) and the conversion of the herbicide 2-(2,4-dichlorophenoxy)ethyl sulfate to 2-(2,4-dichlorophenoxy)ethanol in soil or culture (Vlitos and King, 1953).

THIOLS (RSH): METHYLATION

Several thiols are known to be methylated, e.g., substituted benzenethiols in waste water (Reemtsa *et al.*, 1995) and culture (Drotar and Fall, 1985):

$$ArSH \rightarrow ArSCH_3$$

THIOLS: DIMERIZATION

Several simple thiols are converted to dimers:

$$RSH \rightarrow RSSR$$

For example, 4-mercaptobenzoate in culture is converted to 4,4′-dithiobenzoate (Entsch *et al.*, 1976), and the formation of dimethyl disulfide (CH_3SSCH_3) from methionine in soil probably arises as a consequence of the intermediary formation of methane thiol (CH_3SH) (Banwart and Bremner, 1975). A number of S-containing pesticides are converted to disulfide (-SS-)-containing dimers, presumably after the pesticide is first converted to the thiol, e.g., benthiocarb in culture (Golovleva *et al.*, 1980) and azinphosmethyl (Engelhardt *et al.*, 1981), denmert (Ohkawa *et al.*, 1976), and hinosan (Uesugi *et al.*, 1972) in soil.

THIOETHERS (RSR′): OXIDATION

A number of natural and synthetic compounds of environmental importance are thioethers. Considerable effort has been devoted to studying their transformations because some are pesticides that are used widely. Characteristically these are oxidized to the corresponding sulfoxides and sulfones. A list of thioether-containing insecticides, herbicides, and fungicides in which the S is oxidized to sulfoxide and/or sulfone is given in Table 12.20.

$$RSR' \longrightarrow R\overset{O}{\overset{\|}{S}}R' \longrightarrow R\underset{\underset{O}{\|}}{\overset{\overset{O}{\|}}{S}}R'$$

The same conversion occurs as microorganisms in culture oxidize dibenzothiophene, a nonpesticidal thioether, to its corresponding sulfoxide (Laborde and Gibson, 1977).

Table 12.20

Substrates and Products of Oxidation of Thioethers

Substrate	Products	System	Reference
Aldicarb	Sulfoxide, sulfone	so	Richey *et al.* (1977)
Benthiocarb	Sulfoxide	so	Ishikawa *et al.* (1976)
Carboxin	Sulfoxide, sulfone	so, sw, mc	Chin *et al.* (1970), Wallnöfer *et al.* (1972)
Disulfoton	Sulfoxide, sulfone	so	Clapp *et al.* (1976)
Ethiofencarb	Sulfoxide, sulfone	so, sw	Dräger (1977)
Fensulfoton	Sulfone	so	Chisholm (1974)
Fenthion	Sulfoxide	mc	Wallnöfer *et al.* (1976)
Phorate	Sulfoxide, sulfone	so, mc	Lichtenstein *et al.* (1974), Le Patourel and Wright (1976)
Terbutryn	Sulfoxide	sd	Muir and Yarechewski (1982)

Evidence also exists for the reduction of sulfoxides, e.g., the sulfoxide of phorate is reduced in soil (Getzin and Shanks, 1970) and the sulfoxides of fensulfothion (Timms and MacRae, 1982) and dimethyl sulfide (Zinder and Brock, 1978) are reduced in culture:

$$R\overset{\overset{\displaystyle O}{\|}}{S}R' \longrightarrow RSR'$$

DISULFIDES (RSSR): CLEAVAGE

The disulfide thiram is known to be cleaved. The product of cleavage of this fungicide in soil (Kumarasamy and Raghu, 1976) and culture (Odeyemi and Alexander, 1977) is the reduced form of the monomer:

$$RSSR \rightarrow RSH$$

PHOSPHATE ESTERS: CLEAVAGE

A number of insecticides are phosphate esters. These have the following general structure:

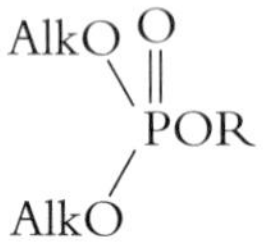

$$(AlkO)_2P(=O)OR$$

where Alk is either a methyl or ethyl group. The structure of the moiety designated R varies greatly. The typical products of their degradation include the monoalkyl derivative (I), a dealkylated compound (II), dialkyl phosphate (III), and/or one or several metabolites generated from the R group:

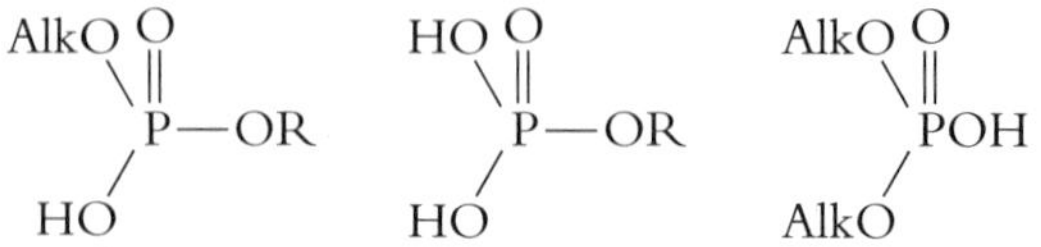

$$\underset{\text{I}}{(AlkO)(HO)P(=O){-}OR} \qquad \underset{\text{II}}{(HO)(HO)P(=O){-}OR} \qquad \underset{\text{III}}{(AlkO)_2P(=O)OH}$$

Such products are generated in the biodegradation of chlorfenvinphos (Beynon and Wright, 1967) and gardona (Beynon and Wright, 1969) in soil and dichlorvos in culture (Lieberman and Alexander, 1983).

PHOSPHOROTHIOATES: CLEAVAGE

A group of phosphorothioates also are potent insecticides:

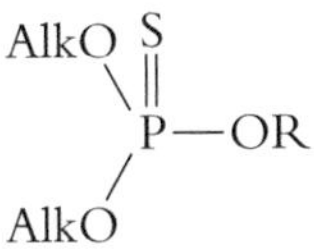

```
AlkO S
    \||
     P—OR
    /
AlkO
```

Here, too, Alk is commonly a methyl or ethyl group, and the structure of R varies widely. The products of degradation may be one or more of the following: the monoalkyl derivative (I), dialkylphosphorothioate (II), and a hydroxy compound derived from the R group (designated ROH):

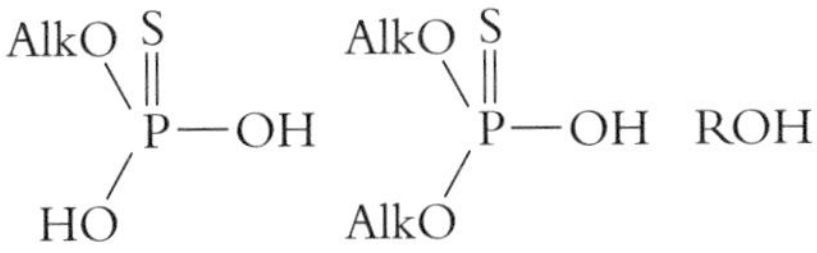

```
AlkO S          AlkO S
    \||             \||
     P—OH            P—OH   ROH
    /               /
  HO            AlkO

   I               II
```

Examples are given in Table 12.21.

In addition, as presented below, the S can be replaced by O so that the products may include the following:

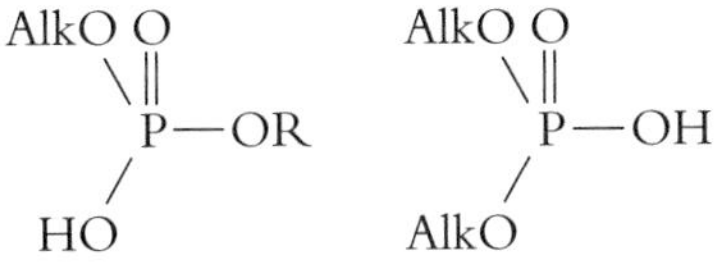

```
AlkO O          AlkO O
    \||             \||
     P—OR            P—OH
    /               /
  HO            AlkO
```

PHOSPHOROTHIOLATES: DEGRADATION

Some insecticides are phosphorothiolates. These have the following structure:

```
AlkO O
    \||
     PSR
    /
AlkO
```

Studies of one, Kitazin P (in which Alk is isopropyl), show that the products in soil and microbial culture (Tomizawa 1975; Uesugi *et al.*, 1972) include the following:

Table 12.21

Products of Degradation of Phosphorothioates

Substrate	Products	System	Reference
Chlorpyrifos	ROH	so	Racke *et al.* (1994)
Cyanox	Monodealkyl derivative, ROH	so	Chiba *et al.* (1976)
Dasanit	Dialkylphosphorothioate	mc	Rosenberg and Alexander (1979)
Diazinon	ROH	so, mc	Sethunathan and Pathak (1972), Sethunathan and Yoshida (1973a)
Diazinon	Dialkylphosphorothioate	so	Konrad *et al.* (1967)
Fenitrothion	Monodealkyl derivative, ROH	so	Takimoto *et al.* (1976)
Fenitrothion	Dialkylphosphorothioate, ROH	mc	Miyamo *et al.* (1966), Baarschers and Heitland (1986)
Fenthion	ROH	mc	Wallnöfer *et al.* (1978)
Methyl parathion	ROH	so	Misra *et al.* (1992)
Parathion	Monodealkyl derivative, ROH	so	Adhya *et al.* (1981), Sethunathan and Yoshida (1973b)
Parathion	Dialkylphosphorothioate, ROH	mc	Munnecke and Hsieh (1976)

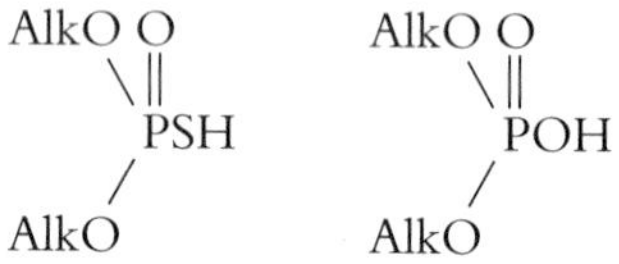

```
AlkO O          AlkO O
    \||             \||
     PSH             POH
    /               /
AlkO            AlkO
```

PHOSPHORODITHIOATES: CLEAVAGE

Some insecticides are phosphorodithioates:

```
AlkO S
    \||
     PSR
    /
AlkO
```

The products formed from malathion (Matsumura and Boush, 1966; Walker and Stojanovic, 1974; Paris *et al.*, 1975), dimethoate (El Beit *et al.*, 1978), and trithion (Rosenberg and Alexander, 1979) in soil, culture, or both include these structures:

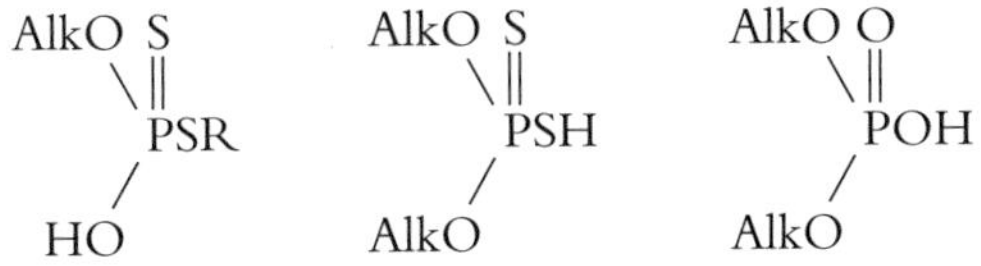

$$\begin{array}{ccc} \text{AlkO}\ \ \text{S} & \text{AlkO}\ \ \text{S} & \text{AlkO}\ \ \text{O} \\ \quad \backslash \| & \quad \backslash \| & \quad \backslash \| \\ \quad\quad \text{PSR} & \quad\quad \text{PSH} & \quad\quad \text{POH} \\ / & / & / \\ \text{HO} & \text{AlkO} & \text{AlkO} \end{array}$$

PHOSPHONATES: CLEAVAGE

Phosphonates are used for several purposes. Some are insecticides, and one of the major herbicides is a phosphonate. Moreover, some of the nerve gases used as chemical weapons are phosphonates. Differing from the organic phosphates, the P in phosphonates is linked directly to C. Because of this C–P bond, their environmental fate is quite different from that of the organic phosphates.

The cleavage of the C–P bond of phosphonates appears to involve a hydrolytic mechanism in the case of the herbicide glyphosate (Shinabarger and Broymer, 1980; Dick and Quinn, 1995), the insecticide trichlorfon (Salama *et al.*, 1975), methyl and ethyl phosphonate (Daughton *et al.*, 1979), and phenyl phosphonate (Cook *et al.*, 1979):

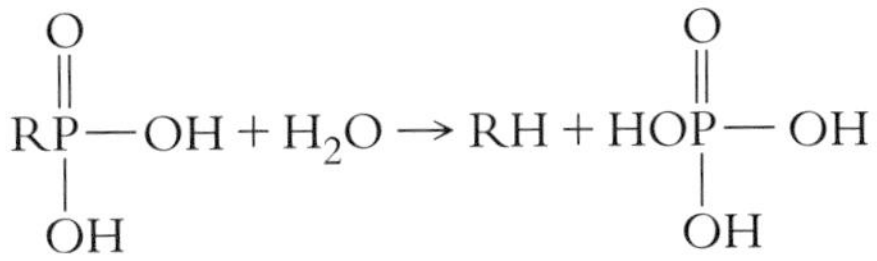

$$\mathrm{R\overset{\overset{\displaystyle O}{\|}}{\underset{\underset{\displaystyle OH}{|}}{P}}{-}OH + H_2O \rightarrow RH + HO\overset{\overset{\displaystyle O}{\|}}{\underset{\underset{\displaystyle OH}{|}}{P}}{-}OH}$$

Interestingly, if R is a phenyl, methyl, or ethyl group, the product is benzene, methane, or ethane.

TRIARYL PHOSPHATES: CLEAVAGE

Triaryl phosphates are used as fire retardants and as additives to lubricants and hydraulic fluids. The compounds whose fate has been studied have two unsubstituted phenyls and either an isopropyl or a *tert*-butyl phenyl group. If the three aryl groups each are represented by Ar, the reaction appears to be:

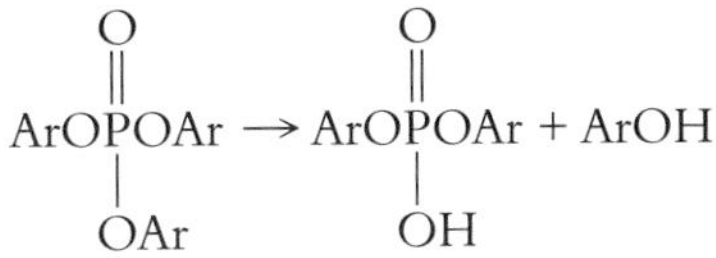

$$\mathrm{ArO\overset{\overset{\displaystyle O}{\|}}{\underset{\underset{\displaystyle OAr}{|}}{P}}OAr \rightarrow ArO\overset{\overset{\displaystyle O}{\|}}{\underset{\underset{\displaystyle OH}{|}}{P}}OAr + ArOH}$$

in sediments (Heitkamp *et al.*, 1984, 1986).

P=S: CONVERSION TO P=O

The P=S of the phosphorothiolate insecticides may be converted to P=O:

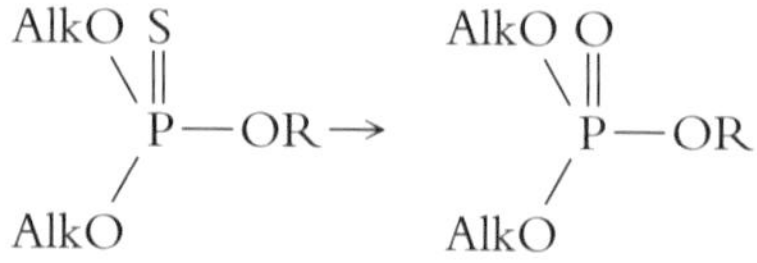

$$(\text{AlkO})_2\text{P}(=\text{S})\text{—OR} \rightarrow (\text{AlkO})_2\text{P}(=\text{O})\text{—OR}$$

Such a reaction has been shown to occur with parathion as the substrate in microbial cultures (Munnecke and Hsieh, 1976) and with parathion (Yu and Sanborn, 1975), chlorpyrifos (Rouchaud *et al.*, 1989), and diazinon (Kim *et al.*, 1989) in soil.

Similarly, the P=S of the phosphorodithioates is converted to the P=O derivative, as shown to occur with dimethoate added to soil (Duff and Menzer, 1973):

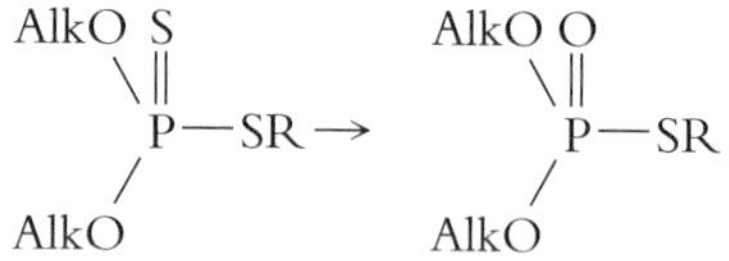

$$(\text{AlkO})_2\text{P}(=\text{S})\text{—SR} \rightarrow (\text{AlkO})_2\text{P}(=\text{O})\text{—SR}$$

These conversions are particularly significant because they represent activations, with the product being far more toxic than the precursors.

ADDITION REACTIONS

Most of the preceding types of reactions are degradative, i.e.; the molecules are becoming less complex. Some microbial transformations, however, result in a product more complex than the substrate, e.g., the *N-N* and *S-S* dimerizations and *N*-nitrosation described previously. Conjugations and oligomerizations, which have been discussed earlier, similarly lead to products more complex than the starting material.

The introduction of a methyl group also represents an addition reaction. Methyl groups may be added to O, N, and S, and *O*-methylations, *N*-methylations, and *S*-methylations were considered earlier. Although the mechanism is unknown and the transformation is probably rare, methyl groups may be added to aromatic rings

$$\text{ArH} \rightarrow \text{ArCH}_3$$

as in the microbial conversion of 4,4′-dichlorobenzophenone to 4-chlorobenzophenone bearing a methyl on the ring (Subba-Rao and Alexander, 1985).

Other elements may also be methylated to give rise to CH_3Hg^+, CH_3HgCH_3, CH_3SeCH_3, CH_3TeCH_3, and methylated As, Pb, Sn, and other elements.

Nitration of aromatic compounds has also been described. Thus, the aromatic ring of the pesticide diethofencarb is converted to 3,4-diethoxy-6-nitrocarbanilate in soil (Sakata *et al.*, 1992)

$$ArH \rightarrow ArNO_2$$

and the pesticide chlorodimeform is converted to a nitro compound in soil (Iwan *et al.*, 1976).

Nitriles ($RC{\equiv}N$) are also sometimes formed, as in the conversion of DDT to bis(4-chlorophenyl)nitrile in sewage (Albone *et al.*, 1978):

$$Ar_2CHCCl_3 \rightarrow Ar_2CHC{\equiv}N$$

Not as well known is the conversion of organic compounds to sulfates, with a hydroxy compound as a likely precursor:

$$ROH + H_2SO_4 \rightarrow ROSO_3H + H_2O$$

This type of process is suggested by the conversion by fungi of pyrene to 1-pyrenyl sulfate (Lange *et al.*, 1994) and 1-hydroxy-8-pyrenyl sulfate (Wunder *et al.*, 1994), a hydroxypyrene likely being formed and then converted to the sulfate ester. Microorganisms in culture also are able to convert biphenyl to a sulfate ester, with the likely intermediate being a hydroxy biphenyl (Golbeck *et al.*, 1983).

Several pesticides are converted to sulfonates (RSO_3H), e.g., the chloromethyl ($-CH_2Cl$) of alachlor in surface and groundwater (Macomber *et al.*, 1992) and propachlor (Lamoureux and Rusness, 1989) and acetochlor (Feng, 1991) in soil:

$$RCH_2Cl \rightarrow RCH_2SO_3H$$

In a few instances, a methylthio (CH_3S-) is added to an aromatic compound, as in the conversion of pentachloronitrobenzene (PCNB) to the pentachlorothioanisole in culture (Mora Torres *et al.*, 1996) and soil (Nakanishi, 1972):

$$ArNO_2 \rightarrow ArSCH_3$$

Some of these various addition reactions are probably multistep processes, but often the intermediates are not known or do not accumulate in culture or in natural or polluted environments.

REFERENCES

Adhya, T. K., Sudhakar-Barik, and Sethunathan, N., *Pestic. Biochem. Physiol.* **16,** 14–20 (1981).
Adrian, N. R., and Suflita, J. M., *Appl. Environ. Microbiol.* **56,** 292–294 (1990).

Adriaens, P., Fu, Q., and Grbić-Galić, D., *Environ. Sci. Technol.* **29,** 2252–2260 (1995).
Aftring, R. P., and Taylor, B. F., *Arch. Microbiol.* **130,** 101–104 (1981).
Albone, E. S., Eglinton, G., Evans, N. C., and Rhead, N. M., *Nature (London)* **240,** 420–421 (1972).
Amblès, A., Parlanti, E., Jambu, P., Mayougou, P., and Jacquesy, J.-C., *Geoderma* **64,** 111–124 (1994).
Aoki, K., Ohtsuka, K., Shinke, R., and Nishira, H., *Agric. Biol. Chem.* **48,** 865–872 (1984).
Arcangeli, J.-P., and Arvin, E., *Biodegradation* **6,** 19–27 (1996).
Asano, Y. Ando, S., Tani, Y., Yamada, H., and Ueno, T., *Agric. Biol. Chem.* **45,** 57–62 (1981).
Ayanaba, A., and Alexander, M., *Appl. Microbiol.* **25,** 862–868 (1973).
Ayanaba, A., and Alexander, M., *J. Environ. Qual.* **3,** 83–89 (1974).
Ayanaba, A., Verstraete, W., and Alexander, M., *Soil Sci. Soc. Am. Proc.* **37,** 565–568 (1973).
Baarschers, W. H., and Heitland, H. S., *J. Agric. Food Chem.* **34,** 707–709 (1986).
Baggi, G., Barbieri, P., Galli, E., and Tollari, S., *Appl. Environ. Microbiol.* **53,** 2129–2132 (1987).
Balba, M. T., Khan, M. R., and Evans, W. C., *Biochem. Soc. Trans.* **7,** 405–407 (1979).
Ball, H. A., Johnson, H. A., Reinhard, M., and Spormann, A. M., *J. Bacteriol.* **178,** 5755–5761 (1996).
Banwart, W. L., and Bremner, J. M., *Soil Biol. Biochem.* **7,** 359–364 (1975).
Bartha, R., *J. Agric. Food Chem.* **16,** 602–604 (1968).
Bartha, R., and Pramer, D., *Science* **156,** 1617–1618 (1967).
Bartha, R., and Pramer, D., *Bull. Environ. Contam. Toxicol.* **4,** 240–245 (1969).
Bartsch, K., and Tebbe, C. C., *Appl. Environ. Microbiol.* **55,** 711–716 (1989).
Basu, S. K., Oleszkiewicz, J. A., and Sparling, R., *Water Res.* **30,** 315–322 (1996).
Baude, F. J., Pease, H. L., and Holt, R. F., *J. Agric. Food Chem.* **22,** 413–418 (1974).
Beam, H. W., and Perry, J. J., *Arch. Mikrobiol.* **91,** 87–90 (1973).
Behki, R. M., and Khan, S. U., *J. Agric. Food Chem.* **42,** 1237–1241 (1994).
Belay, N., and Daniels, L., *Appl. Environ. Microbiol.* **53,** 1604–1610 (1987).
Bestetti, G., DiGennaro, P., Galli, E., Leoni, B., Pelizzoni, F., Sello, G., and Bianchi, D., *Appl. Microbiol. Biotechnol.* **40,** 791–793 (1994).
Beurskens, J. E. M., Toussaint, M., de Wolf, J., van der Steen, J. M. D., Slot, P. C., Commandeur, L. C. M., and Parsons, J. R., *Environ. Toxicol. Chem.* **14,** 939–943 (1995).
Beynon, K. I., Stoydin, G., and Wright, A. N., *Pestic. Biochem. Physiol.* **2,** 153–161 (1972a).
Beynon, K. I., Stoydin, G., and Wright, A. N., *Pestic. Sci.* **3,** 293–305 (1972b).
Beynon, K. I., and Wright, A. N., *J. Sci. Food Agric.* **20,** 250–256 (1969).
Bezalel, L., Hadar, Y., Fu, P. P., Freeman, J. P., and Cerniglia, C. E., *Appl. Environ. Microbiol.* **62,** 2554–2559 (1996).
Bisaillon, J. G., Lepine, F., Beaudet, R., and Sylvestre, M., *Can. J. Microbiol.* **39,** 642–648 (1993).
Blakley, E. R., and Papish, B., *Can. J. Microbiol.* **28,** 1037–1046 (1982).
Blakley, E. R., and Simpson, F. J., *Can. J. Microbiol.* **10,** 175–185 (1964).
Boethling, R. S., Gregg, B., Frederick, R., Gabel, N. W., Campbell, S. E., and Sabljic, A., *Ecotoxicol. Environ. Saf.* **18,** 252–267 (1989).
Boopathy, R., Kulpa, C. F., and Wilson, M., *Appl. Microbiol. Biotechnol.* **39,** 270–275 (1993).
Boopathy, R., Manning, J., Montemagno, C., and Rimkus, K., *Can. J. Microbiol.* **40,** 787–790 (1994).
Bothwell, M. L., Beeton, A. M., and Lech, J. J., *J. Fish. Res. Board Can.* **30,** 1841–1846 (1973).
Bozarth, G. A., and Funderburk, H. H., Jr., *Weed Sci.* **19,** 691–695 (1971).
Bruns-Nagel, D., Breitung, J., von Löw, E., Steinbach, K., Gorontzy, T., Kahl, M., Blotevogel, K.-H., and Gemsa, D., *Appl. Environ. Microbiol.* **62,** 2651–2656 (1996).
Bryant, F. O., *Appl. Microbiol. Biotechnol.* **38,** 276–281 (1992).
Bumpus, J. A., and Tatarko, M., *Curr. Microbiol.* **28,** 185–190 (1994).
Burback, B. L., and Perry, J. J., *Appl. Environ. Microbiol.* **59,** 1025–1029 (1993).
Cain, R. B., and Farr, D. R., *Biochem. J.* **106,** 859–877 (1968).
Cartwright, N. J., and Cain, R. B., *Biochem. J.* **73,** 305–314 (1959).

Castro, C. E., Wade, R. S., and Belser, N. O., *J. Agric. Food Chem.* **31,** 1184–1187 (1983).
Cerniglia, C. E., Freeman, J. P., and Van Baalen, C., *Arch. Microbiol.* **130,** 272–275 (1981).
Cerniglia, C. E., Morgan, J. C., and Gibson, D. T., *Biochem. J.* **180,** 175–185 (1979).
Chang, B.-V., Yeh, L.-N., and Yuan, S.-Y., *Chemosphere* **33,** 303–311 (1996).
Chaudhuri, B. K., and Wiesmann, U., *Appl. Microbiol. Biotechnol.* **43,** 178–187 (1995).
Chauret, C., Mayfield, C. I., and Inniss, W. E., *Can. J. Microbiol.* **41,** 54–63 (1995).
Chen, C., Puhakka, J. A., and Ferguson, J. F., *Environ. Sci. Technol.* **30,** 542–547 (1996).
Chiba, M., Kato, S., and Yamamoto, I., *J. Pestic. Sci.* **1,** 179–191 (1976).
Child, J., and Willetts, A., *Biochim. Biophys. Acta* **538,** 316–327 (1978).
Chin, W.-T., Stone, G. M., and Smith, A. E., *J. Agric. Food Chem.* **18,** 731–732 (1970).
Chisholm, D., *Can. J. Plant Sci.* **54,** 667–671 (1974).
Chouteau, J., Azoulay, E., and Senez, J. C., *Nature (London)* **194,** 576–578 (1962).
Clapp, D. W., Naylor, D. V., and Lewis, G. C., *J. Environ. Qual.* **5,** 207–210 (1976).
Collins, R. F., *Pestic. Sci.* **4,** 181–192 (1973).
Cook, A. M., Daughton, C. G., and Alexander, M., *Biochem. J.* **184,** 453–455 (1979).
Cook, A. M., and Hütter, R., *Appl. Environ. Microbiol.* **43,** 781–786 (1982).
Corbett, M. D., and Corbett, B. R., *Appl. Environ. Microbiol.* **41,** 942–949 (1981).
Corke, C. T., Bunce, N. J., Beaumont, A.-L., and Merrick, R. L., *J. Agric. Food Chem.* **27,** 644–649 (1979).
Corkery, D. M., O'Connor, K. E., Buckley, C. M., and Dobson, A. D. W., *FEMS Microbiol. Lett.* **124,** 23–28 (1994).
Couch, R. W., Gramlich, J. V., Davis, D. E., and Funderburk, H. H., Jr., *Proc. South. Weed Conf.* **18,** 623–631 (1965).
Cserjesi, A. J., and Johnson, E. L., *Can. J. Microbiol.* **18,** 45–49 (1972).
Cullen, W. R., Li, X.-F., and Reimer, K. J., *Sci. Total Environ.* **156,** 27–37 (1994).
Curtis, R. F., Land, D. G., Griffiths, N. M., Gee, M., Robinson, D., Peel, J. L., Dennis, C., and Gee, J. M., *Nature (London)* **235,** 223–224 (1972).
Dalton, R. L., Evans, A. W., and Rhodes, R. C., *Weeds* **14,** 31–33 (1966).
Daughton, C. G., Cook, A. M., and Alexander, M., *FEMS Microbiol. Lett.* **5,** 91–93 (1979).
Dean-Raymond, D., and Alexander, M., *Appl. Environ. Microbiol.* **33,** 1037–1041 (1977).
deKlerk, H., and van der Linden, A. C., *Antonie van Leeuwenhoek* **40,** 7–15 (1974).
deVos, R. H., ten Noever de Brauw, M. C., and Olthof, P. D. A., *Bull. Environ. Contam. Toxicol.* **11,** 567–571 (1974).
Dey, S., and Godbole, S. H., *Indian J. Exp. Biol.* **24,** 29–33 (1986).
Dick, R. E., and Quinn, J. P., *Appl. Microbiol. Biotechnol.* **43,** 545–550 (1995).
DiStefano, T. D., Gossett, J. M., and Zinder, S. H., *Appl. Environ. Microbiol.* **57,** 2287–2292 (1991).
Dräger, G., *Pflanzenschutz-Nachr.* **30,** 18–27 (1977).
Duff, W. G., and Menzer, R. E., *Environ. Entomol.* **2,** 309–318 (1973).
Duffy, J. R., and Wong, N., *J. Agric. Food Chem.* **15,** 457–464 (1967).
Durham, D. R., and Perry, J. J., *J. Gen. Microbiol.* **105,** 39–44 (1978).
Dybas, J. J., Tatara, G. M., and Criddle, C. S., *Appl. Environ. Microbiol.* **61,** 758–762 (1995).
Ecker, S., Widmann, T., Lenke, H., Dickel, O., Fischer, P., Bruhn, C., and Knackmuss, H.-J., *Arch. Microbiol.* **158,** 149–154 (1992).
Egli, C., Scholtz, R., Cook, A. M., and Leisinger, T., *FEMS Microbiol. Lett.* **43,** 257–261 (1987).
El Beit, I. O. D., Wheelock, J. V., and Cotton, D. E., *Int. J. Environ. Stud.* **12,** 212–225 (1978).
Elsner, E., Bieniek, D., Klein, W., and Korte, F., *Chemosphere* **1,** 247–250 (1972).
Engelhardt, G., and Wallnöfer, P. R., *Chemosphere* **7,** 463–466 (1978).
Engelhardt, G., Wallnöfer, P., Fuchsbichler, G., and Baumeister, W., *Chemosphere* **6,** 85–92 (1977).
Engelhardt, G., Ziegler, W., Wallnöfer, P. R., Oehlmann, L., and Wagner, K., *FEMS Microbiol. Lett.* **11,** 165–169 (1981).

Entsch, B., Ballou, D. P., Husain, M., and Massey, V., *J. Biol. Chem.* **251,** 7367–7379 (1976).
Evans, W. C., *Biochem. J.* **41,** 373–382 (1947).
Fahlbusch, K., Hippe, H., and Gottschalk, G., *FEMS Microbiol. Lett.* **19,** 103–104 (1983).
Fathepure, B. Z., and Tiedje, J. M., *Environ. Sci. Technol.* **28,** 746–752 (1994).
Faulkner, J. K., and Woodcock, D., *Nature (London)* **203,** 865 (1964).
Feng, P. C. C., *Pestic. Biochem. Physiol.* **40,** 136–142 (1991).
Finkle, B. J., Lewis, J. C., Corse, J. W., and Lundin, R. E., *J. Biol. Chem.* **237,** 2926–2931 (1971).
Fletcher, C. L., and Kaufman, D. D., *J. Agric. Food Chem.* **27,** 1127–1130 (1979).
Focht, D. D., and Alexander, M., *Science* **170,** 91–92 (1970).
Focht, D. D., and Joseph, H., *Can. J. Microbiol.* **20,** 631–635 (1974).
Fournier, J.-C., and Catroux, G., *C. R. Acad. Sci. D* **275,** 1723–1726 (1972).
Fournier, J.-C., and Salle, J., *Chemosphere* **3,** 77–82 (1974).
Freedman, D. L., and Gossett, J. M., *Appl. Environ. Microbiol.* **55,** 2144–2151 (1989).
Freitag, D., Scheunert, I., Klein, W., and Korte, F., *J. Agric. Food Chem.* **32,** 203–207 (1984).
Funderburk, H. H., Jr., and Bozarth, G. A., *J. Agric. Food Chem.* **15,** 563–567 (1967).
Funtikova, N. S., *Mikrobiologiya* **48,** 57–61 (1979).
Gardin, H., Lebeault, J. M., and Pauss, A., *Meded. Fac. Landbouwkd. Toegepaste Biol. Wet. (Univ. Gent).* **59**(4B), 2155–2160 (1994).
Gardiner, J. A., Rhodes, R. C., Adams, J. B., Jr., and Soboczenski, E. J., *J. Agric. Food Chem.* **17,** 980–986 (1969).
Getzin, L. W., and Shanks, C. H., Jr., *J. Econ. Entomol.* **63,** 52–58 (1970).
Gibson, D. T., Koch, J. R., Schuld, C. L., and Kallio, R. E., *Biochemistry* **7,** 3795–3802 (1968).
Gibson, D. T., Mahadevan, V., Jerina, D. M., Yagi, H., and Yeh, H. J. C., *Science* **189,** 295–297 (1975).
Gibson, D. T., Roberts, R. L., Wells, M. C., and Kobal, V. M., *Biochem. Biophys. Res. Commun.* **50,** 211–219 (1973).
Gibson, S. A., and Suflita, J. M., *Appl. Environ. Microbiol.* **52,** 681–688 (1986).
Gingell, R., *Xenobiotica* **3,** 165–169 (1973).
Golab, T., Althaus, W. A., and Wooten, H. C., *J. Agric. Food Chem.* **27,** 163–179 (1979).
Golab, T., Bishop, C. E., Donoho, A. L., Manthey, J. A., and Zornes, L. L., *Pestic. Biochem. Physiol.* **5,** 196–204 (1975).
Golbeck, J. H., Albaugh, S. A., and Radmer, R., *J. Bacteriol.* **156,** 49–57 (1983).
Golovleva, L A., Golovlev, E. L., Zyakun, A. M., Shurukhin, Yu. V., and Finkelshtein, Z. I., *Izv. Akad. Nauk SSSR Ser. Biol.* (1), 44–51 (1978).
Golovleva, L. A., Klysheva, A. L., Nefedova, M. Yu., Baskunov, B. P., Zyakun, A. M., and Ilyaletdinov, A. N., *Izv. Akad. Nauk. Kaz. SSR Ser. Biol.* (6), 23–30 (1980), Chem. Abstr. **95,** 1742 (1981).
Gorontzy, T., Küver, J., and Blotevogel, K.-H., *J. Gen. Microbiol.* **139,** 1331–1336 (1993).
Grbić-Galić, D., and Vogel, T. M., *Appl. Environ. Microbiol.* **53,** 254–260 (1987).
Grifoll, M., Selifonov, S. A., and Chapman, P. J., *Appl. Environ. Microbiol.* **60,** 2438–2449 (1994).
Gross, D., Laanio, T., Dupuis, G., and Esser, H. O., *Pestic. Biochem. Physiol.* **10,** 49–59 (1979).
Grossenbacher, H., Horn, C., Cook, A. M., and Hütter, R., *Appl. Environ. Microbiol.* **48,** 451–453 (1984).
Gupta, M., Sharma, D., Suidan, M. T., and Sayles, G. D., *Water Res.* **30,** 1377–1385 (1996).
Gutenmann, W. H., Loos, M. A., Alexander, M., and Lisk, D. J., *Soil Sci. Soc. Am. Proc.* **28,** 205–207 (1964).
Haïdour, A., and Ramos, J. L., *Environ. Sci. Technol.* **30,** 2365–2370 (1996).
Hales, S. G., Dodgson, K. S., White, G. F., Jones, N., and Watson, G. K., *Appl. Environ. Microbiol.* **44,** 790–800 (1982).
Hallas, L. E., and Alexander, M., *Appl. Environ. Microbiol.* **45,** 1234–1241 (1983).
Halvorson, H., *Biochem. Biophys. Res. Commun.* **10,** 440–443 (1963).
Haque, A., Weisgerber, I., Kotzinas, D., and Klein, W., *Pestic. Biochem. Physiol.* **7,** 321–331 (1977).

Harbison, K. G., and Belly, R. T., *Environ. Toxicol. Chem.* **1,** 9–15 (1982).

Hasegawa, Y., Obata, H., and Tokuyoma, T., *J. Ferment. Technol.* **58,** 215–220 (1980).

Heitkamp, M. A., Freeman, J. P., and Cerniglia, C. E., *Appl. Environ. Microbiol.* **51,** 316–322 (1986).

Heitkamp, M. A., Huckins, J. N., Petty, J. D., and Johnson, J. L., *Environ. Sci. Technol.* **18,** 434–439 (1984).

Heringa, J. W., Huybregtse, R., and van der Linden, A. C., *Antonie van Leeuwenhoek* **27,** 51–58 (1961).

Hofrichter, M., Bublitz, F., and Fritzsche, W., *J. Basic Microbiol.* **34,** 163–172 (1994).

Holliger, C., Schraa, G., Stams, A. J. M., and Zehnder, A. J. B., *Biodegradation* **1,** 253–261 (1990).

Holliger, C., Schraa, G., Stams, A. J. M., and Zehnder, A. J. B., *Appl. Environ. Microbiol.* **59,** 2991–2997 (1993).

Hsu, J. C., and Camper, N. D., *Soil Biol. Biochem.* **11,** 19–22 (1979).

Hyman, M. R., Page, C. L., and Arp, D. J., *Appl. Environ. Microbiol.,* **60,** 3033–3035 (1994).

Idaka, E., Ogawa, T., Horitsu, H., and Yatome, C., *Eur. J. Appl. Microbiol. Biotechnol.* **15,** 141–143 (1982).

Iizuka, H., Iida, M., and Unami, Y., *J. Gen. Appl. Microbiol.* **12,** 119–126 (1961).

Ikeda, M., Unai, T., and Tomizawa, C., *J. Pestic. Sci.* **11,** 85–96 (1986).

Ishida, M., *in* "Environmental Toxicology of Pesticides" (F. M. Matsumura, G. M. Boush, and T. Misato, eds.), pp. 281–306. Academic Press, New York, 1972.

Ishikawa, K., Nakamura, Y., and Kuwatsuka, S., *J. Pestic. Sci.* **1,** 49–57 (1976).

Iwan, J., Hoyer, G.-A., Rosenberg, D., and Goller, D., *Pestic. Sci.* **7,** 621–631 (1976).

Jain, R. K., Dreisbach, J. H., and Spain, J. C., *Appl. Environ. Microbiol.* **60,** 3030–3032 (1994).

Janota-Bassalik, L., and Wright, L. D., *Nature (London)* **204,** 501–502 (1964).

Johnson, W. G., and Lavy, T. L., *J. Environ. Qual.* **24,** 487–493 (1995).

Joshi, D. K., and Gold, M. H., *Appl. Environ. Microbiol.* **59,** 1779–1785 (1993).

Juengst, F. W., Jr., and Alexander, M., *Mar. Biol.* **33,** 1–6 (1975).

Kaake, R. H., Crawford, D. L., and Crawford, R. L., *Biodegradation* **6,** 329–337 (1995).

Kallio, R. E., *in* "Fermentation Advances" (D. Perlman, ed.), pp. 635–648. Academic Press, New York, 1969.

Katayama, A., Itou, T., and Ukai, T., *J. Pestic. Sci.* **22,** 11–16 (1997).

Kaufman, D. D., Plimmer, J. R., and Iwan, J., Abstr. 162nd Meet., Am. Chem. Soc., Washington, p. 21 (1971).

Kaufman, D. D., Plimmer, J. R., Iwan, J., and Klingebiel, U. I., *J. Agric. Food Chem.* **20,** 916–919 (1972).

Kaufman, D. D., Plimmer, J. R., and Klingebiel, U. I., *J. Agric. Food Chem.* **21,** 127–132 (1973).

Kearney, P. C., Kaufman, D. D., and Sheets, T. J., *J. Agric. Food Chem.* **13,** 369–372 (1965).

Kearney, P. C., and Plimmer, J. R., *J. Agric. Food Chem.* **20,** 584–585 (1972).

Kearney, P. C., Plimmer, J. R., Wheeler, W. B., and Kontson, A., *Pestic. Biochem. Physiol.* **6,** 229–238 (1976).

Keith, C. L., Bridges, R. L., Fina, L. R., Iverson, K. L., and Cloren, J. A., *Arch. Microbiol.* **118,** 173–176 (1978).

Kertesz, M. A., Kölbener, P., Stockinger, H., Beil, S., and Cook, A. M., *Appl. Environ. Microbiol.* **60,** 2296–2303 (1994).

Kiehlmann, E., Pinto, L., and Moore, M., *Can. J. Microbiol.* **42,** 604–608 (1996).

Kim, J. H., Rhee, Y. H., Choi, J. W., and Lee, K. S., *Misaengmul Hakhoechi* **27,** 139–146 (1989), *Chem. Abstr.* **111,** 227209 (1989).

Kloepfer, R. D., Easley, D. M., Haas, B. B., Jr., Deihl, T. G., Jackson, D. E., and Wurrey, C. J., *Environ. Sci. Technol.* **19,** 277–280 (1985).

Knackmuss, H.-J., and Hellwig, M., *Arch. Microbiol.* **117,** 1–7 (1978).

Knoll, G., and Winter, J., *Appl. Microbiol. Biotechnol.* **25,** 384–391 (1987).

Komatsu, T., Omori, T., and Kodama, T., *Biosci. Biotechnol. Biochem.* **57,** 864–865 (1993).

Konrad, J. G., Armstrong, D. E., and Chesters, G., *Agron. J.* **59,** 591–594 (1967).

Kost, A. N., Vorob'eva, L. I., Terent'ev, P. B., Modyanova, L. V., Shibilkina, O. K., and Korosteleva, L. A., *Prikl. Biokhim. Mikrobiol.* **13**(59), 696–703 (1977).

Kozyreva, L. P., Schurukhin, Y. V., Finkelshtein, Z. I., Baskunov, B. P., and Golovleva, L. A., *Mikrobiologiya* **62,** 110–119 (1993).

Kroon, A. G. M., Pomper, M. A., and van Ginkel, C. G., *Appl. Microbiol. Biotechnol.* **42,** 134–139 (1994).

Kumarasamy, R., and Raghu, K., *Chemosphere* **5,** 107–112 (1976).

Kunc, F., *Folia Microbiol. (Prague)* **16,** 41–50 (1971).

Kuwatsuka, S., and Igarishi, S., *Soil Sci. Plant Nutr.* **21,** 405–414 (1975).

Kveštak, R., and Ahel, M., *Arch. Environ. Contam. Toxicol.* **29,** 551–556 (1995).

Laanio, T. L., Kearney, P. C., and Kaufman, D. D., *Pestic. Biochem. Physiol.* **3,** 271–277 (1973).

Laborde, A. L., and Gibson, D. T., *Appl. Environ. Microbiol.* **34,** 783–790 (1977).

Lam, T., and Vilker, V. L., *Biotechnol. Bioeng.* **29,** 151–159 (1987).

Lambert, M., Kremer, S., Sterner, O., and Anke, H., *Appl. Environ. Microbiol.* **60,** 3597–3601 (1994).

Lamberton, J. G., and Claeys, R. R., *J. Agric. Food Chem.* **18,** 92–96 (1970).

Lamoureux, G. L., and Rusness, D. G., *Pestic. Biochem. Physiol.* **34,** 187–204 (1989).

Leadbetter, E. R., and Foster, J. W., *Arch. Biochem. Biophys.* **82,** 491–492 (1959).

Leather, G. R., and Foy, C. L., *Pestic. Biochem. Physiol.* **7,** 437–442 (1977).

Lee, J. K., and Kim, K. C., *Hanguk Nonghwa Hakhoe Chi* **21,** 197–203 (1978), Chem. Abstr. **90,** 99948 (1979).

Lenfant, M., Hunt, P. F., Thérier, L. M., Pinte, F., and Lederer, E., *Biochim. Biophys. Acta* **201,** 82–90 (1970).

Lenke, H., Pieper, D. H., Bruhn, C., and Knackmuss, H.-J., *Appl. Environ. Microbiol.* **58,** 2928–2932 (1992).

Leoni, V., Hollick, C. B., DeLuca, E. D., Collison, R. J., and Merolli, S., *Agrochimica* **25,** 414–426 (1981).

LePatourel, G. N. J., and Wright, D. J., *Comp. Biochem. Physiol. C* **53**(2C), 73–74 (1976).

Li, T., Bisaillon, J.-G., Villemur, R., Létourneau, L., Bernard, K., Lépine, F., and Beaudet, R., *J. Bacteriol.* **178,** 2551–2558 (1996).

Lichtenstein, E. P., Fuhremann, T. W., and Schulz, K. R., *J. Agric. Food Chem.* **22,** 991–996 (1974).

Lichtenstein, E. P., and Schulz, K. R., *J. Econ. Entomol.* **53,** 192–197 (1960).

Lieberman, M. T., and Alexander, M., *J. Agric. Food Chem.* **31,** 265–267 (1983).

Liu, D., Maguire, R. J., Pacepavicius, G. J., and Nagy, E., *Environ. Toxicol. Water Qual.* **7,** 355–372 (1992).

Lu, P.-Y., Metcalf, R. L., and Cole, L. K., *in* "Pentachlorophenol: Chemistry, Pharmacology, and Enviromental Toxicology" (K. Ranga Rao, ed.), pp. 53–63. Plenum Press, New York, 1978.

Lu, P.-Y., Metcalf, R. L., Plummer, N., and Mandel, D., *Arch. Environ. Contam. Toxicol.* **6,** 129–142 (1977).

Lübbe, C., Salcher, O., and Lingens, F., *FEMS Microbiol. Lett.* **13,** 31–33 (1982).

Lusby, W. R., Oliver, J. E., and Kearney, P. C., *J. Agric. Food Chem.* **28,** 641–644 (1980).

Lyons, C. D., Katz, S., and Bartha, R., *Appl. Environ. Microbiol.* **48,** 491–496 (1984).

MacGillivray, A. R., and Shiaris, M. P., *Appl. Environ. Microbiol.* **60,** 1154–1159 (1994).

Macomber, C., Bushway, R. J., Perkins, L. B., Baker, D., Fan, S., and Ferguson, B. S., *J. Agric. Food Chem.* **40,** 1450–1452 (1992).

MacRae, I. C., Alexander, M., and Rovira, A. D., *J. Gen. Microbiol.* **32,** 69–76 (1963).

Madhosingh, C., *Can. J. Microbiol.* **7,** 553–567 (1961).

Maghaddam, A. P., Abbas, R., Fisher, J. W., Stavrono, S., and Lipscomb, J. C. *Biochem. Biophys. Res. Commun.* **228,** 639–645 (1996).

Masaphy, S., Henis, Y., and Levanon, D., *Appl. Environ. Microbiol.* **62,** 3587–3593 (1996).

Masunaga, S., Urushigawa, Y., and Yonezawa, Y., *Chemosphere* **12,** 1075–1082 (1983).

Mathur, S. P., and Saha, J. G., *Soil Sci.* **120,** 301–307 (1975).

Matsumura, F., and Boush, G. M., *Science* **153,** 1278–1280 (1966).

McCall, P. J., Vrona, S. A., and Kelly, S. S., *J. Agric. Food Chem.* **29,** 100–107 (1981).

McCormick, N. G., Cornell, J. H., and Kaplan, A. M., *Appl. Environ. Microbiol.* **35,** 945–948 (1978).

McCormick, N. G., Feeherry, F. E., and Levinson, H. S., *Appl. Environ. Microbiol.* **31,** 949–958 (1976).

McCormick, N. G., Foster, D. M., Cornell, J. H., and Kaplan, A. M., Abstr. Annu. Meet., Am. Soc. Microbiol., p. 196 (1980).

Meng, M., Sun, W.-Q., Geelhaar, L. A., Kumar, G., Patel, A. R., Payne, G. F., Speedie, M. K., and Stacy, J. R., *Appl. Environ. Microbiol.* **61,** 2548–2553 (1995).

Mills, A. L., and Alexander, M., *Appl. Environ. Microbiol.* **31,** 892–895 (1976).

Minard, R. D., Russell, S., and Bollag, J.-M., *J. Agric. Food Chem.* **25,** 841–844 (1977).

Misra, D., Bhuyan, S., Adhya, T. K., and Sethunathan, N., *Soil Biol. Biochem.* **24,** 1035–1042 (1992).

Miyamo, J., Kitagawa, K., and Sato, Y., *Jpn. J. Exp. Med.* **36,** 211–225 (1966).

Modrzakowski, M. C., and Finnerty, W. R., *Arch. Microbiol.* **126,** 285–290 (1980).

Mora Torres, R., Grosset, C., Steiman, R., and Alary, J., *Chemosphere* **33,** 683–692 (1996).

Morikawa, M., Kanemoto, M., and Imanaka, T., *J. Ferment. Bioeng.* **82,** 309–311 (1996).

Moy, P. L., and Ebert, A. G., *J. Pharm. Sci.* **61,** 804–805 (1972).

Muir, D. C. G., Pitze, M., Blouw, A. P., and Lockhart, W. H., *Weed Res.* **21,** 59–70 (1981).

Muir, D. C. G., and Yarechewski, A. L., *J. Environ. Sci. Health B,* **17,** 363–380 (1982).

Munnecke, D. M., and Hsieh, D. P. H., *Appl. Environ. Microbiol.* **31,** 63–69 (1976).

Murado Garcia, M. A., *Rev. Agroquim. Tecnol. Aliment.* **13,** 559–566 (1973), *Chem. Abstr.* **81,** 34208 (1974).

Nakanishi, T., *Ann. Phytopathol. Soc.* **38,** 249–251 (1972).

Narro, M. L., Cerniglia, C. E., van Baalen, C., and Gibson, D. T., *Appl. Environ. Microbiol.* **58,** 1351–1359 (1992).

Neu, H. J., and Ballschmiter, K., *Chemosphere* **6,** 419–423 (1977).

Nicholson, D. K., Woods, S. L., Istok, J. D., and Peek, D. C., *Appl. Environ. Microbiol.* **58,** 2280–2286 (1992).

Niki, Y., and Kuwatsuka, S., *Soil Sci. Plant Nutr.* **22,** 233–245 (1976).

Nishikawa, H., Hosomura, H., Sinoda, Y., and Inagaki, S., *Kenkyu Hokoku-Gifu-Ken Kogyo Gijutsu Senta* **11,** 31–34 (1978), *Chem. Abstr.* **92,** 88933 (1980).

Noguera, K. R., and Friedman, D. L., *Appl. Environ. Microbiol.* **62,** 2257–2263 (1996).

Odeyemi, O., and Alexander, M., *Appl. Environ. Microbiol.* **33,** 784–790 (1977).

Ohisa, N., Yamaguchi, M., and Kurihara, N., *Arch. Microbiol.* **125,** 221–225 (1980).

Ohkawa, H., Shibaike, R., Okihara, Y., Morikawa, M., and Miyamoto, J., *Agric. Biol. Chem.* **40,** 943–951 (1976).

Ohyama, H., and Kuwatsuka, S., *J. Pestic. Sci.* **3,** 401–410 (1978).

Omori, T., and Alexander, M., *Appl. Environ. Microbiol.* **35,** 512–516 (1978).

Otto, S., and Drescher, N., Proc. 7th Brit. Insectic. Fungic. Conf., Brighton, U.K., p. 56 (1973).

Ou, L.-T., Chung, K.-Y., Thomas, J. E., Obreza, T. A., and Dickson, D. W., *J. Nematol.* **27,** 249–257 (1995).

Oyamada, M., and Kuwatsuka, S., *J. Pestic. Sci.* **4,** 157–163 (1979).

Paris, D. F., Lewis, D. L., and Wolfe, N. L., *Environ. Sci. Technol.* **9,** 135–138 (1975).

Parsons, F., Lage, G. B., and Rice, R., *Environ. Toxicol. Chem.* **4,** 739–742 (1985).

Patel, J. C., and Grant, D. J. W., *Antonie van Leeuwenhoek* **35,** 53–64 (1969).

Patel, R. N., Hou, C. T., Laskin, A. I., Felix, A., and Derelanko, P., *Appl. Environ. Microbiol.* **39,** 720–726 (1980a).

Patel, R. N., Hou, C. T., Laskin, A. I., Felix, A., and Derelanko, P., *Appl. Environ. Microbiol.* **39,** 727–733 (1980b).

Patil, K. C., Matsumura, F., and Boush, G. M., *Environ. Sci. Technol.* **6,** 629–632 (1972).

Pennington, J. C., and Patrick, W. H., Jr., *J. Environ. Qual.* **19,** 559–567 (1990).

Pereira, W. E., Short, D. L., Manigold, D. B., and Roscio, P. K., *Bull. Environ. Contam. Toxicol.* **21,** 554–562 (1979).

Pfaender, F. K., and Alexander, M., *J. Agric. Food Chem.* **20,** 842–846 (1972).

Quilt, P., Grossbard, E., and Wright, S. J. L., *J. Appl. Bacteriol.* **46,** 431–442 (1979).

Racke, K. D., Fontaine, D. D., Yoder, R. N. and Miller, J. R., *Pestic. Sci.* **42,** 43–51 (1994).

Ramanand, K., Balba, M. T., and Duffy, J., *Appl. Environ. Microbiol.* **59,** 3266–3272 (1993).

Ramanand, K., and Suflita, J. M., *Appl. Environ. Microbiol.* **57,** 1689–1695 (1991).

Ramos, J. L., Haidour, A., Duque, E., Piñar, G., Calvo, V., and Oliva, J.-M., *Nature Biotechnol.* **14,** 320–322 (1996).

Ranga Rao, D. M., and Murthy, A. S., *J. Agric. Food Chem.* **28,** 1099–1101 (1981).

Raymond, R. L., Jamison, V. W., and Hudson, J. O., *Appl. Microbiol.* **15,** 857–865 (1967).

Redemann, C. T., Meikle, R. W., and Widofsky, J. G., *J. Agric. Food Chem.* **12,** 207–209 (1964).

Resnick, S. M., and Gibson, D. T., *Appl. Environ. Microbiol.* **62,** 4073–4080 (1996).

Rhodes, R. C., *J. Agric. Food Chem.* **28,** 311–315 (1980).

Rhys-Williams, W., Taylor, S. C., and Williams, P. A., *J. Gen. Microbiol.* **139,** 1967–1972 (1993).

Richey, F. A., Jr., Bartley, W. J., and Sheets, K. P., *J. Agric. Food Chem.* **25,** 47–51 (1977).

Rickard, R. W., and Camper, N. D., *Pestic. Biochem. Physiol.* **9,** 183–189 (1978).

Rieck, C. E., *Diss. Abstr. Int.* **B30,** 3945 (1970).

Rogoff, M. H., and Wender, I., *J. Bacteriol.* **74,** 108–109 (1957).

Rosenberg, A., and Alexander, M., *Appl. Environ. Microbiol.* **37,** 886–891 (1979).

Rouchaud, J., Metsue, M., Gustin, F., van de Steene, F., Pelerents, C., Benoit F., Ceustermans, N., Gillet, J., and Vanparys, L., *Toxicol. Environ. Chem.* **23,** 215–226 (1989).

Russel, S., *Zesz. Nauk Szk. Gl. Gospod. Wiejsk-Akad. Roln. Warszawie, Roxp. Nauk* **101** (1979), *Chem. Abstr.* **92,** 1524 (1980).

Sakata, S., Katagi, T., Yoshimura, J., Mikami, N., and Yamada, H., *J. Pestic. Sci.* **17,** 221–230 (1992).

Salama, A. M., Mostafa, I. Y., and El-Zawahry, Y. A., *Acta Biol. Acad. Sci. Hung.* **26,** 1–7 (1975).

Salanitro, J. P., Diaz, L. A., Williams, M. P., and Wisniewski, H. L., *Appl. Environ. Microbiol.* **60,** 2593–2596 (1994).

Sayama, M., Inoue, M., Mori, M. A., Maruyama, Y., and Kozuka, H., *Xenobiotica* **22,** 633–640 (1992).

Schmidt, S., Wittich, R.-M., Erdmann, D., Wilkes, H., Francke, W., and Fortnagel, P., *Appl. Environ. Microbiol.* **58,** 2744–2750 (1992).

Schmitzer, J., Chen, B., Scheunert, I., and Korte, F., *Chemosphere* **18,** 2383–2388 (1989).

Schraa, G., Boone, M. L., Jetten, M. S. M., van Neerven, A. R. W., Colberg, P. J., and Zehnder, A. J. B., *Appl. Environ. Microbiol.* **52,** 1374–1381 (1986).

Sethunathan, N., and Pathak, M. D., J. Agric. Food Chem. **30,** 586–589 (1972).

Sethunathan, N., and Yoshida, T., *Can. J. Microbiol.* **19,** 873–875 (1973a).

Sethunathan, N., and Yoshida, T., *J. Agric. Food Chem.* **21,** 602–604 (1973b).

Seuferer, S. L., Braymer, H. D., and Dunn, J. J., *Pestic. Biochem. Physiol.* **10,** 174–180 (1979).

Seyfried, B., Glod, G., Schocher, R., Tschech, A., and Zeyer, J., *Appl. Environ. Microbiol.* **60,** 4047–4052 (1994).

Shelton, D. R., Sadeghi, A. M., Karns, J. S., and Hapeman, C. J., *Weed Sci.* **43,** 298–305 (1995).

Shinabarger, D. L., and Braymer, H. D., *J. Bacteriol.* **168,** 702–707 (1986).

Skeen, R. S., Gao, J., and Hooker, B. S., *Biotechnol. Bioeng.* **48,** 659–666 (1995).

Skryabin, G. K., Golovleva, L A., and Krupyanko, V. I., *Izv. Akad. Nauk SSSR Ser. Biol.* No. 5, 660–669 (1969).

Sloane, N. H., Untch, K. G., and Johnson, A. W., *Biochim. Biophys. Acta* **78,** 588–593 (1963).

Smith, A. E., *J. Agric. Food Chem.* **22,** 601–605 (1974).

Smith, A. E., *J. Agric. Food Chem.* **25,** 893–898 (1977).

Smith, A. E., *Bull Environ. Contam. Toxicol.* **34,** 150–157 (1985).

Smith, A. E., and Cullimore, D. R., *Can. J. Microbiol.* **20,** 773–776 (1974).

Smith, A. E., and Hayden, B. J., *Weed Res.* **21,** 179–183 (1981).
Smith, A. E., and Milward, L. J., *J. Chromatogr.* **265,** 378–381 (1983).
Smith, R. A., Belles, W. S., Shen, K.-W., and Wood, W. G., *Pestic. Biochem. Physiol.* **3,** 278–288 (1973).
Smolenski, W. J., and Suflita, J. M., *Appl. Environ. Microbiol.* **53,** 710–716 (1987).
Sonier, D. N., Duran, N. L., and Smith. J. B., *Appl. Environ. Microbiol.* **60,** 4567–4572 (1994).
Sotiriou, N., Weisgerber, I., Klein, W., and Korte, F., *Chemosphere* **5,** 53–60 (1976).
Spain, J. C., and Gibson, D. T., *Appl. Environ. Microbiol.* **57,** 812–819 (1991).
Staiger, L. E., and Quistad, G. B., *J. Agric. Food Chem.* **31,** 599–603 (1983).
Staron, T., Allard, C., and Xuong, N. D., *C R Acad. Sci.* **263D,** 81–84 (1966).
Stepp, T. D., Camper, N. D., and Paynter, M. J. B., *Pestic. Biochem. Physiol.* **23,** 256–260 (1985).
Stralka, K. A., and Camper, N. D., *Soil Biol. Biochem.* **13,** 33–38 (1981).
Stromeyer, S. A., Stumpf, K., Cook, A. M., and Leisinger, T., *Biodegradation* **3,** 113–123 (1992).
Subba-Rao, R. V., and Alexander, M., *Appl. Environ. Microbiol.* **33,** 101–108 (1977).
Subba-Rao, R. V., and Alexander, M., *Appl. Environ. Microbiol.* **49,** 509–516 (1985).
Suflita, J. M., Horowitz, A., Shelton, D. R., and Tiedje, J. M., *Science* **218,** 1115–1116 (1982).
Sullivan, J. P., and Chase, H. A., *Appl. Microbiol. Biotechnol.* **45,** 427–433 (1996).
Tafuri, F., Patumi, M., Marucchini, C., and Businelli, M., *Pestic. Sci.* **13,** 665–669 (1982).
Tahara, S., Hafsah, Z., Ono, A., Asaishi, E., and Mizutani, J., *Agric. Biol. Chem.* **45,** 2253–2258 (1981).
Takimoto, Y., Hirota, M., Insui, H., and Miyamoto, J., *J. Pestic. Sci.* **1,** 131–143 (1976).
Taraban, R. H., Berry, D. F., Berry, D. A., and Walker, H. L., Jr., *Appl. Environ. Microbiol.* **59,** 2332–2334 (1993).
Taylor, B. F., and Ribbons, D. W., *Appl. Environ. Microbiol.* **46,** 1276–1281 (1983).
Tewfik, M. C., and Evans, W. C., *Biochem. J.* **99,** 31P–32P (1966).
Tiedje, J. M., and Hagedorn, M. L., *J. Agric. Food Chem.* **23,** 77–81 (1975).
Tillmanns, G. M., Wallnöfer, P. R., Engelhardt, G., Olie, K., and Hutzinger, O., *Chemosphere* **7,** 59–64 (1978).
Timms, P., and MacRae, I. C., *Aust. J. Biol. Sci.* **35,** 661–667 (1982).
Tokieda, T., Niimura, T., Yamaha, T., Hasegawa, T., and Suzuki, T., *Agric. Biol. Chem.* **43,** 25–32 (1979).
Tomizawa, C., *Environ. Qual. Saf.* **4,** 117–127 (1975).
Tweedy, B. G., Loeppky, C., and Ross, J. A., *J. Agric. Food Chem.* **18,** 851–853 (1970).
Uchiyama, H., Nakajima, T., Yagi, O., and Nakahara, T., *Appl. Environ. Microbiol.* **58,** 3067–3071 (1992).
Udod, V. M., *Tr. Vses. Nauch.-Issled. Inst. Vodosnabzh. Kanaliz Gidrotelkh. Sooruzhenii Inzh. Gidrogeol.* **40,** 56–59 (1972), *Chem. Abstr.* **80,** 74052 (1974).
Uesugl, Y., Tomizawa, C., and Murai, T., *in* "Environmental Toxicology of Pesticides" (F. Matsumura, G. M., Boush, and T. Misato, eds.), pp. 327–339. Academic Press, New York, 1972.
Valli, K., Wariishi, H., and Gold, M. H., *J. Bacteriol.* **174,** 2131–2137 (1992).
van Alfen, N. K., and Kosuge, T., *J. Agric. Food Chem.* **24,** 584–588 (1976).
Vanderberg, L. A., Burback, B. L., and Perry, J. J., *Can. J. Microbiol.* **41,** 298–301 (1995).
Vanderberg, L. A., Perry, J. J., and Unkefer, P. J., *Appl. Microbiol. Biotechnol.* **43,** 937–945 (1995).
van der Linden, A. C., *Biochim. Biophys. Acta* **77,** 157–159 (1963).
van Ginkel, C. G., *Biodegradation* **7,** 151–164 (1996).
van Ginkel, C. G., van Dijk, J. B., and Kroon, A. G. M., *Appl. Environ. Microbiol.* **58,** 3083–3087 (1992).
Vannelli, T., and Hooper, A. B., *Appl. Environ. Microbiol.* **58,** 2321–2325 (1992).
Verloop, A., *Residue Rev.* **43,** 55–103 (1972).
Verloop, A., Nimmo, W. B., and DeWilde, P. C., *Dokl. Soobshch.-Mezhdunar Kongr. Zashch. Rast.,* 8th, 4, 43–44 (1975), *Chem. Abstr.* **88,** 184194 (1978).
Vlitos, A. J., and King, L. J., *Nature (London)* **171,** 523 (1953).
Voorman, R., and Penner, D., *Arch. Environ. Contam. Toxicol.* **15,** 595–602 (1986).

Walker, W. W., and Stojanovic, B. J., *J. Environ. Qual.* **3,** 4–10 (1974).
Wallnöfer, P. R., *Weed Res.* **9,** 333–339 (1969).
Wallnöfer, P. R., Koniger, M., Safe, S., and Hutzinger, O., *Int. J. Environ. Anal. Chem.* **2,** 37–43 (1972).
Wallnöfer, P. R., Safe, S., and Hutzinger, O., *J. Agric. Food Chem.* **21,** 502–504 (1973).
Wallnöfer, P. R., Söhlemann, F., and Oehlmann, L., *Pfanzenschutz-Nachr.* **29,** 236–253 (1976).
Wallnöfer, P. R., Ziegler, W., Engelhardt, G., and Rothmeier, H., *Chemosphere* **7,** 967–972 (1978).
Webley, D. M., Duff, R. B., and Farmer, V. C., *Nature (London)* **178,** 1467–1468 (1956).
Webster, G. R. B., and Reimer, G. J., *Weed Res.* **16,** 191–196 (1976).
Wedemeyer, G., *Appl. Microbiol.* **15,** 569–574 (1967).
Wendt, T. M., Cornell, J. H., and Kaplan, A. M., *Appl. Environ. Microbiol.* **36,** 693–699 (1978).
Wheelis, M. L., Palleroni, N. J., and Stanier, R. Y., *Arch. Microbiol.* **59,** 302–304 (1967).
Wiese, M. V., and Vargas, J. M., Jr., *Pestic. Biochem. Physiol.* **3,** 214–222 (1973).
Wilcock, R. J., Corban, G. A., Northcott, G. L., Wilkins, A. L., and Langdon, A. G., *Environ. Toxicol. Chem.* **15,** 670–676 (1996).
Wild, A. P., Winkelbauer, W., and Leisinger, T., *Biodegradation* **6,** 309–318 (1995).
Willetts, A. J., and Cain, R. B., *Biochem. J.* **120,** 28P (1970).
Wilson, B. H., Smith, G. B., and Rees, J. F., *Environ. Sci. Technol.* **20,** 997–1002 (1986).
Wright, S. J. L., and Forey, A., *Soil Biol. Biochem.* **4,** 207–213 (1972).
Wunder, T., Kremer, S., Sterner, O., and Anke, H., *Appl. Microbiol. Biotechnol.* **42,** 636–641 (1994).
Wunder, T., Marr, J., Kremer, S., Sterner, O., and Anke, H., *Arch. Microbiol.* **167,** 310–316 (1997).
Yamada, H., Asano, Y., Hino, T., and Tani, Y., *J. Ferment. Technol.* **57,** 8–14 (1979).
Yordy, J. R., and Alexander, M., *J. Environ. Qual.* **10,** 266–270 (1981).
Yu, C.-C., Hansen, D. J., and Booth, G. M., *Bull. Environ. Contam. Toxicol.* **13,** 280–283 (1975).
Yu, C.-C., and Sanborn, J. R., *Bull. Environ. Contam. Toxicol.* **13,** 543–550 (1975).
Zinder, S. H., and Brock, T. D., *J. Gen. Microbiol.* **105,** 335–342 (1978).

CHAPTER 13

Cometabolism

Microorganisms have long been known to have the ability to transform organic molecules to yield organic products that accumulate in culture media. Such conversions have achieved prominence in industrial microbiology because of the importance of the products, especially pharmaceutical agents, thus generated. The first evidence of analogous transformations with environmentally important chemicals came from a study of chlorinated aliphatic acids. In this early investigation, it was noted that a strain of *Pseudomonas* that grew on monochloroacetate was able to dehalogenate trichloroacetate but not use the latter compound as a C source for growth (Jensen, 1963). This transformation of an organic compound by a microorganism that is unable to use the substrate as a source of energy or of one of its constituent elements is termed *cometabolism* (Alexander, 1967).

The active populations thus derive no nutritional benefit from the substrates they cometabolize. Energy sufficient to fully sustain growth is not acquired even if the conversion is an oxidation and releases energy, and the C, N, S, or P that may be in the molecule is not used as a source, or at least a significant source, of these elements for biosynthetic purposes. Because of the prefix *co,* which often is appended to a word to indicate that something is done jointly or together (as in copilot or cooperate), there have been some semantic disagreements. Specifically, some authorities propose that the term cometabolism should be applied only to circumstances in which a substrate that is not used for growth is metabolized in the presence of a second substrate that is used to support multiplication. According to this view, the transformation of a substance that is not used as a nutrient or energy source but which occurs in the absence of a chemical supporting growth should be designated by another term, for example, *fortuitous metabolism* (Dalton and Sterling, 1982). However, the prefix *co* also has another meaning, namely, the same or similar (as in coconscious). The latter usage implies that the cometabolic transformation is similar to some other metabolic reaction, which is consistent with one explanation for the phenomenon (see the following). Fortuitous metabolism is, indeed, a more attractive term because it suggests an explanation for cometabolism, but the term will be used here as in the original definition, if for no other reason than it has gained wide acceptance. Thus, the term will be used to describe the metabolism of an organic substrate by a microorganism that is unable to use that compound as a souce of energy or an essential nutrient element. Covered by the term will be cases in which the organism is simultaneously growing on a second compound and instances in which multiplication is not occurring at the time the chemical of interest is being metabolized (Horvath, 1972).

The term *cooxidation* is sometimes used in studies of pure cultures of bacteria, this term referring specifically to oxidations of substrates that do not support growth in the presence of a second compound that does support multiplication (Perry, 1979). Cooxidation has historical precedence in the semantic debate (Foster, 1962), but since it is restricted to oxidation, the word does not have sufficient breadth to include many reactions that are not oxidations.

Nevertheless, it should be pointed out that two types of reactions of these sorts take place in pure cultures of bacteria. In one, the cometabolized compound is transformed only in the presence of a second substrate, which indeed may be the compound that supports growth. For heterotrophs, the energy-providing substrate is organic (Malashenko *et al.*, 1976; You and Bartha, 1982; Schukat *et al.*, 1983). For autotrophs, it is inorganic (Vannelli and Hooper, 1992). In the second, the compound is metabolized even in the absence of a second substrate (Horvath and Alexander, 1970a).

Particularly cogent reasons for using the more general definition, and even for maintaining cometabolism as a term apart from bioconversion or biotransformation, are the environmental consequences of cometabolism. Cometabolic reactions have impacts in nature that are different from growth-linked biodegradations, and when the transformations take place, it is usually totally unclear whether the microorganisms do or do not have a second substrate available on which they are growing.

SUBSTRATES AND REACTIONS

A large number of chemicals are subject to cometabolism in culture. Among the compounds thus acted on are cyclohexane (Beam and Perry, 1974), PCBs (Brunner *et al.*, 1985), 3-trifluoromethylbenzoate (Knackmuss, 1981), several chlorophenols (Liu *et al.*, 1991), 3,4-dichloroaniline (You and Bartha, 1982), 1,3,5-trinitrobenzene (Mitchell *et al.*, 1982), such pesticides as propachlor (Novick and

Alexander, 1985), alachlor (Smith and Phillips, 1975), ordram (Golovleva *et al.*, 1978), 2,4-D (Bauer *et al.*, 1979), and dicamba (Ferrer *et al.*, 1985), as well as the compounds listed in Table 13.1. The organisms that carry out these reactions in laboratory media includes species of *Pseudomonas, Acinetobacter, Nocardia, Bacillus, Mycococcus, Achromobacter, Methylosinus, Alcaligenes, Rhodococcus, Mycobacterium, Xanthobacter,* and *Nitrosomonas* among the bacteria and *Penicillium* and *Rhizoctonia* among the fungi. Among cometabolic conversions that appear to involve a single enzyme, the reactions may be hydroxylations, oxidations, denitrations, deaminations, hydrolyses, acylations, or cleavages of ether linkages, but many of the conversions are complex and involve several enzymes. Even a substrate that will, in nature, support growth of microorganisms may be metabolized by some bacteria in culture with no incorporation of the C into their cells (Schmitt *et al.*, 1992).

Some of the cometabolic reactions brought about by bacteria and fungi in culture are given in Table 13.1. Even this incomplete list illustrates the wide range of conversions, reaction types, and products associated with cometabolism. The kinds of transformation come as no surprise in view of the vast array of biological transformations that heterotrophic bacteria and fungi bring about in culture (Kieslich, 1976). The methane monooxygenase of methylotrophic bacteria is able to oxidize alkanes, alkenes, secondary alcohols, di- or trichloromethane, dialkyl ethers, cycloalkanes, and aromatic compounds (Haber *et al.*, 1983), and a single strain of *Nocardia corallina* can cometabolize tri- and tetramethylbenzenes, diethylbenzenes, biphenyl, tetralin, and dimethylnaphthalenes to yield a diversity of products (Jamison *et al.*, 1971).

Cometabolism yields organic products, but the C in the substrate is not converted to typical cell constituents. This is evident in studies of pure cultures and in samples from natural environments. For example, during the metabolism of ^{14}C-labeled 2,5,2′-trichlorobiphenyl, strains of *Alcaligenes* and *Acinetobacter* do not incorporate ^{14}C into cell constituents nor do they generate $^{14}CO_2$ (Furukawa *et al.*, 1978). Similarly, none of the C is assimilated by bacteria that cometabolize propachlor (Novick and Alexander, 1985). During metabolism of a chlorinated disaccharide by two bacteria or microflora from sewage or a lake, none of the substrate C was incorporated into the biomass (Labare and Alexander, 1994), and little of the C from carbofuran was incorporated into the microbial cells in soil and the number of carbofuran-metabolizing bacteria did not increase as this insecticide was being transformed (Robertson and Alexander, 1994). The same lack of use of the C is evident as the natural microflora of sewage cometabolizes the herbicides trifluralin, profluralin, fluchloralin, and nitrofen, the C in the substrate that had been transformed being converted to low-molecular-weight products instead (Jacobson *et al.*, 1980).

Several lines of evidence suggest that many compounds are cometabolized in soils, waters, and sewage. Only one or a few of the following lines of evidence have been obtained for any one chemical, however. (a) The chemical is convertd

Table 13.1

Cometabolism of Various Substrates in Pure Culture

Substrate	Products	Reference
Methyl fluoride	Formaldehyde	Hyman *et al.* (1994)
Dimethyl ether	Methanol	Hyman *et al.* (1994)
Dimethyl sulfide	Dimethyl sulfoxide	Juliette *et al.* (1993)
Nitrapyrin	6-Chloropicolinic acid	Vannelli and Hooper (1992)
Tetrachloroethylene	Trichloroethylene	Fathepure and Boyd (1988)
Benzothiophene	Benzothiophene-2,3-dione	Fedorak and Grbić-Galić (1991)
3-Hydroxybenzoate	2,3-Dihydroxybenzoate	Daumy *et al.* (1980)
Cyclohexane	Cyclohexanol	deKlerk and van der Linden (1974)
3-Chlorophenol	4-Chlorocatechol	Engelhardt *et al.* (1979)
Chlorobenzene	3-Chlorocatechol	Klečka and Gibson (1981)
Bis(tributyltin) oxide	Dibutyl tin	Barug (1981)
3-Nitrophenol	Nitrohydroquinone	Raymond and Alexander (1971)
Trinitroglycerine	1- and 2-Nitroglycerine	Cornell and Kaplan (1977)
Parathion	4-Nitrophenol	Daughton and Hsieh (1977)
4-Chloroaniline	4-Chloroacetanilide	Engelhardt *et al.* (1977)
Metamitron	Desaminometamitron	Engelhardt and Wallnöfer (1978)
Propane	Propionate, acetone	Leadbetter and Foster (1959)
2-Butanol	2-Butanone	Patel *et al.* (1979)
Phenol	*cis,cis*-Muconate	Knackmuss and Hellwig (1978)
DDT	DDD, DDE, DBP	Pfaender and Alexander (1973)
o-Xylene	*o*-Toluic acid	Raymond *et al.* (1967)
2,4,5-T	2,4,5-Trichlorophenol	Rosenberg and Alexander (1980)
4-Fluorobenzoate	4-Fluorocatechol	Clarke *et al.* (1979)
4,4′-Dichlorodiphenyl-methane	4-Chlorophenylacetic acid	Focht and Alexander (1971)
2,3,6-Trichlorobenzoate	3,5-Dichlorocatechol	Horvath and Alexander (1970a)
3-Chlorobenzoate	4-Chlorocatechol	Horvath and Alexander (1970b)
m-Chlorotoluene	Benzyl alcohol	Higgins *et al.* (1979)
Kepone	Monohydrokepone	Orndorff and Colwell (1980)
4-Trifluoromethyl-benzoate	4-Trifluoromethyl-2,3-dihydroxybenzoate	Engesser *et al.* (1988)

to organic products in nonsterile but not in sterile samples of the environment (or is more readily transformed in nonsterile samples), but a microorganism able to use that substrate as a source of energy, C, or another element essential for growth cannot be isolated from that environment. For example, propachlor is converted in sewage and lake water to organic products but not CO_2, and a microorganism able to use it as a sole source of C and energy has not been isolated (Novick and Alexander, 1985). (b) Microorganisms that use other organic molecules as C sources for growth metabolize the chemical in culture to yield products identical to those found in nature (Beam and Perry, 1974). (c) Carbon from the chemical is not incorporated into cell components. The almost quantitative transformation of specific compounds to organic products and the lack of incorporation of ^{14}C from the radioactive substrate into microbial cells are strong lines of evidence for cometabolism (Jacobson *et al.*, 1980). Similar evidence for cometabolism was obtained in a study of the metabolism of ^{14}C-labeled carbon monoxide in soil. Because the ^{14}C is not converted to organic matter in soil (this fraction containing microbial cells), cometabolism is indicated. Populations oxidizing CO to CO_2 in soil apparently do not grow using CO as a C or energy source because prior exposure of the soil to this air pollutant does not result in an enhanced oxidation of later increments of CO; had they grown, the rate presumably should have increased (Bartholomew and Alexander, 1982). A similar argument can be made for the transformation of EPTC in soil since little of the ^{14}C from ^{14}C-labeled EPTC is incorporated into biomass (Moorman *et al.*, 1992). (d) Often but not always, the products known to be generated by cometabolism in culture media also accumulate and persist in nature.

Caution needs to be exercised in concluding that cometabolism is occurring merely because an organism cannot be isolated from an environment in which a chemical is undergoing a biological reaction. The isolation of bacteria acting on specific substrates is usually performed by enriching for the organism in a medium whose only C source is the test chemical, and the agar medium used to plate the enrichments contains that single organic supplement. Yet, many bacteria that are able to grow at the expense of that substrate will not develop in such simple media because they require amino acids, B vitamins, or other growth factors. These essential growth factors are not routinely included in such liquid media, and hence bacteria and fungi needing them fail to proliferate. If the only organisms in the environment able to metabolize a test chemical need these growth factors, no isolate will be obtained, and the conclusion will be reached that the compound is cometabolized; that conclusion may thus be erroneous. If a chemical supports the growth of many species, some will undoubtedly require no growth factors (these organisms are called *prototrophs*), and they will be enriched and ultimately can be isolated. If the compound is acted on by only one species, in contrast, it is likely that the responsible organism will need amino acids, B vitamins, or other growth factors; these species are termed *auxotrophs*. Hence, the failure to isolate a bacterium or fungus capable of using the molecule as the sole C source for growth is not sufficient evidence for cometabolism.

EXPLANATIONS

Several reasons have been advanced to explain cometabolism, that is, why an organic chemical that is a substrate does not support growth but is converted to products that accumulate. Three have experimental support: (a) the initial enzyme or enzymes convert the substrate to an organic product that is not further transformed by other enzymes in the microorganism to yield the metabolic intermediates that ultimately are used for biosynthesis and energy production; (b) the initial substrate is transformed to products that inhibit the activity of late enzymes in mineralization or that suppress growth of the organisms; and (c) the organism needs a second substrate to bring about some particular reaction. It is likely that the first explanation is the most common, especially at concentrations of organic chemicals that are not likely to be metabolized to yield products that have antimicrobial effects. The basis for this explanation is the fact that many enzymes act on several structurally related substrates; thus, an enzyme naturally present in the cell—because it functions in processes characterizing normal growth of the organism on other than synthetic molecules—will catalyze reactions that alter chemicals that are not typical cellular intermediates. These enzymes are not absolutely specific for their substrates. Consider a normal metabolic sequence involving the conversion of A to B by enzyme *a*, B to C by enzyme *b*, and C to D by enzyme *c* in a sequence that ultimately yields CO_2, energy for biosynthetic reactions, and intermediates that are converted to cell constituents.

$$A \xrightarrow{a} B \xrightarrow{b} C \xrightarrow{c} D \rightarrow \rightarrow \rightarrow CO_2 + \text{energy} + \text{cell-C}$$

The first enzyme (*a*) may have a low substrate specifity and act on a molecule structurally similar to A, namely, A′. The product (B′) would differ from B in the same way that A differs from A′. However, if enzyme *b* is unable to act on B′ (because the structural features controlling which substrates it modifies differ from those controlling the substrate specificity of enzyme *a*), B′ will accumulate:

$$A' \xrightarrow{a} B' \not\rightarrow.$$

In addition, CO_2 and energy will not be generated, and because cell-C is not formed, the organisms do not multiply. The formation of B′ is thus entirely fortuitous (Raymond and Alexander, 1971; Alexander, 1979). The initial evidence for this explanation came from studies of the metabolism of 2,4-D. This herbicide is usually converted first to 2,4-dichlorophenol, but the enzyme further metabolizing 2,4-dichlorophenol acts on some but not all the phenols generated by the

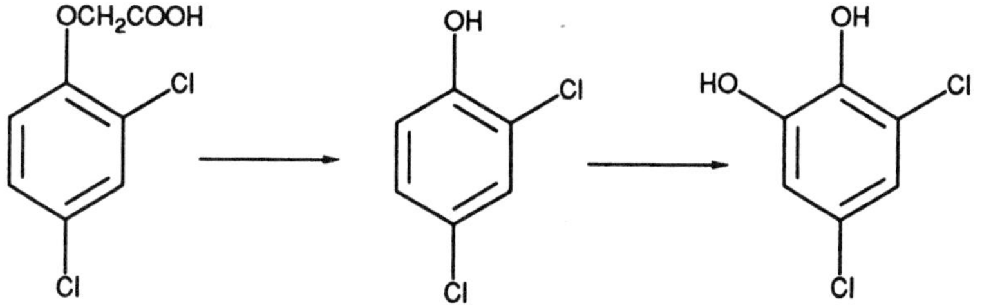

Figure 13.1 Conversion of 2,4-D to 2,4-dichlorophenol and 3,5-dichlorocatechol.

initial enzyme acting on other phenoxyacetic acids (Bollag *et al.*, 1968; Loos *et al.*, 1967) (Fig. 13.1). When this occurs, the product of cometabolism accumulates in almost quantitative yield, at least in pure culture. A typical case is the bacterial conversion of 3-chlorobenzoate to 4-chlorocatechol, the yield of the catechol being 98% of the substrate that is transformed (Fig. 13.2).

In instances in which the chemical concentration is high, cometabolism may result from the conversion of the parent compound to toxic products. In the sequence just depicted, if the rate of reaction catalyzed by enzyme *a* is faster than the process catalyzed by enzyme *b*, B will accumulate because it is not destroyed as readily as it is generated. For example, a strain of *Pseudomonas* that grows on benzoate but not 2-fluorobenzoate converts the latter to fluorinated products that are toxic (Taylor *et al.*, 1979). The inhibitor that accumulates may affect a single enzyme that is important for the further metabolism of the toxin. For example, *Pseudomonas putida* cometabolizes chlorobenzene to 3-chlorocatechol, but the latter is not degraded because it suppresses the enzymes involved in further degradation (Klečka and Gibson, 1981). *Pseudomonas putida* also converts 4-ethylbenzoate to 4-ethylcatechol, and the latter inactivates enzymes necessary for subsequent metabolic steps (Ramos *et al.*, 1987). As a result, growth of this bacterium does not occur on chlorobenzene and 4-ethylbenzoate.

In some instances, in pure culture at least, an organism may not be able to metabolize an organic compound because it needs a second substrate to bring

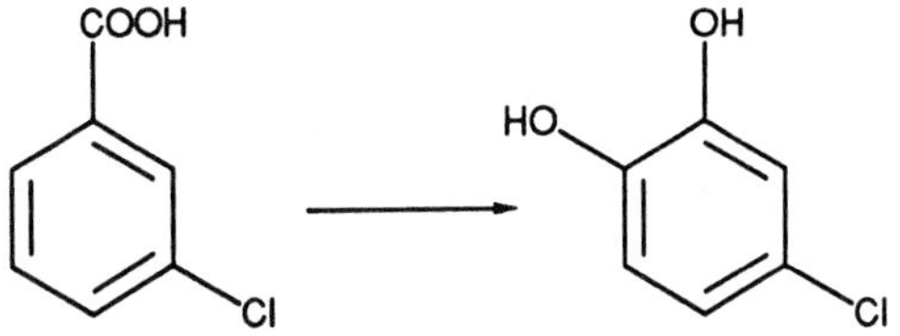

Figure 13.2 Conversion of 3-chlorobenzoate to 4-chlorocatechol by *Arthrobacter* sp. (From Horvath and Alexander, 1970a.)

about a particular reaction. The second substrate may provide something that is present in insufficient supply in the cells for the reaction to proceed, for example, an electron donor for the transformation (Lütjens and Gottschalk, 1980; Schukat *et al.*, 1983).

ENZYMES WITH MANY SUBSTRATES

The first explanation is linked to the existence of enzymes acting on more than a single substrate. Many enzymes are not absolutely specific for a single substrate. As a rule, they act on a series of closely related molecules, but some carry out a single type of reaction on a variety of somewhat dissimilar molecules. The following are examples of single enzymes acting on a range of substrates.

(a) Methane monooxygenase of methylotrophic bacteria. When grown on methane, methanol, or formate, these aerobic bacteria are able to cometabolize a large array of organic molecules, including several major pollutants. Some of the reactions carried out by these bacteria are shown in Fig. 13.3. In each instance, methane monooxygenase is the responsible catalyst. Other chlorinated aliphatic hydrocarbons transformed by one such methylotroph, *Methylosinus trichosporium*, are *cis*- and *trans*-1,2-dichloroethylene, 1,1-dichloroethylene, 1,2-dichloropropane, and 1,3-dichloropropylene (Oldenhuis *et al.*, 1989). Apparently the same enzyme in other bacteria, after growth on methane, will catalyze the oxidation of *n*-alkanes with two to eight C atoms, *n*-alkenes with two to six C atoms, and mono- and dichloroalkanes with five or six C atoms (Imai *et al.*, 1986), as well as dialkyl ethers and cycloalkanes (Haber *et al.*, 1983).

(b) Toluene dioxygenase of a number of aerobic bacteria. This enzyme incorporates both atoms of oxygen from O_2 (hence, it is a dioxygenase) into toluene as it catalyzes the first step in the degradation of toluene by bacteria grown on that aromatic hydrocarbon (Fig. 13.4). However, that enzyme has very low specificity and also is able to bring about the degradation of TCE (Nelson *et al.*,

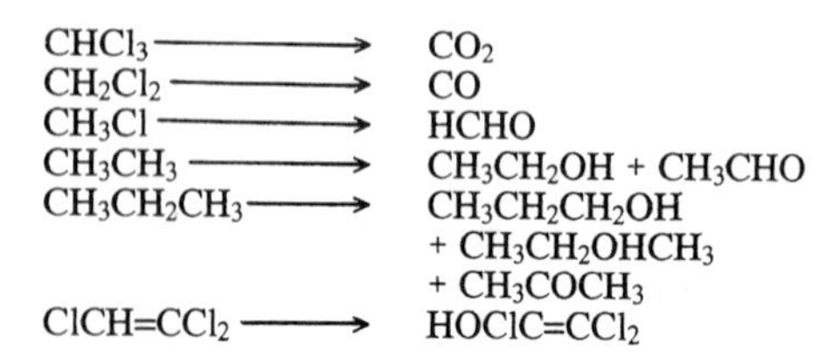

Figure 13.3 Reactions catalyzed by the methane monooxygenase of methylotrophic bacteria. (Reprinted with permission from Haber *et al.*, 1983; Oldenhuis *et al.*, 1989.)

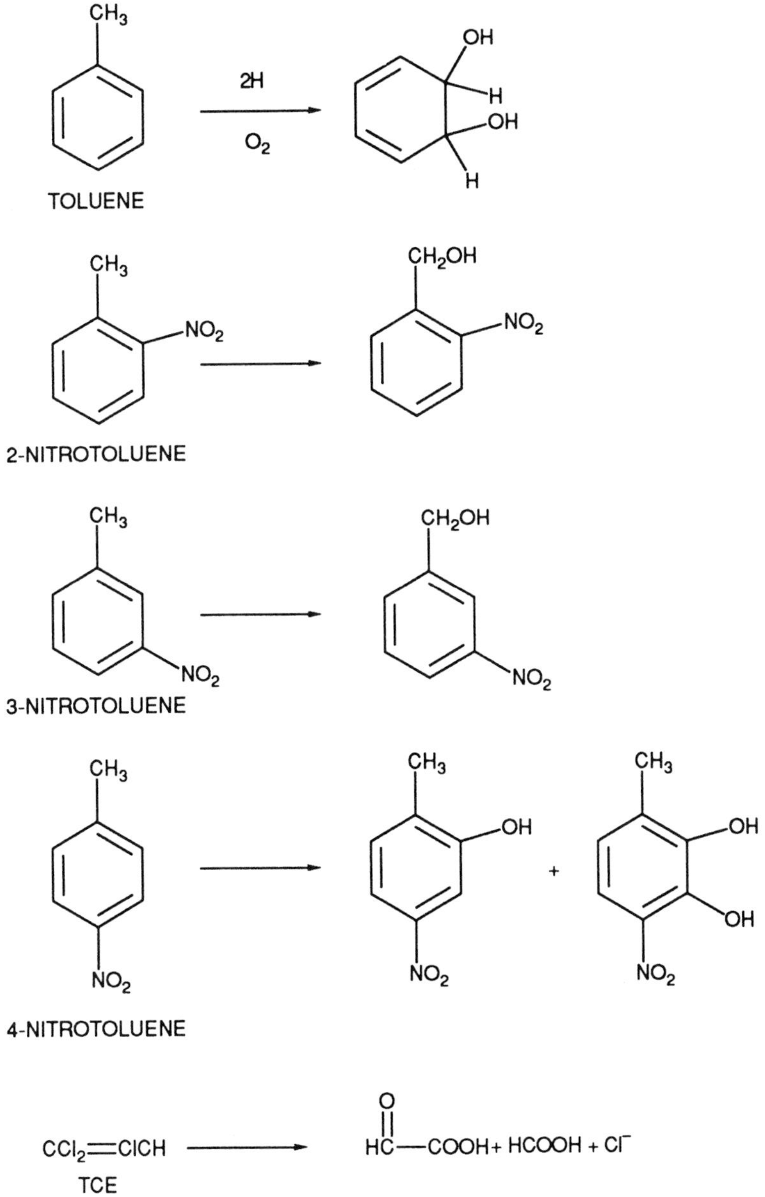

CH_3 — 2H, O_2 → OH, H, OH, H

TOLUENE

CH_3, NO_2 → CH_2OH, NO_2

2-NITROTOLUENE

CH_3, NO_2 → CH_2OH, NO_2

3-NITROTOLUENE

CH_3, NO_2 → CH_3, OH, NO_2 + CH_3, OH, OH, NO_2

4-NITROTOLUENE

$CCl_2{=}ClCH$ → $HC(=O){-}COOH + HCOOH + Cl^-$

TCE

Figure 13.4 Reactions catalyzed by toluene dioxygenase.

1988; Li and Wackett, 1992), to convert 2- and 3-nitrotoluene to the corresponding alcohols, and to hydroxylate the ring of 4-nitrotoluene (Robertson *et al.*, 1992).

(c) Toluene monooxygenase of several aerobic bacteria. Differing from the dioxygenase, this enzyme incorporates only one atom of oxygen from O_2 into toluene to give *o*-cresol (Fig. 13.5). However, because of this enzyme, bacteria can cometabolize TCE, convert 3- and 4-nitrotoluenes to the corresponding benzyl alcohols and benzaldehydes (Delgado *et al.*, 1992), and add hydroxyl groups to other aromatic compounds (Shields *et al.*, 1991).

(d) Oxygenase of propane-utilizing bacteria. Aerobes using propane as C and energy source for growth also have an oxygenase of broad specificity. This enzyme cometabolizes TCE, vinyl chloride, and 1,1-di- and *trans*- and *cis*-1,2-dichloroethylene (Wackett *et al.*, 1989).

(e) Ammonia monooxygenase of *Nitrosomonas europaea*. This bacterium, which is a chemoautotroph whose energy source in nature is NH_3 and whose C source is CO_2, cometabolizes TCE, 1,1-dichloroethylene, various mono- and polyhalogenated ethanes (Rasche *et al.*, 1990, 1991), and a variety of monocyclic aromatic compounds (Keener and Arp, 1994) and thioethers (Juliette *et al.*, 1993), as well as methyl fluoride and dimethyl ether (Hyman *et al.*, 1994).

(f) Halidohydrolase acting on simple halogenated fatty acids. Depending on the specific organism, this enzyme may cleave halogens from fluoro-, chloro-, and iodoacetate (Goldman, 1965), dichloroacetate, 2-chloropropionate, and 2-chlorobutyrate (Goldman *et al.*, 1968), and all monohaloacetates except fluoroacetate (Klages *et al.*, 1983).

(g) Halidohydrolase that removes the halogen from 1-iodomethane, 1-iodoethane, 1-chlorobutane, 1-bromobutane, and 1-chlorohexane to yield the corresponding *n*-alcohols (Scholtz *et al.*, 1987).

(h) A dehalogenase that removes the halogens from CH_2Cl_2, CH_2BrCl, CH_2Br_2, and CH_2I_2 (Kohler-Staub and Leisinger, 1985).

(i) A dehalogenase that acts on 4-chloro-, 4-bromo-, and 4-iodo- but not 4-fluorobenzoate (Thiele *et al.*, 1987).

(j) A catechol dioxygenase that oxidizes catechol, 3- and 4-methylcatechol, and 3-fluoro- but not 3-chlorocatechol (Klečka and Gibson, 1981).

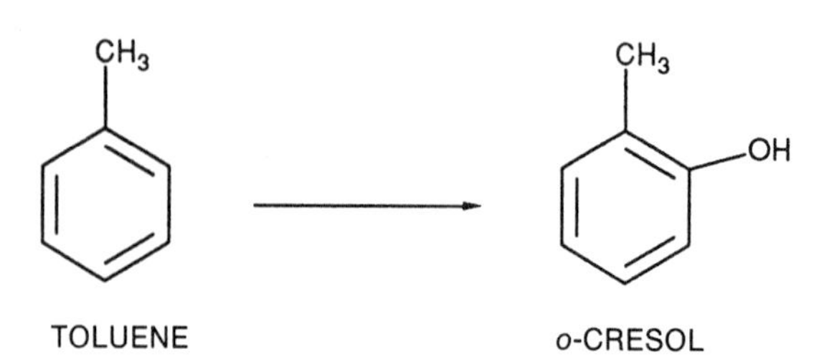

Figure 13.5 The reaction catalyzed by toluene monooxygenase.

(k) A benzoate hydroxylase that metabolizes benzoate and 4-amino-, 4-nitro-, 4-chloro-, and 4-methylbenzoates (Reddy and Vaidyanathan, 1976).

(l) An enzyme that cleaves the nitrile from a number of aromatic nitriles to yield ammonia (Harper, 1977).

(m) A phosphatase that hydrolyzes parathion, paraoxon, diazinon, dursban, and fenitrothion, but not several related insecticides (Munnecke, 1976).

(n) An alcohol dehydrogenase that oxidizes normal aliphatic alcohols containing 1 to 11 carbon atoms (Sperl *et al.*, 1974).

(o) A deaminase cleaving the amine moiety of a number of purines (Sakai and Jun, 1978).

(p) An alkane hydroxylase that hydroxylates a number of alkylbenzenes and linear, branched, and cyclic alkanes (van Beilin *et al.*, 1994).

(q) An alkane monooxygenase that degrades TCE, vinyl chloride, and dichloroethylenes and propylenes (Ensign *et al.*, 1992).

(r) Naphthalene dioxygenase that acts on xylene, isomers of nitrotoluene, and ethylbenzene (Lee and Gibson, 1996).

(s) Biphenyl dioxygenase that transforms several PCB congeners (Hernandez *et al.*, 1995).

The organism containing these enzymes may be able to use one or several of the enzyme's substrates for growth. However, many of the substrates are transformed but do not support growth. The product of the reaction then accumulates.

ENVIRONMENTAL SIGNIFICANCE

In a sense, cometabolism is merely a special type of microbial transformation. As such, it might seem to be of mere academic interest; however, such transformations have considerable importance in nature. These important consequences have been the basis for the great attention given to cometabolism, and it is because of these environmental issues that cometabolism is considered to be a special type of biological transformation. These environmental consequences are readily evident from the characteristics of the process, specifically the inability of the organisms to grow at the expense of the organic compound and the conversion of the substrate to an organic product that often accumulates. Two effects are immediately evident. First, because the size of the population or the biomass of organisms acting on most synthetic chemicals is small in surface and subsurface soils and waters, a chemical subject to cometabolism by these organisms is transformed slowly, and the rate of conversion does not increase with time. This contrasts with chemicals used as C and energy sources because the rate of metabolism of such substrates increases as the responsible organisms multiply (Fig. 13.6). Second, many organic products accumulate as a result of cometabolism, and these products tend to persist.

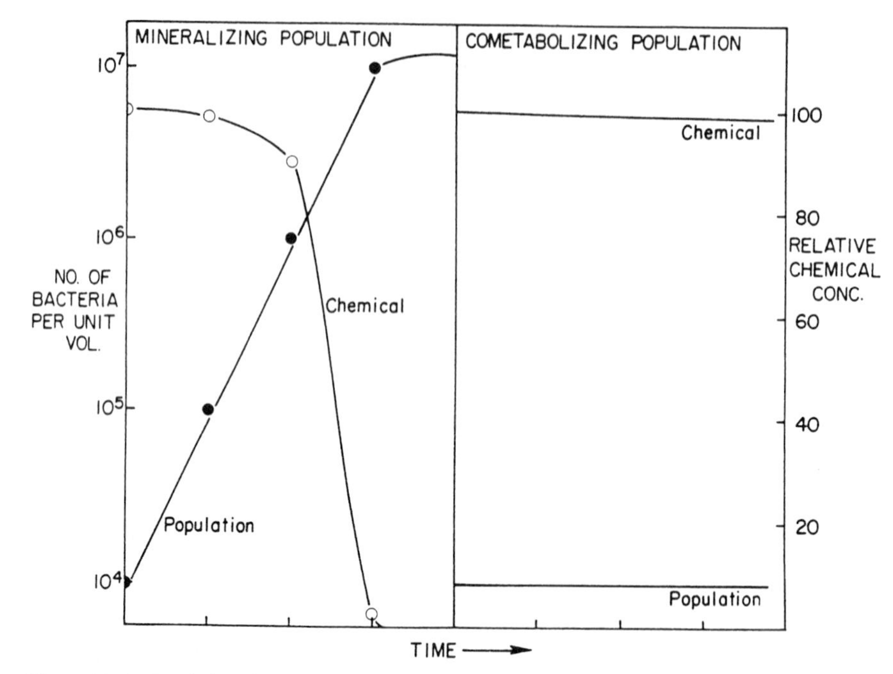

Figure 13.6 Population changes and disappearance of a chemical acted on by (left) bacteria growing logarithmically and using the compound as a C source or (right) bacteria cometabolizing the chemical. (From Alexander, 1981. Reprinted with permission from the American Association for the Advancement of Science.)

This accumulation is an outcome of cometabolism by a single species because it cannot further metabolize the product. Moreover, since the outcome of cometabolism frequently is only a small alteration in the structure of the molecule, a toxic parent compound is often converted to a harmful product (Alexander, 1979).

Few estimates have been made of the numbers or biomass of microorganisms able to cometabolize individual substrates in nature. However, the number of cells in one soil able to cometabolize 2,4-D ranges from 0.3 to 0.8 million per gram (Fournier *et al.*, 1981). In contrast, 20 to 75% of the bacteria isolated from sewage have the ability to cometabolize DDT, and 90 million cells per milliliter of sewage can cometabolize the insecticide (Pfaender and Alexander, 1973).

Although the products of cometabolism accumulate in culture, the same is not necessarily true in nature. Those products may be acted on by a second species and may thereby be either cometabolized or mineralized. Under such circumstances, the initial products may not persist. An example is shown in Fig. 13.7 of a compound, sucralose, that is initially cometabolized to yield organic products, but the latter are subsequently mineralized, presumably by species not

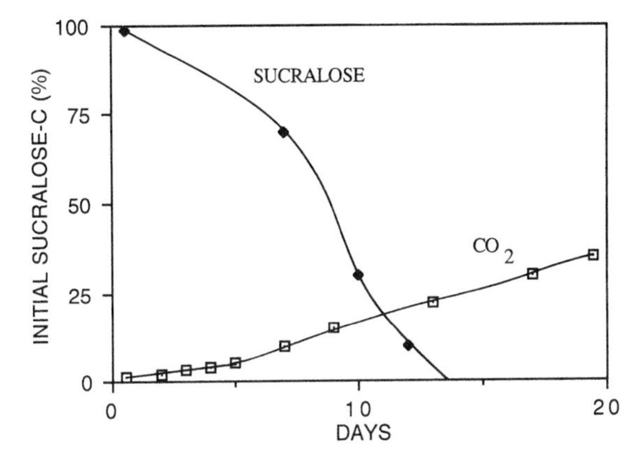

Figure 13.7 Transformation of sucralose in samples of soil. (From M. P. Labare and M. Alexander, unpublished data.)

acting on the original trichlorinated disaccharide. Indeed, if the second population grows by using the cometabolic products of the first population, those products may not be detected at all because the population will grow to the size permitted by the yield of its C source, that is, the cometabolic product.

Several cases have been described in which a second species destroys the metabolites excreted by the first in culture (Fig. 13.8). Six examples will be cited. (a) Parathion is cometabolized by *Pseudomonas stutzeri* to yield 4-nitrophenol and diethyl phosphate, and *Pseudomonas aeruginosa* uses the phenol as a source of C and energy (Daughton and Hsieh, 1977). (b) Cyclohexane is cometabolized to cyclohexanol by one pseudomonad, and cyclohexanol is mineralized by a different species of *Pseudomonas* (deKlerk and van der Linden, 1974). (c) 4,4′-Dichlorobiphenyl is cometabolized to yield 4-chlorobenzoate, and the latter is a C and energy source for an *Acinetobacter* strain (Adriaens *et al.*, 1989). (d) DDT is converted by cometabolism to 4-chlorophenylacetic acid by a strain of *Pseudomonas*, and the product is then used for growth by *Arthrobacter* sp. (Pfaender and Alexander, 1972). (e) 2,4,5-T is cometabolized by *Pseudomonas fluorescens* to 2,4,5-trichlorophenol, which is then further metabolized by other microorganisms (Rosenberg and Alexander, 1980). (f) 4-Chloro-3,5-dinitrobenzoic acid is cometabolized to yield 2-hydroxymuconic semialdehyde, which is then mineralized by *Streptomyces* sp. (Jacobson *et al.*, 1980). In such instances, a single organism able to mineralize the initial substrate would not be obtained in pure culture, yet the product of the first population and the substrate for the second might not be detected in nature. The two phases of the transformation, in effect, are *complementary catabolic pathways*, complementary in the sense that they complete an entire degradative sequence. These complementary catabolic pathways may be exploitable by genetic engineering to construct organisms that are able to mineralize the parent molecule. This

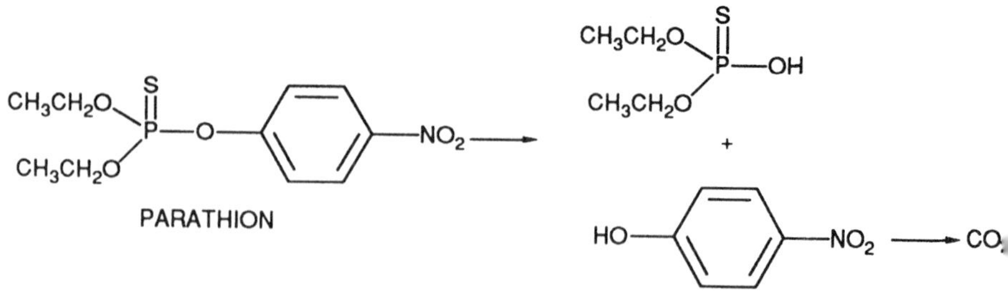

CH_3CH_2O $\overset{S}{\|}$ P—O—(benzene)—NO_2 → $(CH_3CH_2O)_2P(=S)$—OH + HO—(benzene)—NO_2 → CO_2

PARATHION

(cyclohexane) → (cyclohexane)—OH → CO_2

CYCLOHEXANE

CYCLOHEXANOL

Cl—(benzene)—(benzene)—Cl → Cl—(benzene)—COOH → CO_2

4,4'-DICHLOROBIPHENYL

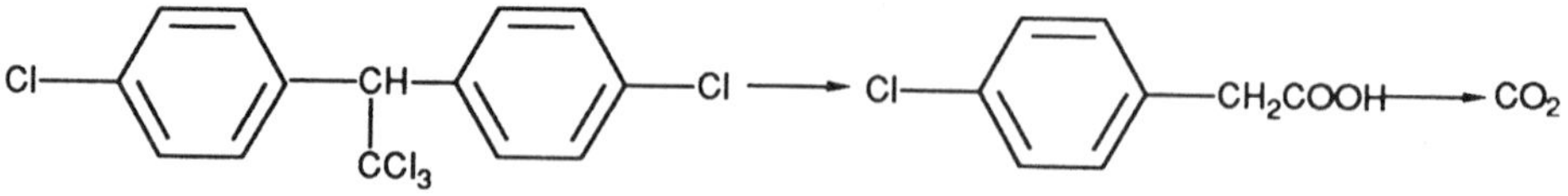

Cl—(benzene)—CH(CCl_3)—(benzene)—Cl → Cl—(benzene)—CH_2COOH → CO_2

DDT

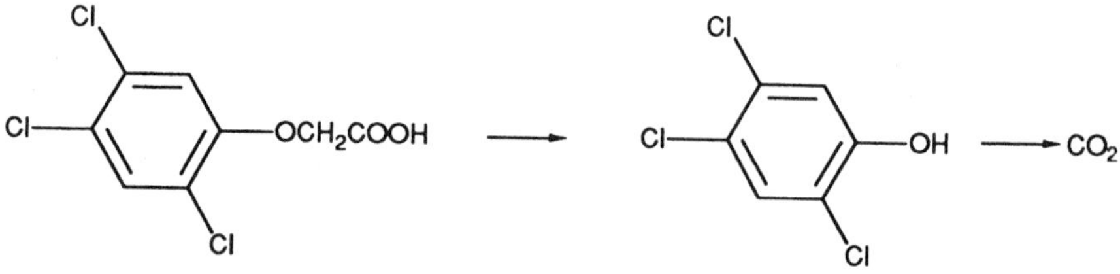

Cl, Cl, Cl—(benzene)—OCH_2COOH → Cl, Cl, Cl—(benzene)—OH → CO_2

2,4,5-T

COOH, O_2N, Cl, NO_2 (benzene) → CHO, COOH, OH → CO_2

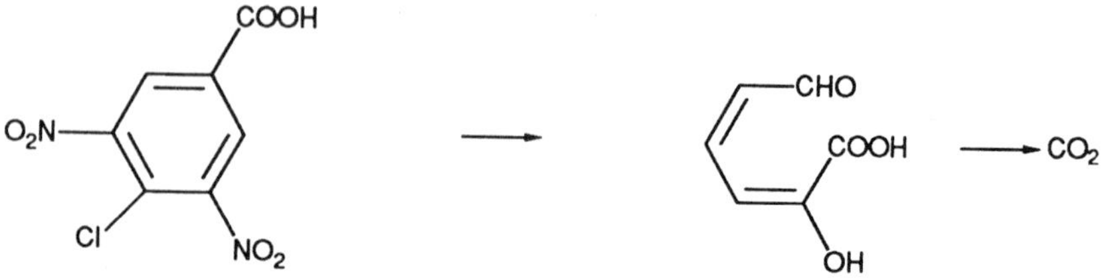

4-CHLORO-3,5-DINITRO-
BENZOIC ACID

might be done by transferring genes into one species so that its cells contain the enzymes bringing about the initial cometabolic sequence but also those enzymes that allow the organism to mineralize and grow on the products of the initial cometabolic sequence (Ramos *et al.*, 1987; Rubio *et al.*, 1986).

A chemical that is cometabolized at one concentration may be mineralized in the same environment at another concentration or it may be cometabolized in one environment and mineralized in another. This suggests that the organic products of cometabolism may accumulate at only certain concentrations or only in some environments. For example, IPC is cometabolized at 1.0 mg/liter but mineralized at 0.4 μg/liter in lake water, and monuron is cometabolized apparently to 4-chloroaniline at 10 mg/liter but mineralized at 10 μg/liter in sewage. Chlorobenzilate is cometabolized in samples from the water column of lakes but is mineralized in the presence of the microflora of freshwater sediments (Wang *et al.*, 1984, 1985). Thus, caution must be exercised in predicting that cometabolism will take place at concentrations or in environments other than those specifically tested.

The kinetics of cometabolism have received scant attention. If the microbial populations are neither growing nor declining and the concentration of substrate for cometabolism is below the K_m of the active organisms, it is likely that the conversion would be first order, as discussed in Chapter 6 on kinetics. The transformation of propachlor may be first or zero order in lake water or sewage (Novick and Alexander, 1985). In a biofilm bioreactor inoculated with methane-oxidizing bacteria, the cometabolism of TCE, 1,1,1-trichloroethane, and *cis*- and *trans*-1,2-dichloroethylene is first order at concentrations up to 1 mg/liter (Arvin, 1991). However, in environments in which the transformations are slow, the C source for growth probably is being depleted, so the kinetic patterns may change with time. Other models have been developed for cometabolism by nongrowing or growing populations (Arcangeli and Ervin, 1995; Chang and Alvarez-Cohen, 1995; Ely *et al.*, 1995a,b; Hill *et al.*, 1996; Alvarez-Cohen and McCarty, 1991; Criddle, 1993).

Because cometabolism generally leads to a slow destruction of the substrate, attention has been given to enhancing its rate. The stimulation of such activities is especially important if the substrate is toxic to humans, agricultural crops, or species in natural ecosystems. The addition of a number of organic compounds to soil or sewage promotes the rate of cometabolism of DDT and a number of chlorinated aromatic compounds and chlorinated fatty acids (Jacobson and Alexander, 1981; Pfaender and Alexander, 1973), but the responses to such additions are not predictable. No relation is known to exist between the metabolic pathways involved in destruction of the added mineralizable substrate and the compound

---

Figure 13.8 Conversions involving cometabolism by one species followed by mineralization of the cometabolic products by a second species.

that is cometabolized in these studies. The added molecules are randomly chosen in these trials, and sometimes they do and sometimes they do not stimulate cometabolism. In instances in which stimulation occurs, the benefit probably results from an unpredicted increase in the biomass of organisms, some of which fortuitously cometabolize the compound of interest.

An alternative approach is to add mineralizable compounds that are structurally analogous to the compound whose cometabolism one wishes to promote. Presumably, the microflora that grows on the mineralizable compound contains enzymes transforming the analogous molecule, that is, the one that is cometabolized. This larger biomass thus has more of the degradative enzyme than is present in the unsupplemented water or soil. This method of *analogue enrichment* has been used to enhance the cometabolism of PCBs by additions of biphenyl. The unchlorinated biphenyl was selected for addition to soil since it is mineralizable, nontoxic, and serves as a C source for microorganisms that are able to cometabolize PCBs (Brunner *et al.,* 1985). A similar approach has been used to enhance the cometabolism of trifluoromethylbenzoates by the addition of alkyl-substituted benzoates (Knackmuss, 1981) and the metabolism of 2,4-dichloroaniline in soil by additions of aniline (You and Bartha, 1982).

Analogue enrichment is a procedure that is similar to the usual means of isolating bacteria that can cometabolize a compound. The enrichment culture contains a C source that supports growth, and the pure cultures thus obtained also cometabolize structurally related compounds that would not support growth. For example, bacteria isolated on diphenylmethane and containing enzymes to degrade it also cometabolize chlorinated diphenylmethanes. Many of the latter do not sustain growth (Focht and Alexander, 1970a,b).

Various approaches have been devised recently to enhance cometabolism as part of actual bioremediation efforts. These will be presented in Chapter 17.

REFERENCES

Adriaens, P., Kohler, H.-P. E., Kohler-Staub, D., and Focht, D. D., *Appl. Environ. Microbiol.* **55,** 887–892 (1989).

Alexander, M., in "Agriculture and the Quality of Our Environment" (N. C. Brady, ed.), pp. 331–342. American Association for the Advancement of Science, Washington, DC, 1967.

Alexander, M., *in* "Microbial Degradation of Pollutants in Marine Environments" (A. W. Bourquin and P. H. Pritchard, eds.), pp. 67–75. U.S. Environmental Protection Agency, Gulf Breeze, FL, 1979.

Alexander, M., *Science* **211,** 132–138 (1981).

Alvarez-Cohen, L., and McCarty, P L., *Environ. Sci. Technol.* **25,** 1381–1387 (1991).

Arcangeli, J.-P., and Arvin, E., *Biodegradation* **6,** 29–38 (1995).

Arvin, E., *Water Res.* **25,** 873–881 (1991).

Bartholomew, G. W., and Alexander, M., *Environ. Sci. Technol.* **16,** 301–302 (1982).

Barug, D., *Chemosphere* **10,** 1145–1154 (1981).

Bauer, S. R., Wood, E. M., and Traxler, R. W., *Int. Biodeterior. Bull.* **15,** 53–56 (1979).
Beam, H. W., and Perry, J. J., *J. Gen. Microbiol.* **82,** 163–169 (1974).
Bollag, J.-M., Helling, C. S., and Alexander, M., *J. Agric. Food Chem.* **16,** 826–828 (1968).
Brunner, W., Sutherland, F. H., and Focht, D. D., *J. Environ. Qual.* **14,** 324–328 (1985).
Chang, H.-L., and Alvarez-Cohen, L., *Environ. Sci. Technol.* **29,** 2357–2367 (1995).
Clarke, K. F., Callely, A. G., Livingstone, A., and Fewson, C. A., *Biochim. Biophys. Acta* **404,** 169–179 (1979).
Cornell, J. H., and Kaplan, A. M., *Abstr. Annu. Meet., Am. Soc. Microbiol.* 276 (1977).
Criddle, C. S., *Biotechnol. Bioeng.* **41,** 1048–1056 (1993).
Dalton, H., and Sterling, D. I., *Philos. Trans. R. Soc. Lond., Ser. B* **297,** 481–495 (1982).
Daughton, C. G., and Hsieh, D. P. H., *Appl. Environ. Microbiol.* **34,** 175–184 (1977).
Daumy, G. O., McColl, A. S., and Andrews, G. C., *J. Bacteriol.* **141,** 293–296 (1980).
deKlerk, H., and van der Linden, A. C., *Antonie van Leeuwenhoek* **40,** 7–15 (1974).
Delgado, A., Wubbolts, M. G., Abril, M. A., and Ramos, J. L., *Appl. Environ. Microbiol.* **58,** 415–417 (1992).
Ely, R. L., Hyman, M. R., Arp, D. J., Guenther, R. B., and Williamson, K. J., *Biotechnol. Bioeng.* **46,** 232–245 (1995a).
Ely, R. L., Williamson, K. J., Guenther, R. B., Hyman, M. R., and Arp, D. J., *Biotechnol. Bioeng.* **46,** 218–231 (1995b).
Engelhardt, G., and Wallnöfer, P. R., *Chemosphere* **7,** 463–466 (1978).
Engelhardt, G., Wallnöfer, P., Fuchsbichler, G., and Baumeister, W., *Chemosphere* **6,** 85–92 (1977).
Engelhardt, G., Rast, H. G., and Wallnöfer, P. R., *FEMS Microbiol. Lett.* **5,** 377–383 (1979).
Engesser, K. H., Rubio, M. A., and Ribbons, D. W., *Arch. Microbiol.* **149,** 198–206 (1988).
Ensign, S. A., Hyman, M. R., and Arp, D. J., *Appl. Environ. Microbiol.* **58,** 3038–3046 (1992).
Fathepure, B. Z., and Boyd, S. A., *Appl. Environ. Microbiol.* **54,** 2976–2980 (1988).
Fedorak, P. M., and Grbić-Galić, D., *Appl. Environ. Microbiol.* **57,** 932–940 (1991).
Ferrer, M. R., del Moral, A., Ruiz-Berraquero, F., and Ramos-Cormenzana, A., *Chemosphere* **14,** 1645–1648 (1985).
Focht, D. D., and Alexander, M., *Science* **170,** 91–92 (1970a).
Focht, D. D., and Alexander, M., *Appl. Microbiol.* **20,** 608–611 (1970b).
Focht, D. D., and Alexander, M., J. *Agric. Food Chem.* **19,** 20–22 (1971).
Foster, J. W., *Antonie van Leeuwenhoek* **28,** 241–274 (1962).
Fournier, J. C., Coddaccioni, P., and Soulas, G., *Chemosphere* **10,** 977–984 (1981).
Furukawa, K., Matsumura, F., and Tonomura, K., *Agric. Biol. Chem.* **42,** 543–548 (1978).
Goldman, P., *J. Biol. Chem.* **240,** 3434–3438 (1965).
Goldman, P., Milne, G. W. A., and Keister, D. B., *J. Biol. Chem.* **243,** 428–434 (1968).
Golovleva, L. A., Golovlev, E. L., Zyakun, A. M., Shurukhin, Y. V., and Finkelshtein, Z. I., *Izv. Akad. Nauk SSSR, Ser. Biol.* **1,** 44–51 (1978).
Haber, C. L., Allen, L. N., Zhao, S., and Hanson, R. S., *Science* **221,** 1147–1153 (1983).
Harper, D. B., *Biochem. J.* **167,** 685–692 (1977).
Hernandez, B. S., Arensdorf, J. J., and Focht, D. D., *Biodegradation* **6,** 75–82 (1995).
Higgins, I. J., Sariaslani, F. S., Best, D. J., Tryhom, S. F., and Davies, M. M., *Soc. Gen. Microbiol. Q.* **6,** 71 (1979).
Hill, G. A., Milne, B. J., and Nawrocki, P. A., *Appl. Microbiol. Biotechnol.* **46,** 163–168 (1996).
Horvath, R. S., *Bacteriol. Rev.* **36,** 146–155 (1972).
Horvath, R. S., and Alexander, M., *Can. J. Microbiol.* **16,** 1131–1132 (1970a).
Horvath, R. S., and Alexander, M., *Appl. Microbiol.* **20,** 254–258 (1970b).
Hyman, M. R., Page, C. L., and Arp, D. J., *Appl. Environ. Microbiol.* **60,** 3033–3035 (1994).
Imai, T., Takigawa, H., Nakagawa, S., Shen, G.-J., Kodama, T., and Minoda, Y., *Appl. Environ. Microbiol.* **52,** 1403–1406 (1986).

Jacobson, S. N., and Alexander, M., *Appl. Environ. Microbiol.* **42,** 1062–1066 (1981).
Jacobson, S. N., O'Mara, N. L., and Alexander, M., *Appl. Environ. Microbiol.* **40,** 917–921 (1980).
Jamison, V. W., Raymond, R. L., and Hudson, J. O., *Dev. Ind. Microbiol.* **12,** 99–105 (1971).
Jensen, H. L., *Acta Agric. Scand.* **13,** 404–412 (1963).
Juliette, L. Y., Hyman, M. R., and Arp, D. J., *Appl. Environ. Microbiol.* **59,** 3718–3727 (1993).
Keener, W. K., and Arp, D. J., *Appl. Environ. Microbiol.* **60,** 1914–1920 (1994).
Kieslich, K., "Microbial Transformations of Non-steroid Cyclic Compounds." Thieme, Stuttgart, 1976.
Klages, U., Krauss, S., and Lingens, F., *Hoppe-Seyler's Z. Physiol. Chem.* **364,** 529–535 (1983).
Klečka, G. M., and Gibson, D. T., *Appl. Environ. Microbiol.* **41,** 1159–1165 (1981).
Knackmuss, H.-J., *in* "Microbial Degradation of Xenobiotics and Recalcitrant Compounds," (T. Leisinger, A. M. Cook, R. Hütter, and J. Nüesch, eds.), pp. 189–212. Academic Press, New York, 1981.
Knackmuss, H.-J., and Hellwig, M., *Arch. Microbiol.* **117,** 1–7 (1978).
Kohler-Staub, D., and Leisinger, T., *J. Bacteriol.* **162,** 676–681 (1985).
Labare, M. P., and Alexander, M., *Appl. Microbiol. Biotechnol.* **42,** 173–178 (1994).
Leadbetter, E. R., and Foster, J. W., *Arch. Biochem. Biophys.* **82,** 491–492 (1959).
Lee, K., and Gibson, D. T., *Appl. Environ. Microbiol.* **62,** 3101–3106 (1996).
Li, S., and Wackett, L. P., *Biochem. Biophys. Res. Commun.* **185,** 443–451 (1992).
Liu, D., Maguire, R. J., Pacepavicius, G., and Dutka, B. J., *Environ. Toxicol. Water Qual.* **6,** 85–95 (1991).
Loos, M. A., Roberts, R. N., and Alexander, M., *Can. J. Microbiol.* **13,** 679–690 (1967).
Lütjens, M., and Gottschalk, G., *J. Gen. Microbiol.* **119,** 63–70 (1980).
Malashenko, Y. R., Romanovskaya, V. A., Sokolov, I. G., and Kryshtab, T. P., *Mikrobiologiya* **45,** 1105–1107 (1976).
Mitchell, W. R., Dennis, W. H., and Burrows, E. P., "Microbial Interactions with Several Munitions Compounds: 1,3-Dinitrobenzene, 1,3,5-Trinitrobenzene, and 3,5-Dinitroaniline." Publ. TR-820.1. U.S. Army Bioengineering Research and Development Laboratory, Ft. Detrick, MD, 1982.
Moorman, T. B., Broder, M. W., and Koskinen, W. C., *Soil Biol. Biochem.* **24,** 121–127 (1992).
Munnecke, D. M., *Appl. Environ. Microbiol.* **32,** 7–13 (1976).
Nelson, M. J. K., Montgomery, S. O., and Pritchard, P. H., *Appl. Environ. Microbiol.* **54,** 604–606 (1988).
Novick, N. J., and Alexander, M., *Appl. Environ. Microbiol.* **49,** 737–743 (1985).
Oldenhuis, R., Vink, R. L. J. M., Janssen, D. B., and Witholt, B., *Appl. Environ. Microbiol.* **55,** 2819–2826 (1989).
Orndorff, S. A., and Colwell, R. R., *Appl. Environ. Microbiol.* **39,** 398–406 (1980).
Patel, R. N., Hou, C. T., Laskin, A. I., Derelanko, P., and Felix, A., *Appl. Environ. Microbiol.* **38,** 219–223 (1979).
Perry, J. J., *Microbiol. Rev.* **43,** 59–72 (1979).
Pfaender, F. K., and Alexander, M., *J. Agric. Food Chem.* **20,** 842–846 (1972).
Pfaender, F. K., and Alexander, M., *J. Agric. Food Chem.* **21,** 397–399 (1973).
Rasche, M. E., Hicks, R. E., Hyman, M. R., and Arp, D. J., *J. Bacteriol.* **172,** 5368–5373 (1990).
Rasche, M. E., Hyman, M. R., and Arp, D. J., *Appl. Environ. Microbiol.* **57,** 2986–2994 (1991).
Ramos, J. L., Wasserfallen, A., Rose, K., and Timmis, K. N., *Science* **235,** 593–596 (1987).
Raymond, D. G. M., and Alexander, M., *Pestic. Biochem. Physiol.* **1,** 123–130 (1971).
Raymond, R. L., Jamison, V. W., and Hudson, J. O., *Appl. Microbiol.* **15,** 857–865 (1967).
Reddy, C. C., and Vaidyanathan, C. S., *Arch. Biochem. Biophys.* **177,** 488–498 (1976).
Robertson, B. K., and Alexander, M., *Pestic. Sci.* **41,** 311–318 (1994).
Robertson, J. B., Spain, J. C., Haddock, J. D., and Gibson, D. T., *Appl. Environ. Microbiol.* **58,** 2643–2648 (1992).
Rosenberg, A., and Alexander, M., *J. Agric. Food Chem.* **28,** 297–302 (1980).
Rubio, M. A., Engesser, K.-H., and Knackmuss, H.-J., *Arch. Microbiol.* **145,** 116–122 (1986).
Sakai, T., and Jun, H.-K., *J. Ferment. Technol.* **56,** 257–265 (1978).

Schmitt, P., Diviès, C., and Cardona, R., *Appl. Microbiol. Biotechnol.* **36,** 679–683 (1992).
Scholtz, R., Schmuckle, A., Cook, A. M., and Leisinger, T. M., *J. Gen. Microbiol.* **133,** 267–274 (1987).
Schukat, B., Janke, D., Krebs, D., and Fritsche, W., *Curr. Microbiol.* **9,** 81–86 (1983).
Shields, M. S., Montgomery, S. O., Cuskey, S. M., Chapman, P. J., and Pritchard, P. H., *Appl. Environ. Microbiol.* **57,** 1935–1941 (1991).
Smith, A. E., and Phillips, D. V., *Agron. J.* **67,** 347–349 (1975).
Sperl, G. T., Forrest, H. S., and Gibson, D. T., *J. Bacteriol.* **118,** 541–550 (1974).
Taylor, B. F., Hearn, W. L., and Pincus, S., *Arch. Microbiol.* **122,** 301–306 (1979).
Thiele, J., Müller, R., and Lingens, F., *FEMS Microbiol. Lett.* **41,** 115–119 (1987).
van Beilen, J. B., Kingma, J., and Witholt, B., *Enz. Microb. Technol.* **16,** 904–911 (1994).
Vannelli, T., and Hooper, A. B., *Appl. Environ. Microbiol.* **58,** 2321–2325 (1992).
Wackett, L. P., Brusseau, G. A., Householder, S. R., and Hanson, R. S., *Appl. Environ. Microbiol.* **55,** 2960–2964 (1989).
Wang, Y.-S., Subba-Rao, R. V., and Alexander, M., *Appl. Environ. Microbiol.* **47,** 1195–1200 (1984).
Wang, Y.-S., Madsen, E. L., and Alexander, M., *J. Agric. Food Chem.* **33,** 495–499 (1985).
You, I.-S., and Bartha, R., *Appl. Environ. Microbiol.* **44,** 678–681 (1982).

CHAPTER 14

Environmental Effects

The microbial populations destroying synthetic chemicals are subject to a variety of physical, chemical, and biological factors that influence their growth, their activity, and their very existence. The environments in which these species function vary enormously, and these differences in environmental properties and characteristics have a profound impact on the resident populations, the rate of biochemical transformations, and the identities and persistence of products of biodegradation.

The great impact of site factors is evident from studies showing that a specific compound is biodegraded in samples from one but not another environment. For example, TCE was found to be metabolized by the indigenous microorganisms in only 1 of 43 samples of water and soil (Nelson *et al.*, 1986), 2,4-D was mineralized in samples from a eutrophic (rich in inorganic nutrients) but not an oligotrophic (nutrient-poor) lake (Rubin *et al.*, 1982), methyl parathion was transformed in sediments but not in samples from the water column of an estuary (Pritchard *et al.*, 1987), IPC was mineralized in samples from only some lakes (Hoover *et al.*, 1986), the anaerobic degradation of some phthalates and chlorobenzoates may occur in sewage digestor sludge but not in municipal landfills (Ejlertsson *et al.*, 1996), and the reductive dehalogenation of aromatic compounds occurred in only some sewage sludges, pond sediments, and aquifer solids under anaerobic conditions (Gibson and Suflita, 1986). Sometimes a compound may be mineralized in one environment but only cometabolized at a different site (Wang *et al.*, 1985) or, even ignoring temperature effects, transformed at one but not another time of year (Rubin *et al.*, 1982). More often than not, the reasons for the sporadic or nonuniversal occurrence of a biodegradative sequence are unknown. In some instances, the random occurrence of biodegradation may be a result of the presence of organisms at only some sites, the existence of fastidious populations whose growth factor requirements are met in not all environments, the presence of toxins, the availability of O_2, or the impact of other environmental characteristics that promote, restrict, or prevent biodegradation. It is inappropriate to assume that a compound biodegraded in one environment is *ipso facto* going to be transformed in another.

A vast amount of information exists on the biochemical activities of bacteria and fungi grown in pure culture at high substrate concentrations in laboratory media. This research has created a foundation for the understanding of the nutrition, genetics, and catabolic potential of microorganisms. Yet, in nature, bacteria and fungi are exposed to enormously different conditions. They may have an insufficient supply of inorganic nutrients, a paucity of essential growth factors, temperatures and pH values at their extremes of tolerance, and toxins that retard their growth or result in loss of viability. They may benefit from the activities of other microorganisms or be consumed by species residing in the same habitat. As a consequence, extrapolations from tests of laboratory-grown pure cultures to nature are fraught with peril. Not only must there be information on the characteristics of the biodegrading species *in vitro,* but there also must be an understanding of those factors in nature that determine the occurrence, rate, and products of biodegradation.

ABIOTIC FACTORS

Every strain of microorganisms has a range of tolerances to ecologically important factors (e.g., temperature, pH, salinity) affecting its growth and activity. That range is bounded by the maximum level tolerated and, for some species, a minimum tolerance level. If a particular environment contains several species able to bring about a particular transformation, the tolerance range often is broader than that of a single species, encompassing the tolerances of all the indigenous populations. Outside of the tolerance ranges of all the inhabitants able to perform the degradation of concern, no activity will occur.

Apart from the supply of nutrients and factors that control the bioavailability of organic compounds, the chief abiotic factors influencing microbial transformations are temperature, pH, moisture level (in the case of soil), salinity in some environments, toxins, and hydrostatic pressure if the compounds are in deep marine sediments or at sites deep below the soil surface. An organic pollutant that is quickly destroyed in one environment will persist at another site if these factors preclude or retard microbial activity.

The prevailing temperature is of paramount importance. If the compound of interest exists near the surfaces of soils or water, the low temperatures of winter and even the time immediately preceding and following the winter season are typically associated with little or no biodegradation of many organic substrates. In the frozen soils of the northern parts of North America, Europe, and Asia, organic molecules will persist for long periods. As the temperature rises with the change of seasons, microbial activity will increase in response to more favorable circumstances. The magnitude of response to a particular increase or decrease in temperature varies with the compound and the environment, which reflects the physiology of the individual populations at the site. To a great degree, the changes in rate of degradation associated with seasons of year are a consequence of the concomitant changes in temperature. On occasion, however, the anticipated increase or decrease in activity with rise or fall in temperature is not evident. This lack of response to the warmer conditions may sometimes be attributable to some other factor becoming limiting during the warmer period; for example, nutrient deficiencies may severely limit degradation of some substances, such as oil, in lakes. At times, the activity in the winter months is not diminished because a factor other than temperature comes into play. This is evident in a study of the degradation of an ester of 2.4-D in a stream in which the transformation rate increased in the winter. The anomalously greater rates in the colder months were a consequence of leaf fall and deposition of leaves in the stream. The surface of the leaves provided abundant sites for microbial colonization, and the enhanced biomass compensated for the otherwise detrimental effects of the cooler water (Lewis *et al.,* 1986).

At extremes of acidity or alkalinity, activity declines. At more moderate pH values, biodegradation tends to be fastest. If a compound in a particular environment can be metabolized by a diverse group of organisms, the range of pH values at which degradation occurs frequently is broader than if only one species can bring about the transformation. Apart from pesticides, however, the effect of pH on biodegradation of polluting chemicals has received scant attention, although it is common practice to add lime to bioremediate acid soils or subsoil materials containing harmful organic compounds.

Microorganisms carrying out a metabolic transformation require adequate moisture for their growth and activity. Moisture obviously is not a limiting factor in oceans, in fresh waters, or in subterranean aquifers. However, an inadequate supply of water can severely restrict biodegradation in surface soils, in which drying to suboptimal water levels is common. In one study, for example, the optimum moisture level for the biodegradation of oily sludges was found to be at 30 to 90% of the soil's water-holding capacity (Dibble and Bartha, 1979). The optimum moisture level will depend on the properties of the soil, the compound in question, and whether the transformation is aerobic or anaerobic. The last factor is of particular significance because excess water displaces air from the pores in soil, and a waterlogged soil soon becomes anaerobic and unfavorable for aerobic processes. Decreasing the moisture content of soil diminishes rates of degradation (Walker,

1976), a result of an inadequate supply of water to sustain proliferation, metabolism, or both.

Salinity sometimes is sufficiently high to become harmful. Soils and inland waters in certain areas of the world are rich in salts, and microbial processes in such environments are inhibited. It is possible that the salinity in estuaries and in oceans may also be detrimental to some species involved in the biodegradation of organic pollutants, but no strong argument can presently be advanced for the salinity in such waters being major deterrents to biodegradation.

Components of oil and other pollutants that have specific gravities greater than that of marine waters will move downward and sink to the deep benthic zone. At these depths, the hydrostatic pressure is notably high. It is likely that a combination of high pressure and low temperatures in the deep ocean will result in low microbial activity, slow biodegradation, and consequently prolonged persistence of substances that reach deep benthic zones (Atlas, 1981).

It is not economically feasible to modify or control some of these factors, yet it is important to understand them, qualitatively at least, in order to predict the likely persistence of organic molecules in environments differing in the intensity of these factors. Understanding the quantitative impacts of these abiotic factors would be far more useful for predictive purposes, but knowledge for attaining such objectives remains scant.

Conversely, some practical technologies to promote biodegradation do entail manipulation of several of these abiotic factors. This is evident at sites in the field encompassing limited areas or in bioreactors. Temperatures may be made more favorable, moisture levels can be improved, and undesirable acidities can be corrected. Optimization is also evident in the controlled microbiological treatment of industrial wastes, a practice that a few prudent chemical companies have followed for many years.

NUTRIENT SUPPLY

To grow, heterotrophic bacteria and fungi require—in addition to an organic compound that serves as a source of C and energy—a group of other nutrient elements and an electron acceptor. That electron acceptor is O_2 for aerobes, but it may be nitrate, sulfate, CO_2, ferric iron, or organic compounds for specific bacteria able to utilize these substances to accept the electrons released in the oxidation of the energy source. Many bacteria and fungi also require low concentrations of one or more amino acids, B vitamins, fat-soluble vitamins, or other organic molecules; these trace organic nutrients are termed growth factors. The absence from a particular environment of any of these essential nutrients will prevent the growth of organisms requiring that substance or prevent any microbial replication if the requisite, such as an inorganic nutrient, is needed by all species.

Soils, sediments, and marine and fresh waters contain low concentrations of readily metabolizable organic matter. This may not seem to be true for soils or sediments, which may contain 1% or more organic matter, but that organic C exists in complex forms that bacteria and fungi either cannot use or utilize only slowly. Typically, the supply of all other nutrient elements exceeds the need of the resident microbial communities given the little readily available C, and hence the limiting nutrient element for heterotrophs in soils, sediments, and natural waters is commonly C.

However, the situation changes markedly if a pollutant that is potentially readily utilizable is introduced into the environment, provided that its concentration is sufficiently high to make one or more previously nonlimiting nutrients into a limiting factor. At very low pollutant concentrations, such a change may not occur. However, even at what might appear to be a low concentration, a pollutant that is in a NAPL or otherwise does not mix throughout the site is, in fact, at high concentrations in the microenvironment in which it is deposited. Thus, at the interface between crude oil, gasoline, or an organic solvent and the surrounding environment, the C concentration is high. Under these circumstances, the supply of one or several nutrients, which previously may have been nonlimiting, may be in concentrations too low to meet the now higher demand. Usually, the nutrients now in short supply are N, P, or both, and a frequent concomitant of the greater growth on the pollutant C is a greater demand for an electron acceptor. For hydrocarbons and many other C compounds, that electron acceptor is O_2. Nearly always, the supply of K, S, Mg, Ca, Fe, and micronutrient elements is greater than the demand.

Oceanic waters, lakes, rivers, soils, and aquifers containing oil, gasoline, or organic solvents from leaking underground storage tanks typically have too low concentrations of inorganic nutrients, O_2, or both at the interface between the water-insoluble pollutants and the aqueous phase to support the activity that is otherwise possible.

The release into marine and estuarine waters of crude oil from leaking tankers and corrosion and the subsequent leakage of petroleum or oil products from underground storage tanks have prompted studies designed to establish means to bioremediate the surface or groundwaters. These investigations show that crude oil degradation in seawater is slow unless both N and P are added. Individually, N or P alone fails to cause appreciable stimulation (Atlas and Bartha, 1972). Similarly, additions of N and P to samples of groundwater contaminated with gasoline stimulate the growth of bacteria (Jamison *et al.*, 1975). The nutrient level for optimal activity varies with the type of oil and the particular water body, but stimulations in seawater have been reported over a range of concentrations (Atlas and Bartha, 1972; Floodgate, 1984; LePetit and N'Guyen, 1976). The concentration of water-soluble salts of N and P introduced into surface waters at or very near the oil—water interface rapidly declines because of turbulence of many waters. Therefore, a number of "oleophilic" fertilizers were developed, and these hydrophobic preparations, after addition, remain associated with the oil and stimulate hydrocarbon-degrading bacteria. N- and P-containing compounds in these early fertilizer materials include octyl phosphate, decyl phosphate, paraffinized urea, and dodecyl urea (Atlas and Bartha, 1972; Olivieri *et al.*, 1978).

The addition of N and P to soil also stimulates the biodegradation of oil and individual hydrocarbons and increases bacterial abundance. The effect of inorganic N and P on the mineralization in subsoil of phenanthrene present in two NAPLs or in soil with no NAPL is shown in Fig. 14.1. The stimulation is sometimes

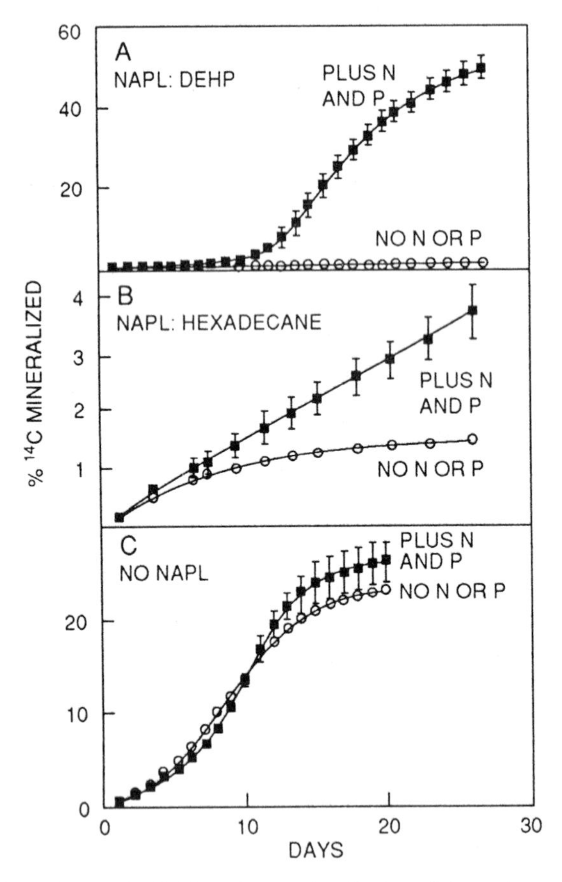

Figure 14.1 The influence of added N and P on mineralization of phenanthrene in a subsoil. The compound was added in di(2-ethylhexyl) phthalate (A) or hexadecane (B) as NAPLs or was added with no NAPL (C). (Reprinted from Efroymson and Alexander, 1994, with kind permission from Pergamon Press Ltd., Headington Hill Hall, Oxford 0X3 OBW, U.K.)

apparent immediately, but it may require some time for a benefit to be evident (Bossert and Bartha, 1984; Jobson *et al.,* 1974). However, fertilizer additions sometimes may be without benefit, possibly because of high N and P levels in the soil, the presence of N and P in the organic pollutants, or the low concentration of chemical whose biodegradation is being determined. Alternately, even if N and P are stimulatory, the rate of degradation may be sufficiently rapid in the absence of fertilizers that their use is not warranted, as suggested for the bioremediation of an experimental spill of oil on a sandy beach (Venosa *et al.,* 1996).

It is widely believed that only one nutrient element is limiting at any one time, and that only when that one deficiency is overcome does another nutrient element become limiting. This view frequently may be incorrect since microbial growth may be simultaneously limited by two nutrients (Egli, 1991). It is not uncommon to find that additions of combinations of inorganic nutrients have a greater effect on biodegradation than single nutrients (Swindoll *et al.,* 1988), although such responses may often result from the second nutrient element becoming limiting as the deficiency of the first is overcome by the supplement.

Even in the absence of added N and P, biodegradation continues in waters, soils, and sediments, albeit at a slow rate. This probably is a consequence of nutrient regeneration, that is, a recycling of the elements as they are first assimilated into microbial cells and then are converted back to the inorganic forms as the cells lyse or are consumed by predators or parasites, both of which release some of the N and P contained in their prey or hosts. Under such circumstances, the rate of biodegradation will be governed by the rate at which the limiting nutrient is recycled. Protozoa probably are especially important for nutrient regeneration in oceans and lakes, and possibly also in soil.

The variation in rate of biodegradation with time is often the result of diurnal or seasonal changes in temperature; however, it may have other causes. For example, the concentration of N and P in lake and river water varies with rainfall as drainage in the watershed carries soil materials into streams, rivers, and lakes, thereby making the water more fertile. In turn, the rate of a N- or P-limited transformation in the water may be enhanced, as has been observed for 4-nitrophenol mineralization (Zaidi *et al.,* 1988).

The concentration of N and P needed for the biodegradation of oil or other materials present at high concentrations, either throughout the environment or within the NAPL that is the oil itself, is usually assumed to reflect the amount of those elements that must be incorporated into the biomass that would be formed as the microorganisms use the organic materials as C sources for growth. For example, consider the mineralization of 1000 g of organic C. If the active organisms assimilate 30% of the substrate-C to make 300 g of biomass-C and elemental analysis of those cells shows that they contain a C : N ratio of 10 : 1 and a C : P ratio of 50 : 1, then the amount of N and P needed to be incorporated into the biomass is 30 g of N and 6 g of P. This is a convenient assumption and is probably appropriate for predicting the N and P needed to totally destroy the C source, but it is probably not valid for predicting the concentration of N and P to support the maximum rate of degradation. It is important to distinguish between the optimum nutrient level for the extent of degradation and that needed for the highest rate. This argument of rate versus extent is particularly important in considering explanations for the occasional need for high N or P concentrations to degrade organic molecules whose level is too low to give large biomasses.

It is not yet clear why phosphate or inorganic N stimulates biodegradation of chemicals present at concentrations appreciably below 1.0 mg/liter, for example, of 4-nitrophenol at concentrations of 2–200 μg/liter in lake water. With no added P, the degradation was observed to be slow or failed to proceed (Jones and Alexander, 1988a,b). The slow biodegradation of comparably low levels of chlorophenols by marine plankton communities (Kuiper and Hanstveit, 1984), of phenols in lake water (Rubin and Alexander, 1983), and of IPC at 400 ng/liter and 2,4-D at 200 ng/liter (Wang *et al.,* 1984) was also limited by the supply of inorganic nutrients, and the rates were enhanced by providing those nutrients. The requirement for high concentrations of P and N in waters may be related not to the amount of these elements needed to be incorporated into the biomass but rather to the K_s value for P or N. As with C compounds, microbial growth at P or N concentrations below the K_s (of the P or N source) is slower than at higher concentrations. If the K_s value for the rate of P or N utilization is high, the maximum rate of degradation would require a high P or N concentration. The K_s values for P for different microorganisms may range from as low as 0.4 to as high as 500 μg/liter (Owens and Legan, 1987). Alternatively, the need for high P levels may result from nonbiological reactions that reduce phosphate availability. These reactions could be the precipitation of phosphate as insoluble salts of Ca, Fe, or Mg. However, phosphate in solution may not be represented solely by $H_2PO_4^-$ and HPO_4^{2-} because Ca, Fe, and other metallic salts are also present in soluble form, and the dependency of microbial nutrition on the solution chemistry of inorganic P is unexplored.

Calcium and Mg are abundant in many inland waters, and reactive Ca, Fe, and Mg exist in soils and sediments. These cations alter the availability of P. Moreover, pH affects the identities of the Ca, Fe, and other salts of P in the aqueous phase and also alters the relative abundance of $H_2PO_4^-$ and HPO_4^{2-}. These changes in solution chemistry of P may explain why a strain of *Pseudomonas* requires high P concentrations for phenol mineralization at pH 8.0 but only low concentrations at pH 5.2 (Robertson and Alexander, 1992).

Rarely do the additions of elements other than P, N, and O_2 stimulate biodegradation in natural or polluted environments. However, Fe may sometimes limit the rate of microbial destruction of oil in seawater, in which the available forms of that element are often present at very low concentrations.

Nevertheless, the disappearance of many pesticides in soil and probably of many other chemicals at low concentrations in soils and waters is not known to be stimulated by supplementary N and P. The reason is either the presence of an adequate supply of these nutrients to support growth of the active species or the existence of some other limiting factor, for example, sorption.

Little attention has been given to the possible role of growth factors in controlling microbial activity. In an environment containing several species able to degrade a particular compound, it is likely that both auxotrophs and prototrophs will coexist, and the absence of growth factors will not affect the transformation because the prototrophs will flourish. However, in environments containing only one or two species active on the compound of concern, it is likely that the supply or rate of excretion of growth factors will limit the rate of degradation since one or both species are quite possibly auxotrophs. The abundance of auxotrophs is evident from findings that about 90% of the bacteria in marine waters, 75–80% of the bacteria in marine sediments (Skerman, 1963), a high percentage of those in lakes (Fondén, 1969), and more than 90% of the bacteria in soil (Rouatt and Lochhead, 1955) need one or more B vitamins, amino acids, or other growth factors to multiply. In these environments, the growth factors would be excreted by bacteria, fungi, or algae, or they would be generated as these organisms are grazed by protozoa or higher animals or are parasitized. The rates of such excretions are unknown, but their importance to auxotrophic populations must be great.

Growth factors may also affect the threshold concentration of a C source for growth and biodegradation. Thus, the concentration of glucose below which a bacterium would not multiply was lowered by a mixture of amino acids (Law and Button, 1977), and the threshold for phenol mineralization by lake water bacteria was reduced by a single amino acid (Rubin and Alexander, 1983).

Because bacteria and fungi that cometabolize organic compounds need a substrate for growth, it is not surprising that additions of organic materials or individual chemicals to natural environments often stimulate degradation. Some examples are presented in Table 14.1. However, the mechanism by which such stimulations occur is rarely known. In the case in which the addition of biphenyl to soil promotes PCB transformation, the effect may be the result of the larger population of biphenyl degraders, which grow using biphenyl as C source but can cometabolize PCBs, since biphenyl is an analogue of the chlorinated biphenyls. However, most of the stimulatory organic amendments are not analogues of the compounds being cometabolized (such as DDT, heptachlor, endrin, and BHC) so any benefit must be nonspecific, for example, by increasing the biomass of organisms that only coincidentally carry out a cometabolic reaction. In some instances, the effect may result from the added material causing a depletion of O_2, at least when the transformation is favored by anaerobiosis. Moreover, some of the substrates listed in Table 14.1 presumably are acted on by growth-linked and not cometabolic processes (e.g., MCPA, *m*-cresol, and 4-chlorophenol) so that

Table 14.1

Stimulation of Biodegradation of Test Substrates by Additions of Individual Compounds or Complex Organic Materials

Substrate	Environment	Amendment	Reference
BHC	Soil suspension	Peptone	Ohisa and Yamaguchi (1978)
m-Cresol, 4-chlorophenol	Lake water	Amino acids	Shimp and Pfaender (1985)
DDT	Sewage	Glucose	Pfaender and Alexander (1973)
DDT, heptachlor	Flooded soil	Plant residues	Guenzi *et al.* (1971)
2,6-Dichlorobenzamide	Soil	Benzamide	Fournier (1975)
Malathion	Soil	Heptadecane	Merkel and Perry (1977)
MCPA	Soil	Straw	Duah-Yentumi and Kuwatsuka (1980)
Naphthalene	Soil	Salicylate	Ogunseitan and Olson (1993)
PCB	Soil	Biphenyl	Focht and Brunner (1985)
TCE	Aquifer	Methane	Kane *et al.* (1997)
TCE	Bioreactor	Phenol	Shih *et al.* (1996)

the explanation may not be one associated with cometabolism. However, some organic amendments may reduce the rate of degradation (Subba-Rao *et al.*, 1982).

In many environments, the supply of the electron acceptor is not sufficient to meet the need, especially if the microflora places a large demand on it because of an abundance of organic substrates. For hydrocarbons and several other chemical classes, the only or preferred electron acceptor is O_2, and the transformations are only aerobic or the most rapid conversions are carried out by obligate aerobes. This need for O_2 is especially evident in the degradation of crude oil and individual hydrocarbons, particularly where O_2 diffusion from the atmosphere to replenish the supply is restricted or physically prevented. In groundwaters contaminated with gasoline or oil, the O_2 initially in the aqueous phase is rapidly consumed, and the subsequent degradation either is extremely slow or does not occur. As a result, remediation strategies typically involve the introduction of O_2 from forced air, pure O_2, or H_2O_2. The biodegradation of fractions of crude oil that sink to the sediments of marine and freshwater environments often is limited because the small amount of O_2 dissolved in the pore water or the immediately overlying water column is consumed by aerobic bacteria that act on the hydrocarbons. In soils, the O_2 dissolved in the liquid phase and in the gas-filled pores is also quickly consumed if much hydrocarbon or oil is present, and the rate of O_2 entry from the overlying air is too slow to sustain appreciable transformation. Although some bacteria metabolize hydrocarbons anaerobically, such reactions in most natural ecosystems proceed very slowly, if at all (Atlas, 1981). In contrast, the supply of

O_2 rarely limits biodegradation of hydrocarbons at the surfaces of marine and fresh waters because O_2 from the atmosphere is accessible and diffuses readily into the top of the water column, especially if there is turbulence from wave action.

The biodegradation of many organic compounds is independent of the O_2 supply, and anaerobic conversions are common. Indeed, the transformation of some substrates is more rapid or only occurs in anoxic environments. In many instances, the electron acceptor is an organic molecule, but sometimes it is nitrate, sulfate, CO_2/bicarbonate, or possibly ferric iron. However, the supply of nitrate or sulfate may be totally consumed, so that further conversions may stop or be governed by the reentry of additional electron acceptors.

MULTIPLE SUBSTRATES

Laboratory studies are typically conducted with individual organic substrates, but natural and polluted environments characteristically contain a multiplicity of organic compounds that can be used by one or more of the indigenous bacteria or fungi. These substrates may be synthetic compounds, discrete natural products, the complex materials associated with humic fractions of soil or sediments, or the dissolved organic C (DOC) of natural waters. Their concentrations may be either quite high or extremely low, and the levels may be sufficiently high to be toxic or so low that they will not support growth. Because of the number of coexisting species and compounds, biodegradation of individual substrates will differ from the transformations brought about by a single species acting on one chemical.

Several organic substrates may be used at the same time. Such simultaneous metabolism has been reported for mixtures containing widely different classes of compounds. For example, marine bacteria degrade linear alkanes with 16 to 30 carbons simultaneously in oil-contaminated marine sediments and waters (Kator, 1973), the rate of metabolism of glucose in activated sludge is unaffected by the ongoing degradation of acetate (Painter *et al.*, 1968), mixed cultures simultaneously degrade 2,4-D and mecoprop (Hallberg *et al.*, 1991), *Pseudomonas putida* metabolizes phenol and glucose at the same time (Wang *et al.*, 1996), and an ultramicrobacterium metabolizes glucose and alanine simultaneously (Schut *et al.*, 1995). Such a simultaneous use of two C sources is usually assumed to occur only at low substrate concentrations in microbial cultures, but it sometimes occurs even at high concentrations. Bacteria in laboratory media may sometimes metabolize three or more C sources, and more than one N source or electron acceptor may sometimes be transformed (Egli, 1995).

Frequently one substrate enhances the rate of degradation of a second. Such stimulations occur in environmental samples, bioreactors, mixtures containing two organisms, or pure culture. Thus, salicylate stimulates naphthalene mineralization in soil (Ogunseiten and Olson, 1993), fluorene promotes carbazole mineralization in groundwater samples (Millette *et al.*, 1995), the addition of glucose to a sludge bioreactor enhances the anaerobic transformation of pentachlorophenol (Hendriksen *et al.*, 1992), the addition of glucose promotes the biodegradation of 2,4-dinitrophenol by an actinomycete and a strain of *Janthinobacterium* (Hess *et al.*, 1990), and toluene stimulates the degradation of benzene and *p*-xylene by a pseudomonad (Alvarez and Vogel, 1991).

The converse may also occur, namely, one substrate may slow the degradation of a second. This is evident in the suppression of caprolactam degradation by benzylamine in a culture of two bacteria (Steffensen and Alexander, 1995) and the reduced rates of pentachlorophenol utilization by an enrichment culture in solutions containing phenol or 2,4,5-trichlorophenol, both of which could be utilized (Klečka and Maier, 1988), and of the degradation of low concentrations of acetate by a *Pseudomonas* strain also provided with methylene chloride as a substrate (LaPat-Polasko *et al.*, 1984). In some instances, the suppression by one compound of the metabolism of a second is manifested in a sequential use of the substrates, one disappearing only after the second is largely or wholly destroyed, as in the sequential destruction of linear alkanes by *Cladosporium resinae* (Lindley and Heydeman, 1986). Such sequential utilization is similar to the common finding that some components of oil or prepared mixtures of hydrocarbons are degraded by microorganisms in natural waters, soils, or enrichments before others (Mechalas *et al.*, 1973; Raymond *et al.*, 1976), although the differences in rates of disappearance in these instances probably are frequently, but not always, a result of the differences in intrinsic resistance of the various hydrocarbons to biodegradation.

Sequential utilization of substrates by pure cultures is frequently the result of diauxie, and it is evident when one substrate delays utilization of a second. However, in diauxie, the first substrate is being utilized during the period of apparent inhibition of the second conversion. Diauxie, which has been considered previously as a cause of acclimation, occurs in cultures of a number of bacterial genera in media containing high concentrations of organic substrates, and it sometimes has been noted in enrichment cultures that probably are dominated by a single bacterial species (Gaudy *et al.*, 1964; Stumm-Zollinger, 1966). The substrate that supports the faster growth is generally the one that is used first when two growth-supporting compounds are included in culture media. For molecules for which individual species exhibit diauxie at high substrate concentrations, the two compounds are used simultaneously when their concentrations are low (Harder and Dijkhuizen, 1982). Similarly, one P source may be used in preference to another in a diauxie-type relationship, as in the preferential use of inorganic phosphate over methylphosphonate by *Pseudomonas testosteroni* (Daughton *et al.*, 1979). It is not certain whether diauxie occurs in environments with many different species, and it is likely that a compound not being transformed because of a diauxic effect on one species will be metabolized by a second species in such environments.

The explanations for the effect of one substrate on the biodegradation of a second in natural and polluted environments are largely unknown and have scarcely been explored. The reasons for the absence of an influence of one compound on the metabolism of a second are likewise uncertain. The absence of an effect in nature probably can be attributed frequently to the action of two different species functioning independently on the substrates of concern. The two may act independently unless they are limited by some common factor (e.g., grazing by protozoa or a deficiency of O_2 or an inorganic nutrient). Alternatively, if a single species is degrading the two compounds, their concentrations may be too low for diauxie to come into play. Because diauxie involves repression of synthesis of the enzymes catalyzing degradation of the second substrate as the first is being metabolized (Harder *et al.,* 1984), diauxie may not be important if the catabolic pathways for the two C sources or the enzyme-regulatory mechanisms in the organisms are not subject to control by the physiological processes associated with diauxie.

A number of hypotheses have been advanced to account for the stimulation of biodegradation of one compound by a second, but few have experimental support. A likely cause in many instances is the greater population size or biomass arising because of the additional C source; if the resulting organisms can act readily on both substrates, growth on the second C source would enhance destruction of the first. If the second compound is only degraded cometabolically, then the first would obviously be beneficial inasmuch as a large mass of cometabolizing cells is produced. In some instances, if the organisms acting on one substrate are auxotrophs, the stimulation may result from the excretion of growth factors by the population acting on the second. Alternatively, the second compound may be beneficial because it induces enzymes necessary for catabolizing the other molecule. Should one of the two chemicals be at concentrations below the threshold for growth of the requisite bacteria or fungi, the second may serve as an energy source and thus facilitate destruction of the trace contaminant (Bouwer and McCarty, 1984).

More attention has been given to explanations of how second compounds inhibit biodegradation. (a) Undoubtedly, the suppression in many highly polluted sites results from toxicity of the second compound—a toxicity that slows or prevents growth or that diminishes activity of the microorganisms. If both compounds individually are at levels just below those that are toxic, a combination of the two could then exceed the tolerance of the active microorganisms (Smith *et al.,* 1991). (b) One substrate could be converted to products that are detrimental to the population acting on the second, as with the products of 4-nitrophenol metabolism by a pseudomonad, which inhibit phenol oxidation by a different bacterium (Murakami and Alexander, 1989). (c) Studies of the biodegradation of two substrates by two bacterial species, each of which can metabolize only one of the molecules, show that competition between the organisms for limiting concentrations of P may be reflected in a reduction in the rate of biodegradation of one or both of the compounds as compared to media with only one substrate. The two bacteria are competing for an inadequate supply of a limiting factor, and this competition is manifested in an effect on the transformation (Steffensen and Alexander, 1995). (d) Similarly, competition for O_2 or another electron acceptor, if present in amounts insufficient for the microbial demand, may be the reason why the organisms degrading one substance apparently have an effect on the utilization of a second. (e) The number of bacterial cells will be greater if two rather than a single C source is present, and this larger population would result in more intense grazing in environments in which protozoa are active; a likely outcome would be a lower rate or extent (or both) of biodegradation of a compound when a second substrate is being degraded by a different bacterial species. (f) Should a single species be responsible for the biodegradation of both organic molecules, the inhibition may result from a repression of further synthesis of enzymes needed for the catabolism of one substrate by an intermediate formed in the catabolism of the second (catabolite repression), the inhibition by an intermediate of the activity of already existing enzymes, or by an interference by one substrate in the uptake by the cell of the second substrate (Harder and Dijkhuizen, 1982).

The effects of one compound on the biodegradation of a second are frequent and occur in many environments. However, given the number of compounds, the undefined populations causing their destruction, and the multitude of chemical mixtures, generalizations on whether there will or will not be an effect, whether the effect will be stimulatory or inhibitory, and the reasons for the enhancement or suppression are premature.

SYNERGISM

Many biodegradations require the cooperation of more than a single species. These interactions may be necessary for the initial step in the conversion, a later phase of the transformation, or the mineralization of the compound. These various interactions represent several types of *synergism,* in which two or more species carry out a transformation that one alone cannot perform or in which the process carried out by the multispecies mixture is more rapid than the sums of the rates of reactions effected by each of the separate species. Thus, some reactions take place in mixtures of species but not in pure culture or take place more readily in multispecies associations.

Several examples of synergism will serve as adequate illustrations of the phenomenon. Isolates of *Arthrobacter* and *Streptomyces* together are able to mineralize diazinon, but neither bacterium alone produces CO_2 from this insecticide (Gunner and Zuckerman, 1968). A mixture of *Pseudomonas* and *Arthrobacter* together degrades the herbicide silvex, although neither alone has this capacity (Ou and Sikka, 1977). Synergism is also shown by the more rapid degradation of dodecyl-1-

decaethoxylate, a surfactant, by a mixture of two bacteria than by either organism alone (Watson and Jones, 1979).

A number of mechanisms for synergistic relationships have been described, but undoubtedly other mechanisms have yet to be discovered. (a) One or more species provide B vitamins, amino acids, or other growth factors to one or more of the other organisms. (b) One species grows on the test compound and carries out an incomplete degradation to yield one or several organic products, and the second species mineralizes the products that otherwise would accumulate. The second species commonly grows on the intermediate in the sequence. (c) The initial species cometabolizes the test compound to yield a product that it can no longer metabolize, and the second species destroys that product. As an illustration, the transformation of a PCB without the accumulation of chlorinated aromatic products is brought about by two bacteria, one converting the PCB to chlorine-containing benzoates and the other mineralizing the chlorobenzoates (Fava, 1996). This mechanism differs from the second in the type of activity of the initial population, that is, whether it uses the parent compound as a C source for growth or only cometabolizes it. (d) The first species converts the substrate to a toxic metabolite that then slows the transformation, but the reaction proceeds rapidly if the second member of the association destroys the inhibitor. This detoxication may sometimes be a consequence of the use of the inhibitor as a C source for growth, so that it is somewhat analogous to mechanism (b). However, it may involve interspecies hydrogen transfer, in which H_2 or reducing equivalents generated by one population are used by another.

Many bacteria and fungi that are capable of degrading toxicants require one or more growth factors. These auxotrophs will not grow in liquid media with the test substrate as sole C source because they need a single or several B vitamins, amino acids, purines, pyrimidines, or more complex growth factors. This inability of an organism to grow in simple media does not necessarily mean that the organism is unimportant in nature because a high percentage of the heterotrophic microorganisms in soils, sediments, and waters excrete growth factors (Burkholder, 1963; Lochhead, 1958), thereby permitting proliferation of the auxotrophs responsible for biodegradation. In such environments, there exist both a continuous formation and a continuous utilization of growth factors, resulting in a constant turnover or flux of trace nutrients essential for the auxotrophic inhabitants. Indeed, two fastidious species may coexist because each excretes and provides its associate with the required growth factor, a mutual feeding that is known as *syntrophy*.

Many growth factors are responsible for synergistic interactions, but B vitamins and amino acids are most often implicated. Excretion of vitamin B_{12}, for example, is required for a bacterium to grow on and dechlorinate trichloroacetic acid (Jensen, 1957), and growth factors excreted by a strain of *Pseudomonas* appear to be needed by an isolate of *Xanthomonas* to grow and degrade dodecyltrimethylammonium bromide (Dean-Raymond and Alexander, 1977).

A large number of synthetic molecules are converted to organic products, and little or no mineralization is evident. Even aerobic bacteria and fungi in pure culture often convert little or sometimes none of their C and energy source to CO_2. However, such species in nature often coexist with others that use the products of their associates as sources of C and energy for growth. Several examples are presented in Fig. 14.2. The first species in each instance converts the initial compound to a product that accumulates in a one-membered culture, but that compound is destroyed if the second species is present.

Cometabolizing microorganisms do not mineralize their organic substrates and hence are responsible for the accumulation of organic products. However, many of those products are C sources for other organisms and support their growth; hence, the accumulation is transitory. Several examples are depicted in Fig. 14.3. Each of these illustrations comes from studies of defined cultures. However, the same types of relationships undoubtedly occur in nature, so that the intermediate in the conversion would be at low levels or undetectable in natural habitats, and the parent compound is converted to CO_2. Thus, although cyclohexane is appar-

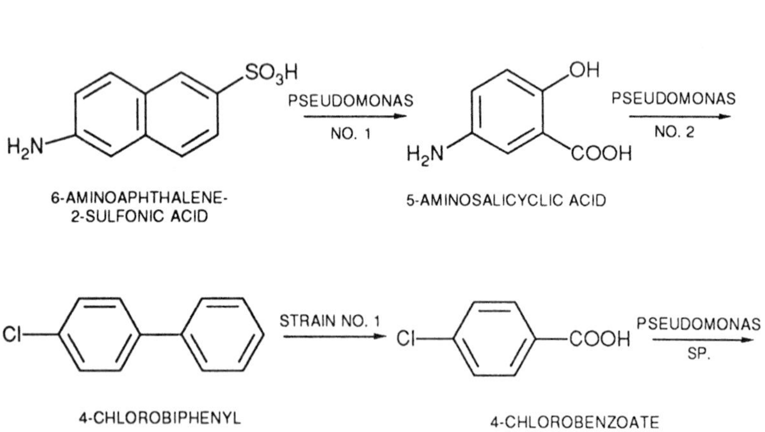

Figure 14.2 Synergistic associations leading to the complete destruction of dodecylcyclohexane (Feinberg *et al.*, 1980), 6-aminonaphthalene-2-sulfonic acid (Nörtemann *et al.*, 1986), and 4-chlorobiphenyl (Sylvestre *et al.*, 1985).

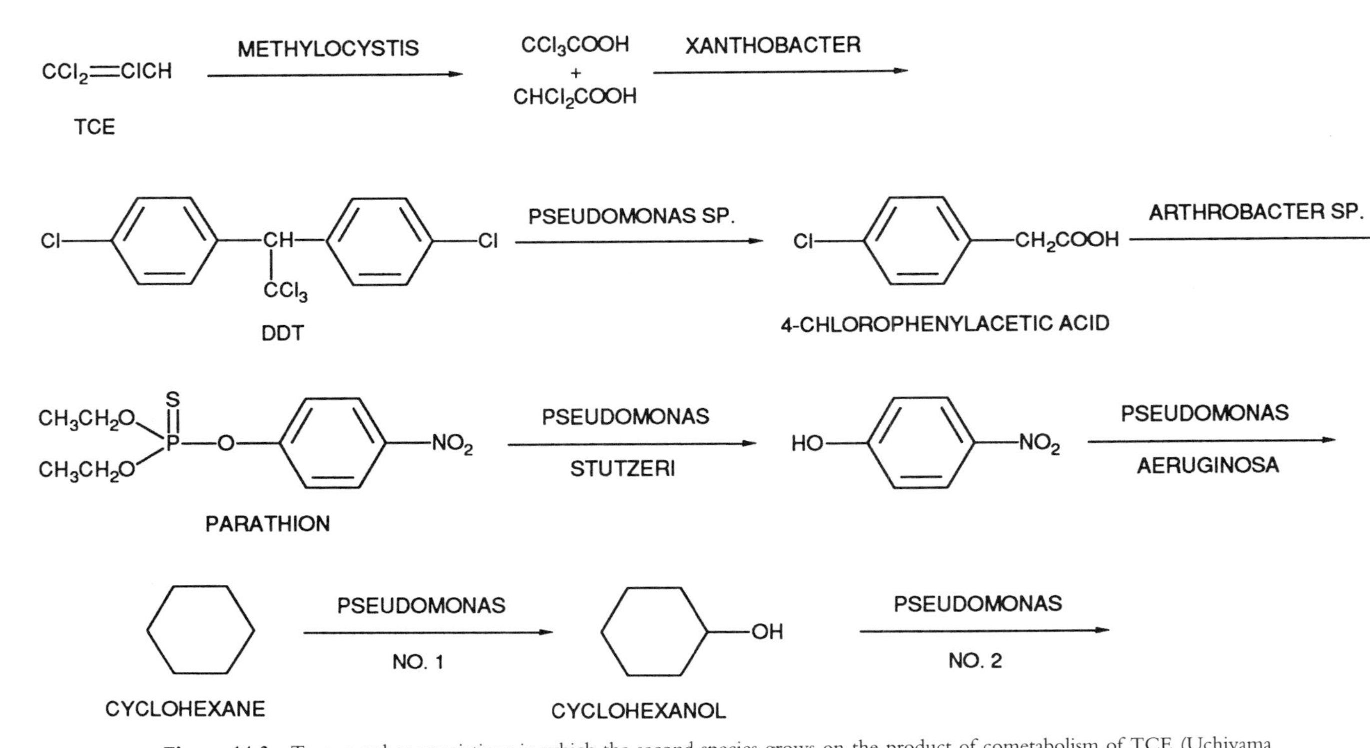

Figure 14.3 Two-member associations in which the second species grows on the product of cometabolism of TCE (Uchiyama *et al.*, 1992), DDT (Pfaender and Alexander, 1972), parathion (Daughton and Hsieh, 1977), and cyclohexane (deKlerk and van der Linden, 1974).

ently acted on by cometabolism to yield cyclohexanone, the latter can be used as a C source by aerobic bacteria; as a consequence, cyclohexane is mineralized in samples of soil and marine sediments (Beam and Perry, 1973, 1974). Nevertheless, the second species in some associations, as well as the first, may act only by cometabolism, so products of the second accumulate. This topic is also considered in Chapter 13.

The basis for some synergisms is the destruction of a toxin that is formed by one species. In some biodegradative reactions, the population acting on the parent chemical forms a metabolite that inhibits either the initial species or a species that is involved in a later step in the metabolic sequence. However, if the toxin producer has an organism nearby that destroys the inhibitor, the sensitive species continues to function without hindrance. In most of such synergisms that have been investigated, the first species generates a molecule that is harmful to itself, and the second species uses that compound as a source of C, energy, or both. However, the second organism may act in a somewhat different manner, as in the conversion of *N*-(3,4-dichlorophenyl)propionamide (the herbicide propanil) by *Penicillium piscarium* to 3,4-dichloroaniline, a compound that inhibits further propanil biodegradation, but the toxicity is relieved by *Geotrichum candidum,* a fungus that dimerizes the aniline derivative to yield 3,3′,4,4′-tetrachloroazobenzene (Bordeleau and Bartha, 1971). Species that metabolize nitro compounds commonly produce nitrite, which is toxic to many micro- as well as macroorganisms, but many bacteria and fungi are able to destroy nitrite—by converting it to ammonium, N oxides, N_2, or sometimes nitrate.

Interspecies hydrogen transfer represents a unique type of synergism. In part, such an association also relies on the destruction by the second species of a toxin produced by the first. In this instance, the inhibitor is H_2, which metabolically represents reducing power not needed by the initial population. The reactions carried out by each of the two anaerobic bacteria are:

$$2CH_3CH_2OH + 2H_2O \rightarrow 2CH_3COOH + 4H_2$$

$$4H_2 + CO_2 \rightarrow CH_4 + 2H_2O$$

The net effect of these two organisms is to make methane and acetate from ethanol, and the H_2 toxicity to the first organism is relieved by the second (Reddy *et al.,* 1972):

$$2CH_3CH_2OH + CO_2 \rightarrow 2CH_3COOH + CH_4$$

Other substrates may be acted on synergistically by mixtures of anaerobes in sequences that also involve interspecies hydrogen transfer (Laube and Martin, 1981; McCarty and Smith, 1986). A three-membered association has also been described in which 3-chlorobenzoate is converted anaerobically to methane. The different bacteria convert (a) 3-chlorobenzoate to benzoate; (b) benzoate to acetate, H_2, and CO_2; and (c) $H_2 + CO_2$ to methane. The reducing power for the reductive dehalogenation of 3-chlorobenzoate comes from the organism responsible for the second step (Dolfing and Tiedje, 1986).

PREDATION

An environment with a high density of bacteria or a large fungal biomass usually will also contain microorganisms that act as predators or parasites and some that will cause lysis. These predatory, parasitic, or lytic inhabitants may affect the biodegradation carried out by the bacteria and fungi. The impact is often deleterious, but it may be beneficial.

Among the predators and parasites found in soils, sediments, and surface and groundwaters are protozoa, bacteriophages, viruses affecting fungi, *Bdellovibrio,* mycobacteria, Acrasiales, and organisms that excrete enzymes that destroy cell walls of fungi and bacteria and thereby cause their lysis. Of these several groups, only the protozoa are known to affect biodegradation. This does not mean that the other groups are not important, only that evidence for their role has not been obtained.

Protozoa typically multiply by feeding on bacteria. In environments in which these microscopic animals are abundant, their grazing may markedly reduce the number of bacteria since 10^3 to 10^4 bacteria may be consumed to permit the division of a single protozoan. However, not only may protozoa affect bacterial activity by grazing but they may facilitate the cycling of limiting inorganic nutrients (especially P and N) and excrete essential growth factors. In some environments, protozoa are sparse and not particularly active, so that their role is highly dependent on prevailing conditions.

Active grazing requires a prey density greater than 10^6 to 10^7 bacterial cells per milliliter or, for nonaqueous environments, per cubic centimeter. Below this threshold density of bacteria, protozoan feeding is inconsequential. Similarly, when a substrate is provided in high concentrations in enrichment cultures containing protozoa, the predators will feed on the bacteria to bring their density down to about 10^6 cells per milliliter. Thus, if one assumes that approximately 1 pg of a C source will support 1 bacterial cell, a population size of 10^8 bacteria per milliliter would be expected in an enrichment containing 100 μg of an organic substrate per milliliter (100 mg/liter); instead, the cell density is sometimes only about 10^6 cells/ml (DiGeronimo *et al.,* 1979). However, the 10^6 cells/ml threshold is not for a single bacterial species but rather for all prey, so the abundance of a single species may be reduced below the threshold when other species are at high cell densities. For example, if species A is present at 10^8 cells/ml, species B is at 10^6/ml, and species C is at 10^3/ml, nonselective grazing would reduce the three populations to a final density of about 10^6/ml, but the mixture would contain 10^6 of A, 10^4 of B, and 10 of C per milliliter (Mallory *et al.,* 1983).

In environments in which they are active and abundant, the impact of protozoa depends on their grazing rate and the rate of biodegradation or, for transformations that proceed parallel to growth, bacterial multiplication. If grazing is slow and bacterial multiplication is rapid, protozoa will have little or no effect. If the predation rate is rapid (as occurs when the community of all bacteria is large) and the growth of the particular bacterial species causing the degradation is slow, protozoa may have a large impact. Such slow multiplication characterizes bacteria growing on a low concentration of an organic molecule, typically at levels below the K_s. At these low concentrations, the density of the biodegradative species may even decline despite the presence of a substrate that it alone, of the organisms present, can utilize. This is illustrated by the finding that indigenous protozoa have no effect on 4-nitrophenol mineralization at 50, 75, or 100 mg/liter in lake water samples inoculated with a nitrophenol-utilizing *Corynebacterium,* but they markedly suppress the transformation and prevent growth of the bacterium when the compound is added at 26 mg/liter (Zaidi *et al.,* 1989). However, a sufficient number of cells may survive the attack by protozoa so that, when the period of active predation is over, bacteria with the requisite metabolic capacity can grow and destroy the chemical (Ramadan *et al.,* 1990). Active grazing typically ends when the protozoa have reduced the density of all susceptible bacteria to approximately 10^6 cells/ml.

As discussed in Chapter 3, in sewage or other wastewaters containing many protozoa and a large number of indigenous bacteria to serve as prey, the acclimation period prior to active biodegradation may be the result of feeding by these unicellular animals. The acclimation phase lasts as long as the grazing is intense. Once the predation rate declines, usually because of the decrease in total number of bacteria to serve as prey, the remaining cells of bacteria able to degrade the chemical of interest begin to multiply and the active period of biodegradation commences (Wiggins and Alexander, 1988; Wiggins *et al.,* 1987).

Conversely, protozoa may sometimes stimulate microbial activity. This is evident, for example, in the enhanced decomposition of crude oil by a bacterial mixture in the presence of the ciliate *Colpidium colpoda* (Rogerson and Berger, 1983). Similar stimulatory effects on the rate of degradation of plant constituents or particulate matter have been observed with a number of flagellates and ciliates. A likely reason for this enhancement is the regeneration of P, N, or both. In environments in which the concentrations of inorganic P or N are so low that they limit microbial growth, the P or N is assimilated by bacteria, algae, and fungi, and little is available for the species important in a specific biodegradation; hence, the rate of microbial transformation is slow. However, the grazers consume part of that microbial biomass and excrete some of the P and N in the material they consume. That P and N are then available for use by bacteria and fungi active in biodegradation. Such a regeneration of P and N, which represents P and N mineralization, is believed to be important in soil and fresh and marine waters

(Anderson *et al.*, 1986; Cole *et al.*, 1978; Johannes, 1968). Protozoa also excrete growth factors as they ingest and digest bacteria, and they may thereby enhance biodegradation by auxotrophs (Huang *et al.*, 1981), which rely on other species for the vitamins, amino acids, and other growth factors they need.

GROWING PLANTS

Soil immediately surrounding the roots of growing plants, a site that is known as the rhizosphere, is a zone of intense microbial activity. This activity is a consequence of the large number of bacteria that utilize the simple organic compounds continuously excreted by the roots of plants during their active stages of development. In view of the large and metabolically active bacterial community of the rhizosphere, it is not unexpected that the rates of biodegradation, of some compounds at least, are more rapid in rhizosphere than adjacent nonrhizosphere soil or in soil under vegetation than in comparable fallow soil. For example, the mineralization of several surfactants is 1.1- to 1.9-fold faster in the rhizosphere of several plants than in nonrhizosphere soil (Knaebel and Vestal, 1992), and more benz(*a*)anthracene, chrysene, benzo(*a*)pyrene, and dibenz(*a,h*)anthracene disappear from soils supporting deep-rooted prairie grasses than from fallow soil (Aprill and Sims, 1990). Moreover, because TCE is more readily destroyed in rhizosphere than nonrhizosphere soil, the use of growing plants to promote bioremediation of TCE-contaminated soils has been proposed (Walton and Anderson, 1990).

In the last few years, biodegradation in the rhizosphere has attracted widespread interest. The reason is the possibility of practically exploiting the active microbial community around roots to destroy toxicants. That this possibility is real is evident from a number of investigations. For example, stimulations have been noted in the biodegradation of aliphatic hydrocarbons by growth of ryegrass (*Lolium perenne*) (Günther *et al.*, 1996), the destruction of naphthalene in a field test of buffalo grass (*Buchloe dactyloides*) (Qiu *et al.*, 1997), and the mineralization of benzo(*a*)pyrene by tall fescue (*Festuca arundinacea*) (Epuri and Sorensen, 1997) and PCP by crested wheatgrass (*Agropyron desortorum*) (Ferro *et al.*, 1994). However, sometimes the presence of plants has no effect, as in the case of benzene degradation in soils planted to alfalfa (*Medicago sativa*) (Ferro *et al.*, 1997).

The use of plants for bioremediation in the field is discussed in Chapter 16.

ANAEROBIC BIODEGRADATION

Biodegradation under aerobic conditions has been the subject of intense inquiry, but it was only in recent years that anaerobic transformations have begun to receive their long overdue attention. These more recent studies have demonstrated that bacteria that function under anaerobiosis are frequently highly versatile and they can destroy a variety of compounds. Many of these substrates can also be metabolized in the presence of O_2, often more rapidly but sometimes more slowly than under anaerobiosis. With certain molecules, however, the only known transformations occur when O_2 is not present, and such compounds persist in aerobic sites but disappear, albeit often slowly, under anaerobic conditions.

Some environments characteristically are devoid of O_2. In others, the O_2 is depleted by the aerobes that initially act on the organic materials. Aerobic bacteria and fungi are then displaced, and species able to function with other electron acceptors become prominent. These new populations may use organic compounds as electron acceptors, but many sites contain bacteria that use nitrate, sulfate, ferric iron, or CO_2 as electron acceptors. As a result of the use of the inorganic acceptors, N_2 and N_2O, sulfide, ferrous iron, or methane are generated. When the level of pollution is high, provided that the pollutants are not highly toxic to microorganisms, the processes of concern would thus be anaerobic. Not only are such conversions important in biodegradation, but several bioremediation technologies are specifically designed to exploit anaerobic activities, especially for reactions that only occur, or only take place rapidly, when the system is O_2 free.

Individual species of anaerobes rarely bring about an extensive conversion of most compounds to CO_2. As a rule, a single species carries out only part of the sequence of steps necessary to mineralize organic molecules, but the species responsible for the initial transformation frequently coexists with other anaerobes that carry out the later steps. In some cases, three different bacteria may be involved, as in the destruction of 3-chlorobenzoate cited earlier as an example of synergism. Several of the compounds metabolized anaerobically are listed in Table 14.2. Some of the chemicals, however, are degraded anaerobically in one environment but not in another, at least in the time periods tested.

Of particular interest are compounds that are metabolized anaerobically but not by bacteria or fungi when O_2 is present. For these molecules, only anaerobic transformations will result in destruction of the molecules and will be the basis of bioremediation. The substrates include highly chlorinated PCBs, hexachlorobenzene (Mohn and Tiedje, 1992), 2,6-dinitrotoluene, 3,5-dinitrobenzoic acid (Hallas and Alexander, 1983), RDX (McCormick *et al.*,1980), and DDT (Parr and Smith, 1974). It is likely that some of the substrates now believed to be degraded only anaerobically will ultimately be found to be metabolized by some aerobe, but that possible discovery will not change the conclusion that the process is clearly favored when no O_2 is available. In some instances, as with hexachlorocyclohexane (Ohisa and Yamaguchi, 1978), the conversion proceeds with or without O_2, but the anaerobic conversion is more rapid.

Although the initiation of anaerobic metabolism of some compounds is detected almost immediately after their introduction into a suitable environment,

Table 14.2

Compounds Degraded under Anaerobic Conditions

Chloroalkanes and alkenes	Benzoates
Carbon tetrachloride	Benzoate
Chloroform	2-, 3-, and 4-Chlorobenzoate
Vinyl chloride	3,4- and 3,5-Dichlorobenzoate
1,2-Dichloroethane	Aromatic hydrocarbons
1,1,1-Trichloroethane	Toluene
Trichloroethylene	Ethylbenzene
1,1,2,2-Tetrachlorethane	*o*- and *m*-Xylene
Tetrachloroethylene	Others
Phenols	Highly chlorinated PCBs
Phenol	Dimethyl phthalate
2- and 3-Chlorophenol	Pyridine
2,4- and 2,5-Dichlorophenol	Quinoline
Trichlorophenols	*m*- and *p*-Cresol
Tetrachlorophenols	2,4-D
Pentachlorophenol	2,4,5-T
2-, 3-, and 4-Nitrophenol	Diuron
	Linuron

extremely long periods of acclimation are required before a detectable disappearance of others is evident. This acclimation may even be of several months' duration, and the reductive dechlorination of chlorinated benzoates or benzenes may have acclimation phases as long as 6 months (Mohn and Tiedje, 1992). In the case of benzoate or 2- or 3-hydroxybenzoates, an anaerobic enrichment acting on these compounds only developed after 18 months (Sahm *et al.*, 1986).

Should an anaerobic bioremediation process lead to the accumulation of organic products, as is frequently true, it is likely that those products can be destroyed aerobically. Thus, monochlorobenzene that is generated anaerobically from hexachlorobenzene or the compounds that accumulate in the metabolism of PCBs under anoxic conditions can be transformed aerobically (Bédard *et al.*, 1987; Mohn and Tiedje, 1992). Such two-stage processes involving an initial anaerobic phase followed by a final aerobic phase represent promising means for the mineralization of certain persistent pollutants.

Because of the mammalian toxicity and persistence of many highly chlorinated molecules and their susceptibility to anaerobic bacteria, considerable attention has been given to the reductive dechlorinations that frequently represent the first and critical metabolic steps. Reductive dechlorination or, more generically, reductive dehalogenation may lead to the replacement of one or two halogens with one or two hydrogens (Fig. 14.4). In the third equation in Fig. 14.4, the substrates are typically alkyl halides, and the halogens are removed from adjacent

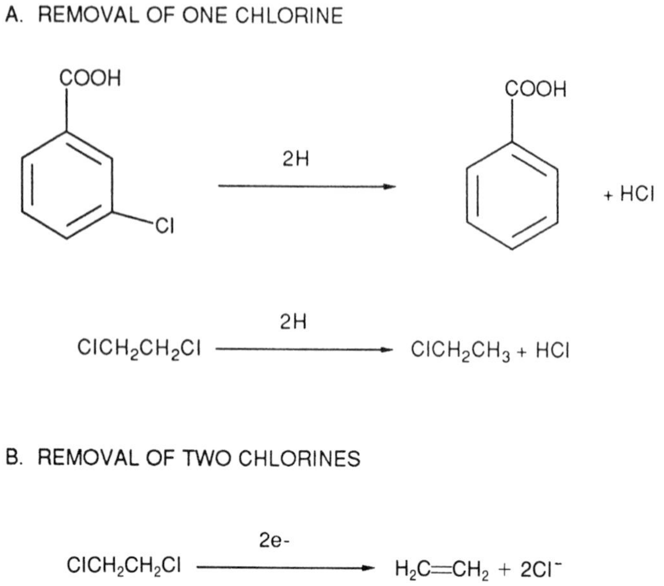

A. REMOVAL OF ONE CHLORINE

$$ClCH_2CH_2Cl \xrightarrow{2H} ClCH_2CH_3 + HCl$$

B. REMOVAL OF TWO CHLORINES

$$ClCH_2CH_2Cl \xrightarrow{2e-} H_2C{=}CH_2 + 2Cl^-$$

Figure 14.4 Typical reductive dehalogenations.

C atoms with the formation of a double bond between the two C atoms. The conversion requires microbial activity, but it is not certain whether microorganisms catalyze the reduction enzymatically or only generate a reductant that functions nonenzymatically to bring about halogen removal. Reductive dechlorination is evident in anoxic soils, sediments, and sludges and is responsible for initial phases in the metabolism of highly chlorinated PCBs, halogenated alkanes and ethylenes (e.g., chloroform, methyl chloride, and tri- and tetrachloroethylene), persistent chlorinated pesticides (including DDT, dieldrin, toxaphene, and lindane), and other halogenated molecules. Reductive dechlorination is usually slow and requires anaerobiosis. Often, the process is carried out by consortia or mixtures of bacteria, and pure cultures are typically obtained with difficulty, if at all. In some instances, the dechlorination reaction itself functions as an electron acceptor, allowing for the growth of the organism (Tiedje *et al.*, 1994; Mohn and Tiedje, 1992).

An inadequate supply of electron acceptors often limits anaerobic conversions. However, some communities of anaerobic microorganisms are able to use nitrate, sulfate, CO_2/bicarbonate, or ferric iron as electron acceptors to destroy compounds they would not otherwise degrade. For example, a mixture of microorganisms was found to be capable of degrading benzoate and 4-chlorobenzoate anaerobically but only if nitrate was present (Genthner *et al.*, 1989). These reactions not only destroy the organic molecules but reduce the electron acceptors nitrate, sulfate, and CO_2/bicarbonate to N_2 and N_2O, sulfide, and methane, respectively. If all three electron acceptors coexist, nitrate characteristically is metabolized first and disappears, sulfate reduction follows next, and finally methane is formed from CO_2. Ferric iron may serve as an electron acceptor, allowing some organisms to oxidize benzoate, phenol, several other simple aromatic compounds, and vinyl chloride anaerobically (Lovley *et al.*, 1989; Bradley and Chapelle, 1996). It appears that some bacteria are also able to use humic acid as an electron acceptor in the absence of O_2 (Lovley *et al.*, 1996a). However, sulfate may sometimes inhibit anaerobic conversions, for example, the reductive dehalogenation of aromatic compounds (Gibson and Suflita, 1986), and nitrate may also have harmful effects. Nevertheless, the supply of sulfate and nitrate in anaerobic environments is often limited, and the formation of additional amounts requires aerobic organisms; hence, if the supply is depleted, as would occur when the quantity of readily available C at a site is large, the anaerobic conversion may stop.

Recent research has thus disclosed the previously unrecognized potential of anaerobes for the decomposition of many pollutants, and these bacteria may be particularly important for compounds not metabolized by aerobes. In most instances, these reports come from investigations of pure or mixed cultures and not of natural or polluted environments, and commonly the degradation occurs with nitrate, sulfate, ferric iron, or sometimes CO_2 as required electron acceptors. Because the supply of nitrate and sulfate, in particular, in most anaerobic environments is limited, the extent of such transformations in nature remains uncertain. Such anaerobic conversions have been shown for benzene in sediment samples in the presence of sulfate, CO_2, or ferric iron as electron acceptors (Kazumi *et al.*, 1997; Lovley *et al.*, 1996b), toluene and xylenes in aquifer sediment and groundwater with a supply of nitrate or sulfate (Ball and Reinhard, 1996), naphthalene in sediment columns in the presence of nitrate (Langenhof *et al.*, 1996), pristane in aquifer sediments in the presence of nitrate (Bregnard *et al.*, 1997), some alkanes in microbial cultures supplemented with sulfate or nitrate (Holliger and Zehnder, 1996), and, to a limited extent, methylnaphthalenes, biphenyl, and anthraquinone in enrichment cultures (Genthner *et al.*, 1997).

However, many organic molecules persist in anaerobic environments, whether they be natural or polluted. Some of these are known to be potentially degradable or have actually been shown to be metabolized in nature or at some polluted site but, for reasons as yet unclear, the transformations are not ubiquitous, possibly because of the sparse distribution of organisms, the absence of suitable electron acceptors, the limited supply of nitrate and sulfate in anoxic environments, toxins present at individual locations, or the need for O_2 not as an electron acceptor but rather because O_2 is a reactant in the actual oxidative step itself. Several of the compounds reported to be resistant to anaerobic degradation are listed in Table 14.3. Even for these examples, however, some transformation but possibly not mineralization may occur, or the reaction may proceed if periods long enough for

Table 14.3

Compounds Reported to be Resistant to Anaerobic Degradation

Compounds	Reference
Benzene	Langenhoff *et al.* (1996), Ball and Reinhard (1996)
Three- to five-ring PAHs	Holliger and Zehnder (1996), Genthner *et al.* (1997)
Benzothiophene, benzofuran	Licht *et al.* (1996)
Anthracene	Bauer and Capone (1985)
2- and 4-Chlorobenzoate	Horowitz *et al.* (1983)
Chlorobenzene	Acton and Barker (1992)
Aniline, 4-toluidine	Hallas and Alexander (1983)
1- and 2-Naphthol, pyridine	Fox *et al.* (1988)
3,3′-Dichlorobenzidine	Boyd *et al.* (1984)
Saturated alkanes	Zehnder and Svensson (1986)

extended acclimations are allowed to elapse. Still, the reactions are slow, incomplete, or not ubiquitous. Nevertheless, because pure cultures have been isolated or enrichments have been established that bring about the destruction of organic compounds that persist anaerobically in nature, it should be possible to develop practical bioremediation techniques that lead to the anaerobic destruction of otherwise long-lived pollutants.

REFERENCES

Acton, D. W., and Barker, J. F., *J. Contam. Hydrol.* **9,** 325–332 (1992).

Alvarez, P. J. J., and Vogel, T. M., *Appl. Environ. Microbiol.* **57,** 2981–2985 (1991).

Anderson, O. K., Goldman, J. C., Caron, D. A., and Dennett, M. R., *Mar. Ecol.: Prog. Ser.* **31,** 47–55 (1986).

Aprill, W., and Sims, R. C., *Chemosphere* **20,** 253–265 (1990).

Atlas, R. M., *Microbiol. Rev.* **45,** 180–209 (1981).

Atlas, R. M., and Bartha, R., *Biotechnol. Bioeng.* **14,** 309–318 (1972).

Ball, H. A., and Reinhard, M., *Environ. Toxicol. Chem.* **15,** 114–122 (1996).

Bauer, J. E., and Capone, D. G., *Appl. Environ. Microbiol.* **50,** 81–90 (1985).

Beam, H. W., and Perry, J. J., *Arch. Microbiol.* **91,** 87–90 (1973).

Beam, H. W., and Perry, J. J., *J. Gen. Microbiol.* **82,** 163–169 (1974).

Bédard, D. L., Wagner, R. E., Brennan, M. J., Haberl, M. L., and Brown, J. F., Jr., *Appl. Environ. Microbiol.* **53,** 1094–1102 (1987).

Bordeleau, L. M., and Bartha, R., *Soil Biol. Biochem.* **3,** 281–284 (1971).

Bossert, I., and Bartha, R., *in* "Petroleum Microbiology" (R. M. Atlas, ed.), pp. 435–473. Macmillan, New York, 1984.

Bouwer, E. J., and McCarty, P. L., *Ground Water* **22,** 433–440 (1984).

Boyd, S. A., Kao, C.-W., and Suflita, J. M., *Environ. Toxicol. Chem.* **3,** 201–208 (1984).
Bradley, P. M., and Chapelle, F. H., *Environ. Sci. Technol.* **30,** 2084–2086 (1996).
Bregnard, T. P.-A., Häner, A., Höhener, P., and Zeyer, J., *Appl. Environ. Microbiol.* **63,** 2077–2081 (1997).
Burkholder, P. R., *in* "Symposium on Marine Microbiology" (C. H. Oppenheimer, ed.), pp. 133–150. Thomas, Springfield, IL, 1963.
Cole, C. V., Elliott, E. T., Hunt, H. W., and Coleman, D. C., *Microb. Ecol.* **4,** 381–387 (1978).
Daughton, C. G., and Hsieh, D. P. H., *Appl. Environ. Microbiol.* **34,** 175–184 (1977).
Daughton, C. G., Cook, A. M., and Alexander, M., *Appl. Environ. Microbiol.* **37,** 605–609 (1979).
Dean-Raymond, D., and Alexander, M., *Appl. Environ. Microbiol.* **33,** 1037–1041 (1977).
deKlerk, H., and van der Linden, A. C., *Antonie van Leeuwenhoek* **40,** 7–15 (1974).
Dibble, J. T., and Bartha, R., *Appl. Environ. Microbiol.* **37,** 729–739 (1979).
DiGeronimo, M. J., Nikaido, M., and Alexander, M., *Appl. Environ. Microbiol.* **37,** 619–625 (1979).
Dolfing, J., and Tiedje, J. M., *FEMS Microbiol. Ecol.* **38,** 293–298 (1986).
Duah-Yentumi, S., and Kuwatsuka, S., *Soil Sci. Plant Nutr.* **26,** 541–549 (1980).
Efroymson, R. A., and Alexander, M., *Environ. Toxicol. Chem.* **13,** 405–411 (1994).
Egli, T., *Antonie van Leeuwenhoek* **60,** 225–234 (1991).
Egli, T., *Adv. Microb. Ecol.* **14,** 305–386 (1995).
Ejlertsson, J., Johansson, E., Karlsson, A., Meyerson, U., and Svensson, B. H., *Antonie van Leeuwenhoek* **69,** 67–74 (1996).
Epuri, V., and Sorensen, D. L., *in* "Phytoremediation of Soil and Water Contaminants" (E. L. Kruger, T. A. Anderson, and J. R. Coats, eds.), pp. 200–222. American Chemical Society, Washington, DC, 1997.
Fava, F., *Chemosphere* **32,** 1477–1483 (1996).
Feinberg, E. L., Ramage, P. I. N., and Trudgill, P. W., *J. Gen. Microbiol.* **121,** 507–511 (1980).
Ferro, A., Kennedy, J., Douette, W., Nelson, S. Jauregui, G., McFarland, B., and Bugbee, B., *in* "Phytoremediation of Soil and Water Contaminants" (E. L. Kruger, T. A. Anderson, and J. R. Coats, eds.) pp. 223–237. American Chemical Society, Washington, DC, 1997.
Ferro, A. M., Sims, R. C., and Bugbee, B., *J. Environ. Qual.* **23,** 272–279 (1994).
Floodgate, G. D., *in* "Petroleum Microbiology" (R. M. Atlas, ed.), pp. 354–397. Macmillan, New York, 1984.
Focht, D. D., and Brunner, W., *Appl. Environ. Microbiol.* **50,** 1058–1063 (1985).
Fondén, R., *Oikos* **20,** 373–383 (1969).
Fournier, J.-C., *Chemosphere* **4,** 35–40 (1975).
Fox, P., Suidan, M. T., and Pfeffer, J. T., *J. Water Pollut. Control Fed.* **60,** 86–92 (1988).
Gaudy, A. F., Jr., Komolrit, K., and Gaudy, E. T., *Appl. Microbiol.* **12,** 280–286 (1964).
Genthner, B. R. S., Price, W. A., II, and Pritchard, P. H., *Appl. Environ. Microbiol.* **55,** 1472–1476 (1989).
Genthner, B. R. S., Townsend, G. T., Lantz, S. E., and Mueller, J. G., *Arch. Environ. Contam. Toxicol.* **32,** 99–105 (1997).
Gibson, S. A., and Suflita, J. M., *Appl. Environ. Microbiol.* **52,** 681–688 (1986).
Guenzi, W. D., Beard, W. E., and Viets, F. G., Jr., *Soil Sci. Soc. Am. Proc.* **35,** 910–913 (1971).
Gunner, H. B., and Zuckerman, B. M., *Nature (London)* **217,** 1183–1184 (1968).
Günther, T., Dornberger, U., and Fritsche, W., *Chemosphere* **33,** 203–215 (1996).
Hallas, L. E., and Alexander, M., *Appl. Environ. Microbiol.* **45,** 1234–1241 (1983).
Hallberg, K. B., Kelly, M. P., and Tuovinen, O. H., *Curr. Microbiol.* **23,** 65–69 (1991).
Harder, W., and Dijkhuizen, L., *Philos. Trans. R. Soc. London, Ser. B* **297,** 459–479 (1982).
Harder, W., Dijkhuizen, L., and Veldkamp, H., *in* "The Microbe" (D. P. Kelly and N. G. Carr, eds.), Part II, pp. 51–95. Cambridge Univ. Press, Cambridge, UK, 1984.
Hendriksen, H. V., Larsen, S., and Ahring, B. K., *Appl. Environ. Microbiol.* **58,** 365–370 (1992).
Hess, T. F., Schmidt, S. K., Silverstein, J., and Howe, B., *Appl. Environ. Microbiol.* **56,** 1551–1555 (1990).

Holliger, C., and Zehnder, A. J. B., *Curr. Opin. Biotechnol.* **7,** 326–330 (1996).
Hoover, D. G., Borgonovi, G. E., Jones, S. H., and Alexander, M., *Appl. Environ. Microbiol.* **51,** 226–232 (1986).
Horowitz, A., Suflita, J. M., and Tiedje, J. M., *Appl. Environ. Microbiol.* **45,** 1459–1465 (1983).
Huang, T.-C., Chang, M.-C., and Alexander, M., *Appl. Environ. Microbiol.* **41,** 229–232 (1981).
Hutchinson, D. H., and Robinson, C. W., *Appl. Microbiol. Biotechnol.* **29,** 599–604 (1988).
Jamison, V. W., Raymond, R. L., and Hudson, J. O., Jr., *Dev. Ind. Microbiol.* **16,** 305–312 (1975).
Jensen, H. L., *Can. J. Microbiol.* **3,** 151–164 (1957).
Jobson, A., McLaughlin, M., Cook, F. D., and Westlake, D. W. S., *Appl. Microbiol.* **27,** 166–171 (1974).
Johannes, R. E., *in* "Advances in Microbiology of the Sea" (M. R. Droop and E. J. F. Wood, eds.), Vol. 1, pp. 203–213. Academic Press, London, 1968.
Jones, S. H., and Alexander, M., *FEMS Microbiol. Lett.* **52,** 121–126 (1988a).
Jones, S. H., and Alexander, M., *Appl. Environ. Microbiol.* **54,** 3177–3179 (1988b).
Kane, A. C., Wilson, T. P., and Fisher, J. M., *in* "In Situ and On-Site Bioremediation," Vol. 3, pp. 115–123. Battelle Press, Columbus, OH, 1997.
Kator, H., *in* "The Microbial Degradation of Oil Pollutants" (D. G. Ahearn and S. P. Meyers, eds.), pp. 47–65. Louisiana State University, Center for Wetland Resources, Baton Rouge, 1973.
Kazumi, J., Caldwell, M. E., Suflita, J. M., Lovley, D. R., and Young, L. Y., *Environ. Sci. Technol.* **31,** 813–818 (1997).
Klečka, G. M., and Maier, W. J., *Biotechnol. Bioeng.* **31,** 328–333 (1988).
Knaebel, D. B., and Vestal, J. R., *Can. J. Microbiol.* **38,** 643–653 (1992).
Kuiper, J., and Hanstveit, A. O., *Ecotoxicol. Environ. Saf.* **8,** 15–33 (1984).
Langenhoff, A. A. M., Zehnder, A. J. B., and Schraa, G., *Biodegradation* **7,** 267–274 (1996).
LaPat-Polasko, L. T., McCarty, P. L., and Zehnder, A. J. B., *Appl. Environ. Microbiol.* **47,** 825–830 (1984).
Laube, V. M., and Martin, S. M., *Appl. Environ. Microbiol.* **42,** 413–420 (1981).
Law, A. T., and Button, D. K., *J. Bacteriol.* **129,** 115–123 (1977).
LePetit, J., and N'Guyen, M.-H., *Can. J. Microbiol.* **22,** 1364–1373 (1976).
Lewis, D. L., Freeman, L. F., III, and Watwood, M. E., *Environ. Toxicol. Chem.* **5,** 791–796 (1986).
Licht, D., Ahring, B. K., and Arvin, E., *Biodegradation* **7,** 83–90 (1996).
Lindley, N. D., and Heydeman, M. T., *Appl. Microbiol. Biotechnol.* **23,** 384–388 (1986).
Lochhead, A. G., *Bacteriol. Rev.* **22,** 145–153 (1958).
Lovley, D. R., Baedecker, M. J., Lonergan, D. J., Cozzarelli, I. M., Phillips, E. J. P., and Siegel, D. J., *Nature (London)* **339,** 297–300 (1989).
Lovley, D. R., Coates, J. D., Blunt-Harris, E. L., Phillips, E. J. P., and Woodward, J. C., *Nature (London)* **382,** 445–448 (1996a).
Lovley, D. R., Woodward, J. C., and Chapelle, F. H., *Appl. Environ. Microbiol.* **62,** 288–291 (1996b).
Mallory, L. M., Yuk, C.-S., Liang, L.-N., and Alexander, M., *Appl. Environ. Microbiol.* **46,** 1073–1079 (1983).
McCarty, P. L., and Smith, D. P., *Environ. Sci. Technol.* **20,** 1200–1206 (1986).
McCormick, N. G., Foster, D. M., Cornell, J. H., and Kaplan, A. M., *Abstr., Annu. Meet., Am. Soc. Microbiol.*, p. 196 (1980).
Mechalas, B. J., Meyers, T. J., and Kolpack, R. L., *in* "The Microbial Degradation of Oil Pollutants" (D. G. Ahearn and S. P. Meyers, eds.), pp. 67–79. Louisiana State University, Center for Wetland Resources, Baton Rouge, 1973.
Merkel, G. J., and Perry, J. J., *J. Agric. Food Chem.* **25,** 1011–1012 (1977).
Millette, D., Barker, J. F., Comeau, Y., Butler, B. J., Frind, E. O., Clement, B., and Samson, R., *Environ. Sci. Technol.* **29,** 1944–1952 (1995).
Mohn, W. W., and Tiedje, J. M., *Microbiol. Rev.* **56,** 482–507 (1992).
Murakami, Y., and Alexander, M., *Biotechnol. Bioeng.* **33,** 832–838 (1989).

Nelson, M. J. K., Montgomery, S. O., O'Neill, E. J., and Pritchard, P. H., *Appl. Environ. Microbiol.* **52,** 383–384 (1986).

Nörtemann, B., Baumgarten, J., Rast, H. G., and Knackmuss, H.-J., *Appl. Environ. Microbiol.* **52,** 1195–1202 (1986).

Ogunseitan, O. A., and Olson, B. H., *Appl. Microbiol. Biotechnol.* **38,** 799–807 (1993).

Ohisa, N., and Yamaguchi, M., *Agric. Biol. Chem.* **42,** 1983–1987 (1978).

Olivieri, R., Robertiello, A., and Degen, L., *Mar. Pollut. Bull.* **9,** 217–220 (1978).

Ou, L. T., and Sikka, H. C., *J. Agric. Food Chem.* **25,** 1336–1339 (1977).

Owens, J. D., and Legan, J. D., *FEMS Microbiol. Rev.* **46,** 419–432 (1987).

Painter, H. A., Denton, R. S., and Quarmby, C., *Water Res.* **2,** 427–447 (1968).

Parr, J. F., and Smith, S., *Soil Sci.* **118,** 45–52 (1974).

Pfaender, F. K., and Alexander, M., *J. Agric. Food Chem.* **20,** 842–846 (1972).

Pfaender, F. K., and Alexander, M., *J. Agric. Food Chem.* **21,** 397–399 (1973).

Pritchard, P. H., Cripe, C. R., Walker, W. W., Spain, J. C., and Bourquin, A. W., *Chemosphere* **16,** 1509–1520 (1987).

Qiu, X., Leland, T. W., Shah, S. I., Sorensen, D. L., and Kendall, E. W., *in* "Phytoremediation of Soil and Water Contaminants" (E. L. Kruger, T. A. Anderson, and J. R. Coats, eds.), pp. 186–199. American Chemical Society, Washington, DC, 1997.

Ramadan, M. A., El-Tayeb, O. M., and Alexander, M., *Appl. Environ. Microbiol.* **56,** 1392–1396 (1990).

Raymond, R. L., Hudson, J. O., and Jamison, V. W., *Appl. Environ. Microbiol.* **31,** 522–535 (1976).

Reddy, C. A., Bryant, M. P., and Wolin, M. J., *J. Bacteriol.* **110,** 126–132 (1972).

Robertson, B. K., and Alexander, M., *Appl. Environ. Microbiol.* **58,** 38–41 (1992).

Rogerson, A., and Berger, J., *J. Gen. Appl. Microbiol.* **29,** 41–50 (1983).

Rouatt, J. W., and Lochhead, A. G., *Soil Sci.* **80,** 147–154 (1955).

Rubin, H. E., and Alexander, M., *Environ. Sci. Technol.* **17,** 104–107 (1983).

Rubin, H. E., Subba-Rao, R. V., and Alexander, M., *Appl. Environ. Microbiol.* **43,** 1133–1138 (1982).

Sahm, H., Brunner, M., and Schoberth, S. M., *Microb. Ecol.* **12,** 147–153 (1986).

Schut, F., Jansen, M., Pedro Gomes, T. M., Gottschal, J. C., Harder, W., and Prins, R. A., *Microbiology* **141,** 351–361 (1995).

Shih, C.-C., Davey, M. E., Zhou, J., Tiedje, J. M., and Criddle, C. S., *Appl. Environ. Microbiol.* **62,** 2953–2960 (1996).

Shimp, R. J., and Pfaender, F. K., *Appl. Environ. Microbiol.* **49,** 394–401 (1985).

Skerman, T. M., *in* "Symposium on Marine Microbiology" (C. H. Oppenheimer, ed.), pp. 685–698. Thomas, Springfield, IL, 1963.

Smith, M. R., Ewing, M., and Rutledge, C., *Appl. Microbiol. Biotechnol.* **34,** 536–538 (1991).

Steffensen, W. S., and Alexander, M., *Appl. Environ. Microbiol.* **61,** 2859–2862 (1995).

Stumm-Zollinger, E., *Appl. Microbiol.* **14,** 654–664 (1966).

Subba-Rao, R. V., Rubin, H. E., and Alexander, M., *Appl. Environ. Microbiol.* **43,** 1139–1150 (1982).

Swindoll, C. M., Aelion, C. M., and Pfaender, F. K., *Appl. Environ. Microbiol.* **54,** 212–217 (1988).

Sylvestre, M., Massé, R., Ayotte, C., Messier, F., and Fauteux, J., *Appl. Microbiol. Biotechnol.* **21,** 192–195 (1985).

Tiedje, J. M., Fries, M., Chee-Sanford, J., and Cole, J., *in* "Trans. 15th World Congr. Soil Sci.," Vol. 4a, pp. 364–374, 1994.

Uchiyama, H., Nakajima, T., Yagi, O., and Nakahara, T., *Appl. Environ. Microbiol.* **58,** 3067–3071 (1992).

Venosa, A. D., Suidan, M. T., Wrenn, B. A., Strohmeier, K. L., Haines, J. R., Eberhart, B. L., King, D., and Holder, E., *Environ. Sci. Technol.* **30,** 1764–1775 (1996).

Walker, A., *Proc. Br. Crop. Prot. Conf.-Weeds, 13th, 1976,* Vol. 2, 635–642 (1976).

Walton, B. T., and Anderson, T. A., *Appl. Environ. Microbiol.* **56,** 1012–1016 (1990).

Wang, K.-W., Baltzis, B. C., and Lewandowski, G. A., *Biotechnol. Bioeng.* **51,** 87–94 (1996).

Wang, Y.-S., Subba-Rao, R. V., and Alexander, M., *Appl. Environ. Microbiol.* **47,** 1195–1200 (1984).
Wang, Y.-S., Madsen, E. L., and Alexander, M., *J. Agric. Food Chem.* **33,** 495–499 (1985).
Watson, G. K., and Jones, N., *Soc. Gen. Microbiol. Q.* **6,** 78 (1979).
Wiggins, B. A., and Alexander, M., *Can. J. Microbiol.* **34,** 661–666 (1988).
Wiggins, B. A., Jones, S. H., and Alexander, M., *Appl. Environ. Microbiol.* **53,** 791–796 (1987).
Zaidi, B. R., Murakami, Y., and Alexander, M., *Environ. Sci. Technol.* **22,** 1419–1425 (1988).
Zaidi, B. R., Murakami, Y., and Alexander, M., *Environ. Sci. Technol.* **23,** 859–863 (1989).
Zehnder, A. J. B., and Svensson, B. H., *Experientia* **42,** 1197–1205 (1986).

CHAPTER 15

Inoculation

Microorganisms with a phenomenal array of catabolic activities are widespread. Soils, sediments, fresh and marine waters, and industrial and municipal waste-treatment systems possess large and often highly diverse microbial communities that potentially can exhibit many degradative capacities, and when these capacities are expressed fully and rapidly, organic chemicals are readily destroyed. Nevertheless, many synthetic compounds persist for some time in these same environments, even though these molecules are biodegradable, and the question has been asked whether inoculation might appreciably enhance the decomposition of these compounds. Such inoculation is sometimes called *bioaugmentation*.

In polluted sites in which the period for destruction of the chemicals is not important, it is likely that inoculation is not warranted because the initially small population will multiply to destroy the unwanted chemical. However, when rapid destruction is important, it may not be appropriate to rely on the natural response of members of the indigenous community. For example, a slow biodegradation may result in the uptake by plants of toxicants present in soil, the movement of chemicals through soil to underlying groundwater, the transport of pollutants through a contaminated plume of groundwater to enter waters used for human consumption, or the dissemination of unwanted compounds through a biological treatment system to the receiving water and from that water to distant rivers or lakes, from which human, animal, or plant uptake may occur. The indigenous species may act, but often not sufficiently rapidly to prevent the spreading of a local problem.

It also is now clear that microorganisms acting on certain pollutants are absent from particular sites. A compound that is metabolized by many species will likely encounter one or several species in all microbial communities that can transform it. However, certain synthetic compounds are apparently transformed by very few species, and it is thus likely that not a single one of the very few species with the requisite enzymes may be present in a particular site. This view is in line with the frequent observation that some organic compounds are mineralized or otherwise metabolized in samples from one but not another environment and that active organisms can only be isolated from some environments.

Inoculation may also markedly reduce the acclimation period. If the time for the community to reach full activity is but a day or two, attempts to establish an organism probably would be pointless. However, if the acclimation period is weeks or months, as it often is, and the risk of human, animal, or plant exposure increases as the persistence of the toxicant increases, some form of intervention to enhance decomposition is called for.

Finally, inoculation may be necessary because conditions at the site preclude members of the resident community from functioning rapidly. Thus, when the unwanted chemical is present at a concentration high enough to suppress the native biodegrading species, when the temperature is too high, or the circumstances are otherwise stressful, the addition of a species able to destroy the chemical and also to tolerate the stress may be highly beneficial.

The approach to inoculation must be prudent. If there is an indigenous flora capable of carrying out the reaction, conditions favor its multiplication, and rapid destruction is not essential, additions of inocula are not needed. If these conditions do not pertain, intervention is called for. The lack of need for supplementation with microorganisms is well illustrated in waters and soils contaminated with oil. Such environments contain bacteria able to grow on and destroy a variety of hydrocarbons, and the persistence of components of oil is not a consequence of the absence of organisms but rather the absence of the full set of conditions necessary for the resident species to function rapidly (Atlas, 1977). In a typical study, the addition of a mixture of hydrocarbon-degrading bacteria to a marine-water microcosm did not enhance the degradation of crude oil polluting the seawater, and the indigenous microflora degraded the oil (Tagger *et al.*, 1983). Similarly, the addition of soil with a large population of hydrocarbon degraders to soil freshly contaminated with oil reduced the acclimation period, but the indigenous population soon multiplied and carried out the desired transformation (DeBorger *et al.*, 1978).

The preceding comments apply to field sites. In contrast, inoculation is very frequently successful in bioreactors. The conditions in these bioreactors are quite different from those in nature, and frequently few and sometimes no microorganisms having the needed biodegradative activity are present. Hence, the addition of such organisms is often beneficial and sometimes essential. Furthermore, because bioreactors are engineered systems whose conditions are readily altered or optimized for particular processes, they can be designed to promote the multiplication and activity of the inoculated species—in contrast with field sites.

SUCCESSES

As indicated earlier, it is important to distinguish between microorganisms added to, or allowed to grow in, engineered systems and those added to natural environments. Aboveground bioreactors of many types have microorganisms added to them—either pure cultures, enrichments, or microbial mixtures—and these organisms usually develop and destroy the compounds on which they were grown or enriched. These engineered bioreactors, for example, may be industrial waste-treatment systems involving immobilized cells or biofilms designed for specific compounds or waste streams. The record of success in these instances is good, e.g., a full-scale sand filter used for the treatment of 1,2-dichloroethane-contaminated groundwater (Stucki and Thüer, 1995), an *in situ* fixed-bed bioreactor (inoculated with *Methylosinus trichosporium*) created in a field test of the destruction of TCE in groundwater (Duba *et al.*, 1996), laboratory evaluations of soil slurries for the aerobic biodegradation of phenanthrene in NAPLs (Birman and Alexander, 1996) and anthracene (Gray *et al.*, 1994) and for the anaerobic dechlorination of 3-chlorobenzoate by *Desulfomonile tiedjei* (El Fantroussi *et al.*, 1997), and a biofilter designed to destroy dimethyl sulfide (Smet *et al.*, 1996).

In contrast, the record of success in enhancing biodegradation in soils, aquifers, and surface waters *in situ* is spotty. On the one hand, the initiation or enhancement of degradation has often been reported following the addition to natural environments (or, far more commonly, to samples of these environments brought to the laboratory) of bacteria and fungi that can metabolize and grow on specific organic compounds in culture. On the other hand, failures frequently have been reported also.

The usual way of obtaining a population for subsequent inoculation is to prepare an enrichment culture. This is typically done by adding a sample of soil, sewage, or natural water into a solution containing the organic compound and the variety of inorganic salts necessary for bacterial growth. The C source is usually added at a level far higher than exists in nature, so that a high cell yield is obtained. The pH is maintained near neutrality, and the mixture is incubated in the dark. When growth or chemical disappearance is evident, a subculture is added to a sterile portion of the same medium. This procedure may be repeated several times to increase the number of bacteria active on the test substrate relative to other organisms, and then the mixture usually is plated on an agar medium containing the test chemical as well as inorganic salts. The procedure provides isolates active on many compounds. However, the procedure favors organisms that grow well at high substrate concentrations, require no growth factors, multiply at pH values near 7.0, grow quickly, are not resistant to light inactivation, and multiply at levels of N and P in the enrichment medium that are far higher than prevail in the natural environment. Such approaches are based on other needs for isolates: to study the pathways of metabolism or the physiology of organisms catabolizing particular substrates. They are ideal for organisms active at high nutrient levels in the absence of biotic or abiotic stresses, but they do not fit in well with circumstances in natural ecosystems or polluted sites. It is unfortunately not common to make enrichments that are more likely to function well at the low levels of substrate, N, and P characteristic of the polluted environment or tolerate the pH, photochemical, or other stresses likely encountered in the environment in which rapid biodegradation is sought.

First, let us examine reports that inoculation enhances the destruction of pollutants in soil. These studies have centered on pesticides and either oil or specific hydrocarbon constituents of oil. Such studies are commonly carried out with 10 to 30 g of soil in depths of no more than a few centimeters, a method that must be kept in mind in extrapolation to fields where the depths of concern are measured in meters and the adequate mixing of inocula with soil is difficult.

a. *Parathion*. This insecticide is readily destroyed in soil inoculated with a mixture containing *Pseudomonas stutzeri* and *Pseudomonas aeruginosa*. The first bacterium converts parathion to 4-nitrophenol, and the second grows on and destroys 4-nitrophenol. In soil contaminated with 5.0 g of parathion per kilogram, more than 90% is destroyed within 3 weeks as a direct result of the inoculation. These tests were conducted using 10-g soil samples (Daughton and Hsieh, 1977). Addition of the same bacteria to soil in the field also results in the destruction of the insecticide. The latter study was done by adding the bacteria to soil contained within pipes (3.2-cm diameter) that were driven 10 cm into the ground (Barles *et al.*, 1979). The reason for stating the size of the containers or the amount of soil will be presented in the following.

b. *IPC*. A mixture of bacteria added in a large volume of liquid to flats of soil (10 cm deep) in the greenhouse destroys this herbicide, as well as two related herbicides, when present initially at levels equivalent to 5, 10, and 15 kg/ha (McClure, 1972). Similarly, treating 57-g samples of soil in petri dishes with *Arthrobacter* sp. able to grow in culture on IPC leads to inactivation of the herbicide (Clark and Wright, 1970). In these investigations, the action of the microorganisms was assessed by regular bioassay of the toxicity of the herbicide-treated soil to susceptible plants.

c. *Chlorpropham*. The addition to 25-g quantities of soil of either of two chlorpropham-utilizing *Pseudomonas* species results in destruction of this herbicide (Milhomme *et al.*, 1989).

d. *PCP*. Fungi of the genera *Irpex, Bjerkandera,* and *Trametes* destroy this compound rapidly when inoculated into small samples of soil (Leštan and Lamar,

1996), and a strain of *Sphingomonas* mineralizes high concentrations in a soil inoculated with the bacterium (Colores *et al.*, 1995). Mixing an inoculum of *Rhodococcus chlorophenolicus* with 50-g quantities of soil brings about mineralization of PCP (Middledorp *et al.*, 1990). In earlier studies, it was found that introduction of a PCP-metabolizing strain of *Arthrobacter* into 250-g samples of PCP-contaminated soil that is mixed with the inoculum results in destruction of the chemical. The same bacterium also decomposes the chemical in soil placed 10-cm deep on a concrete floor, but more PCP is destroyed in soil that is mixed daily than if it is not mixed (Edgehill and Finn, 1983). Similar results are obtained when a PCP-degrading *Flavobacterium* is added to 100-g samples of soil supplemented with 100 mg of PCP per kilogram, although marked PCP mineralization in this soil is evident after 10 days as a result of the action of the indigenous community (Crawford and Mohn, 1985).

e. *Oil.* Inoculation of 100 g of soil with *Candida guillermondii* promotes the rate of degradation of crude petroleum and hydrocarbons (Ismailov, 1985). In contrast, the addition of a mixture of oil-utilizing bacteria in field plots in an earlier study was found to cause only a slight enhancement in the rate of breakdown of *n*-alkane constituents with chain lengths of 20 to 25 carbons (Jobson *et al.*, 1974).

f. *2,4-D.* Amendment of samples of forest soil with wood chips colonized by *Phanerochaete chrysosporium* enhances mineralization of this herbicide (Entry *et al.*, 1996). Inoculation of columns of soil with *Burkholderia cepacia* leads to 2,4-D disappearance (Cattaneo *et al.*, 1997). The typical acclimation phase prior to the onset of mineralization of the herbicide is almost wholly abolished if the soil is amended with a suspension containing organisms acting on this compound (Kunc and Rybarova, 1983).

g. *PCBs.* Addition of a *Pseudomonas* strain and biphenyl to soil samples results in PCB loss (Hickey *et al.*, 1993). Inoculation of *Acinetobacter* sp. stimulates the rate of PCB mineralization and favors the destruction of the more highly chlorinated PCBs (Focht and Brunner, 1985).

h. *Pyrazon.* A gram-negative coccus able to degrade this herbicide in liquid medium also enhances its breakdown in soil (Engvild and Jensen, 1969).

i. *2,4,5-T.* The introduction of a 2,4,5-T-utilizing strain of *Pseudomonas cepacia* into test tubes of soil containing 1.0 g of 2,4,5-T per kilogram of soil results in destruction of much of the herbicide as measured by bioassays with sensitive plants (Karns *et al.*, 1984). Tests involving addition of the fungus *Phanerochaete chrysosporium* to 1.0 g of soil amended with 4.0 g of ground corn cob show that 33% of the 2,4,5-T is mineralized in 30 days (Ryan and Bumpus, 1989).

j. *Lindane and chlordane. Phanerochaete chrysosporium* mineralizes both of these chlorinated hydrocarbon insecticides when the fungus is inoculated into a mixture containing 1.0 g of sterile soil and 4.0 g of corn cobs (Kennedy *et al.*, 1990).

k. *Dicamba.* Inoculation of soil with microorganisms degrading this herbicide results in destruction of the chemical and protection of seedlings from phytotoxicity (Krueger *et al.*, 1991).

l. *Nitrophenols.* Inoculation of a flooded soil with a microbial mixture increases the rate of disappearance of 2-, 3-, and 4-nitrophenol and 2,4-dinitrophenol as compared to samples not so treated (Sudhakar-Barik and Sethunathan, 1978).

m. *Atrazine.* The mineralization of the herbicide is brought about in samples of soil inoculated with a strain of *Pseudomonas* (Kontchou *et al.*, 1995) or *Phanerochaete chrysosporium* (Entry *et al.*, 1996).

n. *Carbofuran.* A bacterium active on this insecticide in culture also degrades the compound when added to soil samples (Duquenne *et al.*, 1996).

Such findings show that it is often not difficult to demonstrate that bioaugmentation works in shallow layers of soil under laboratory conditions. The same is far from true with the depths of soil contaminated in the field, not only because of depth and problems of effectively mixing an inoculum with large volumes of soil but also because of stresses imposed on nonindigenous species by environmental variables not as readily controlled in the field as in the laboratory.

Soils are three-dimensional environments. From the point of inoculation, bacteria or fungi will have to be transported or migrate through an environment with pores ranging from macro- to microscopic in size. The cells or hyphae will not be able to penetrate into many micropores, and they will not be able to pass through the narrow necks between even some larger pores. Bacteria will also become sorbed to surfaces of the soil particles. From this viewpoint, experiments involving gram quantities or several centimeters of depth of soil do not serve as models for a three-dimensional environment composed of a solid matrix; they are merely two-dimensional trials. This is not to say that such inoculations will fail under more realistic conditions, rather it should be taken as an argument that realistic evaluations need to be conducted.

Success with bioaugmentation has also been reported for samples of aquifer materials containing groundwater. Thus, repeated addition of *Burkholderia cepacia* enhanced microbial metabolism when phenol was provided as a C source (Munakata-Marr *et al.*, 1996).

Inoculation has also been reported to result in the destruction of organic chemicals in natural waters. For example, a hydrocarbon-degrading bacterium obtained from an estuary enhances the biodegradation of oil spilled into a saline pond (Atlas and Busdosh, 1976). Similarly, data from laboratory studies indicate that oil-degrading bacteria added to seawater destroy a substantial part of the crude oil that has been added to the water (Miget *et al.*, 1969). In both of the preceding tests, the water had been supplemented with inorganic nutrients. The effect of an inoculum is evident even when the concentration of the organic compound is low; thus, addition of a benzoate-utilizing bacterium to lake water stimulates the rate of biodegradation of benzoate present at initial concentrations of 5 and 50 μg/liter (Subba-Rao *et al.*, 1982). In beach sands contaminated with oil from ocean pollution, an inoculum may also enhance the elimination of the unwanted oil (Ahlfeld and LaRock, 1973). However, a strain of a 4-nitrophenol-utilizing *Corynebacterium* had little activity in destroying that compound in lake water at 26 μg/liter, although it was active at higher concentrations (Fig. 15.1). Biodegradation is also observed when a strain of *Mycobacterium* active on PAHs is added to a mixture of 180 ml of reservoir water with 20 g of sediment, and the bacterium is capable of mineralizing pyrene (Heitkamp and Cerniglia, 1989).

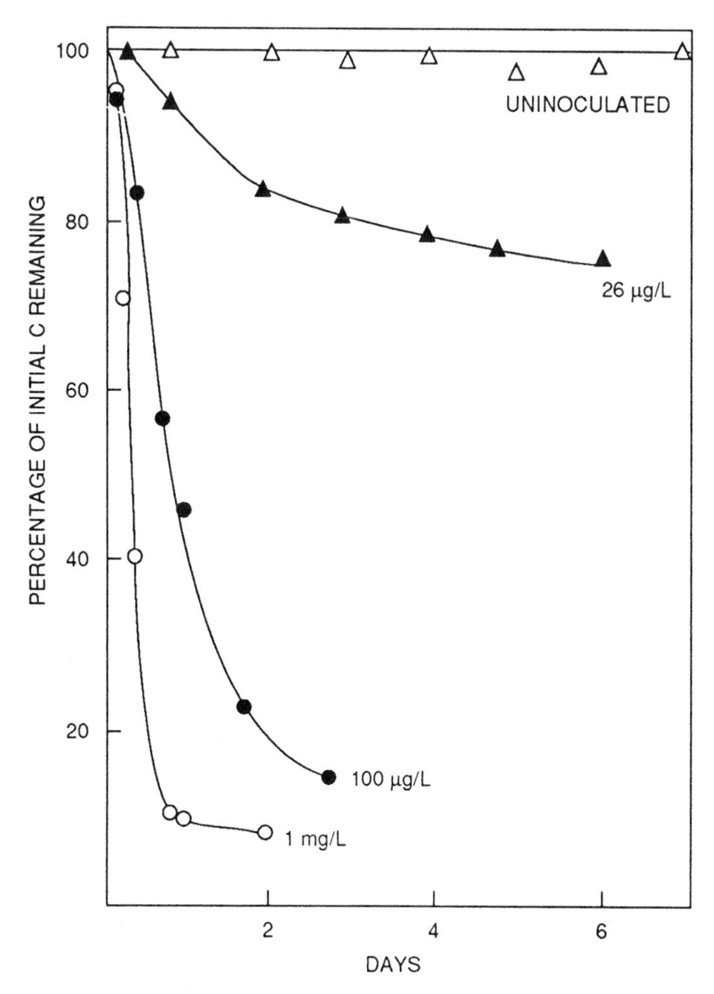

Figure 15.1 Mineralization of 1.0 mg of 4-nitrophenol per liter in uninoculated lake water and of 26 μg, 100 μg, and 1.0 mg of 4-nitrophenol per liter in lake water inoculated with *Corynebacterium* sp. (From Zaidi *et al.*, 1988a. Reprinted with permission from the American Chemical Society.)

Again, the spatial issue should be raised. A small water sample is not a lake or a reasonable volume of the oceans. Although physical obstructions to microbial movement do not exist in open waters, the question can be raised about the environmental relevancy of very small-scale tests that are not followed by pilot-scale experimentation, especially in a nonturbulent environment.

Activated-sludge systems for treating sewage are well mixed in actual practice, and here many of the spatial concerns are of less relevancy. Although possessing a large and metabolically diverse microbial community, sewage has long been known to exhibit an acclimation period prior to the initiation of rapid decomposition of many synthetic compounds. Thus, it is not surprising that addition of a 4-nitrophenol-metabolizing strain of *Pseudomonas* to samples of sewage enhances the mineralization of this compound (Goldstein *et al.*, 1985).

Evidence does exist, however, of the successful outcome of inoculation to bring about pollutant destruction in the field. For example, either of two species of *Phanerochaete* added to a field site containing soil contaminated with PCP, which was used as a wood preservative, brings about destruction of about 90% of the PCP in less than 7 weeks, even under less than optimal conditions for these fungi. However, unidentified complexes are formed in the transformation (Lamar and Dietrich, 1990). A subsequent study of field plots containing soil dug from a pile of contaminated material revealed that PCP concentrations were substantially reduced after 20 weeks as a result of inoculation with *Phanerochaete sordida* (Lamar *et al.*, 1994). Several field tests have also shown that the degradation of CCl_4 in an aquifer is enhanced by augmentation with *Pseudomonas stutzeri* (Dybas *et al.*, 1997) and that the destruction of TCE in groundwater is effected by addition of *Methylosinus trichosporium* (Duba *et al.*, 1996) or *Burkholderia cepacia* (Bourquin *et al.*, 1997). PCP-contaminated soil was also remediated at a field site by an inoculum containing a microbial mixture. In the latter instance, the pollutant was removed from the soil by a washing procedure, and the PCP in the wash solution was degraded by the added microorganisms in an aboveground bioreactor (Compeau *et al.*, 1991). The latter success is not the result of inoculation of the soil *in situ* but rather in an engineered system.

FAILURES

These findings of success have promoted considerable optimism among some investigators. All one needs to do, or so it seemed, is to set up an enrichment culture, isolate a bacterium or fungus able to use the unwanted chemical as a C source or cometabolize it, grow the organism in culture to get a large cell biomass, and then add the organism to the natural environment containing the substance whose destruction is sought. That this unbridled optimism was premature soon became apparent as evidence was collected that such inocula often failed to carry out in environmental samples the transformations they effected in laboratory media.

Failure of bioaugmentation is evident in a well-designed evaluation in the field of bioremediation of oil. In this instance, a mixture of microorganisms was added to randomized plots on a sandy beach deliberately contaminated with light crude oil, but the inoculum failed to have a significant effect (Venosa *et al.*, 1996). Inoculation with a mixture of microorganisms also did not bring about a statistically significant loss of hydrocarbons in a field investigation of contaminated soil (Walter *et al.*, 1997).

In many laboratory tests of samples from soil, subsoil, marine and lake waters, and sewage, bioaugmentation has also not been found to be necessary. The substances tested include crude, fuel, and diesel oils, Cr(VI), and such individual compounds as toluene, phenanthrene, 2,4-dichlorophenol, 4-nitrophenol, and metolachlor. The environmental samples and organisms used are presented in Table 15.1. In addition, it has been observed that periodic bioaugmentation of aquifer samples with *B. cepacia* for TCE metabolism did not result in establishment of the bacterium (Munakata-Marr *et al.*, 1997).

EXPLANATIONS FOR FAILURES

Such negative results come as no surprise to ecologists or agricultural scientists. The possession of one agronomically or ecologically useful trait is not sufficient

Table 15.1

Laboratory Studies in which Inoculation Did Not Stimulate or Cause Biodegradation

Test substance	Environmental sample	Microorganism	Reference
Fuel oil	Soil	Hydrocarbon degraders	Lehtomaki and Niemala (1975)
Diesel oil	Soil	Commercial preparation	Møller *et al.* (1975)
Cr(VI)	Soil	*Pseudomonas maltophila*	Cifuentes *et al.* (1995)
Toluene	Soil	*Corynebacterium variabilis*	Fuller *et al.* (1996)
2,4-Dichlorophenol	Soil	*Pseudomonas* sp.	Goldstein *et al.* (1985)
Metolachlor	Soil	*Streptomyces* sp.	Liu *et al.* (1990)
Diesel oil	Subsoil	Mixed culture	Margesin and Schinner (1997)
Phenanthrene	Soil slurry	*Pseudomonas* sp.	Weir *et al.* (1995)
Crude oil	Sea water	Mixed culture	Tagger *et al.* (1983)
Benzoate	Lake water	Bacterium	Subba-Rao *et al.* (1982)
4-Nitrophenol	Lake water	*Pseudomonas* sp.	Goldstein *et al.* (1985)
2,4-Dichlorophenol	Sewage	*Pseudomonas* sp.	Goldstein *et al.* (1985)

to guarantee success. Without question, an organism having a substrate uniquely available to it has a distinct advantage, yet that advantage may not be sufficient to compensate for many other traits that are also necessary for survival, no less multiplication, in a natural ecosystem. Possessing the requisite enzymes to metabolize a novel compound is a *necessary* attribute for the organism to carry out that transformation in a natural ecosystem, but it is not *sufficient* for the organism to succeed. Populations of microorganisms are subject to a variety of abiotic and biotic stresses, and these must be overcome for an introduced organism to be able to express its beneficial traits. A simple analogy is evident in agriculture: the development by plant breeders of a new crop variety that has desirable characteristics and shows vigor and outstanding growth potential in the greenhouse. The novice might think that the plant wil be successful because it has these beneficial traits, but both the agricultural scientist and any good farmer realizes that there are a variety of soil fertility and soil structural problems, plant pathogens, weeds, and insects that must be controlled in order for the introduced species to be successful. In the same way, a microorganism that has beneficial traits for biodegradation must also be able to overcome the biotic and nonbiological stresses in the environment in which it is to be introduced.

The reasons for the frequent failures of inocula to function in nature are many, even for species selected because they can rapidly grow on and mineralize pollutants present at a site. These reasons for failure often reflect ecological constraints on the introduced organism. These constraints are of several types, and to be able to colonize a particular environment, the added species must be able to cope with each. Some are immediately obvious to the laboratory scientist, others are not. In the next few pages are given some of the constraints and reasons for failure.

Limiting Nutrients

The added population has a nutrient uniquely available to it, namely, the organic compound whose destruction is sought, but it must also obtain N, P, O_2, other inorganic nutrients, and possibly growth factors from the environment into which it is introduced. The supply of these nutrients is frequently less than the demand, particularly where organic pollution is extensive, so that the members of the microbial community compete for the limiting inorganic nutrients. A species that grows slowly will not be as effective a competitor as its rapidly growing neighbors, and in instances where the chemical of interest is at a concentration below the K_s value, the added organism will probably grow slowly. Obviously, the addition of N, P, aeration to supply O_2, or other nutrient elements will reduce or eliminate the stress of competition for these requisites. Competition for inorganic nutrients may explain why an inoculum destroys synthetic chemicals in sterile but not in identical but nonsterile environmental samples, as has been observed in studies of *Streptomyces* sp. introduced into soil to degrade metolachlor (Liu *et al.*, 1990), *Flavobacterium* sp. added to soil to destroy 4-(2,4-DB) (MacRae and Alexander, 1965), or *Pseudomonas* sp. added to mineralize 2,4-dichlorophenol (Goldstein *et al.*, 1985). These bacteria decomposed the chemicals in sterile soil.

Even apart from the competition for inorganic nutrients in insufficient supply, a problem may exist because of the low concentration of inorganic nutrients. Probably for every inorganic nutrient, one can describe the growth rate of a microorganism as a function of nutrient concentration, as previously described for Monod kinetics applied to the organic nutrient. At high levels of the inorganic nutrient, growth rate is independent of its concentration. At low levels, the growth rate is directly dependent on its concentration and would be slower at progressively lower concentration. A threshold probably also exists below which growth does not occur. Thus, a strain to be used for inoculation would function slowly if its K_s value for a particular nutrient is higher than the prevailing level in the environment of interest, and it would not function at all if the prevailing concentration is below its threshold. Clearly, the level of the nutrient would be reduced due to competition with other species. Alternatively, the effect of low nutrient concentration may be compounded by grazing protozoa; that is, at concentrations of the nutrient near or below K_s, the inoculum strain would grow slowly, and the bacteria may not multiply fast enough to replace the cells consumed by predators. Low concentrations of P, N, and possibly other elements may in this way be the cause of failure of inoculation in natural ecosystems.

Suppression By Predators and Parasites

Soils, natural waters, sewage, and sediments contain predators and parasites that may suppress not only the indigenous bacteria but also an added species. Especially prominent among the predators are the protozoa, which often are abundant in these environments. *Bdellovibrio,* bacteriophages, myxobacteria, cellular slime molds, and species producing lytic enzymes may also be numerous, although little evidence exists that they markedly suppress particular species or control the activities of bacterial communities in such environments. Protozoa feed on many bacterial genera, and although their grazing requires a bacterial cell density greater than about 10^6 per milliliter or gram (Alexander, 1981), such densities are common in nature. Should a bacterial species be multiplying rapidly, the cells lost to protozoan grazing, especially if grazing is not intense, may be replaced as new cells are formed. However, if a bacterium is multiplying slowly, as is likely the case for a species added to destroy a chemical at concentrations near or below the K_s value, the cells eliminated by protozoan grazing will not be replaced. Hence, that species is suppressed or eliminated even as the total bacterial community is maintained (Mallory *et al.*, 1983; Wiggins and Alexander, 1988).

Direct evidence that protozoa affect the ability of inocula to carry out biodegradation comes from studies of the biodegradation of 4-nitrophenol by nitrophenol-utilizing bacteria. A *Corynebacterium* with this capacity in culture mineralizes little of the compound present at low concentrations in lake water, and its population declines; however, the bacterium grows and the compound is mineralized in lake water amended with cycloheximide, an inhibitor of protozoa and other eucaryotes but not of the bacterium (Zaidi *et al.*, 1989). An investigation using low but environmentally realistic levels of inoculation by another 4-nitrophenol-degrading bacterium revealed that the organism fails to cause appreciable mineralization of the nitro compound unless the protozoa are suppressed by eucaryotic inhibitors (Fig. 15.2). Large inocula do effect mineralization, but population estimates show that some survivors remain after protozoan feeding on the cells in the large inoculum. These survivors then multiply and metabolize 4-nitrophenol when grazing pressure by the protozoa is reduced. Nevertheless, inhibition of protozoa also enhances activity by the large inoculum (Ramadan *et al.*, 1990). Hence,

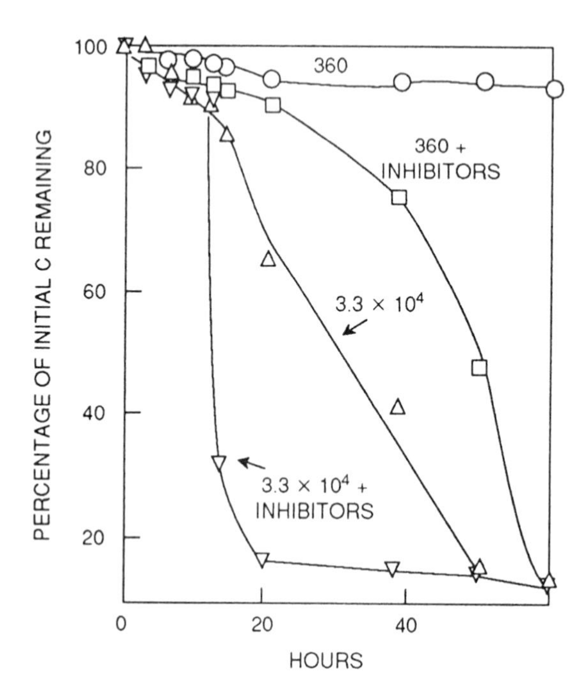

Figure 15.2 Effect of cycloheximide and nystatin, two eucaryotic inhibitors, on the mineralization by *Pseudomonas cepacia* of 1.0 mg of 4-nitrophenol per liter of lake water. The inoculum was added to give 360 or 3.3×10^4 cells per milliliter. (From Ramadan *et al.*, 1990. Reprinted with permission from the American Society for Microbiology.)

protozoa represent a major deterrent to successful remediation by introduced bacteria in natural environments containing active protozoan communities.

Protozoa are also active in suppressing the development of bacteria introduced into soil (Acea *et al.*, 1988). Therefore, they may have a similar role in soil as in lake water in determining the outcome of inoculation to effect biodegradation.

Inability of Bacteria to Move Appreciably through Soil

The finding of both successes and failures as a result of inoculation of soil might seem to be anomalous. However, in studies in which the introduced bacteria or fungi brought about biodegradation, the test system usually contained only 1 to 250 g of soil in centrifuge tubes, test tubes, petri dishes, or occasionally pots, the soil was in depths of 10 cm, or the soil sample was mixed with the added bacteria. These procedures probably result in penetration or mixing of the bacteria with the body of soil to an extent that the organisms encounter much of their substrate. However, in a soil in which channels are absent, movement is very limited, as is evident in the finding that a phenanthrene-metabolizing strain of *Pseudomonas* is able to transform only the compound present very close to the soil surface (Devare and Alexander, 1995). Similarly, when a bacterium able to grow on 2,4-dichlorophenol or another able to grow on 4-nitrophenol is added to the surface of sterile soil (in which there is no competition, predation, or parasitism), the bacteria mineralize little of either of the two phenols, although they are active if mixed with the sterile soil (Fig. 15.3). The latter findings probably result from the lack of appreciable movement of the microorganisms through the soil matrix to degrade the chemical at a distance from the point of inoculation. Presumably, only the chemical very near the sites of bacterial introduction is destroyed, as would occur in shallow soil depths or if bacteria are well mixed with the soil. Mixing the soil following inoculation may promote biodegradation, as observed with carbofuran-containing samples of soil that were inoculated (Duquenne *et al.*, 1996), but such mixing in the field is not sufficient to distribute the microorganisms to a high percentage of the microsites penetrated by the pollutant and is not economically feasible for subsurface material.

Such findings are not surprising in view of the many observations, in both the laboratory and the field, that bacteria do not move appreciably through soil. It is not enough for individual cells to move through macropores, channels, or openings previously made by roots, root hairs, earthworms, or other soil animals because much of the chemical in soil is at a distance from these channels. The inoculated bacterium must be distributed to nearly all sites immediately adjacent to the chemical. The inability of bacteria, even in sterile soil with no competition or predation, to destroy a chemical only a short distance from the point of their addition presumably results from the lack of movement of the cells through the

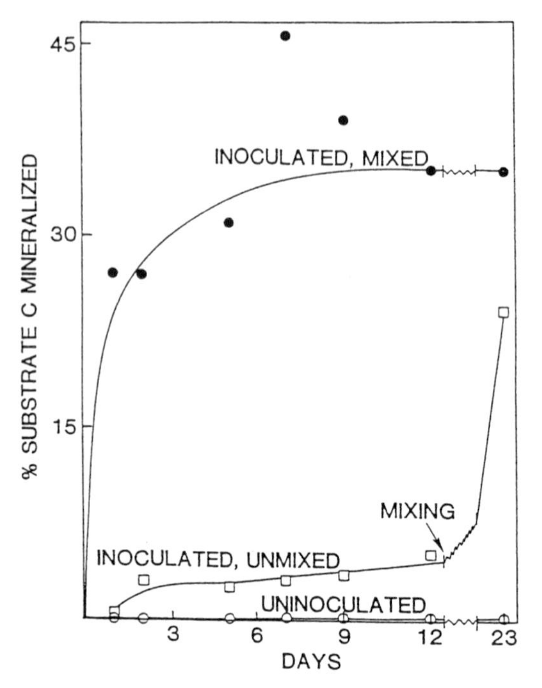

Figure 15.3 Mineralization of 5.0 μg of 4-nitrophenol per gram in sterile soil that received no inoculum or that was inoculated at the surface with a 4-nitrophenol-utilizing *Pseudomonas* and either mixed or left unmixed. The arrow indicates when the unmixed soil was shaken vigorously. (From Goldstein *et al.*, 1985. Reprinted with permission from the American Society for Microbiology.)

porous matrix (Goldstein *et al.*, 1985). The lack of movement is a consequence of (a) physical filtration, in which the cells are mechanically blocked by the solids and are unable to penetrate and move through small pores, and (b) adsorption of the cells by soil particles. If the soil is made into a slurry with water or if an economically feasible way exists to mix soil intimately with an inoculum in the field, the failures to achieve remediation because of poor movement of the inoculum may be overcome.

Limited movement through soil in the field or laboratory has been demonstrated frequently. For example, some mobility of naphthalene- and phenanthrene-metabolizing bacteria has been reported for field sites (Madsen *et al.*, 1996). However, it is likely that much of this movement results from the passage of cells through macropores, channels created by earthworms or other invertebrates, and cracks in the soil. This view is supported by the finding that nearly all cells of an inoculum strain of *Pseudomonas fluorescens* were found in channels of preferential flow rather than in the soil matrix (Natsch *et al.*, 1996). Typically, few of the bacteria applied to the surface penetrate past the first 5 cm (Edmonds, 1976), and most fecal coliforms, even when applied in liquid derived from a sewage-lagoon effluent, do not pass beyond the surface 8 cm (Bell and Bole, 1978). Although the extent of movement is probably greater in sandy soils because of their larger pores, bacterial transport with water is probably very limited in nonsandy soils. For example, the movement of *Pseudomonas putida* and *Bradyrhizobium japonicum* through a soil does not exceed 3 cm; although earthworm activity or the presence of growing roots increases vertical transport, the bacteria still did not move far (Madsen and Alexander, 1982). With enormous amounts of water and considerable time, good movement does occur through sand (Robeck *et al.,* 1962), but the amount of time required and the fact that soils do not have as large pores as sand make extensive movement highly unlikely. However, because the degree of movement does vary among bacteria (Weiss *et al.,* 1995; Wong and Griffin, 1976; Gannon *et al.,* 1991) and among the spores of certain fungi (Hepple, 1960), it may be possible to find species that are susceptible to transport with water and hence that are more suitable for soil inoculation. Nevertheless, one must consider that dispersal of cells through soil to destroy chemicals more than a few centimeters from the point of inoculation is a major obstacle to success.

It is not clear whether bacteria that are deliberately injected into groundwater will be dispersed sufficiently or will move with a contaminant plume to an extent that they will destroy contaminants in aquifers. A recent investigation shows that most cells of a TCE-degrading strain of *B. cepacia* became attached within a few centimeters of the site of their introduction, but an adhesion-deficient strain of the same species was moved readily (Malusis *et al.,* 1997). Most other evidence, largely obtained with fecal coliforms but sometimes with other bacteria, indicates that bacteria do not move far through septic fields, but successful transport for somewhat more than 10 m is possible (Reneau and Peittry, 1975; Viraraghavan, 1978). Conversely, distant dissemination of bacteria does occur through fractured bedrock, at least for 30 m (Morrison and Allen, 1972), and distant transport also probaby occurs through a variety of underground cracks and fissures. Nevertheless, because appreciable biodegradation in groundwaters requires extensive dispersal of the organisms added at discrete points of belowground injection, it is not yet clear whether such approaches are feasible.

Use of Other C Sources

An organism isolated because of its capacity to metabolize an unwanted chemical is able to grow on a number of other organic substrates, some of which may be present in the environment of interest. Following its addition to that environment, the organism may multiply by preferentially using one or more of these other C sources, leaving unmetabolized the unwanted compound (Goldstein *et al.*, 1985).

Concentration of Organic Substrate too Low to Support Multiplication

For many practical problems, the initial population density of the organisms added for the purposes of biodegradation will be small because the volume of material to be freed of the compound will be large. Hence, to effect a significant loss of the compound, the organism will have to multiply. However, because microorganisms will not grow on organic molecules below a threshold concentration and the selective advantage of an introduced species is its ability to grow on the unwanted molecule, it cannot degrade that substrate to a significant degree if its selective advantage is lost. Evidence for such a problem is found in the reports that a benzoate-utilizing bacterium fails to mineralize 34 and 350 ng of benzoate per liter although it destroys benzoate at higher levels following inoculation into lake water (Subba-Rao *et al.*, 1982). It is possible that the inability to function at low substrate concentrations may sometimes not be a consequence of concentrations below the threshold for growth but rather reflect the inability of the inoculated organism to grow sufficiently rapidly at the low substrate concentration to replace cells consumed by protozoan grazing.

Need for C Source to Support Growth

A species that acts on an organic substrate by cometabolism must have a C source for growth. It is unlikely that the site of interest will have a sufficient supply of such a C source to support the introduced cells because, even if present, the indigenous species would likely use it more readily than an introduced organism. Hence, without additions of such an organic nutrient, the inoculum will not carry out the desired transformation. This possible cause of failure is illustrated by studies of a strain of *Acinetobacter* that cometabolizes a number of PCBs. Inoculation of soil with this bacterium does not enhance PCB mineralization unless the soil is also amended with biphenyl, a nonchlorinated analogue of PCBs that supports proliferation of the bacterium. In this instance, *Acinetobacter* only carries out the first step in the mineralization, and indigenous populations convert the products of PCB cometabolism to CO_2 (Brunner *et al.*, 1985). However, some bioremediation strategies are based on additions of growth substrates to support the cometabolizing species.

Temperature

It is common to isolate bacteria using enrichment cultures incubated at 25 to 37°C. Such temperatures are characteristic of some natural and man-made environments, but the temperatures of most are significantly lower. Many species that multiply well at these common laboratory temperatures do not multiply at lower temperatures. However, enrichments could just as well be established and organisms isolated at temperatures similar to those that prevail at the sites of interest.

pH

It is also common to maintain enrichment cultures near neutrality and thus to isolate microorganisms that grow well or that only multiply at pH values close to 7.0. Nevertheless, the oceans and many inland waters are at higher pH values, and many soils have lower values. Thus, an organism may fail to destroy an organic compound in nature simply because its pH range for growth does not include the pH at the site of interest. Overcoming or preventing such failure is simple: the enrichment should be maintained at pH values similar to those at the site of concern, and the isolate thus will be capable of development at those values (Zaidi *et al.*, 1988b).

Salinity

Some waters and soils have modest to occasionally high levels of salts. The oceans and estuaries are more saline than nearly all inland waters, and some soils are also reasonably rich in salts. As with temperature and pH, the inoculum strain must be able to multiply at the prevailing levels of salts, and its inability to do so will result in the failure of the organism to function. Conversely, an organism isolated from a salt-rich habitat may not degrade a chemical in fresh water; for example, Atlas and Busdosh (1976) found that a hydrocarbon-utilizing bacterium isolated from an estuary enhanced degradation of oil spilled in a saline but not in a freshwater pond.

Toxins

Natural inhibitors affecting bacteria are present in some unpolluted waters and soils. Although their identities are largely unknown, these toxins prevent the growth and may affect the survival of a species introduced into an environment in which it is not native. Polluted soils and surface and ground waters usually contain many inhibitors harmful to microorganisms. To function at a site in which natural or synthetic inhibitors are present at injurious levels, the organism to be used must be resistant to the toxins.

Overcoming some of these constraints is easy. Overcoming others will be difficult or, in some instances, impossible. Bacteria or fungi able to cope with many of these stresses can be isolated, or the site containing the unwanted chemical could be modified by aeration, mixing, nutrient supplementation, etc. Frequent success will then be attained when the identities of the constraints are defined and means are devised to overcome or minimize their importance.

GENETICALLY ENGINEERED MICROORGANISMS

Molecular biology has provided highly useful techniques to modify the genetic composition of microorganisms and thus to allow for the potential construction of new organisms having the capacity to carry out catabolic sequences not possible in existing organisms or under conditions not suitable for existing organisms. These genetic modifications thus provide a highly important and potentially very useful approach to effect the bioremediation of compounds not otherwise destroyed rapidly by microbiological means or under conditions in which microbial transformations would otherwise be too slow to be practical. The constraints affecting introduced organisms, to be sure, are the same as those that apply to existing organisms, and an inoculated genetically engineered microorganism must be able to cope with the ecological and environmental stresses that apply to any nonindigenous species. However, the constructed microorganism still is of special significance because of its new characteristics.

A variety of problems in the future may be solved by use of genetically engineered microorganisms. (a) Constructing microorganisms able to grow on and mineralize pollutants that presently are only cometabolized. Substrates that are transformed solely by cometabolism are biodegraded slowly and thus persist, and they yield products that are often toxic, long-lived, or both. However, by combining in one organism the genes encoding for the enzymes that cometabolize the compound of concern with the genes encoding for enzymes that allow an organism to grow on and mineralize what otherwise would be the end product of cometabolism, the engineered organism would use the parent molecule as C source and bring about its mineralization. The new organism would thus have two catabolic sequences that complement one another, and such complementary catabolic sequences would then effect a conversion not otherwise possible. Such approaches could result in a biodegradation that does not yield a persistent product or does not result in the formation of a compound that inhibits the bacteria destroying the target compound. (b) Creation of new catabolic pathways to effect transformations not presently carried out efficiently or rapidly, such as by altering the range of substrates used by a particular microorganism. (c) Increasing the amount or activity of specific enzymes in a microorganism. This increase might be useful in enhancing the rate of degradation brought about by a microorganism deliberately added to a polluted site or in providing highly active bacteria for use as immobilized cells or for the preparation of immobilized enzymes. (d) Construction of microorganisms that not only can destroy target pollutants but also are resistant to inhibitors at the site that prevent degradation by indigenous microorganisms. Many industrial sites, possibly most, that have high levels of synthetic organic compounds also contain high concentrations of heavy metals or other substances that suppress microbial development, or the concentration of the target compound itself may be too high to allow for activity of existing bacteria or fungi. (e) Creating microorganisms able to act on a broader array of polluting compounds. (f) Developing strains that have low adherence to the surfaces of the solids present in soils or aquifers so that they are more capable of being transported to sites at some distance from the point of inoculation (Timmis *et al.*, 1994; Pieper *et al.*, 1996).

Most of the genetic determinants of bacteria are on a single, circular chromosome. In addition, many bacteria have far smaller genetic elements known as plasmids that bear some of the genetic determinants of the organisms; the plasmids are considered to be not crucial for the survival of the cell, but they do have special significance in many catabolic sequences. The enzymes catalyzing the degradation of a particular compound may be encoded by chromosomal genes, by genes on a plasmid, or partly by chromosomal and partly by plasmid genes.

Bacteria may exchange genetic material in three ways: transduction, transformation, or conjugation. Gene exchange in transduction is mediated by a bacteriophage, whereas transformation entails the release of DNA by lysis of one bacterium and its uptake by a second. In conjugation, the DNA is transferred from one cell to another through a conjugal tube joining the two cells. Construction of novel bacterial strains by transformation, transduction, or conjugation is termed *in vivo* genetic engineering, that is, the genetic rearrangement occurs in living organisms. *In vitro* genetic engineering, in contrast, may involve separation of DNA from the cell, its treatment with a specific restriction endonuclease to cleave the DNA molecule, the rejoining of DNA fragments with DNA ligase to give a new sequence of nucleotide bases, and the reintroduction of this hybrid molecule into a suitable bacterial cell in which it will replicate and be expressed. Protoplast fusion and transposon-mediated gene manipulation may also be used to construct organisms with new characteristics. Bacterial protoplasts are cells with the rigid, outer peptidoglycan layer removed enzymatically, and fusion of such protoplasts may lead to genetic rearrangements. Transposons are short sequences of DNA bases that can be inserted *in vivo* into many sites in replicating DNA molecules.

Plasmids have been of great attraction as a means to construct novel bacteria. This attraction results from the many plasmid-borne genes that encode enzymes important in biodegradation. These are known as catabolic plasmids, and they give the bacterium containing them the ability to degrade certain compounds. The type of plasmid that is of particular interest is the one that can be transferred from one organism to another; some of these can only be transferred between closely related strains (narrow host-range plasmids), but others are transferred freely between different species and genera. The latter, the broad host-range plasmids, replicate in the cells in which they have been introduced, and genetic information thereby may be transmitted to quite dissimilar species. Catabolic plasmids have been discovered that encode enzymes catalyzing the degradation of ABSs, benzoate and chlorobenzoates, biphenyl and 4-chlorobiphenyl, chloroacetate, *p*-cresol, 2,4-D, naphthalene, octane, parathion, phenanthrene, styrene, toluene, and other compounds, and they have been found in species of *Pseudomonas, Alcaligenes, Acinetobacter, Flavobacterium, Beijerinckia, Klebsiella, Moraxella* and *Arthrobacter* (Sayler *et al.*, 1990). Some of the plasmid-encoded degradative activities that have been transferred from one bacterial species to a second are listed in Table 15.2. The recipient of the plasmid in some instances is a different species in the same genus as the source of the plasmid, but sometimes the transfer involves different genera. In several instances, the plasmid confers on the new host the capacity to metabolize the compound but not use it as a C source for growth. In other instances, however, the recipient acquires the ability to grow on the molecule it previously was unable to metabolize.

By means of transmissible plasmids or by the use of other genetic techniques, bacteria have been constructed that have activities different from those of the original organisms. One of the first of such new organisms was a bacterium that had acquired the ability to destroy the herbicide 2,4,5-T (Kellogg *et al.*, 1981). A

Table 15.2

Plasmid-Borne Genes That Have Been Transferred from One Bacterium to Another

Activity encoded by plasmid that is transferred	Activity of bacterial recipient of plasmid	Reference
2,4-D Degradation	Metabolism but not growth on 2,4-D	Friedrich *et al.* (1983)
Benzene metabolism	Growth on benzene	Irie *et al.* (1987)
Haloacetate dehalogenase	Metabolism but not growth on chloroacetate	Kawasuki *et al.* (1981)
Naphthalene metabolism	Metabolism but not growth on naphthalene	Oh *et al.* (1985)
3-Chlorobenzoate metabolism	Growth on chlorophenols	Reineke *et al.* (1982)
Five catabolic pathways	Growth on chlorobenzoates	Rojo *et al.* (1987)
TCE cometabolism	Degradation of TCE	Winter *et al.* (1989)

strain of *Pseudomonas putida* has been constructed that extensively metabolizes pentachloroethane in culture (Wackett, 1995). A derivative of *P. putida* has also been made that is able to simultaneously mineralize benzene, toluene, and *p*-xylene (Lee *et al.*, 1995). A strain of *Pseudomonas* was constructed that is able to use TNT as a C and N source and thus grow on the compound; it was derived from a culture using the compound only as a N source (Duque *et al.*, 1993). In addition, an isolate of *P. putida* has been constructed that is capable of extensively degrading TCE (Fujita *et al.*, 1995). The issue of obtaining bacteria active in biodegradation and also tolerant of the heavy metals found at many hazardous waste sites was addressed in a study in which a strain of *Alcaligenes eutrophus* was constructed that degraded 2,4-D and di- and trichlorobiphenyl isomers and also tolerated high concentrations of Ni and Zn (Springael *et al.*, 1993). Of special significance is the finding that genes encoding for a metabolic activity (the conversion of halogen-containing aromatic compounds to halogenated catechols) can be cloned in a bacterium that is able to grow on the product of cometabolism formed by the first species, the result being a new organism that is able to mineralize the original halogenated aromatic compound (Reineke, 1986). Other constructed bacteria are listed in Table 15.3.

Thus, genetic engineering does indeed promise to provide organisms with novel and presently nonexisting biodegradative activities. However, the possession of these new catabolic traits will only lead to successful bioremediations if these unique organisms can cope with the stresses in the environments in which they will

Table 15.3

Degradative Activity of Genetically Engineered Bacteria

Activity of parent cultures	Compounds metabolized by constructed bacteria	Reference
Biphenyl-grown *Acinetobacter*, 3-chlorobenzoate-grown *Pseudomonas*	3-Chlorobiphenyl	Adams *et al.* (1992)
Pseudomonas using 4-chloro-2-nitrophenol for N, *Alcaligenes* using haloaromatics	4-Chloro-2-nitrophenol used as C source	Bruhn *et al.* (1988)
Toluene-grown *Pseudomonas putida*, benzoate-grown *P. alcaligenes*	1,4-Dichlorobenzene	Kröckel and Focht (1987)
Aniline-degrading *Pseudomonas*, chlorocatechol-degrading *Pseudomonas*	Chloroanilines	Latorre *et al.* (1984)
4-Chlorophenol-utilizing *Pseudomonas*, phenol-utilizing *Alcaligenes*	2- and 3-Chlorophenols	Schwien and Schmidt (1982)
Biphenyl-grown *Pseudomonas putida*, 4-chlorobenzoate-grown *P. cepacia*	Dichlorobiphenyls	Havel and Reineke (1991)

be introduced. To date, only a few tests have been conducted with environmental samples. The trials that have been conducted are of a strain of *P. putida* that degrades 2,3,2′,5′-tetrachlorobiphenyl (a PCB) in soil slurries (Lajoie *et al.,* 1994) and a constructed strain of *B. cepacia* that destroys TCE in groundwater and aquifer sediment samples (Krumme *et al.,* 1993). Assessments under realistic field conditions are still required. Nevertheless, genetically engineered organisms that are found to be successful ecologically and catabolically should greatly enhance society's ability to destroy pollutants in natural environments. In bioreactors, in which such stresses are less important, however, the availability of constructed organisms tailor-made for the compounds of concern should have a more immediate impact.

REFERENCES

Acea, M., Moore, C. R., and Alexander, M., *Soil Biol. Biochem.* **20,** 509–515 (1988).

Adams, R. H., Huang, C.-M., Higson, F. K., Brenner, V., and Focht, D. D., *Appl. Environ. Microbiol.* **58,** 647–654 (1992).

Ahlfeld, T. E., and LaRock, P. A., *in* "The Microbial Degradation of Oil Pollutants" (D. G. Ahearn and S. P. Meyers, eds.), pp. 199–203. Louisiana State University, Center for Wetland Studies, Baton Rouge, 1973.

Alexander, M., *Annu. Rev. Microbiol.* **35,** 113–133 (1981).

Atlas, R. M., *CRC Crit. Rev. Microbiol.* **5,** 371–386 (1977).

Atlas, R. M., and Busdosh, M., *in* "Proceedings of the Third International Biodegradation Symposium" (J. M. Sharpley and A. M. Kaplan, eds.), pp. 79–85. Applied Science Publishers, London, 1976.

Barles, R. W., Daughton, C. G., and Hsieh, D. P. H., *Arch. Environ. Contam. Toxicol.* **8,** 647–660 (1979).

Bell, R. G., and Bole, J. B., *J. Environ. Qual.* **7,** 193–196 (1978).

Birman, I., and Alexander, M., *Appl. Microbiol. Biotechnol.* **45,** 267–272 (1996).

Bourquin, A. W., Mosteller, D. C., Olsen, R. L., Smith, M. J., and Reardon, K. F., *in* "In Situ and On-Site Bioremediation," Vol. 4, pp. 513–518. Battelle Press, Columbus, OH, 1997.

Bruhn, C., Bayly, R. C., and Knackmuss, H.-J., *Arch. Microbiol.* **150,** 171–177 (1988).

Brunner, W., Sutherland, F. H., and Focht, D. D., *J. Environ. Qual.* **14,** 324–328 (1985).

Cattaneo, M. V., Masson, C., and Greer, C. W., *Biodegradation* **8,** 87–96 (1997).

Cifuentes, F. R., Lindemann, W. C., and Barton, L. L., *Soil Sci.* **161,** 233–241 (1996).

Clark, C. G., and Wright, S. J. L., *Soil Biol. Biochem.* **2,** 19–26 (1970).

Colores, G. M., Radehaus, P. M., and Schmidt, S. K., *Appl. Biochem. Biotechnol.* **54,** 271–275 (1995).

Compeau, G. C., Mahaffey, W. D., and Patras, L., *in* "Environmental Biotechnology for Waste Treatment" (G. S. Sayler, R. Fox, and J. W. Blackburn, eds.), pp. 91–109. Plenum, New York, 1991.

Crawford, R. L., and Mohn, W. W., *Enzyme Microb. Technol.* **7,** 617–620 (1985).

Daughton, C. G., and Hsieh, D. P. H., *Bull. Environ. Contam. Toxicol.* **18,** 48–56 (1977).

DeBorger, R., Vanloocke, R., Verlinde, A., and Verstraete, W., *Rev. Ecol. Biol. Sol* **15,** 445–452 (1978).

Devare, M., and Alexander, M., *Soil Sci. Soc. Am. J.* **59,** 1316–1320 (1995).

Duba, A. G., Jackson, K. J., Jovanovich, M. C., Knapp, R. B., Shah, N. N., and Taylor, R. T., *Environ. Sci. Technol.* **30,** 1982–1989 (1996).

Duque, E., Haidour, A., Godoy, F., and Ramos, J. L., *J. Bacteriol.* **175,** 2278–2283 (1993).

Duquenne, P., Parekh, N. R., Catroux, G., and Fournier, C., *Soil Biol. Biochem.* **28,** 1805–1811 (1996).

Dybas, M. J., Bezborodinikov, S., Voice, T., Wiggert, D. C., Davies, S., Tiedje, J., Criddle, C. S., Kawka, O., Barcelona, M., and Mayotte, T., *in* "In Situ and On-Site Bioremediation," Vol. 4, pp. 507–512. Battelle Press, Columbus, OH, 1997.

Edgehill, R. U., and Finn, R. K., *Appl. Environ. Microbiol.* **45,** 1122–1125 (1983).

Edmonds, R. L., *Appl. Environ. Microbiol.* **32,** 537–546 (1976).

El Fantroussi, S., Mahillon, J., Naveau, H., and Agathos, S. N., *Appl. Environ. Microbiol.* **63,** 806–811 (1997).

Engvild, K. C., and Jensen, H. L., *Soil Biol. Biochem.* **1,** 295–300 (1969).

Entry, J. A., Donnelly, P. K., and Emmingham, W. H., *Appl. Soil Ecol.* **3,** 85–90 (1996).

Focht, D. D., and Brunner, W., *Appl. Environ. Microbiol.* **50,** 1058–1063 (1985).

Friedrich, B., Meyer, M., and Schlegel, H. G., *Arch. Microbiol.* **134,** 92–97 (1983).

Fujita, M., Ike, M., Hioki, J.-I., Kataoka, K., and Takeo, M., *J. Ferment. Bioeng.* **79,** 100–106 (1995).

Fuller, M. E., Mu, D. Y., and Scow, K. M., *Microb. Ecol.* **29,** 311–325 (1995).

Gannon, J. T., Mingelgrin, U., Alexander, M., and Wagenet, R. J., *Soil Biol. Biochem.* **23,** 1155–1160 (1991).

Goldstein, R. M., Mallory, L. M., and Alexander, M., *Appl. Environ. Microbiol.* **50,** 977–983 (1985).

Gray, M. R., Banerjee, D. K., Fedorak, P. M., Hashimoto, A., Masliyah, J. H., and Pickard, M. A., *Appl. Microbiol. Biotechnol.* **40,** 933–940 (1994).

Havel, J., and Reineke, W., *FEMS Microbiol. Lett.* **78,** 163–169 (1991).

Heitkamp, M. A., and Cerniglia, C. E., *Appl. Environ. Microbiol.* **55,** 1968–1973 (1989).

Hepple, S., *Trans. Br. Mycol. Soc.* **43,** 73–79 (1960).

Hickey, W. J., Searles, D. B., and Focht, D. D., *Appl. Environ. Microbiol.* **59,** 1194–1200 (1993).

Irie, S., Shirai, K., Doi, S., and Yorifuji, T., *Agric. Biol. Chem.* **51,** 1489–1493 (1987).

Ismailov, N. M., *Mikrobiologiya* **54,** 835–841 (1985).

Jobson, A., McLaughlin, M., Cook, F. D., and Westlake, D. W. S., *Appl. Microbiol.* **27,** 166–171 (1974).

Karns, J. S., Kilbane, J. J., Chaterjee, D. K., and Chakrabarty, A. M., *in* "Genetic Control of Environmental Pollutants" (G. S. Omenn and A. Hollaender, eds.), pp. 3–21. Plenum, New York, 1984.

Kawasuki, H., Tone, N., and Tonomura, K., *Agric. Biol. Chem.* **45,** 29–34 (1981).

Kellogg, S. T., Chatterjee, D. K., and Chakrabarty, A. M., *Science* **214,** 1133 (1981).

Kennedy, D. W., Aust, S. D., and Bumpus, J. A., *Appl. Environ. Microbiol.* **56,** 2347–2353 (1990).

Kontchou, C. Y., and Gschwind, N., *J. Agric. Food Chem.* **43,** 2291–2294 (1995).

Kröckel, L., and Focht, D. D., *Appl. Environ. Microbiol.* **53,** 2470–2475 (1987).

Krueger, J. P., Butz, R. G., and Cork, D. J., *J. Agric. Food Chem.* **39,** 1000–1003 (1991).

Krumme, M. L., Timmis, K. N., and Dwyer, D. F., *Appl. Environ. Microbiol.* **59,** 2746–2749 (1993).

Kunc, F., and Rybarova, J., *Soil Biol. Biochem.* **15,** 141–144 (1983).

Lajoie, C. A., Layton, A. C., and Sayler, G. S., *Appl. Environ. Microbiol.* **60,** 2826–2833 (1994).

Lamar, R. T., Davis, M. W., Dietrich, D. M., and Glaser, J. A., *Soil Biol. Biochem.* **26,** 1603–1611 (1994).

Lamar, R. T., and Dietrich, D. M., *Appl. Environ. Microbiol.* **56,** 3093–3100 (1990).

Latorre, J., Reineke, W., and Knackmuss, H.-J., *Arch. Microbiol.* **140,** 159–165 (1984).

Lee, J.-Y., Jung, K.-H., Choi, S. H., and Ki, H.-S., *Appl. Environ. Microbiol.* **61,** 2211–2217 (1995).

Lehtomäki, M., and Niemelä, S., *Ambio* **4,** 126–129 (1975).

Leštan, D., and Lamar, R. T., *Appl. Environ. Microbiol.* **62,** 2045–2052 (1996).

Liu, S.-Y., Lu, M.-H., and Bollag, J.-M., *Biodegradation* **1,** 9–17 (1990).

Madsen, E. L., and Alexander, M., *Soil Sci. Soc. Am. J.* **46,** 557–560 (1982).

Madsen, E. L., Thomas, C. T., Wilson, M. S., Sandoli, R. L., and Bilotta, S. E., *Environ. Sci. Technol.* **30,** 2412–2416 (1996).

Mallory, L. M., Yuk, C.-S., Liang, L.-N., and Alexander, M., *Appl. Environ. Microbiol.* **46,** 1073–1079 (1983).

Malusius, M. A., Adams, D. J., Reardon, K. F., Shackelford, C. D., Mosteller, D. C., and Bourquin, A. W., *in* "In Situ and On-Site Bioremediation," Vol. 4, pp. 559–564. Battelle Press, Columbus, OH, 1997.

Margesin, R., and Schinner, F., *Appl. Microbiol. Biotechnol.* **47,** 462–468 (1997).
McClure, G. W., *J. Environ. Qual.* **1,** 177–180 (1972).
Middledorp, P. J. M., Briglia, M., and Salkinoja-Salonen, M. S., *Microb. Ecol.* **20,** 123–139 (1990).
Miget, R., Oppenheimer, C. H., Kator, H. I., and LaRock, P. A., *in* "Proceedings of the Joint Conference on Prevention and Control of Oil Spils," pp. 327–331. American Petroleum Institute, New York, 1969.
Milhomme, H., Vega, D., Marty, J.-L., and Bastide, J., *Soil Biol. Biochem.* **21,** 307–311 (1989).
Møller, J., Gaarn, H., Steckel, T., Wedebye, E. B., and Westermann, P., *Bull. Environ. Contam. Toxicol.* **54,** 913–918 (1995).
Morrison, S. M., and Allen, M. J., "Bacterial Movement through Fractured Rock." Environmental Resources Center, Colorado State University, Fort Collins, 1972.
Munakata-Marr, J., Matheson, V. G., Forney, L. J., Tiedje, J. M., and McCarty, P. L., *in* "In Situ and On-Site Bioremediation," Vol. 4, pp. 501–506. Battelle Press, Columbus, OH, 1997.
Munakata-Marr, J., McCarty, P. L., Shields, M. S., Reagin, M., and Francesconi, S. C., *Environ. Sci. Technol.* **30,** 2045–2052 (1996).
Natsch, A., Keel, C., Troxler, J., Zala, M., von Albertini, N., and Defago, G., *Appl. Environ. Microbiol.* **62,** 33–40 (1996).
Oh, S., Quensen, J., Matsumura, F., and Momose, H., *Environ. Toxicol. Chem.* **4,** 21–27 (1985).
Pieper, D. H., Timmis, K. N., and Ramos, J. L., *Naturwissenschaften* **83,** 201–213 (1996).
Ramadan, M. A., El-Tayeb, O. M., and Alexander, M., *Appl. Environ. Microbiol.* **56,** 1392–1396 (1990).
Reineke, W., *J. Basic Microbiol.* **26,** 551–567 (1986).
Reineke, W., Wessels, S. W., Rubio, M. A., Latorre, J., Schwien, U., Schmidt, E., Schlömann, M., and Knackmuss, H.-J., *FEMS Microbiol. Lett.* **14,** 291–294 (1982).
Reneau, R. B., Jr., and Peittry, D. E., *J. Environ. Qual.* **4,** 41–44 (1975).
Robeck, G. G., Bryant, A. R., and Woodward, R. L., *J. Am. Water Works Assoc.* **54,** 75–82 (1962).
Rojo, F., Pieper, D. H., Engesser, K.-H., Knackmuss, H.-J., and Timmis, K. N., *Science* **238,** 1395–1398 (1987).
Ryan, T. P., and Bumpus, J. A., *Appl. Microbiol. Biotechnol.* **31,** 302–307 (1989).
Sayler, G. S., Hooper, S. W., Layton, A. C., and King, J. M. H., *Microb. Ecol.* **19,** 1–20 (1990).
Schwien, U., and Schmidt, E., *Appl. Environ. Microbiol.* **44,** 33–39 (1982).
Smet, E., Chasaya, G., Van Langenhove, H., and Verstraete, W., *Appl. Microbiol. Biotechnol.* **45,** 293–298 (1996).
Springael, D., Diels, L., Hooyberghs, L., Kreps, S., and Mergeay, M., *Appl. Environ. Microbiol.* **59,** 334–339 (1993).
Stucki, G., and Thüer, M., *Environ. Sci. Technol.* **29,** 2339–2345 (1995).
Subba-Rao, R. V., Rubin, H. E., and Alexander, M., *Appl. Environ. Microbiol.* **43,** 1139–1150 (1982).
Sudhakar-Barik and Sethunathan, N., *J. Environ. Qual.* **7,** 349–352 (1978).
Tagger, S., Bianchi, A., Julliard, M., LePetit, J., and Roux, B., *Mar. Biol. (Berlin)* **78,** 13–20 (1983).
Timmis, K. N., Steffan, R. J., and Unterman, R., *Annu. Rev. Microbiol.* **48,** 525–557 (1994).
Venosa, A. D., Suidan, M. T., Wrenn, B. A., Strohmeier, K. L., Haines, J. R., Eberhart, B. L., King, D., and Holder, E., *Environ. Sci. Technol.* **30,** 1764–1775 (1996).
Viraraghavan, T., *Water Air Soil Pollut.* **9,** 355–362 (1978).
Wackett, L. P., *Environ. Health Perspect.* **103**(Suppl. 5), 45–48 (1995).
Walter, M. V., Nelson, E. C., Firmstone, G., Martin, D. G., Clayton, M. J., Simpson, S., and Spaulding, S., *J. Soil Contam.* **6,** 61–77 (1997).
Weir, S. C., Dupuis, S. P., Providenti, M. A., Lee, H., and Trevors, J. T., *Appl. Microbiol. Biotechnol.* **43,** 946–951 (1995).
Weiss, T. H., Mills, A. L., Hornberger, G. M., and Herman, J. S., *Environ. Sci. Technol.* **29,** 1737–1740 (1995).
Wiggins, B. A., and Alexander, M., *Can. J. Microbiol.* **34,** 661–666 (1988).

Winter, R. B., Yen, K.-M., and Ensley, B. D., *Bio/Technology* **7,** 282–285 (1989).
Wong, P. T. W., and Griffin, D. M., *Soil Biol. Biochem.* **8,** 215–218 (1976).
Zaidi, B. R., Murakami, Y., and Alexander, M., *Environ. Sci. Technol.* **22,** 1419–1425 (1988a).
Zaidi, B. R., Stucki, G., and Alexander, M., *Environ. Toxicol. Chem.* **7,** 143–151 (1988b).
Zaidi, B. R., Murakami, Y., and Alexander, M., *Environ. Sci. Technol.* **23,** 859–863 (1989).

Bioremediation
In Situ & Solid Phase

CHAPTER 16

Bioremediation Technologies: *In Situ* and Solid Phase

Recent years have witnessed an enormous growth in the controlled, practical use of microorganisms for the destruction of chemical pollutants. These various technologies rely on the biodegradative activities of microorganisms and focus on enhancing existent but slow biodegradation processes in nature or technologies that bring chemicals into contact with microorganisms in some type of reactor that allows for rapid transformation. In many instances, the focus of attention is on existing sites of pollution, and such technologies are encompassed by the term *bioremediation*. The term is an apt one because a remedy is being applied to a problem. Approaches that deal with waste streams from industry are often considered under the purview of bioremediation, although they might be more appropriately termed bioprophylaxis.

Bioremediation of contaminated sites is a new field of endeavor and many new or altered technologies are appearing; nevertheless, the utilization of microbial processes to destroy chemicals is neither a novel idea nor a new technology. Such processes have been used for decades for the elimination of chemicals from waste streams of industries that had biological-treatment systems for their effluents and, knowingly or unknowingly, for the breakdown of chemicals from households or industries serviced by municipal waste-treatment systems. The fact that many compounds were not so destroyed was not necessarily the result of the absence of a biodegradative microflora but rather that the systems were optimized for different purposes.

The goal of bioremediation is to degrade organic pollutants to concentrations that are either undetectable or, if detectable, to concentrations below the limits established as safe or acceptable by regulatory agencies. Bioremediation is being used for the destruction of chemicals in soils, groundwater, wastewater, sludges, industrial-waste systems, and gases. The list of compounds that may be subject to biological destruction by one or another bioremediation system is long. However, because they are widespread, represent health or ecological hazards, and are susceptible to microbial detoxication, most interest has been directed to oil and oil products, gasoline and its constituents, polycyclic aromatic hydrocarbons, chlorinated aliphatics such as trichloroethylene (TCE) and tetrachloroethylene (also called perchloroethylene or PCE), and chlorinated aromatic hydrocarbons. Although they are not biodegraded, metals are of interest in bioremediation because they can be altered and rendered less harmful by microorganisms. The remediation of metals is considered in Chapter 18.

Certain criteria must be met for bioremediation to be seriously considered as a practical means for treatment. (a) Microorganisms must exist that have the needed catabolic activity. (b) Those organisms must have the capacity to transform the compound at reasonable rates and bring the concentration to levels that meet regulatory standards. (c) They must not generate products that are toxic at concentrations likely to be achieved during the remediation. (d) The site must not contain concentrations or combinations of chemicals that are markedly inhibitory to the biodegrading species, or means must exist to dilute or otherwise render innocuous the inhibitors. (e) The target compound(s) must be available to the microorganisms. (f) Conditions at the site or in a bioreactor must be made conducive to microbial growth or activity, e.g., an adequate supply of inorganic nutrients, sufficient O_2 or some other electron acceptor, favorable moisture content, suitable temperature, and a source of C and energy for growth if the pollutant is to be cometabolized. (g) The cost of the technology must be less or, at worst, no more expensive than other technologies that can also destroy the chemical. None of these criteria is trivial, and none is platitudinous. The failure to meet any one has resulted in a rejection of a biodegradative approach or the inability to meet the cleanup goals that were established.

A major reason for bioremediation being the technology of choice for eliminating organic pollutants is its cost. Although any procedure for removing or destroying pollutants is expensive, biological procedures tend to be the least expensive. Citing potential costs is always difficult because they will vary from year to year and will depend on soil characteristics, identities of the contaminants, fluctuations in labor and equipment costs, the specific location, and other factors. However, a comparison of the values is worthwhile. Some typical costs per metric ton (10^3 kg or 1.1 U.S. ton): land farming, $39 to $88; composting, $44 to $110; soil pile, $99 to $110; and slurry treatment, $88 to $165 (Anderson, 1995). Assuming 1.0 m^3 of soil weighs 1760 kg, the estimate of Cho *et al.* (1997) gives a value of $40 to treat a metric ton of JP-4 jet fuel-contaminated soil by a procedure involving air injection, vacuum extraction, and biodegradation. In comparison, the expenditures for nonbiological processes are estimated to be $220 to $330 for disposal off-site and $330 to $2200 for incineration off-site (Anderson, 1995).

An appreciation of recent activities, in the United States at least, can be obtained from the results of a survey conducted by the U.S. Environmental Protection Agency (Environmental Protection Agency, 1995). The survey includes only those sites described to that agency, and many cleanups were probably not reported. Of the reported sites, 368 were soils, 204 were groundwaters, 33 were sediments, 15 contained sludges, and 15 were surface waters. The types of contaminants (and number of sites containing them) were petroleum (313), solvents (60), wood-preserving wastes (51), pesticides (15), and munitions (4). The technologies (and number of sites using them) were bioventing (108), *in situ* groundwater treatment (103), *in situ* bioremediation of soil (103) and sediment (11), air sparging (40), intrinsic bioremediation (11), *ex situ* treatment with bioreactors (including aerated lagoons, prepared beds, soil piles, and compositing) (116), and *ex situ* treatment with bioreactors (73). The survey included activities that were completed, in operation, or being installed or planned. In addition to the many unreported cleanups, bioremediation at many more locations has been planned, designed, and begun or completed since the survey was conducted.

A variety of different technologies and procedures are currently being used, and a number of new and promising approaches have been suggested or have reached advanced stages of development. Some of these technologies are *in situ* treatments, in which soil is not removed from the field, materials from beaches contaminated with oil are not removed from the site, or groundwater is not pumped for aboveground treatment. *In situ* biomediation has the advantage of relatively low cost but the disadvantage of being less subject to rigorous control. Other bioremediation technologies require removal of the contaminated material in some manner from its original location. These technologies represent *ex situ* treatments. Such removals increase the costs modestly or appreciably, but the processes are more subject to control.

LAND FARMING, PREPARED BEDS, AND SOIL PILES

Microorganisms in soil have a broad array of catabolic activities, and a simple way of destroying pollutants is to add them to the soil and rely on the indigenous microflora. This procedure, often called *land farming* or land treatment, has been used frequently by the oil industry to destroy oily wastes, and it is a procedure that has been utilized for many years. Such an approach has been employed to biodegrade sludges of various types, wastes from manufactured-gas plants, creosote-containing soils, and wastes from industries involved in food processing, pulp and paper production, and leather tanning. It is also used for oily or hydrocarbon-rich materials that are inadvertently spilled on soil. A typical example is the treatment of 127,000 m^3 of soil at a terminal that stored and distributed petroleum. The soil initially contained 2000–75,000 mg petroleum hydrocarbons/kg, but the concentration had declined in two seasons of treatment to an extent that 60% of the land was suitable for residential use, and an additional season presumably would render all the land acceptable for human habitation (Baldwin *et al.*, 1997).

The considerable amount of C added in these wastes has the potential to support a large biomass, but the soil has too little N and P—and possibly other inorganic nutrients—to support such large biomasses, so N and P are added to the soil, often in the form of commercial fertilizers. Furthermore, the O_2 demand of the microflora increases with the added organic C, and the rate of diffusion of O_2 from the overlying air into the soil is too slow to sustain the aerobic bacteria that are chiefly, or solely, responsible. The need for supplemental O_2 is satisfied by mixing the soil in some way, sometimes by simple plowing, sometimes by more thorough mixing. Another common limiting factor for rapid microbial transformation is moisture because surface soil often dries out, so arrangements are made to provide water to maintain optimum moisture levels for aerobic organisms. The pH of the soil is sometimes of concern, and often the desired pH is between 6.0 and 8.0, especially for hydrocarbon degradation. The transformations may still proceed at somewhat more acid conditions. Because the nitrification of N-containing fertilizers increases the soil acidity, the soil pH must be monitored to prevent undesirable conditions for the responsible microorganisms. The effectiveness of this type of remediation in a soil containing diesel oil is shown in Fig. 16.1. Remediation by these means is limited to times of the year when the soil temperature is in a range that permits reasonably rapid microbial growth and activity, and little or no biodegradation occurs during the cold parts of the year in the temperate zone.

For petroleum hydrocarbons, oily materials, and other N- and P-deficient substances for which bioremediation is anticipated, the quantity of N and P to be added is calculated from the amount of C in the material to be degraded. For example, if it is assumed that 30% of the C is assimilated into the biomass of cells carrying out the bioremediation and that the resulting biomass has a C : N : P ratio of 50 : 5 : 1, the amount of N and P to be added would be equivalent to 3 and 0.6% of the C; i.e., for 100 units of substrate-C, 30 units of biomass-C would be formed, and 3 units of N and 0.6 units of P would be needed by that biomass. Such calculations often considerably overestimate the need for N and P because (a) the biomass is itself decomposed, which renders the N and P available once again to further enhance the biomediation, and (b) the soil, sediment, or water will contain some or considerable available N and P for microbial use. Overuse of fertilizer N and P results in an unwarranted expense and may also result in nitrate pollution of ground or surface waters. Hence, an initial laboratory study is often performed to determine the appropriate amount of fertilizer to add. That assessment may include determination of the amount of available N (typically

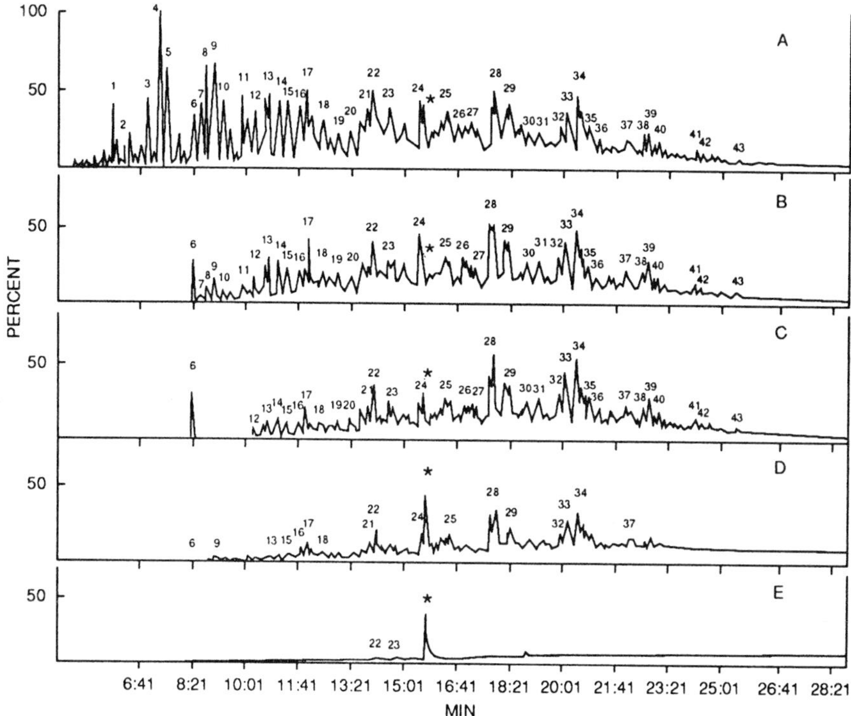

Figure 16.1 Gas chromatograms of the aromatic fraction of diesel oil residue in contaminated soil. Zero time (A); 2 weeks, without (B) and with (C) bioremediation; 12 weeks, without (D) and with (E) bioremediation. The numbers over the peaks designate different PAHs, and the asterisks designate a compound added as a standard for analytical purposes. (From Wang *et al.,* 1990. Reprinted with permission from the American Chemical Society.)

ammonium and nitrate) and available (but not total) P in the soil, sediment, or water. It is not uncommon to find that the biodegradation sometimes proceeds rapidly even without fertilizer, as in some beaches or salt marsh mesocosms contaminated with oil (Venosa *et al.,* 1996; Wright *et al.,* 1997) or that only one of the two fertilizer elements is stimulatory in soil (Chang *et al.,* 1996).

The addition of nutrients to soil to enhance biodegradation, which is sometimes called *biostimulation,* is not always beneficial. On occasion, in laboratory tests, the addition of N inhibits the mineralization of aromatic and aliphatic hydrocarbons (Manilal and Alexander, 1991; Morgan and Watkinson, 1990). The frequency and the explanations for the reduced rate of mineralization are unknown. However, it is possible that more substrate-C is incorporated into the biomass in the presence of higher levels of N and hence less CO_2 is produced; if this explanation is correct, the rate of loss of the substrate may not have diminished but rather the flow of C from substrate may be changed to yield more cells and less CO_2.

The efficacy of land treatment for spills of oil and oil products has been confirmed in carefully controlled experiments in the laboratory and in the field. Thus, the hydrocarbons in gasoline, jet fuel, and heating oil were found to be extensively degraded in soils in the laboratory that were treated with fertilizer, lime, and simulated tilling (Song *et al.*, 1990), and crude oil, crankcase oil, jet fuel, heating oil, and diesel oil disappeared at an enhanced rate in field plots receiving fertilizer, lime, and simulated tilling as compared to soil not receiving these treatments (Raymond *et al.*, 1976; Wang and Bartha, 1990). Similarly, an investigation of 120-m^2 fertilizer plots in the field showed 80% reduction in the amount of oil within 15 months in deliberately contaminated soil (Al-Awadhi *et al.*, 1996), and a study of 135-m^2 plots of soil showed extensive destruction of alachlor present in the soil or applied as a spray to the soil (Dzantor *et al.*, 1993). The results of many unpublished studies are summarized by Loehr and Webster (1997). Such experiments are worth citing because they contain untreated controls and are thus scientifically more convincing. Controls are not included in actual remediations because the issue is not to convince scientists of the response to the treatments but rather to remove the unwanted materials. However, the efficacy if not the scientific rigor is evident in analyses conducted in a field inundated with approximately 1.9 million liters of kerosene. After the initial emergency cleanup operation, the oil content of the soil was 0.87% in the top 30 cm and about 0.7% at 30–45 cm. However, in soil that received 200 kg N and 20 kg P as well as lime, the oil content had declined to $<0.1\%$ in the upper 30 cm, although only to 0.3% at the lower depth (Dibble and Bartha, 1979). On the other hand, some hydrocarbon-rich materials are not readily destroyed by land farming, e.g., sediment contaminated with PCP, creosote (Mueller *et al.*, 1991a), and bunker C oil (Song *et al.*, 1990). A major advantage of land farming is its low cost—for equipment, construction, and operation. However, land farming is slow and requires a reasonably large area. Moreover, organic pollutants (and toxic metals, if present) may leach from the site into the underlying groundwater, and volatile emissions may pose hazards in the vicinity of the site; such leaching or volatilization of toxic constituents or products, if they are present, must be controlled or prevented, or else regulatory authorities might ban the procedure.

Laboratory studies with a view to implementing land farming have focused on problems in soils that have special constraints on bioremediation. These include low temperature and salinity. Low temperatures characterize sites in the arctic and in mountainous areas. These investigations show that bioremediation of diesel oil and jet fuel will take place in soils at temperatures of 10°C, although the process is slow. Nitrogen and P are often but not always needed, with the lack of a response resulting from the availability of sufficient N and P to sustain the slow degradation of the oil constituents (Braddock *et al.*, 1997; Margesin and Schinner, 1997; Walworth and

Reynolds, 1995). The presence of high concentrations of salt in soil has been observed to slow appreciably the bioremediation of oil (Rhykerd *et al.*, 1995).

In a similar technology, more engineering controls are included. The additions include systems to provide irrigation water and nutrients, a liner at the bottom of the soil, and a means to collect leachate. This is termed a *prepared bed reactor* (Fig. 16.2). Either clay or a synthetic material is the liner. These reactors are used at many Superfund sites in which bioremediation is being used, and often the contaminants are polycyclic aromatic hydrocarbons, BTEX (benzene, toluene, ethylbenzene, and xylene), or both (Ryan *et al.*, 1991). The liner and a system to collect leachate are included because of concern that conventional land treatment may result in contamination of the underlying groundwater with the parent compounds or products of microbial transformation that are carried downward with percolating water. The level of sophistication, and consequently the cost, will vary enormously. In some instances, perforated pipes are placed above the liner to collect the leachate, and sand is placed on the liner and over the pipes to improve drainage and hence the ultimate collection of the leachate. The leachate is removed for subsequent treatment, which may be in an adjacent bioreactor. Water and nutrients may be dispensed through an overhead spray-irrigation system, and the entire operation may be enclosed in a plastic greenhouse if volatile hazardous products may be emitted.

This is essentially the method that was used for the treatment of 115,000 m^3 of soil contaminated with bunker C fuel oil (Compeau *et al.*, 1991) and 23,000 m^3 of soil contaminated with gasoline and fuel oil (Block *et al.*, 1990). The treatments were fertilizer, lime, irrigation, and mixing the soil to provide O_2 to the aerobic populations able to degrade the unwanted material. Sometimes the soil is placed in piles that extend laterally for some distance (Hildebrandt and Wilson, 1991). In another illustration of this technology, a soil containing tars was treated in a 150 × 60-m bed that was placed on an impermeable clay liner, and

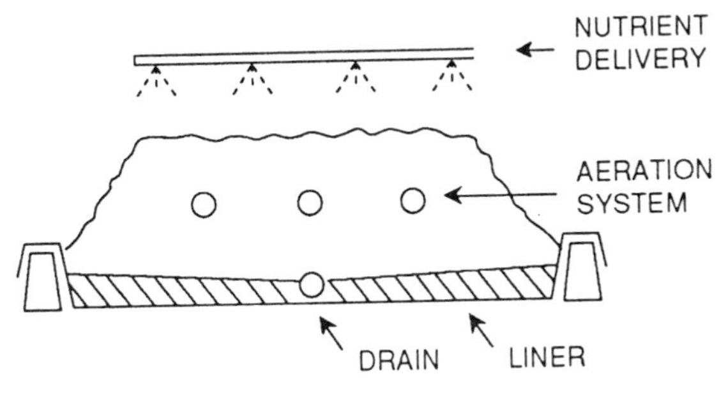

Figure 16.2 Diagram of a prepared bed bioreactor for treatment of excavated contaminated soil. (Reprinted with permission from Fogel *et al.*, 1989. Copyright CRC Press, Inc., Boca Raton, Florida.)

the remediation of PAHs and benzo(*a*)pyrene resulted in a change from average initial concentrations of 1000 and 100 mg/kg to less than 100 and less than 10 mg/kg, respectively (Hyzy and Schepart, 1995). A bench-scale evaluation showed that the TNT level in treated soil was reduced from ca. 4000 mg/kg to <100 and sometimes <1 mg/kg in 12 months when the contaminated samples were alternatively flooded with a dilute solution containing molasses and then exposed to aerobic conditions (Widrig *et al.*, 1997). Many other bench-scale experiments have shown that such technologies are useful for the remediation of oil-contaminated soils (Loehr and Webster, 1997).

Land treatment is also a means to dispose of contaminated water. This is illustrated by a system in which contaminated water from a facility originally designed for treating wood with creosote was introduced into a soil to destroy PAHs. The soil was placed on a polyethylene liner, and the treatment unit was provided with a means to collect leachate. In the first year of operation in the field, 60% of the extractable hydrocarbons, more than 95% of the two- and three-ring PAHs, and more than 70% of the four- and five-ring PAHs disappeared; this destruction occurred chiefly in the first 90 days when the temperatures were still warm (Lynch and Genes, 1989).

A high percentage of the bioremediations that have been completed recently are land treatments by one of these means (Devine, 1992). These successes supplement the several decades of use of conventional land treatment for the destruction of oily and petroleum wastes.

A major concern of the practicing engineer is the scaleup from the bench-scale or even pilot-scale assessment to field conditions. Spatial heterogeneity is the rule rather than the exception in the field, and these heterogeneities in contaminant localization, presence of NAPLs, variations in availability of O_2 and nutrients, and permeability as well as difficulties of mixing large masses of soil with amendments or inocula are not included—and indeed are often scrupulously avoided—in small-scale assessments of the feasibility of land farming or the use of prepared bed reactors. Some laboratory tests are conducted with 5 or 10 g of soil in a tube or flask, but problems of scaleup are issues even with 5- or 10-kg samples (Sturman *et al.*, 1995). Many novel ideas, which appear to be attractive based on these laboratory trials, fail because problems or pitfalls associated with scaleup were not considered.

Microorganisms are sometimes added to soils that are treated by land farming or in prepared bed reactors—the so-called bioaugmentation. The topic has been reviewed in Chapter 15. The practice has been verified as useful in fully engineered bioreactors under controlled conditions, but land farming and even prepared beds are subject to variations in field conditions and have problems of scale not typically considered in the small-scale laboratory tests of inoculation. Moreover, many of the reports of successful remediations associated with bioaugmentation in land farming or prepared bed reactors have lacked an essential experimental component, namely a comparison of all treatments (e.g., aeration, irrigation, mixing, and fertilization) without and with inoculation; therefore, when the remediation was successful, an objective reviewer cannot assess whether the addition of microorganisms was, or was not, beneficial. Some field studies that have included parallel inoculated and uninoculated treatments have failed to show a stimulation resulting from the added inocula, e.g., in field plots of a beach contaminated with crude oil (Venosa *et al.*, 1996). Similarly, adding oil-degrading isolates to a soil receiving a simulated oil spill did not promote hydrocarbon degradation (Radwan *et al.*, 1997). Some data exist that remediation of PCP was enhanced in field plots inoculated with *Phanerochaete sordida* (Lamar *et al.*, 1994), but the same fungus, although able to destroy DDT in culture, did not eliminate the insecticide when tested in pans containing soil naturally contaminated with DDT (Safferman *et al.*, 1995). At the present time, therefore, little convincing evidence exists to show that inoculation aids in soil treatment by land farming or in prepared beds.

A somewhat more sophisticated approach is associated with *soil piles*, which are sometimes called *biopiles*. The soil containing the contaminants is dug up and placed on an impermeable liner that retains contaminated leachate. A piping system is placed in the pile, and air or O_2 is introduced or a vacuum is pulled to enhance aerobic decomposition of the pollutant. A solution containing nutrients is applied to the soil surface to stimulate microbial activity, and the leachate may be collected and recycled through the pile. If the compounds being treated are volatile or if volatile products that are toxic may be formed, the gaseous emissions are collected in some manner, as by the use of activated carbon. Soil piles have been employed for the bioremedation of soils contaminated with hydrocarbons (Stefanoff and Garcia, 1995) and PCP (Buchanan *et al.*, 1997), and 4-m^3 soil piles have been found to be an effective means of destroying RDX and HMX in munitions-contaminated soil (Greer *et al.*, 1997).

A number of bench-scale investigations have shown that low concentrations of certain surfactants, usually anionic or nonionic compounds, stimulate the biodegradation of hydrocarbons (Aronstein *et al.*, 1991; Aronstein and Alexander, 1993; Jahan *et al.*, 1997) or DDT (You *et al.*, 1996) sorbed in soil. Recent field studies with guanidinium cocoate have demonstrated that this anionic surfactant stimulates the biodegradation of hydrocarbons in soil (Nelson *et al.*, 1996; Walter *et al.*, 1997). Nevertheless, the inclusion of surfactants in a bioremediation must be approached carefully because some are toxic if used at high concentrations (Deschênes *et al.*, 1996), some may increase the need for O_2 because of their own biodegradability (Sanseverino *et al.*, 1994), and some fail to be stimulatory at economically reasonable concentrations.

An approach similar to that used in land farming can be employed to bioremediate wetlands polluted with oil. Evidence for the utility of the microbial elimination of this type of pollution comes from the cleanup of an experimental spill of light crude oil added to wetland sediments. The slow biodegradation resulting from the indigenous microflora was stimulated by the addition of nutrients and was further enhanced by the addition of nitrate as supplemental electron acceptor (Mills *et al.*, 1997).

Nearly all evaluations of the effectiveness of bioremediation rely on a reduction in the concentration of the contaminant. However, a diminution or elimination in toxicity or uptake by test species has also been established. For example, the toxicity of a heavy, medium, and light oil to the earthworm *Eisenia foetida* in soil was progressively reduced and eventually disappeared as bioremediation proceeded (Salanitro *et al.*, 1997). Similar results were obtained as a consequence of the bioremediation of a PAH-contaminated soil (Hund and Traunspurger, 1994). The toxicity to wheat and oats growing in the soil or to seed germination is also reduced or abolished as a result of the treatment (Salanitro *et al.*, 1997; Hund and Traunspurger, 1994; Wang and Bartha, 1990). However, metabolic products formed during the bioremediation may sometimes be toxic, but this toxicity commonly disappears as the degradation proceeds further (Wang and Bartha, 1994).

PHYTOREMEDIATION

A technology for remediation that has attracted considerable attention involves the use of higher plants that, directly or indirectly, result in the removal or degradation of organic pollutants. This technology, which has been of interest for contaminants in soil and, to a lesser extent, for chemicals in shallow sediments, is known as *phytoremediation*. Phytoremediation includes processes that may involve uptake of the contaminant by the plant or biodegradation by microorganisms colonizing the root or the soil immediately next to the root system (Cunningham *et al.*, 1996). However, the present discussion will only be concerned with phytoremediation resulting from microorganisms on root surfaces or in the adjacent soil.

The portion of soil intimately associated with the roots of growing plants is known as the *rhizosphere*. The rhizosphere includes the immediate surfaces of the roots, which are extensively colonized by microorganisms, especially bacteria; this region is sometimes termed rhizoplane, but the term rhizosphere is commonly considered to include the root surface as well as the adjacent soil. The rhizosphere does not extend far from the root surface, however, because the physical and chemical factors that make this zone unique are spatially limited. Probably the chief factor that distinguishes the rhizosphere is the continuous supply of low-molecular-weight organic compounds excreted by the roots. These compounds serve as a source of readily available carbon and energy that sustains a large community of bacteria. In contrast, the carbon sources in nonrhizosphere soil are high-molecular-weight, poorly available substances that are only slowly utilized and support a community of bacteria and fungi that are not as metabolically active. In addition to being a large and metabolically more active community of bacteria continuously receiving excretions to support their growth and activity, the rhizosphere microflora resides in a region with different O_2 and inorganic nutrient concentrations and dissimilar physical properties than encountered by the soil microflora. Of particular importance in considering the plant species for phytoremediation is the fact that the size, activity, and species composition of the rhizosphere community as well as the volume occupied by the rhizosphere vary appreciably with the plant species.

Several studies under controlled conditions have shown the beneficial influence of plants. In one investigation, for example, a defined mixture of alkanes and PAHs was added to pots containing 400 g of soil, which were then planted with ryegrass (*Lolium perenne*). The rate and extent of loss of total hydrocarbons were enhanced by the ryegrass (Table 16.1), although no marked beneficial effect on PAH degradation was evident (Günther *et al.*, 1996). However, following the deliberate addition of a mixture of benzo(*a*)anthracene, chrysene, benzo(*a*)pyrene, and dibenz(*a,h*)anthracene to a soil and the mixed seeding of eight deep-rooted prairie grasses, more of these PAHs disappeared from the planted than the unplanted soil (Aprill and Sims, 1990). Similarly, the rate of degradation of pyrene added to soil in the greenhouse was enhanced by the presence of growing fescue, sudan grass, switch grass, or alfalfa (Schwab and Banks, 1994), and several species of plants enhanced the disappearance of 2-chlorobenzoic acid when tested in growth chambers (Siciliano and Germida, 1997). In a similar study in which pyrene and anthracene were the test compounds and plants were cultivated in pots of soil, the concentrations of the two PAHs declined from the initial value of 100 to <14 mg/kg in 4 weeks in the presence or absence of the plants, but the subsequent degradation was promoted by the plants. No differences in the effects of the plants were evident after 24 weeks (Reilley *et al.*, 1996).

Table 16.1

Effect of Ryegrass on Hydrocarbon Degradation in Soil[a]

	Total hydrocarbons (mg/kg)	
Weeks	Unplanted	Planted
0	4330	4330
5	3690	2140
12	2150	605
17	1270	223
22	792	112

[a] From Günther *et al.* (1996).

Mineralization of ^{14}C-labeled compounds has also been used to demonstrate the beneficial effects. The earliest tests were conducted with pesticides. Hsu and Bartha (1979), for example, found that 12.9 and 17.9% of [^{14}C]diazinon and [^{14}C]parathion were mineralized in 1 month in soil supporting the growth of bush beans, whereas 5.0 and 7.8%, respectively, were mineralized in vegetation-free soil. Enhancement of parathion mineralization also occurs during the growth of rice, both in flooded and nonflooded conditions (Reddy and Sethunathan, 1983). More recent investigations have confirmed the enhancement of mineralization of ^{14}C-labeled compounds. Thus, in a 20-week period, more PCP was mineralized in the presence of crested wheatgrass (*Agropyron desortorum*) than in the absence of the plants (Fig. 16.3). With ^{14}C-surfactants as the test compounds, the rate of mineralization was 1.1- to 1.9-fold faster in the rhizosphere of corn and soybeans grown in the laboratory than in nonrhizosphere soil (Knaebel and Bestal, 1992).

Further evidence comes from a study in which the addition of a 2,5-dichlorobenzoate-metabolizing strain of *Pseudomonas fluorescens* caused the disappearance of that compound (initially added to give 10 mg/kg) in 3 days in the presence of beans (*Phaseolus vulgaris*), whereas appreciable amounts remained in the inoculated soil not supporting the plants (Crowley *et al.*, 1996). Phytoremediation may even be feasible in aquatic sediments, as suggested by a laboratory study in which it was observed that mineralization of two ^{14}C-labeled surfactants (a linear alkylbenzene sulfonate and a linear alcohol ethoxylate) was more rapid in the rhizosphere of cattails (*Typha latifolia*) grown in sediment than in root-free sediments (Federle and Schwab, 1989). A number of assays have been conducted in which the rates of mineralization of 2,4-D, 2,4,5-T (Boyle and Shann, 1995), atrazine (Perkovich

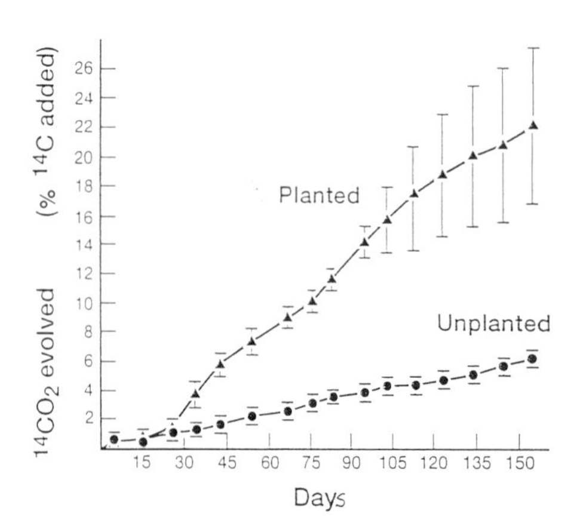

Figure 16.3 Mineralization of ^{14}C-PCP in the presence and absence of crested wheat. (Reprinted with permission from Ferro *et al.*, 1994.)

et al., 1996), and other herbicides (Anderson *et al.*, 1994) have been compared in soil taken from rhizosphere and nonrhizosphere soil. However, the practical significance of such information is difficult to assess because the plants were no longer present, and thus the effect of plant growth and root excretion cannot be evaluated.

Plants may also favor the survival of bacteria introduced into soil for the purposes of destroying unwanted compounds. This benefit is evident from the better survival in the corn rhizosphere of a mixture of three bacterial species that mineralized atrazine (Alvey and Crowley, 1996) and the larger numbers of a 2,5-dichlorobenzoate-utilizing strain in the bean rhizosphere than in unplanted soil (Crowley *et al.*, 1996).

The influence of growing plants is sometimes marked, but sometimes the stimulation is small. However, sometimes a benefit in having a plant is not evident. An illustration is the lack of stimulation by corn of the mineralization of atrazine by indigenous microorganisms or by a mixture of three bacteria that converted this herbicide to CO_2 (Alvey and Crowley, 1996).

Although data on phytoremediation come largely from laboratory and greenhouse tests, field tests are being conducted (Schnoor *et al.*, 1995). The promise of this technology is suggested by a 3-year field study in which it was found that naphthalene concentrations declined more rapidly and to a greater extent in soil planted to buffalo grass (*Buchloe dactyloides*) than in fallow soil. The stimulation was restricted to the surface layer, however (Qiu *et al.*, 1997). Final decisions on the utility of phytoremediation must await the outcome of additional field investigations.

The reasons for the observed enhancements of biodegradation in the rhizosphere are as yet unknown. Possibly the greater rate is a result of the larger bacterial mass in soil near roots than at a distance away or it may be a consequence of the greater metabolic activity of the individual organisms in the rhizosphere. However, the simple organic compounds that are constantly excreted by the roots may serve as supplementary carbon and energy sources to promote the activity of species that are either capable of growing on the compounds of interest or, for chemicals that are cometabolized, such as diazinon (Hsu and Bartha, 1979), serve to supply the microorganisms with the nutrients they require to grow while metabolizing the substrate that is cometabolized.

Plants vary in the extent to which they are stimulatory. Some appear to have a positive effect, but others provide little or no benefit. Undoubtedly some will depress biodegradation. Thus, the choice of the correct plant species is of paramount importance. Unfortunately, little information exists on which to base choices. Clearly, the plant must be able to grow in the presence of pollutants that may be deleterious to many other types of vegetation and must tolerate local conditions at the site, e.g., toxic metals, NAPLs, poor drainage, high salt concentrations, and extremes of pH. The ability to grow rapidly and being a perennial rather than an annual would also offer advantages. Of special importance is the pattern of rooting because it would seem particularly desirable for the root system to explore a large soil volume, which could be achieved with deep roots and by a dense, fibrous root system rather than by shallow tap roots. Should the contaminants be at a depth below which there is a low density of roots, phytoremediation probably would not be a useful technology, and root density is typically high near the soil surface and diminishes markedly with depth.

A major advantage of phytoremediation is its low cost compared to many other biological technologies and its far lower cost than nonbiological approaches. It is believed to offer promise for sites where the contamination is near the surface, possibly only in the top 1–2 m or possibly somewhat deeper (Cunningham *et al.*, 1996; Schnoor *et al.*, 1995). The types of compounds that are more likely to be destroyed by phytoremediation are still unclear, but pollutants that are strongly sorbed or that have become aged or sequestered may be resistant. This technology probably will not be useful if phytotoxicity prevents the plants from rooting extensively, if the contaminants are not readily bioavailable or leach quickly out of the rooting zone, or if the site under consideration is O_2 deficient. Some of these limitations may be overcome if the high pollutant concentrations are first removed or active microbial bioremediation proceeds to a point at which O_2 diffusion allows the return of aerobiosis or the phytotoxicity is reduced before introducing the plants. The time for cleanup may be a major limitation because phytoremediation will usually be slow, but if time is not a constraint, the low operating costs may make phytoremediation the technology of choice.

BIOVENTING AND BIOSPARGING

A technology that has attracted considerable attention is *bioventing*. This approach to solid-phase treatment of contaminants relies on methods of introducing air into soil above the water table (the *vadose* zone or the unsaturated portion of soil), which thereby provides the O_2 needed as the terminal electron acceptor for aerobic bacteria. The air is introduced either by a vacuum extraction method or by forcing air into the soil under positive pressure. Appropriate withdrawal or air-injection wells are the ways by which the vacuum or positive pressure is applied to the soil. The procedure is attractive because it operates *in situ* and because of the little equipment required.

Bioventing has found application particularly for hydrocarbon remediation. However, compounds with high vapor pressures may volatilize too quickly and thus would not be degraded. Soils of low permeability generally are not suitable because the air may not move through the soil sufficiently rapidly to provide enough O_2 to sustain active metabolism by aerobes. Moreover, care must be exercised when air injection is used to prevent the introduced air from spreading volatile compounds into portions of soil not already contaminated.

Bioventing, where useful, can be performed at reasonably low cost, particularly if there is little or no need to treat offgases released from the site. For example, the cost of treatment of a large diesel spill was estimated to be \$10/m^3 of contaminated soil, although the treatment would be somewhat more expensive per unit volume if the amount of contaminated soil was small (Downey *et al.*, 1995).

A full-scale evaluation demonstrated the efficacy of bioventing for the remediation of a spill of diesel oil that contaminated 11,500 m^3 of soil to a depth of 20 m. In the most contaminated locations, the concentration of diesel oil had been reduced 55–60% in a 2-year period. Some loss of the contaminants was the result of volatilization, but >90% was attributed to microbial action (Downey *et al.*, 1995). Other assessments have confirmed the usefulness of the approach for the degradation of fuel oil, motor oil, and individual aromatic hydrocarbons, both in the field and in the laboratory (Gan and Wright, 1995; Gruiz and Kriston, 1995; Baker *et al.*, 1994).

Biosparging entails a similar approach, but the air is introduced into the saturated zone, i.e., below the water table. The purpose is not only to provide O_2 but also to transfer volatile pollutants into the overlying unsaturated (vadose) zone, which usually contains a population of microorganisms able to degrade those compounds. In addition, some of the biodegradation will occur within the aquifer in response to the added O_2. As with bioventing, the flow rate of air should not be so great that volatile organic compounds pass through the vadose zone without decomposition since they will then enter the atmosphere.

Biosparging with vapor extraction was found to be successful in lowering the concentration of JP-4 jet fuel at a site with contaminated soil and groundwater to a depth of 12 m, and the concentrations of fuel oil in the groundwater and soil were reduced up to 97 and 46%, respectively (Strzempka *et al.*, 1997). In sites with BTEX in the groundwater, biosparging represents a way of enhancing the microbial destruction of these aromatic compounds (Griffin *et al.*, 1993; Strzempka *et al.*, 1997). Modifications of the usual methods of sparging with air include the use of O_3 in place of air (Nelson *et al.*, 1997) and sparging with steam to increase the temperature in the vadose zone and the upper part of the groundwater (Dablow *et al.*, 1997), both of which enhance biodegradation.

COMPOSTING

In composting as a treatment procedure, the polluted material is mixed together in a pile with a solid organic substance that is itself reasonably readily degraded, such as fresh straw, wood chips, wood bark, or straw that had been used for livestock bedding. The pile is often supplemented with N, P, and possibly other inorganic nutrients. The material is placed in a simple heap, it is formed in long rows known as windrows, or it is introduced into a large vessel equipped with some means of aeration. Moisture must be maintained, and aeration is provided either by mechanical mixing or by some aeration device. Aeration may be provided by a simple blower or by introducing a more complicated air-distribution system beneath the open pile. If the aeration results in the release of volatile toxicants, they must be trapped so that no such compounds are emitted to the atmosphere. A contained vessel is desirable when the compost contains hazardous chemicals. Heat released during microbial growth on the solid organic material is not adequately dissipated and hence the temperature rises. The higher temperatures (50–60°C) are often more favorable to biodegradation than the lower temperatures that are maintained in some composts. With some hazardous materials, however, the temperature does not reach 50°C.

Composting has been used as a means of treating soil contaminated with chlorophenols. The concentrations of the various chlorophenols present in the composted material, which is placed in windrows in the field, decline markedly during the summer months, when the temperature in the compost is high, but the conversion is slow during the cold part of the year (Valo and Salkinoja-Salonen, 1986). A field-scale demonstration has also shown that the concentrations of TNT, RDX, and HMX in contaminated sediments placed into composts declined to a marked extent as these three explosives were biodegraded (Ziegenfuss *et al.*, 1991). A laboratory-scale test of the compositing of TNT suggests that an initial anaerobic treatment before the aerobic phase of composting is desirable because of the more complete destruction of the explosive (Breitung *et al.*, 1996). Bench-scale assessments also show extensive degradation of PAHs containing two, three, and four rings but not five- and six-ring PAHs (Potter *et al.*, 1997).

IN SITU GROUNDWATER BIORESTORATION

A common procedure for *in situ* bioremediation entails the introduction of nutrients and O_2 into subsurface aquifers, relying on the indigenous microflora to destroy the unwanted molecules. This process is sometimes called *biorestoration*. Most of the contaminated sites treated so far contain petroleum hydrocarbons as the contaminants. Leakages from underground storage tanks containing gasoline result in the appearance of benzene, toluene, ethylbenzene, and xylenes. Although these BTEX compounds are initially in the gasoline phase, particular attention is given to them because they are toxic and because they enter the aqueous phase in the form of a sustained release. This sustained release and the amounts present in the aqueous phase are consequences of their reasonable water solubilities and the constant partitioning of these compounds from the gasoline to the aqueous phases. Groundwaters contaminated with diesel fuel and JP-4 aviation fuel are also treated in similar manners.

Initially, as much of the free oil or hydrocarbon as possible is removed by one of several physical means. Bioremediation without such removal is pointless because the bulk source would continue adding new chemical to the groundwater. Laboratory tests are also conducted to determine the optimal amount of nutrients to add; this is especially important to avoid too little or too much being supplied. Too little will result in a slow transformation, and too much might clog the aquifer because of the large biomass that is formed, thereby causing cessation of the remediation. The three nutrients that are commonly required for optimal activity are N, P, and O_2. These are typically the factors that limit the activity of the indigenous microflora. The N and P salts are usually dissolved in the groundwater that is circulated through the contaminated site. A common procedure is to add the nutrients in solution through injection wells into the saturated zone or through infiltration galleries into the unsaturated or surface-soil zone (Fig. 16.4). The water is recovered from production wells, and that water is again amended with nutrients and recirculated. The concentrations of contaminants and nutrients are often measured on a regular basis by taking samples from wells that are installed between the points of injection and removal. In some instances, the water is not recirculated, but, instead, is disposed at the surface (Thomas and Ward, 1989).

The rapid biodegradation of hydrocarbons typically is carried out by aerobic bacteria, and their activity must be sustained. This poses major problems because even under the best conditions, little O_2 is present in groundwater and natural sources provide O_2 at exceedingly slow rates. The same problem of poor O_2 solubility applies to the water added during the remediation itself. For example, for the degradation of a 4000-liter spill of hydrocarbons, 5000 kg of O_2 is needed. An enormous volume of water mixed with air to give up to 8 mg O_2/liter would have to be pumped through the aquifer to provide the requisite O_2. Even if the introduced water is saturated with pure O_2 (to give 40 mg O_2/liter) rather than air, approximately 110 million liters of water would be needed for the 4000-liter spill. As a result, H_2O_2 is sometimes added to the nutrient solution as a source of O_2 (McCarty, 1988). H_2O_2 is highly water soluble and it slowly breaks down in the aquifer to give free O_2. However, care needs to be taken because H_2O_2 is toxic to some species at 100 to 200 mg/liter. The problem of toxicity can be minimized or avoided if a low H_2O_2 concentration is added initially, possibly 50 mg/liter, and then the amounts can be stepwise increased to give 1000 mg/liter (Thomas and Ward, 1989; Wilson and Ward, 1987). The method of enhancing the activity of aerobes may create difficulties, however, whether the enhancement is provided by air, O_2, or H_2O_2. Inorganic Fe or Mn may precipitate because of the increased oxidation–reduction potential associated with the method of promoting aerobic activity or the biomass will become larger, both of which may reduce permeability of the aquifer. Alternatively, the decomposition of H_2O_2 may result

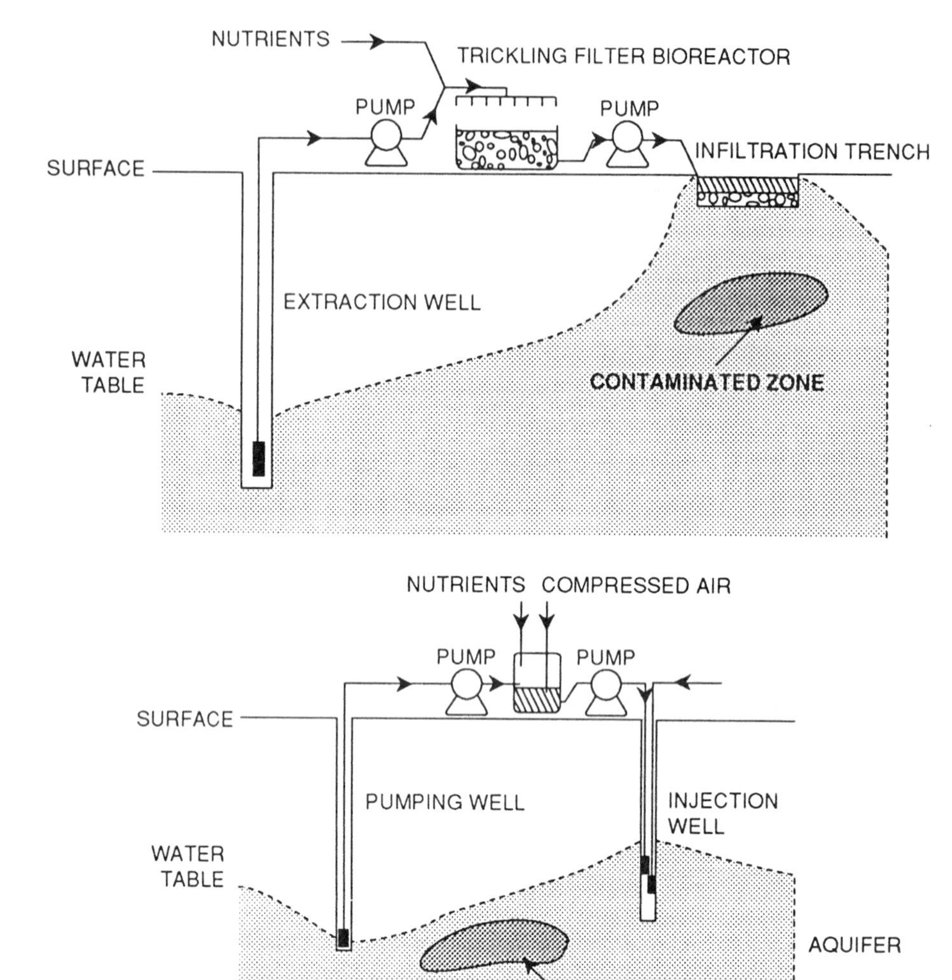

Figure 16.4 *In situ* groundwater treatment using an infiltration trench (top) and an injection well (bottom). Note the use of an aboveground trickling-filter bioreactor. (From Morgan and Wilkinson, 1989. Reproduced with permission from Elsevier Science Publishers.)

in gas formation within the aquifer, which may also reduce the permeability of the aquifer (Wiesner *et al.,* 1996).

The success of biorestoration depends on the hydrogeology of the site. If the hydrogeology is complex, success is problematic, and bioremediation sometimes will be of dubious value. Adequate procedures to characterize many sites are not currently available, and thus the likelihood of success at complex sites is questionable. Moreover, the subsurface environment must be sufficiently permeable to permit the transport of the added N, P, and O_2 to the microorganisms situated at the various subsurface sites containing the contaminants. This water movement, referred to as hydraulic conductivity, is critical to a positive outcome (McCarty, 1991; Thomas *et al.*, 1992). Flow rates of groundwater $>10^{-4}$ cm/sec (or >30 m/year) are considered to be conducive to successful bioremediation, but such values of hydraulic conductivities necessary for treatment are not absolute.

Such procedures have successfully destroyed a number of pollutants at a variety of sites. For example, in a 3-year treatment of groundwater contaminated by an oil- and gas-production facility, the average BTEX and benzene concentrations fell from 210 to 4 and from 115 to 0.6 μg/liter, respectively (Chang *et al.*, 1997). Similarly, the groundwater at a former retail gasoline facility was essentially freed of petroleum hydrocarbons after 10 months of treatment with solutions containing H_2O_2 and ammonium and phosphate salts (Fogel *et al.*, 1989). A site contaminated with ethylene glycol was freed of detectable levels in 26 days (Jerger and Flatham, 1990), and a 94% or greater decline in PCP concentration in the top 10 m was achieved in a pilot-scale test for the remediation of groundwater at a site where PCP had previously been used to treat wood and wood products (Fu and O'Toole, 1990). The microbial populations and their activity may remain high in treated areas even 2 years after the *in situ* bioremediation has ended (Thomas *et al.*, 1990).

Similar procedures have been used to enhance the biodegradation of TCE in the field. In one case, methanol as an energy source for TCE-degrading microorganisms as well as inorganic nutrients were added in a system that contained injection and extraction trenches, and the resulting rate of TCE degradation was ca. 10% per month (Litherland and Anderson, 1997). In another case, methane and O_2 were added alternately at concentrations of 20 and 32 mg/liter, respectively, the two additions serving to stimulate indigenous CH_4-oxidizing bacteria (methanotrophs) that cometabolize TCE. Other nutrients were not added. From 20 to 30% of the TCE was destroyed in the test period (P. L. McCarty, cited by Thomas and Ward, 1989).

In situ treatments involving nutrients and one or another source of O_2 are utilized in many field operations (Devine, 1992). A number of new methods for providing O_2 have been devised recently. Some rely on the slow decomposition of magnesium peroxide (MgO_2), calcium peroxide (CaO_2), or urea hydrogen peroxide, the O_2 thereby being provided at a slow and sustained rate. For example, MgO_2 placed in some solid phase will slowly break down according to this reaction:

$$MgO_2 + H_2O \rightarrow Mg(OH)_2 + \frac{1}{2} O_2$$

In a typical case, wells are installed in the path of flow of the groundwater containing the pollutants. Concrete briquets containing MgO_2 are then introduced into those wells. In operation, this approach resulted in a decline in BTEX concentration from 17 to 3.4 mg/liter as the groundwater moved through the permeable material. Nevertheless, some operational problems remain to be resolved (Borden *et al.*, 1997; Chapman *et al.*, 1997). Systems using CaO_2 (Waltz and Ricotta, 1997) and MgO_2 are being evaluated for bioremediation of petroleum hydrocarbons.

Some bacteria are able to use, in addition to O_2, nitrate as an electron acceptor, and they thus grow and degrade a number of substrates in anoxic waters provided with nitrate. Nitrate is attractive because of its high solubility in water and low cost, although caution must be exercised because nitrate, if present in drinking water at levels in excess of 10 mg/liter (as N), is a pollutant itself. In addition, bubbles of the gaseous end product of the complete reduction of nitrate (N_2) may displace water in pores within the aquifer and thus reduce hydraulic conductivity. However, nitrate has been used successfully as an electron acceptor for the bioremediation of BTEX in an O_2-deficient aquifer contaminated with fuel. Groundwater removed from the aquifer was supplemented with KNO_3 (as well as N and P as nutrient sources) and reintroduced to the site of contamination through an infiltration gallery. The concentration of BTEX in an extraction well declined by almost 99% as a result of the stimulation of bacteria able to use the nitrate as an alternate acceptor (denitrifiers) when O_2 levels were low or had fallen to zero (Gersberg *et al.*, 1995). A field test was also conducted at a location in Michigan contaminated with JP-4 jet fuel to determine the efficacy of nitrate. Although the concentration of each constituent of BTEX was reduced from 760, 4500, and 840 to <1, <1, and 6 μg/liter for benzene, toluene, and ethylbenzene but only to 20–40 μg/liter for total xylene isomers, the design of the test did not permit separating the degree of biodegradation associated with O_2 and nitrate as electron acceptors (Hutchins *et al.*, 1991). Nitrate may also be supplied in permeable briquets placed in trenches situated perpendicular to the direction of groundwater flow, and the electron acceptor would then be released slowly. A laboratory-scale assessment showed biodegradation of toluene, ethylbenzene, and xylenes but not benzene under the denitrifying conditions that were established (Kao and Borden, 1997).

Sulfate and ferric iron also serve as electron acceptors for some bacteria, and in the presence of one or the other of the ions, certain organic substrates are metabolized. Both are important in the biodegradation of natural products and probably also in the destruction of synthetic compounds. Nevertheless, the potential use of sulfate for an engineered bioremediation is limited because the end product of its reduction, H_2S, is toxic to microorganisms, humans, higher animals, and plants. The use of ferric iron as an electron acceptor for field-scale cleanups has not been explored because of the problems of dissemination of this insoluble material through a contaminated aquifer.

As with land farming, inoculation/bioaugmentation is sometimes practiced for *in situ* groundwater restoration. In nearly every instance, controls were not performed so the efficacy of the practice cannot be evaluated. One might say that because it causes no harm and possibly could be beneficial, why not do it? The arguments to be made against such an approach include the additional cost, abetting the dishonesty associated with some of the claims, and the intellectual uneasiness associated with the use of procedures that have rarely been evaluated appropriately. Nevertheless, some short-term tests show stimulation of biodegradation of TCE and carbon tetrachloride in small-scale field tests. However, even some of these tests show problems with the approach, including plugging of aquifers by high cell densities, the lack of movement of appreciable numbers of the cells away from the point of their introduction, and the disappearance of the beneficial effect of inoculation after several weeks (Bourquin *et al.*, 1997; Duba *et al.*, 1995; Dybas *et al.*, 1997).

INTRINSIC BIOREMEDIATION

For almost five decades, it was known that certain synthetic compounds added to soil disappeared with time and that this activity was the result of microbial action. Similar information was available for surface waters. If the rate of disappearance was sufficiently rapid that humans or higher animals and plants were not exposed, concern with pollution was not aroused. The rate of disappearance may have been slow or rapid, but the absence of exposure of higher organisms to the chemical was considered to represent an acceptable way for the elimination of the chemical. However, the emission of industrial pollutants in vast quantities, the leakage of large volumes of liquids from storage tanks, and the deliberate or inadvertent dumping of wastes into soils and waters brought about the current emphasis on engineered remediation. Nevertheless, provided that potential receptor populations of higher organisms are not exposed, the natural microbial processes remain a perfectly reasonable option for eliminating toxicants.

Recent years have witnessed a marked rise in interest in these natural processes. In line with the vogue to create new terms, these natural phenomena are variously termed *intrinsic bioremediation, natural bioremediation, passive bioremediation,* or *bioattenuation*. All the terms refer to the same thing, allowing indigenous microorganisms to use the supply of nutrients and electron acceptors in the ambient environment and grow on, or possibly cometabolize, the pollutants and cause their destruction. The obvious advantage: low cost.

For intrinsic bioremediation to be an appropriate option for cleanup, a risk assessment—either formal or informal—is needed. It must be shown that adverse effects on humans, higher plants or animals, and natural ecosystems are not likely to occur. Intrinsic bioremediation may require that the source of the pollutant is first removed to prevent additional entry and spread into the surroundings or that some system is instituted to prevent further migration of a contaminant plume in an aquifer. A long-term monitoring program is often required to confirm that biodegradation is, in fact, taking place and that further movement of the pollutants is not occurring. This program may entail measurements of the contaminants as well as demonstrating microbial activity, as by showing that the levels of dissolved O_2 or nitrate are less than the background concentration or that products of biodegradation are appearing (Hooker and Skeen, 1996). The option of intrinsic bioremediation thus results in low costs, but it does require considerable effort, action, and moderate expenses.

MARINE OIL SPILLS

Interest in the cleanup of oil spilled in marine, estuarine, and fresh waters and in the use of microorganisms for freeing the adjacent shorelines of oil has existed for many years. Early studies showed that hydrocarbon-oxidizing bacteria were widespread, that they were limited by N and P when oil was introduced into the water, and that formulations containing oleophilic fertilizers were particularly beneficial. However, it was not until the tanker Exxon Valdez was grounded on a reef in Prince William Sound, Alaska, that a major bioremediation of oil in surface waters was undertaken. The March 24, 1989 grounding of this oil tanker released 42 million liters into the sound, contaminating both the water and adjacent beaches. As a result, a program was immediately mounted to determine, both in the laboratory and in the field, how to promote the biodegradation. The subsequent cleanup represents the largest field bioremediation ever undertaken.

Laboratory investigations conducted shortly after the spill confirmed the early studies that N and P were limiting and that almost all of the alkanes in the Alaskan oil and an appreciable amount of the PAHs were metabolized in 6 weeks with an inorganic salts solution or an oleophilic fertilizer containing N and P. Far less hydrocarbon loss was evident with no added N and P (Tabak *et al.*, 1991). Field tests confirmed the abundance of hydrocarbon-degrading bacteria. Specific N and P fertilizers were chosen to be added to the beaches because they would remain associated with the oil, a critical issue in Prince William Sound due to water movement associated with tidal activity and the occasional strong storms. The oleophilic fertilizer chosen was a liquid containing urea in oleic acid as the N source and tri(laureth-4)-phosphate as the P source. Although this oleophilic preparation remained with the oil on the beach surface, little penetrated to the underlying sediment; therefore, an encapsulated formulation of N and P was added to stimulate the subsurface microflora. Within 2 weeks, differences in the quantities of oil were visually evident between fertilizer-treated and untreated beaches, and subsequent quantitative measurements revealed that 60–70% of the oil was degraded within 16 months of the spill. Even in the absence of the fertilizer, biodegradation occurred, but the added N and P promoted the process. In general, however, the oil was biodegraded three to more than five times faster as a result of fertilizer additions (Lessard *et al.*, 1995). As a result of the positive response to N and P, more than 117 km of shoreline was treated in 1989, additional locations were fertilized in 1990, and those isolated areas still containing oil were treated in 1991. Although oil constituents were degraded with no added N and P as the indigenous bacteria used the N initially present and that subsequently made available as the initial populations died, an enhancement in rate is especially important in a region, such as Alaska, in which only a few months are warm enough for appreciable microbial decomposition to occur (Prince, 1992; Pritchard and Costa, 1991). On the beaches that were treated, O_2 deficiencies were not deemed to be of practical significance because of wave action with aerated water in the sound. The data from the inadvertent spill in Alaska were confirmed in a controlled study in which crude oil was deliberately added to plots established on beaches on the shore of Delaware Bay, the results showing the benefit resulting from fertilization (Venosa *et al.*, 1996).

The response of microorganisms to N, P, and O_2 or the effectiveness of biodegradation is simple to determine in the laboratory. It is also easy to assess in a bioreactor in which measurements can be made of the inflow to the reactor and the outflow from it. The same is not true in an *in situ* bioremediation. The pollutants may disappear from the site as a result of volatilization from the soil or water or merely by dilution in flowing water. A number of methods have been proposed to show that biodegradation has occurred or that a deliberate bioremediation has been successful in the field (Madsen, 1991). The issue of confirming effectiveness of a bioremediation was particularly acute at Prince William Sound because of the enormous cost for the cleanup of the large affected area.

One way of assessing microbial activity in the field is to determine the changes in concentration with time of two substances that have similar physical and chemical behaviors in nature but only one of which is degraded reasonably quickly. Both might be subject to abiotic degradation or volatilization, but a finding that the more biodegradable compound disappears more readily strongly suggests microbial degradation. This assessment may be accomplished by comparing the rate of disappearance of linear alkanes having 17 or 18 C atoms with highly branched alkanes of similar molecular weights, such as pristane and phytane. The linear alkanes usually disappear far more quickly than the highly branched hydrocarbons. Tests following the Exxon Valdez spill showed that the ratio of the straight-chain 18-C alkane to phytane did indeed change with time (Fig. 16.5). However, once biodegradation is extensive, even most of the branched alkanes are destroyed by the microflora. Hence, another internal standard for bioremediation is necessary. For this purpose, a still less degradable standard is needed, and a highly resistant five-ring saturated hydrocarbon was chosen, namely $17\alpha(H),21\beta(4)$-hopane. Hopane is not one of the PAHs, which are, by definition, unsaturated. Tests with hopane as an internal standard confirmed biodegradation of the oil from the spill in Prince

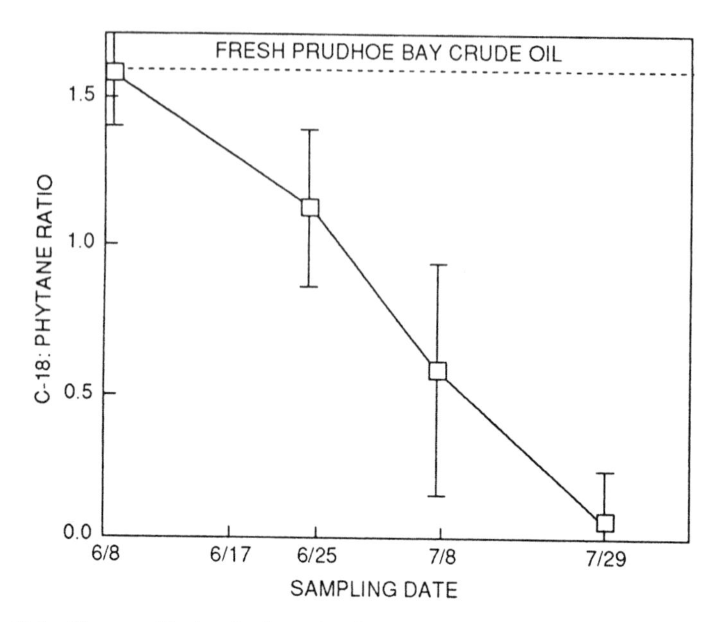

Figure 16.5 Change with time in the ratio of octadecane to phytane (on a mg:g weight basis) following the application of an oleophilic fertilizer on June 8, 1989. (Reproduced with permission from Pritchard and Costa, 1991.)

William Sound (Lessard *et al.*, 1995) as well as crude oil deliberately added to an experimental site (Le Dréau *et al.*, 1997).

Other means have been proposed to confirm *in situ* biodegradation. One makes use of the facts that the ratio of stable C isotopes (^{13}C and ^{12}C) in the CO_2 produced biologically in mineralization differs from that associated with abiotic processes and that the relative rates of utilization of ^{12}C and ^{13}C in organic compounds differ in biotic and abiotic processes (Aggarwal and Hinshee, 1991). Field use has not been made of the method employing stable C isotopes.

Nearly all of the successful bioremediations using the techniques described earlier rely on the actions of indigenous microorganisms. Although proprietary concerns in commercial practice usually preclude the release of information on procedures and sometimes of results, little information exists in the peer-reviewed literature suggesting that inoculation (or bioaugmentation) is needed for the previously discussed technologies. The claims by some entrepreneurs of remarkable successes by undescribed organisms under undefined operating conditions must be taken as just that—claims. Undoubtedly, some such inoculations are indeed beneficial, but better information is needed before conclusions can be reached. However, no question exists that some of the treatments carried out in bioreactors benefit, modestly or enormously, from the deliberate introduction of specific microorganisms.

WHITE-ROT FUNGI

Phanerochaete and related fungi that have the ability to attack wood possess a powerful extracellular enzyme that, differing from many enzymes, acts on a broad array of organic compounds. The enzyme is a peroxidase that, with H_2O_2 produced by the fungus, catalyzes a reaction that cleaves a surprising number of compounds. For example, cultures of *Phanerochaete chrysosporium,* the most widely studied of these fungi for its biodegradative capacity, can degrade a number of PAHs [including benzo(*a*)pyrene, benz(*a*)anthracene, and pyrene], di- and trichlorobenzoic acids, several PCBs, 2,3,7,8-tetrachlorodibenzo-*p*-dioxin, DDT, lindane, chlordane, and explosives such as TNT, RDX, and HMX (Starr and Aust, 1994) that few other microorganisms are capable of decomposing.

Bioremediation technologies using this fungus thus have considerable promise, especially for compounds not acted on readily, if at all, by bacteria. The transformations by the fungus are slow, and a test of the biodegradation of DDT in soil pans failed to show an effect of *Phanerochaete sordida* in promoting bioremediation (Safferman *et al.,* 1995). However, the addition of large inocula (10% dry weight inoculum pus growth substrate on a soil basis) of this fungus to plots (21.3 $\times$ 30.5 m) containing soil to a depth of 25 cm resulted in an enhanced degradation of PCP as well as three- and four- but not five- and six-ring PAHs (Lamar *et al.,* 1994). Another field test with 1 $\times$ 1-m plots containing soil to a 25-cm depth similarly showed a stimulation in the degradation of PCP by two species of *Phanerochaete* (Glaser and Lamar, 1995).

REFERENCES

Aggarwal, P. K., and Hinchee, R. E., *Environ. Sci. Technol.* **25,** 1178–1180 (1991).

Al-Awadhi, N., Al-Daher, R., ElNawawy, A., and Balba, M. T., *J. Soil Contam.* **5,** 243–260 (1996).

Alvey, S., and Crowley, D. E., *Environ. Sci. Technol.* **30,** 1596–1603 (1996).

Anderson, T. A., Kruger, E. L., and Coats, J. R., *Chemosphere* **28,** 1551–1557 (1994).

Anderson, W. C., ed., "Innovative Site Remediation Technology." American Academy of Environmental Engineers, Annapolis, MD, 1995.

Aprill, W., and Sims, R. C., *Chemosphere* **20,** 253–265 (1990).

Aronstein, B. N., and Alexander, M., *Appl. Microbiol. Biotechnol.* **39,** 386–390 (1993).

Aronstein, B. N., Calvillo, Y. M., and Alexander, M., *Environ. Sci. Technol.* **25,** 1728–1731 (1991).

Baker, R. S., Ghaemghami, J., Simkins, S., and Mallory, L. M., *in* "In-Situ Remediation: Scientific Basis of Current and Future Technologies" (G. W. Gee and N. R. Wing, eds.), Vol. 1, pp. 259–277. Battelle Press, Columbus, OH, 1994.

Baldwin, J. J., Swider, K. T., Scalzi, M., and Nowak, J., *in* "In Situ and On-Site Bioremediation, Fourth Symposium," Vol. 5, pp. 419–424. Battelle Press, Columbus, OH, 1997.

Barr, D. P., and Aust., S. D., *Rev. Environ. Contam. Toxicol.* **138,** 49–72 (1994).

Block, R. N., Clark, T. P., and Bishop, M., *in* "Petroleum Contaminated Soils" (P. T. Kostecki and E. J. Calabrese, eds.), Vol. 3, pp. 167–175. Lewis Publishers, Chelsea, MI, 1990.

Borden, R. C., Goin, R. T., and Kao, C.-M., *Ground Water Monitoring Remediation.* 17(1), 70–80 (1997).

Bourquin, A. W., Mosteller, D. C., Olsen, R. L., Smith, M. J., and Reardon, K. F., *in* "In Situ and On-Site Bioremediation, Fourth Symposium," Vol. 4, pp. 513–518. Battelle Press, Columbus, OH, 1997.

Boyle, J. J., and Shann, J. R., *J. Environ. Qual.* **24,** 782–785 (1995).

Braddock, J. F., Ruth, M. L., Catterall, P. H., Walworth, J. L., and McCarthy, K. A., *Environ. Sci. Technol.* **31,** 2078–2084 (1997).

Breitung, J., Bruns-Nagel, D., Steinbach, K., Kaminski, L., Gemsa, D., and von Löw, E., *Appl. Microbiol. Biotechnol.* **44,** 795–800 (1996).

Buchanan, L. R., Nolen, C. H., Bourquin, A. W., and Duaime, T. E., *in* "In Situ and On-Site Bioremediation, Fourth Symposium," Vol. 1, pp. 311–317. Battelle Press, Columbus, OH, 1997.

Chang, Z. Z., Weaver, R. W., and Rhykerd, R. L., *J. Soil Contam.* **5,** 215–224 (1996).

Chapman, S. W., Byerley, B. T., Smyth, D. J. A., and Mackay, D. M., *Ground Water Monitoring Remediation* 17(2), 93–105 (1997).

Chiang, C. Y., Petkovsky, P. D., and Rouse, S. J., *in* "In Situ and On-Site Bioremediation, Fourth Symposium," Vol. 5, pp. 413–418. Battelle Press, Columbus, OH, 1997.

Cho, J. S., DiGuilio, D. C., and Wilson, J. T., *Environ. Progr.* **16**(1), 35–42 (1997).

Compeau, G. C., Mahaffey, W. D., and Patras, L., *in* "Environmental Biotechnology for Waste Treatment" (G. S. Sayler, R. Fox, and J. W. Blackburn, eds.), pp. 91–109. Plenum Press, New York, 1991.

Crowley, D. E., Brennerova, M. V., Irwin, C., Brenner, V., and Focht, D. D., *FEMS Microbiol. Ecol.* **20,** 79–89 (1996).

Cunningham, S. D., Anderson, T. A., Schwab, A. P., and Hsu, F. C., *Adv. Agron.* **56,** 55–114 (1996).

Dablow, J., Rodgers, D., and Morris, H., *in* "In Situ and On-Site Bioremediation, Fourth Symposium," Vol. 5, pp. 439–444. Battelle Press, Columbus, OH, 1997.

Deschênes, L., Lafrance, P., Villeneuve, J.-P., and Samson, R., *Appl. Microbiol. Biotechnol.* **46,** 638–646 (1996).

Devine, K., "Bioremediation Case Studies: An Analysis of Vendor Supplied Data." Publ. EPA/600/R-92/043. Office of Engineering and Technology Demonstration, U.S. Environmental Protection Agency, Washington, DC, 1992.

Dibble, J. T., and Bartha, R., *Soil Sci.* **128,** 56–60 (1979).

Downey, D. C., Guest, P. R., and Ratz, J. W., *Environ. Progr.* **14,** 121–125 (1995).

Duba, A. G., Jackson, K. J., Jovanovich, M. C., Knapp, R. B., Shah, N. N., and Taylor, R. T., *Environ. Sci. Technol.* **30,** 1982–1989 (1996).

Dybas, M. J., Bezborodinikov, S., Voice, T., Wiggert, D. C., Davies, S., Tiedje, J., Criddle, C. S., Kawka, O., Barcelona, M., and Mayotte, T., *in* "In Situ and On-Site Bioremediation, Fourth Symposium," Vol. 4, pp. 507–512. Battelle Press, Columbus, OH, 1997.

Dzantor, E. K., Felsot, A. S., and Beck, M. J., *Appl. Biochem. Biotechnol.* **39/40,** 621–630 (1993).

Environmental Protection Agency, "Bioremediation in the Field." EPA/540/N-95/500. U.S. Environmental Protection Agency, Washington, DC, August 1995.

Federle, T. W., and Schwab, B. S., *Appl. Environ. Microbiol.* **55,** 2092–2094 (1989).

Ferro, A. M., Sims, R. C., and Bugbee, B., *J. Environ. Qual.* **23,** 272–279 (1994).

Fogel, S., Findlay, M., and Moore, A., *in* "Petroleum Contaminated Soils" (E. J. Calabrese and P. T. Kostecki, eds.), Vol. 2, pp. 201–209. Lewis Publishers, Chelsea, MI, 1989.

Fu, J. K., and O'Toole, R., *in* "Gas, Oil, Coal, and Environmental Biotechnology II" (C. Akin and J. Smith, eds.), pp. 145–169. Institute of Gas Technology, Chicago, 1990.

Gan, D. R., and Wright, C. C., *Proc. 68th Annu. Conf. Water Environ. Fed.* **2,** 457–467 (1995).

Gersberg, R. M., Korth, K. G., Rice, L. E., Randall, J. D., Bogardt, A. H., Dawsey, W. J., and Hemmingsen, B. B., *World J. Microbiol. Biotechnol.* **11,** 549–558 (1995).

Glaser, J. A., and Lamar, R. T., *in* "Bioremediation: Science and Applications" (H. D. Skipper and R. F. Turco, eds.), pp. 117–133. Soil Science Society of America, Madison, WI, 1995.

Greer, C. W., Godbout, J. G., Zilber, B., Labelle, S., Sunahara, G., Hawari, J., Ampleman, G., Thiboutot, S., and Dubois, C., *in* "In Situ and On-Site Bioremediation, Fourth Symposium," Vol. 5, pp. 393–398. Battelle Press, Columbus, OH, 1997.

Griffin, C. J., Kampbell, D., and Blaha, F. A., *Hydrocarbon Contam. Soils* **3,** 351–361 (1993).

Gruiz, K., and Kriston, E., *J. Soil Contam.* **4,** 163–173 (1995).

Günther, T., Dornberger, U., and Fritsche, W., *Chemosphere* **33,** 203–215 (1996).

Hildebrandt, W. W., and Wilson, S. B., *J. Pet. Technol.* **43,** 18–22 (1991).

Hooker, B. S., and Skeen, R. S., *Curr. Opin. Biotechnol.* **7,** 317–320 (1996).

Hsu, T. S., and Bartha, R., *Appl. Environ. Microbiol.* **37,** 36–41 (1979).

Hund, K., and Traunspurger, W., *Chemosphere* **29,** 371–390 (1994).

Hutchins, S. R., Downs, W. C., Wilson, J. T., Smith, G. B., Kovacs, D. A., Fine, D. D., Douglass, R. H., and Hendrix, D. J., *Ground Water* **29,** 571–580 (1991).

Hyzy, J. B., and Schepart, B. S., *in* "Bioremediation of Pollutants in Soil and Water" (B. S. Schepart, ed.), pp. 61–74. American Society for Testing and Materials, Philadelphia, PA, 1995.

Jahan, K., Ahmed, T., and Maier, W. J., *Water Environ. Res.* **69,** 317–325 (1997).

Jerger, D. E., and Flatham, P. E., *in* "Gas, Oil, Coal, and Environmental Biotechnology II" (C. Akin and J. Smith, eds.), pp. 67–81. Institute of Gas Technology, Chicago, 1990.

Kao, C.-M., and Borden, R. C., *J. Environ. Eng.* **123,** 18–24 (1997).

Knaebel, D. B., and Vestal, J. R., *Can. J. Microbiol.* **38,** 643–653 (1992).

Lamar, R. T., Davis, M. W., Dietrich, D. M., and Glaser, J. A., *Soil Biol. Biochem.* **26,** 1603–1611 (1994).

Le Dréau, Y., Gilbert, F., Doumenq, P., Asia, L., Bertrand, J.-C., and Mille, G., *Chemosphere* **34,** 1663–1672 (1997).

Lessard, R. R., Wilkinson, J. B., Prince, R. C., Bragg, J. R., Clark, J. R., and Atlas, R. M., *in* "Bioremediation of Pollutants in Soil and Water" (B. S. Schepart, ed.), pp. 207–225. American Society for Testing and Materials, Philadelphia, PA, 1995.

Litherland, S. T., and Anderson, D. W., *in* "In Situ and On-Site Bioremediation, Fourth Symposium," Vol. 5, pp. 425–430. Battelle Press, Columbus, OH, 1997.

Loehr, R. C., and Webster, M. T., *in* "Environmentally Acceptable Endpoints in Soil" (D. G. Linz and D. V. Nakles, eds.), pp. 137–386. American Academy of Environmental Engineers, Annapolis, MD, 1997.

Lynch, J., and Genes, B. R., *in* "Petroleum Contaminated Soils" (P. T. Kostecki and E. J. Calabrese, eds.), Vol. 1, pp. 163–174. Lewis Publishers, Chelsea, MI, 1989.

Madsen, E. L., *Environ. Sci. Technol.* **25,** 1662–1673 (1991).

Manilal, V. B., and Alexander, M., *Appl. Microbiol. Biotechnol.* **35,** 401–405 (1991).

Margesin, R., and Schinner, F., *Appl. Microbiol. Biotechnol.* **47,** 462–468 (1997).

McCarty, P. L., *in* "Environmental Biotechnology" (G. S. Omenn, ed.), pp. 143–162. Plenum Press, New York, 1988.

McCarty, P. L., *J. Hazard. Mater.* **28,** 1–11 (1991).

Mills, M. A., Bonner, J. S., McDonald, T. J., and Autenrieth, R. L., *in* "In Situ and On-Site Bioremediation, Fourth Symposium," Vol. 3, pp. 355–360. Battelle Press, Columbus, OH, 1997.

Morgan, P., and Watkinson, R. J., *FEMS. Microbiol. Rev.* **63,** 277–299 (1989).

Mueller, J. G., Lantz, S. E., Blattmann, B. O., and Chapman, P. J., *Environ. Sci. Technol.* **25,** 1045–1055 (1991).

Nelson, C. H., Seaman, M., Peterson, D. M., Nelson, S., and Buschbom, R., *in* "In Situ and On-Site Bioremediation, Fourth Symposium," Vol. 3, pp. 457–462. Battelle Press, Columbus, OH, 1997.

Nelson, E. C., Walter, M. V., Bossert, I. D., and Martin, D. G., *Environ. Sci. Technol.* **30,** 2406–2411 (1996).

Perkovich, B. S., Anderson, T. A., Kruger, E. L., and Coats, J. R., *Pestic. Sci.* **46,** 391–396 (1996).

Potter, C. L., Glaser, J. A., Dosani, M. A., Krishnan, S., and Krishnan, E. R., *in* "In Situ and On-Site Bioremediation, Fourth Symposium," Vol. 2, pp. 85–90. Battelle Press, Columbus, OH, 1997.

Prince, R. C., *in* "Microbial Control of Pollution" (J. C. Fry, G. M. Gadd, R. A. Herbert, C. W. Jones, and I. A. Watson-Craik, eds.), pp. 19–34. Cambridge Univ. Press, Cambridge, UK, 1992.

Pritchard, P. H., and Costa, C. F., *Environ. Sci. Technol.* **25,** 372–379 (1991).

Qiu, X., Leland, T. W., Shah, S. I., Sorensen, D. L., and Kendall, E. W., *in* "Phytoremediation of Soil and Water Contaminants" (E. L. Kruger, T. A. Anderson, and J. R. Coats, eds.), pp. 186–199. American Chemical Society, Washington, DC, 1997.

Radwan, S. S., Sorkhoh, N. A., El-Nemr, I. M., and El-Desouky, A. F., *J. Appl. Microbiol.* **83,** 353–358 (1997).

Raymond, R. L., Hudson, J. O., and Jamison, V. W., *Appl. Environ. Microbiol.* **31,** 522–535 (1976).

Reddy, B. R., and Sethunathan, N., *Appl. Environ. Microbiol.* **45,** 826–829 (1983).

Reilley, K. A., Banks, M. K., and Schwab, A. P., *J. Environ. Qual.* **25,** 212–219 (1996).

Rhykerd, R. L., Weaver, R. W., and McInnes, K. J., *Environ. Pollut.* **90,** 127–130 (1995).

Ryan, J. R., Loehr, R. C., and Rucker, E., *J. Hazard. Mater.* **28,** 159–169 (1991).

Safferman, S. I., Lamar, R. T., Vonderhaar, S., Neogy, R., Haught, R. C., and Krishnan, E. R., *Toxicol. Environ. Chem.* **50,** 237–251 (1995).

Salanitro, J. P., Dorn, P. B., Huesemann, M. H., Moore, K. O., Rhodes, I. A., Jackson, L. M. R., Vipond, T. E., Western, M. M., and Wisniewski, H. L., *Environ. Sci. Technol.* **31,** 1769–1776 (1997).

Sanseverino, J., Graves, D. A., Leavitt, M. E., Gupta, S. K., and Luthy, R. G., *in* "Remediation of Hazardous Waste Contaminated Soils" (D. L. Wise and D. J. Trantolo, eds.), pp. 345–372. Dekker, New York, 1994.

Schnoor, J. L., Licht, L. A., McCutcheon, S. C., Wolfe, N. L., and Carreira, L. H., *Environ. Sci. Technol.* **29,** 318A–323A (1995).

Schwab, A. P., and Banks, M. K., *in* "Bioremediation through Rhizosphere Technology" (T. A. Anderson and J. R. Coats, eds.), pp. 132–141. American Chemical Society, Washington, DC, 1994.

Siciliano, S. D., and Germida, J. J., *Environ. Toxicol. Chem.* **16,** 1098–1104 (1997).

Song, H.-G., Wang, X., and Bartha, R., *Appl. Environ. Microbiol.* **56,** 652–656 (1990).

Stefanoff, J. G., and Garcia, M. B., Jr., *Environ. Progr.* **14,** 104–110 (1995).

Strzempka, C. P., Woodhull, P. M., Vasser, T. M., and Jerger, D. E., *in* "In Situ and On-Site Bioremediation, Fourth Symposium," Vol. 1, pp. 245–250. Battelle Press, Columbus, OH, 1997.

Sturman, P. J., Stewart, P. S., Cunningham, A. B., Bouwer, E. J., and Wolfram, J. H., *J. Contam. Hydrol.* **19,** 171–203 (1995).

Tabak, H. H., Haines, J. R., Venosa, A. D., Glaser, J. A., Desai, S., and Nisamaneepong, W., *in* "Gas, Oil, Coal and Environmental Biotechnology III" (C. Akin and J. Smith, eds.), pp. 3–38. Institute of Gas Technology, Chicago, 1991.

Thomas, J. M., Gordy, V. R., Fiorenza, S., and Ward, C. H., *Water Sci. Technol.* **22**(6), 53–62 (1990).

Thomas, J. M., Marlow, H. J., Ward, C. H., and Raymond, R. L., *in* "Fate of Pesticides and Chemicals in the Environment" (J. L. Schnoor, ed.), pp. 211–227. Wiley (Interscience), New York, 1992.

Thomas, J. M., and Ward, C. H., *Environ. Sci. Technol.* **23,** 760–766 (1989).

Valo, R., and Salkinoja-Salonen, M., *Appl. Microbiol. Biotechnol.* **25,** 68–75 (1986).

Venosa, A. D., Suidan, M. T., Wrenn, B. A., Strohmeier, K. L., Haines, J. R., Eberhart, B. L., King, D., and Holder, E., *Environ. Sci. Technol.* **30,** 1764–1775 (1996).

Walter, M. V., Nelson, E. C., Firmstone, G., Martin, D. G., Clayton, M. J., Simpson, S., and Spaulding, S., *J. Soil Contam.* **6,** 61–77 (1997).
Waltz, M. D., and Ricotta, A. C., *in* "In Situ and On-Site Bioremediation, Fourth Symposium," Vol. 5, pp. 489–493. Battelle Press, Columbus, OH, 1997.
Walworth, J. L., and Reynolds, C. M., *J. Soil Contam.* **4,** 299–310 (1995).
Wang, X., and Bartha, R., *Soil Biol. Biochem.* **22,** 501–505 (1990).
Wang, X., and Bartha, R., *in* "Remediation of Hazardous Waste Contaminated Soils" (D. L. Wise and D. J. Trantolo, eds.), pp. 175–197. Dekker, New York, 1994.
Wang, X., Yu, X., and Bartha, R., *Environ. Sci. Technol.* **24,** 1086–1089 (1990).
Widrig, D. L., Boopathy, R., and Manning, J. F., Jr., *Environ. Toxicol. Chem.* **16,** 1141–1148 (1997).
Wiesner, M. R., Grant, M. C., and Hutchins, S. R., *Environ. Sci. Technol.* **30,** 3184–3191 (1996).
Wilson, J. T., and Ward, C. H., *Dev. Ind. Microbiol.* **27,** 109–116 (1987).
Wright, A. L., Weaver, R. W., and Webb, J. W., *Water Air Soil Pollut.* **95,** 179–191 (1997).
You, G., Sayles, G. D., Kupferle, M. J., Kim, I. S., and Bishop, P. L., *Chemosphere* **32,** 2269–2284 (1996).
Ziegenfuss, P. S., Williams, R. T., and Myler, C. A., *J. Hazard. Mater.* **28,** 91–99 (1991).

CHAPTER 17

Bioremediation Technologies: *Ex Situ* and Bioreactors

A variety of bioremediation technologies have been developed to treat toxic chemicals not at the place where they exist but rather after their removal or transfer from that location. Some of these techniques are used for waste-disposal sites, and because the contaminated material is removed from its location in soil, sediment, or aquifer, those procedures are considered *ex situ* technologies. The line of demarcation between *in situ* and *ex situ* procedures is not always clear because often, as in the case of pump-and-treat approaches to groundwater remediation, the treatment is performed immediately adjacent to or above the site of contamination. The methods used in many of the *ex situ* as well as pump-and-treat operations are quite similar and thus are considered together. In addition, similar technologies are designed not for contaminated sites but rather for industrial discharges, and they too are considered in this chapter.

The bioremediation is commonly effected by the use of a *bioreactor*. Bioreactors of one or another sort have been developed for the treatment of liquids, solids, or gases. Such systems are often used for the destruction of residual chemicals, by-products, and other wastes from manufacturing facilities, and they are attractive because of the frequently low cost compared to other means of waste disposal. However, *ex situ* treatment is usually more expensive than an *in situ* approach, but it is used because properties of the waste or of the site make *in situ* treatment impractical. The higher costs of an *ex situ* vs *in situ* treatment come from the expenses of moving the contaminated material, the capital required for constructing the needed equipment for the particular cleanup, frequently additional labor, the need for power, or a combination of these factors.

Some of the technologies represent old practices, such as activated-sludge treatment of waste streams containing synthetic chemicals; others have been introduced into practice quite recently; and still others exist only in laboratory or sometimes in pilot-plant treatment units. When coupled with an *in situ* treatment, the reactors may be mobile units taken to field sites and used to treat groundwater pumped out of a contaminated aquifer or wash water used to remove chemicals from a contaminated soil. Such processes may be continuous or may be conducted in batches with contaminated material added and treated material removed intermittently. A typical example is the mobile unit used to degrade PCP-containing groundwater at a facility in which preservatives were once added to wood. Groundwater was pumped up from a well, the pH was adjusted, and nutrients were added to the liquid. The bioreactor, which had a means for temperature control, brought about extensive biodegradation of PCP (Stinson *et al.,* 1991). In another case, groundwater containing dichloromethane, which came from a broken underground pipeline, was passed through an aboveground bioreactor, and the effluent from the treatment unit was then injected back belowground; the result was a reduction in the dichloromethane concentration in monitoring wells from initial values of up to several thousand milligrams per liter to values usually <1 and sometimes <0.005 mg/liter (Flatham *et al.,* 1992).

BIOREACTORS

Many types of bioreactors are used for the microbial production of antibiotics, solvents, amino acids, and other products of fermentation. Only a few of these have been applied to the destruction of pollutants in contaminated sites or to prevent the discharge into waters or soils of industrial effluents. Most of these reactors are designed for aerobic treatment, and the transfer of O_2 throughout the liquid at a rate sufficient to maintain continuous aerobiosis represents a major cost of operation. Anaerobic reactors have also been designed for several types of pollutants.

Reactors may be of two general types, relying on either suspended growth of the microorganisms or immobilized cells. In the first type, the microorganisms are kept in suspension continuously and they grow freely in the water or attached to soil or sediment that is maintained in suspension. These kinds of reactors may be stirred by a mechanical device or by continuous sparging of air. Some examples are slurry reactors, aerated lagoons, and activated sludge. Reactors in which the cells are immobilized in some fashion are of many types. In these designs, the cells are fixed on some type of support and so are not removed as the effluent leaves the reactor, except if desired. The support may be fixed or, if small particulate matter is the support, it may be maintained in suspension (designated *fluidized*). If the support is fixed, the waste stream may simply enter the top and pass out the bottom of the reactor. If the support is fluidized, the particles containing the biomass must be sufficiently heavy that they are not washed out of the reactor with the stream of water containing the compounds being treated. Many microorganisms naturally adhere to the solids used in these reactors, and thus the cells become immobilized.

In many bioremediations, the operation is *continuous,* i.e., the contaminated liquid or suspension enters and treated liquid is removed continuously. Some treatments are done by *batch* operations. Batch systems are common for the treatment of soil, sludge, or other solids and for biodegradations that require long periods of time. A large volume of liquid can be treated by a continuous procedure, but only small volumes are treated by batch systems. Some bioreactors are operated by *semibatch* processes, as in the sequencing batch reactor.

Bioremediation can be effected by a variety of procedures in which contaminated solids are mixed constantly with a liquid in a *slurry-phase treatment.* The system may be reasonably unsophisticated and entail introduction of the contaminated soil, sludge, or sediment into a lagoon that has been constructed with a liner or it may be a sophisticated reactor in which the contaminated materials are mixed (Fig. 17.1). In many ways the operation resembles the activated-sludge procedure that is common for the treatment of municipal wastes, and it allows for aeration, adequate mixing, and control of many of the factors affecting biodegradation. Some designs allow for the capture of volatile organic products that may be generated. The level of dissolved O_2, the pH, and the concentration of inorganic nutrients may be monitored and controlled. Some bioreactors are inoculated with a single species or a mixture of microorganisms able to function effectively under the controlled conditions. In locations where biodegradation slows and sometimes ceases because of low temperatures during part of the year, as is common in many land-treatment systems, the temperature in slurry reactors is maintained in ranges suitable for rapid biodegradation.

A full-scale slurry bioreactor was used to treat TNT-contaminated soil under anaerobic conditions. The 75,000-liter reactor was mounted on a trailer that was brought to the site and filled with 23 m^3 of soil. The soil was made into a 1:1 slurry with water. To make the system anaerobic, starch was added, and indigenous bacteria removed the dissolved O_2 as they metabolized the starch. The treatment resulted in a fall in TNT concentration from 3000 to <1 mg/kg of soil in 5 months (Funk *et al.*, 1995). A slurry reactor maintained aerobically was the means to remediate hydrocarbon-contaminated soils following their excavation (Stefanoff and García, 1995). Many laboratory evaluations have shown the efficacy of slurry reactors for the destruction of PAHs, with extensive degradation occurring of the two-, three-, and often four-ring PAHs (Pinelli *et al.*, 1997; Loehr and Webster, 1997). PAHs, heterocycles, and phenols in creosotes are also quickly destroyed in 3 to 5 days. However, some of the higher-molecular-weight PAHs are only slowly destroyed (Mueller *et al.*, 1991).

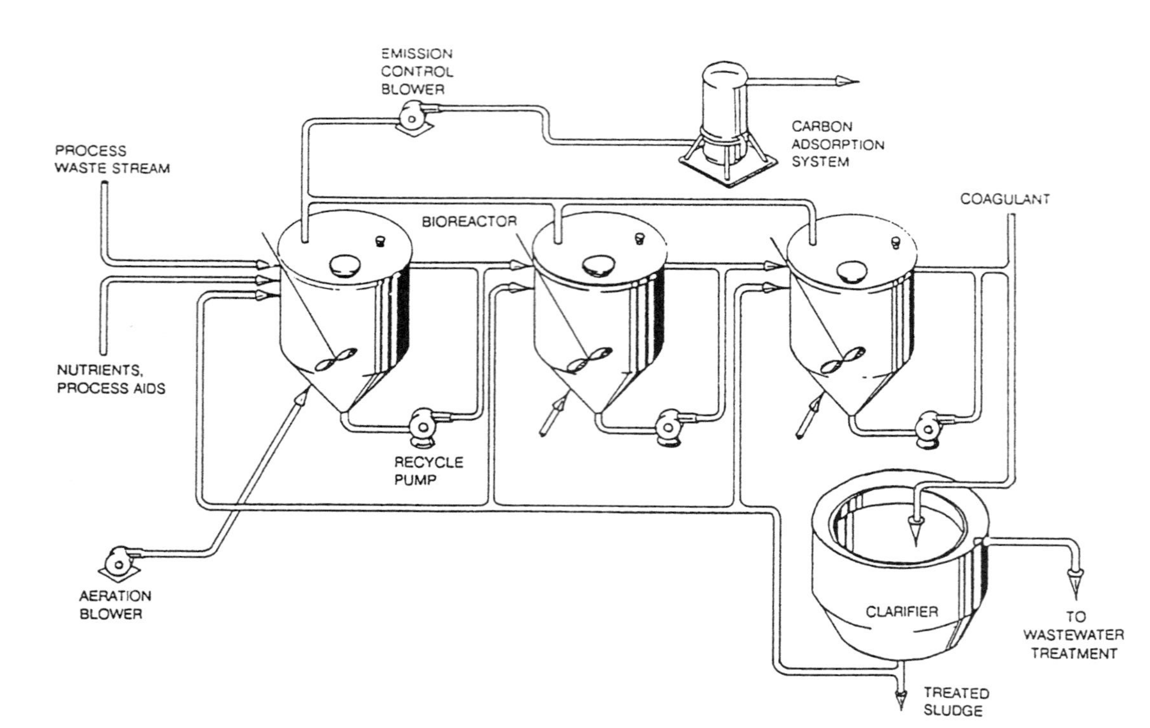

Figure 17.1 A system for bioremediation by slurry-phase treatment. (From Ryan *et al.*, 1991. Reproduced by permission from Elsevier Science Publishers.)

The rate of destruction of compounds sorbed to soil or of oil in soil may be increased markedly by the addition of appropriate surfactants or dispersants to the slurries of soil (Aronstein and Alexander, 1992; Rittmann and Johnson, 1989; Van Hoof and Jafvert, 1996). However, some surfactants do not appear to be beneficial in laboratory tests, and surfactants at high concentrations are often toxic and inhibit the degradation (Laha and Luthy, 1992; Van Hoof and Jafvert, 1996). Laboratory studies also suggest that the addition of relatively inexpensive materials that are quickly degraded may result in anaerobic conditions that enhance the degradation of specific compounds, such as dinoseb (Kaake *et al.*, 1992).

Slurry-phase procedures may be combined with a washing technique to remove the contaminant from soil. Such a soil washing was used for a field-scale bioremediation at a wood-treating facility that contained up to 8000 mg of PCP per kilogram of soil. The soil was initially washed to remove the PCP, and then the wash solution was introduced into a slurry-phase reactor. This two-phase procedure reduced the PCP concentration in the soil to the target cleanup level at the site, namely, less than 0.5 mg/kg of soil (Compeau *et al.*, 1991).

Even a pool or lagoon may be used as a suspended-growth reactor. Aeration and mixing can be accomplished by mechanical stirring. Aerated lagoons have been employed to bioremediate refinery wastes and other materials containing petrochemicals. One such cleanup was designed for a wastewater-sludge mixture from a refinery. Aeration, N, and P were provided, but the temperature was not controlled. The 7000 m^3 of oil- and grease-rich sludge contained 2000 mg of PAHs and 200 mg of BTEX per kilogram. The extent of biodegradation is depicted in Table 17.1. Wastes from refineries and petrochemical manufacturers were also treated successfully in a lagoon that was aerated and amended with phosphate and nitrate (Woodward and Ramsden, 1990).

A common way of bringing about biodegradation is to use reactors in which microbial cells become attached as a film to some matrix. A solution containing the chemicals is passed over the resulting biofilm, which brings about a rapid biodegradation because of the high cell density. A modification of fixed-film treatment employs immobilized or strongly sorbed cells. The cells are immobilized by firmly attaching the organisms or physically embedding them in the solid matrix. The cells may thus be retained in or on a variety of materials, including alginate beads, diatomaceous earth, hollow glass fibers, polyurethane foam, activated C, and polyacrylamide beads. Common to many of these systems is the greater tolerance to high chemical concentrations of the cells that are in the films or that are immobilized than cells in suspension. The greater resistance may be associated with sorption of the substrate to the solid or immobilizing material, thereby reducing the amount available to suppress the microorganisms, or to some other mechanism.

Several bacteria and fungi as well as microbial mixtures have been used as biofilms or in immobilized cell systems, and a number of compounds are readily biodegraded by these procedures (Table 17.2). The immobilized cells are contained

Table 17.1

Bioremediation of Sludge Rich in Oil and Grease in an Aerated Lagoon[a]

Compound	Concentration (mg/kg)	
	Initially	At end
Benzene	64.4	1.19
Toluene	19.4	1.14
Ethylbenzene	32.4	0.32
Naphthalene	290	ND[b]
Phenanthrene	150	ND
Pyrene	540	0.03
Anthracene	20	0.02
Benzoanthracene	91	ND
Chrysene	20	ND
Benzopyrene	100	<0.01

[a] From Vail (1991).
[b] Not detected.

Table 17.2

Biodegradation of Organic Compounds by Immobilized Cells or Strongly Sorbed Cells

Compound	Immobilizing material or solid phase	Microorganism	Reference
Acrylonitrile	Alginate	*Brevibacterium* sp.	Hwang and Chang (1989)
Aniline, chloroanilines	Diatomaceous earth	Mixed culture	Livingston and Willacy (1991)
Anthracene	Alginate beads	*Trichoderma harzianum*	Ermisch and Rehm (1989)
2-Chlorophenol	Alginate beads	*Phanerochaete chrysosporium*	Lewandowski *et al.* (1990)
Glyphosate	Diatomaceous earth	Mixed culture	Hallas *et al.* (1992)
4-Nitrophenol	Diatomaceous earth	*Pseudomonas* sp.	Heitkamp *et al.* (1990)
PCP	Alginate beads	*Arthrobacter* sp.	Lin and Wang (1991)
Phenol	Activated carbon	*Pseudomonas putida*	Ehrhardt and Rehm (1989)
TCE	Alginate beads	*Methylocystis* sp.	Uchiyama *et al.* (1995)
Trichloropyridinol	Diatomaceous earth	*Pseudomonas* sp.	Feng *et al.* (1997)

in one of several different types of reactors to facilitate rapid biodegradation. Such technologies are particularly useful for waste streams from chemical-manufacturing facilities, as illustrated by a pilot plant used for the high-volume treatment of low concentrations of glyphosate. Wastewater containing this compound was introduced at 45 liter/min into a column containing cells configured as an upflow reactor. The microorganisms were found to bring about rapid and extensive degradation (Hallas *et al.*, 1992).

The types of reactors that rely on biofilms or immobilized cells include trickling filters, fixed-film or fixed-bed bioreactors, fluidized-bed reactors, and rotating bioreactors.

The *trickling filter* is a technology that has been used for wastewater treatment for about 100 years. The solution is allowed to trickle downward over the packing material, the biomass having developed on this solid phase. The organisms metabolize the contaminants in the water during the time of passage, and the trickling filter, where applicable, is highly useful and cost effective.

In the *fixed-film bioreactor*, the reactor contains some support on which bacteria grow and become attached. That support typically has a large surface area to allow for the development of an extensive and exposed biomass. The contaminant-containing liquid forms a thin layer over the cell mass, and the compounds diffuse from the liquid to the microbial film, where they are degraded. Such an approach has been used in the field and has been effective in removing >99% of the BTEX and most PAHs from groundwater polluted by a former manufactured-gas plant (Leahy *et al.*, 1997).

Frequently, the solid phase is granular activated carbon, which not only sorbs many organic chemicals but also is a base on which a highly active microbial film appears, frequently after inoculation with a microbial mixture or sometimes individual bacteria. The microorganisms destroy many of the contaminants in the water but also extend the life of the activated carbon, which otherwise would have to be regenerated sooner. The process of regeneration to allow the reuse of the activated carbon is expensive. The effectiveness of such an approach has been demonstrated on a field scale for the degradation of 1,2-dichloroethane (Stucki and Thüer, 1994, 1995) and chlorobenzenes and hexachlorocyclohexane (Feidieker *et al.*, 1995) in groundwater.

Biofilms are also the basis for the bioremediation effected by *rotating-disk bioreactors* (or contactors). The reactor contains several disks that are attached to a shaft that is placed horizontally (Fig. 17.2). The disks may have a diameter of 1.5 to 3.5 m. The biofilm develops on those disks. The shaft, which is typically placed immediately above the contaminated liquid, slowly rotates, and thus a high percentage of the biofilm adhering to the surfaces of the disks are within the liquid phase. Because the disks are rotating, the portion above the liquid phase is aerated, but the adhering biomass is kept moist even when above the water because of the adhering water film. The cylinders and horizontal shaft are in an elongated tank

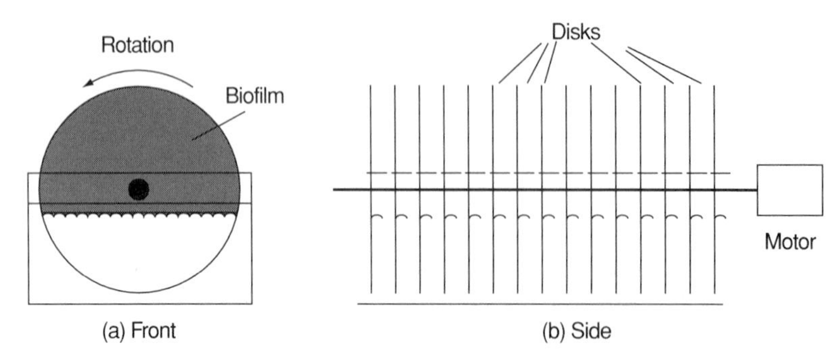

Figure 17.2 A rotating-disk bioreactor. (From Armenante, 1993. Reproduced with permission of the McGraw-Hill Companies.)

or series of tanks through which the contaminated water passes. A somewhat similar approach has been employed in a laboratory study designed to evaluate the effectiveness of fungi, such as *Trametes versicolor* and *Phanerochaete chrysosporium,* to degrade PCB; in this instance, the fungi were allowed to colonize steel mesh placed within tubes that were slowly rotated to allow for the wetting and aeration characteristic of the conventional rotating-disk bioreactor (Alleman *et al.,* 1995).

A common means of treatment of contaminated waters involves the *fixed-bed reactor.* In this type of system, a vessel is filled with some loose packing material containing a surface area sufficiently large for the attachment of a large biomass. That packing may be diatomaceous earth, porous silica, hollow fibers, plastic particles, or sometimes merely pebbles. A sketch of a packed-bed reactor is presented in Fig. 17.3. The vessel may be open to incoming liquid or a gas stream or it may be sealed. The fluid may flow up through the vessel (upflow mode) or downward (downflow mode). Some fixed-bed reactors contain granular activated carbon as the packing material. Such systems have been used for the field-scale treatment of water containing atrazine and simazine following inoculation with strains of *Rhodococcus* and *Acinetobacter* capable of degrading these herbicides (Feakin *et al.,* 1995) and for the anaerobic degradation of 1,1,1-trichloroethane (de Best *et al.,* 1997).

Widely used for the degradation of liquids containing synthetic chemicals is the *fluidized-bed reactor.* The bacteria became attached to and immobilized on small particles within the vessel, and these particles are suspended (or fluidized) by continuously mixing or recycling the liquid in the reactor. The particles may be activated carbon, sand, a synthetic resin, a porous plastic, or stainless-steel mesh. In some instances, the fluidization is brought about by a gas or a combination of an aqueous solution and a gas. The fluidized bed is not subject to the mechanical blockage by the large biomass that occasionally appears in fixed beds. Fluidized-bed reactors have been used at a pilot scale for the treatment of groundwater containing BTEX and PAHs (Leahy *et al.,* 1997) and wastes from a former

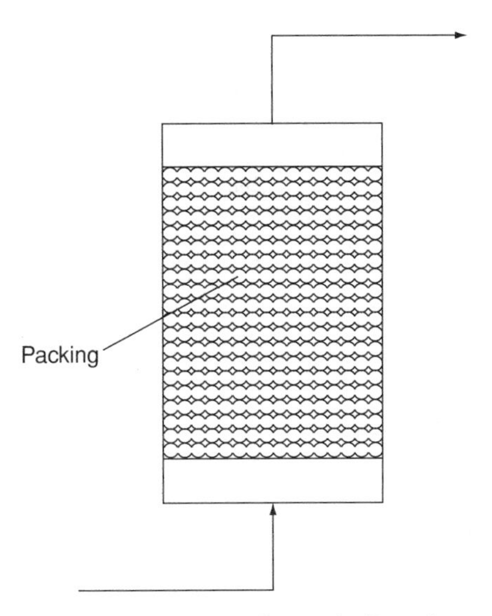

Figure 17.3 A fixed-bed reactor operating in an upflow mode. (From Armenante, 1993. Reproduced with permission of the McGraw-Hill Companies.)

manufactured-gas plant site (Rajan *et al.,* 1997), in both cases with removal of >99% of the target compounds. Laboratory-scale evaluations have shown the feasibility of such approaches for the destruction of 1-naphthylamine (Ro *et al.,* 1997) and chlorophenols (Järvinen *et al.,* 1994). Fluidized beds have also been investigated for the anaerobic treatment of PCP (Khodadoust *et al.,* 1997) and phenols (Suidan *et al.,* 1996).

Sequencing batch reactors have been the basis for some approaches to bioremediation. At the start of processes involving these systems, a reactor containing an acclimated mixture of microorganisms is filled with the contaminated liquid and biodegradation is allowed to proceed. When the degradation is deemed to have gone sufficiently far, the biomass is permitted to settle, the treated liquid is removed, and additional contaminated water is introduced to start a new cycle. The reactor may be aerated, agitated, or both during the period of treatment. Sequencing batch reactors have been studied for potential use in the treatment of tetrachloroethylene (PCE) (Hirl and Irvine, 1997) and have been applied to the destruction of BTEX and other components of JP-4 jet fuel found in groundwater (Yocum *et al.,* 1995).

In the *bubble column reactor,* air or possibly some other gas is sparged through the liquid, and no mechanical device for agitation is provided. The sparger is placed at the bottom of the vessel to allow the gas bubbles to pass through the entire column of liquid. This type of system was used in laboratory-scale evaluations showing the efficacy for the degradation of cyanuric acid by a strain of *Pseudomonas* adsorbed on clay (Ernst and Rehm, 1995) and toluene by *Pseudomonas putida* immobilized on celite particles (Choi *et al.*, 1995).

The *airlift reactor* also uses a pneumatic system for mixing. In its usual design, the reactor is a deep cylinder in which a vertical cylindrical tube is placed. That tube, which is open at both ends, is situated right above the place where the air is introduced into the reactor. The bubbles in the air stream pass through the liquid in the central tube and escape at the top. However, the movement of air brings about an upward movement of liquid in the central tube, which leads to a flow of the liquid in the portion of the fluid outside that tube, as shown in Fig. 17.4. The *airlift loop reactor* is a modification of the airlift reactor in which a pump is used for mechanical agitation to recirculate the liquid faster than the rate at which the contaminated liquid is introduced into the reactor. An airlift loop reactor containing *Rhodococcus rhodochrous* and *Hyphomicrobium* sp. has been evaluated for the decomposition of dichloromethane, 2-propanol, and methanol used as solvents for paint stripping (Vanderberg-Twary *et al.*, 1997).

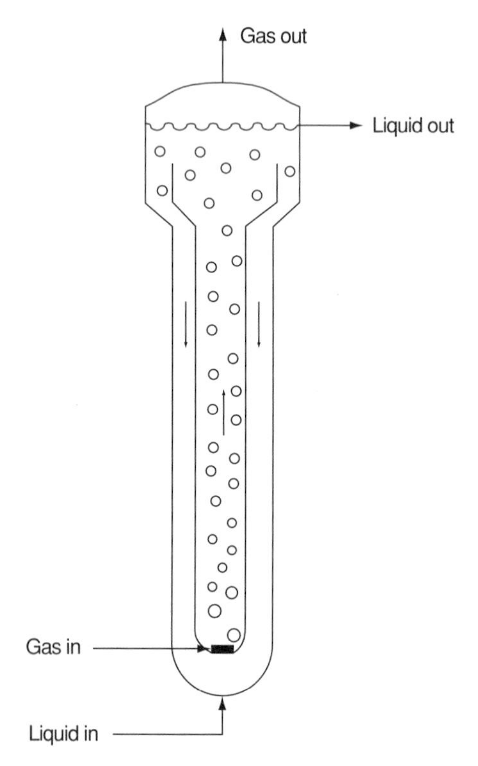

Figure 17.4 An airlift reactor. (From Armenante, 1993. Reproduced with permission of the McGraw-Hill Companies.)

Inoculation of bioreactors with individual bacteria, known mixtures of species, or a mixed culture is commonly practiced. It is also usually successful in these engineered systems, in which the design favors the specific organisms in a liquid and usually well mixed unit. This is in marked contrast to the frequently negative or sometimes questionable outcome of inoculation/bioaugmentation for *in situ* bioremediation. For *in situ* treatment of polluted soils and sediments, the factors limiting the growth of *individual* bacterial or fungal species are not known or are rarely known. The presence of an organic substrate (the target pollutant) overcomes one limiting factor and adding N, P, and O_2 overcomes these limitations (if relevant), but lack of mobility, sorption of cells, susceptibility to protozoan predation and to parasitism, inability to make use of sorbed or sequestered organic substrate, and microbial susceptibility to toxins may limit the success of an inoculum for in situ remediation. In the bioreactor, in contrast, many of these limitations do not exist or are overcome; if not, the treatment system is modified (which cannot be easily or cheaply done for *in situ* treatments) to mitigate or bypass the limitation. Success in treatment after inoculation has been shown in a full-scale granular activated column with a bed volume of 10 m^3 to treat groundwater contaminated with 1,2-dichloroethane (Stucki and Thüer, 1994), a 2000-liter static reactor used to treat dinoseb (Roberts *et al.*, 1993), a biofilter designed to destroy dimethyl sulfide (Smet *et al.*, 1996), and a laboratory-scale slurry bioreactor for PCP degradation (Otte *et al.*, 1994).

BIOFILTERS

Microorganisms are also being used to destroy a variety of volatile compounds. In such technologies, the microorganisms are allowed to grow on some solid support, and a stream of gas containing the unwanted molecules is passed through the solid support. The resulting microbial action leads to destruction of the contaminants. This process, which is termed *biofiltration,* is common in Germany, the Netherlands, Japan, and a few other countries. The trickle-bed reactor, in which bacteria are fixed on a column, may be designed for volatile compounds, with the chemical to be degraded (or "scrubbed") being passed through and dissolved in a solution. Somewhat similar to biofilters are *bioscrubbers,* in which the gases and O_2 are usually first passed into a unit in which the volatiles dissolve in water, and then the solution is introduced into a system, usually an activated-sludge system, in which the organic compounds are degraded by microorganisms dispersed in the aqueous phase. These treatment systems are attractive because of the little energy needed, the comparatively low cost, the simplicity of operation, and the frequent ability to destroy compounds at low concentrations.

The solid phase of the biofilter may be peat, soil, composted organic matter, sawdust, bark chips, activated carbon, clay particles, diatomaceous earth, or porous glass. Soil and compost are popular solid phases, and a volatile organic compound passing through a soil or compost bed is first sorbed by the solid phase, which must be kept moist to maintain biological activity, and then the sorbed molecule is metabolized by the film of microorganisms adhering to the solid. An unsophisticated but effective biofilter is simply a bed of soil placed on top of a system of pipes through which the volatiles pass.

A large number of volatile compounds can be degraded in biofilters, including naphthalene, acetone, propionaldehyde, volatile S compounds, toluene, benzene, dichloromethane, and vinyl chloride. Particular attention has been given to volatiles with offensive odors, and a number of systems have been designed to remove H_2S, methane thiol, dimethyl disulfide (Shoda, 1991), SO_2 (Dasu and Sublette, 1989), and dimethyl sulfide (Hirai *et al.,* 1990). Among the organic pollutants, waste air containing the components of BTEX can be treated to destroy the aromatic compounds in the vapor phase by passing the air through soil. Such an approach offers promise for the destruction of the organic volatiles generated when gasoline-contaminated groundwaters are treated by air stripping or when soil venting is employed to eliminate volatiles emitted from the vadose zone. Alternative treatments to remove the volatiles, such as incineration or sorption by activated carbon, are costly (Canter *et al.,* 1989; Miller and Canter, 1991). Propane and *n*- and isobutane are also readily degraded when exposed to the microorganisms present in soil (Kampbell *et al.,* 1987), as are volatile organics present in fumes from aviation gasoline (Kampbell and Wilson, 1991) and carbon monoxide (Frye *et al.,* 1992). Ammonia, esters, and amines can also be removed by use of biofilters (Swanson and Loehr, 1997). Biofilters have been employed to reduce volatile emissions from wastewater-treatment plants and composting facilities. In addition, such processes have been employed in iron foundries and in industries concerned with manufacturing chemicals, fish processing, flavorings and fragrances, slaughtering animals, and tobacco processing.

A conceptual design of a biofilter is presented in Fig. 17.5. The unit should have adequate inorganic nutrients, and some means of preventing acidification may be important. The solid phase should have a large surface area to maximize sorption of the compounds to be treated and to permit the development of a large biomass. It should also have high porosity and retain moisture. Inoculation is often not necessary because the solid support (e.g., compost or soil) may have a sufficiently active population or one that will develop rapidly to yield a large biomass, but the solids may be inoculated with activated sludge, a mixed culture, or sometimes even individual species if the compound is not utilized by many different species, if the acclimation phase would otherwise be long, or if the solid phase is not naturally rich in microorganisms (such as granular-activated carbon or diatomaceous earth) (Swanson and Loehr, 1997). Difficulties may arise, however, because of the fall in pH that is characteristic of some microbial transformations or because the biomass clogs the pores of the solid phase.

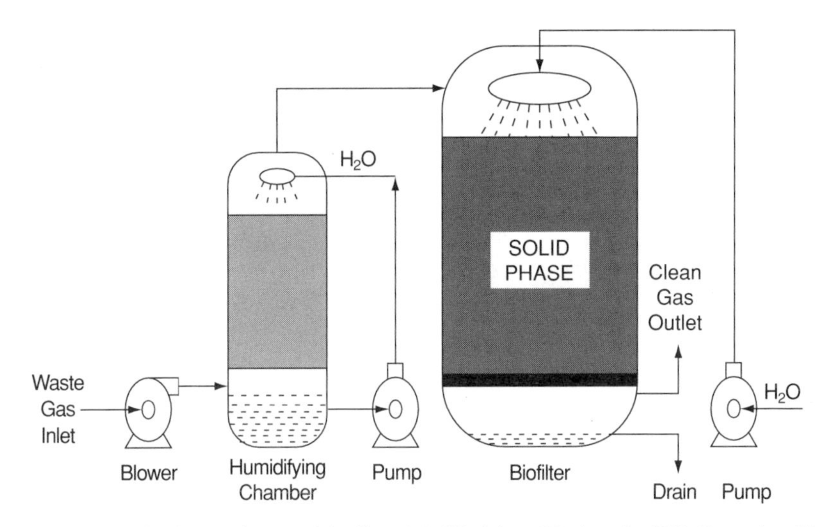

Figure 17.5 The design of a typical biofilter. (Modified from Wani *et al.*, 1997. Reproduced by courtesy of Marcel Dekker Inc.)

The use of biofiltration for the treatment of volatile compounds in soil is often linked with an initial vacuum or vapor extraction technique to move the compounds from the soil to the biofilter. Such systems have been used in the field for the bioremediation of sites contaminated with gasoline from leaking underground storage tanks with biofilters containing pine bark chips (Lesley and Rangan, 1997) or a mixture of diatomaceous earth and composted manure (Jutras *et al.*, 1997) or sites containing offgases from petroleum processing and remediation equipment with a biofilter containing peat (Leson and Smith, 1997). Pilot-scale evaluations have also shown the efficacy of destroying BTEX and other constituents of gasoline vapors with a compost-containing biofilter (Wright *et al.*, 1997) or a biofilter containing celite pellets (Sorial *et al.*, 1997) and methyl, ethyl, and *n*-propyl alcohols with a peat-containing biofilter (Kiared *et al.*, 1997). Several innovations have been suggested based on laboratory tests. These include the use of microporous hollow fiber membranes for the degradation of toluene (Ergas and McGrath, 1997) and the removal of ammonia with a biofilter containing nitrifying autotrophs (Weckhuysen *et al.*, 1994).

Biofilters or bioscrubbers can also be operated to destroy volatile chlorinated compounds. Thus, 98% of the dichloromethane was removed with a biofilter with compost as a solid phase (Ergas *et al.*, 1994). It may also be accomplished by organisms cometabolizing the pollutants, for example, in systems in which gaseous methane or propane is passed through a column of an aquifer material. Perfusing a liquid containing TCE and 1,1,1-trichloroethane leads to their cometabolism by populations growing on the methane or propane (Wilson *et al.,* 1987). A similar type of degradation of TCE occurs with a muck soil as the solid phase (Kampbell *et al.,* 1987). Biofilters containing activated carbon and fed with phenol have also been found capable of destroying all of the TCE and PCE in gas streams (Kim, 1997). Dichloromethane and vinyl chloride in a gas stream can also be metabolized, and most of the compounds are destroyed in trickle-bed filters inoculated with species of *Mycobacterium* or *Hyphomicrobium,* respectively (Diks and Ottengraf, 1991; Hartmans *et al.,* 1985).

Trickling filters have also been designed for the treatment of gas streams. Typically, air containing the compounds of concern is introduced into water, which is supplemented with inorganic nutrients and possibly an inoculum, and that liquid is sprayed on or allowed to trickle over the solid phase, on which an active biofilm develops. Such trickling filters have been employed for the degradation of toluene (Pedersen and Arvin, 1995), methyl ethyl ketone (Chou and Huang, 1997), amines, mercaptans, various other organic compounds, and CS_2 (van Groenestijn and Hesselink, 1993).

COMETABOLISM

Several systems have been designed to bioremediate compounds that are transformed by cometabolizing microorganisms. These technologies require that a C and energy source is provided so that the organisms can cometabolize the target pollutant. TCE has been the compound that has received the most attention, and processes have been devised for aerobic bioremediation in the field using phenol, toluene, methane, or methanol as sources of C and energy for the bacteria. In some instances, phenol or toluene as well as O_2 are injected into the groundwater, and TCE as well as *cis*-1,2-dichloroethylene are metabolized extensively (Hopkins and McCarty, 1995; Hopkins *et al.,* 1993). A full-scale *in situ* system containing alternating extraction and injection trenches has shown that approximately 10% of the TCE in groundwater was cometabolized per month following the introduction of methanol (Litherland and Anderson, 1997). TCE in groundwater was degraded extensively in an experimental trial in which a methanotrophic bacterium (*Methylosinus trichosporium*), methane, and phosphate were injected at 27 m into a contaminated aquifer; the TCE level initially fell markedly, but then slowly increased to background levels (Duba *et al.,* 1996).

Bench-scale studies have also indicated the feasibility of cometabolic bioremediation of TCE. One system relies on an airlift reactor in which gases containing TCE are introduced into an aboveground reactor, and 90 to 95% of the TCE in the gas phase is destroyed with phenol serving as the C and energy source (Ensley and Kurisko, 1994). A membrane biofilter has been evaluated in which TCE-contaminated air is passed through microporous hollow fibers of the biofilter with toluene serving as the C and energy source (Parvatiyar *et al.*, 1996). A fixed-bed reactor provided with a number of different C and energy sources led to the conversion of TCE to *cis*-1,2-dichloroethylene (Schöllhorn *et al.*, 1997), and a bioreactor containing *Methylocystis* sp., a methane-oxidizing bacterium, immobilized in Ca alginate has also been proposed (Uchiyama *et al.*, 1995). Columns containing granular activated carbon colonized by phenol-oxidizing bacteria and fed with a solution containing phenol and inorganic nutrients were found to destroy TCE and PCE; such biofilters are useful for the treatment of TCE- and PCE-containing gases released when groundwater is subjected to air stripping or when soil is treated by vapor extraction (Kim, 1997).

Bioremediation technologies for several other cometabolized compounds have also been investigated. For example, an anaerobic fixed-bed reactor supplied with acetate has been shown to support the transformation of 1,1,1-trichloroethane to 1,1-dichloroethane and chloroethane by methanogenic bacteria (de Best *et al.*, 1997), and a slurry reactor provided with corn steep liquor was found to be useful for the cometabolic transformation of RDX (Young *et al.*, 1997). DDT is cometabolized anaerobically, but the products of the cometabolic transformation are metabolized by aerobes (Pfaender and Alexander, 1972). This partnership is the basis for a method proposed for the degradation of DDT. The two-organism system is made up of *Enterobacter cloacae* that anaerobically dechlorinates DDT and forms 4,4′-dichlorodiphenylmethane as it grows on and ferments lactose and a strain of *Alcaligenes* that grows aerobically on diphenylmethane but cometabolizes 4,4′-dichlorodiphenylmethane. When immobilized in Ca alginate, the two bacteria are able to carry out the two steps in a single reactor (Beunink and Rehm, 1988).

ANAEROBIC PROCESSES

Nearly all bioremediations in practical use are aerobic. Although most early research on the decomposition of benzene, toluene, ethylbenzene, and PAHs dealt with metabolism in the presence of aerobes, many of these compounds can be decomposed in the absence of O_2, often when nitrate, sulfate, CO_2, or ferrous iron is available as an alternate electron acceptor. However, engineered bioremediation technologies have not relied extensively on electron acceptors other than O_2 as the basis for controlled processes. Nevertheless, it is now clear that anaerobic bacteria are able to catalyze many reactions and destroy many compounds that are resistant to aerobes. For example, dinoseb in contaminated soil was destroyed in 10 to 15 days in a 6000-liter anaerobic reactor containing approximately 2000 kg of soil, starch, a phosphate supplement, and 2000 liters of water (Roberts *et al.*, 1993).

Particular attention in recent years has been given to chlorinated molecules, not only because of their toxicity but also because of their persistence in polluted environments. Anaerobes can reductively dehalogenate chlorinated molecules that persist and that are rarely attacked by aerobic bacteria (Mohn and Tiedje, 1992). A simple dechlorination is evident in the proposed anaerobic fluidized bed containing granular activated carbon used to degrade 2-chlorophenol present in industrial wastewater (Suidan *et al.*, 1996). Highly chlorinated PCBs, carbon tetrachloride, PCE, and many other chlorinated products, some of which are ubiquitous as well as persistent, are converted to less highly chlorinated compounds. Metabolism of the latter could probably be best accomplished by aerobes because aerobic conversions of the less chlorinated molecules are often faster. An example is the initial anaerobic dechlorination of PCE in groundwater and soil vapor linked with the injection of methanol followed by the aerobic metabolism of the dichloroethylene and other products that had been formed (Gerritse *et al.*, 1995; Spuij *et al.*, 1997). Another example is a two-stage anaerobic–aerobic biofilm reactor proposed for treating groundwater and industrial effluents containing highly chlorinated molecules. In the anaerobic phase, the mixed culture reductively dechlorinates TCE, chloroform, and hexachlorobenzene in the presence of acetate to yield partially dehalogenated products. These less chlorinated products are then introduced into an aerobic reactor, with the net result being that more than 93% of the three substrates are converted to nonvolatile products and CO_2 (Fathepure and Vogel, 1991). A laboratory study also showed that TCE was anaerobically cometabolized to *cis*-1,2-dichloroethylene in the presence of acetate and that the dichloroethylene was then degraded in the presence of O_2 (Schanke *et al.*, 1997). Another laboratory evaluation suggested that the PCB concentration could be reduced by 70% in soil slurries exposed first to anaerobiosis and then to aerobic conditions (Evans *et al.*, 1996). One-carbon halogenated compounds are nearly completely transformed anaerobically following the onset of sulfate reduction in a biofilm column reactor inoculated with sewage microorganisms, suggesting a possible way of enhancing the transformation (Cobb and Bouwer, 1991).

The possible use of a sequential anaerobic–aerobic system for the bioremediation of TNT-contaminated soil has attracted considerable attention. The approaches have involved soil placed in layers and flooded with shredded plant material to create anaerobiosis as the aerobes consume O_2 (Widrig *et al.*, 1997), use of compost flooded with water in the initial phase (Bruns-Nagel *et al.*, 1997), and the addition of simple organic substrates in the initial phase (Roberts *et al.*, 1996). In addition, azo dyes and effluents from bleaching operations and the textile industry have been subject to a treatment sequence of anaerobiosis followed by aerobiosis (Field *et al.*, 1995).

COMBINED MICROBIAL AND NONBIOLOGICAL TREATMENT

A technology that relies solely on bioremediation often fails to effectively or extensively destroy individual organic pollutants or one or many in a mixture of compounds in a waste stream. The reasons are several: (a) the single compound of concern may be intrinsically nonbiodegradable; (b) the waste stream contains a number of organic substances, some of which are not readily biodegradable; (c) the individual substance or the mixture of compounds is biodegradable, but the concentrations are sufficiently high that species having the capacity to metabolize those substances are inhibited; or (d) an organic product accumulates because it is not further metabolized by the organisms present. The existence of these difficulties has been the basis for a number of technologies that couple a nonbiological treatment with biodegradation. The nonbiological step usually, but not always, comes first (Scott and Ollis, 1995).

The abiotic phase may involve treatment with O_3, H_2O_2, ultraviolet light, sunlight, or Fenton's reagent (Fe salts plus H_2O_2). Typically, the approach is applied to wastewater from industry or contaminated groundwater, but some tests have been done with soil. Nearly all the information comes from laboratory studies.

The concepts underlying these approaches and the basis for interest in them reflect the four reasons cited earlier. If the compound is intrinsically nonbiodegradable, it may be converted abiotically to a product that can be metabolized. This is the basis for studies with both low-molecular-weight compounds that are not metabolized at significant rates, if at all, as well as polymers, many of which, in the macromolecular form, cannot be transformed biologically. If the waste stream contains biodegradable as well as poorly utilized or nonutilizable components, the concentrations of the refractory constituents may be sufficiently high that bioremediation is not an acceptable technology. In addition, should some chemical treatment or sorption be needed for this multicomponent waste stream, a greater degree of biodegradation would reduce the cost of that subsequent treatment and/or increase the life of the sorbent. If the initial waste is toxic, then abiotic treatment may reduce that toxicity to an extent that bioremediation is feasible. Finally, if an organic product accumulates, it may have to be removed or destroyed in some manner, thereby increasing the cost of treatment.

A common approach involves an initial treatment with O_3. This strong oxidizing agent has been applied for the destruction of atrazine in pesticide waste streams in both pilot- and laboratory-scale assessments. The ozonation converted the herbicide to chlorodiamino-*s*-triazine, and then *Klebsiella terragena* further degraded that intermediate to CO_2 (Leeson *et al.*, 1993; Hapeman *et al.*, 1995). An O_3 pretreatment also appears to be feasible for the initial phase of treatment of a number of otherwise slowly biodegraded or nonbiodegradable chlorinated and nitro-containing benzenes and phenols (Stockinger *et al.*, 1995; Adams *et al.*, 1997). Ultraviolet irradiation prior to biodegradation has been reported to be a way of destroying many PCB congeners in solution (Shimura *et al.*, 1996), polystyrene (Guillet *et al.*, 1974), and other compounds (Katayama and Matsumura, 1991). Sunlight, often with a catalyst, has been used as the first stage prior to the biodegradation of PCP and other chlorophenols (Manilal *et al.*, 1992; Hwang *et al.*, 1986). Similarly, the use of Fenton's reagent has been the first step in a proposed treatment of a number of chlorinated molecules (Koyama *et al.*, 1994; Lee and Carberry, 1992).

In some instances, the initial stage is biodegradation, and the second is a nonbiological procedure. The first step is a relatively inexpensive procedure to reduce the amount of material that has to be destroyed in the second step. For example, after aerobic bacteria metabolize trifluoromethylbenzoate, the resistant breakdown product was subsequently destroyed by exposure of the liquid to sunlight (Taylor *et al.*, 1993). Such a two-stage approach has been evaluated in laboratory microcosms containing samples of soil, and the data suggest that 50% of the PAHs remaining after bioremediation were eliminated by Fenton's reagent (Stokley *et al.*, 1997).

ENZYMATIC CONVERSIONS

Proposals have been made to use enzyme preparations to destroy individual pollutants or toxic chemicals. Some of the enzymes are quite stable during storage and thus could be used in emergency responses to spills since the catalysts would be immediately available in active form. It has also been suggested that they can be used to decontaminate soils having unwanted pesticides or other toxicants. They might even be useful for converting some persistent pollutants to products that are harmless. Both soluble and immobilized enzymes have been suggested for these purposes, but most research has focused on immobilized enzymes.

An enzyme immobilized on porous glass or porous silica beads was found to hydrolyze a number of organophosphate insecticides in solution (Munnecke, 1979, 1981), and a phosphotriesterase from *Escherichia coli* was immobilized on nylon membrane, powder, and tubing (Caldwell and Raushel, 1991). Both enzymes were suggested as means for detoxifying pesticides. A foam-based enzyme that hydrolyzes a number organophosphorus compounds used as nerve agents in chemical warfare has been proposed for field use as a means of decontamination (Cheng and Calomiris, 1996). Similarly, a phosphotriesterase in a polyurethane foam, which had good storage characteristics, has been proposed as a way of degrading organophosphorus nerve agents (LeJeune *et al.*, 1997). Enzymes have also been added to soil, where they sorb to particulate matter, and the sorbed catalysts have been suggested as a means of hydrolyzing insecticides such as diazinon. Indeed, parathion hydrolase added to soil converts more than 90% of diazinon at 1.0 g/kg to nontoxic products in 4 h (Honeycutt *et al.*, 1984). It has also been suggested that toxic aromatic compounds in water may be rendered innocuous by converting them to less soluble, high-molecular-weight products, such as by the addition of peroxidase and H_2O_2 (Maloney *et al.* 1986) or by the use of laccase to convert 2,4-dichlorophenol to water-insoluble oligomers and products that do not leach through soil (Ruggiero *et al.*, 1989; Shannon and Bartha, 1988). A peroxidase immobilized on magnetite has been suggested as a means of eliminating all of a variety of chlorophenols from solution (Tatsumi *et al.,* 1996).

Practical technologies based on these innovations have yet to be applied.

REFERENCES

Adams, C. D., Cozzens, R. A., and Kim, B. J., *Water Res.* **31,** 2655–2663 (1997).

Alleman, B. C., Logan, B. E., and Gilbertson, R. L., *Water Res.* **29,** 61–67 (1995).

Armenante, P. M., *in* "Biotreatment of Industrial and Hazardous Waste" (M. A. Levin and M. A. Gealt, eds.), pp. 65–112, McGraw Hill, New York, 1993.

Aronstein, B. N., and Alexander, M., *Environ. Toxicol. Chem.* **11,** 1227–1233 (1992).

Banerjee, D. K., Fedorak, P. M., Hashimoto, A., Masliyah, J. H., Pickard, M. A., and Gray, M. R., *Appl. Microbiol. Biotechnol.* **43,** 521–528 (1995).

Beunink, J., and Rehm, H.-J., *Appl. Microbiol. Biotechnol.* **29,** 72–80 (1988).

Caldwell, S. R., and Raushel, F. M., *Appl. Biochem. Biotechnol.* **31,** 59–73 (1991).

Canter, L. W., Streebin, L. E., Arguiaga, M. C., Carranza, F. E., Miller, D. E., and Wilson, B. H., "Innovative Processes for Reclamation of Contaminated Subsurface Environments." Publ. EPA/600/S2-90/017. Kerr Laboratory, U.S. Environmental Protection Agency, Ada, OK, 1989.

Cheng, T.-C., and Calomiris, J. J., *Enz. Microb. Technol.* **18,** 597–601 (1996).

Choi, Y.-B., Lee, J.-Y., and Kim, H.-S., *J. Microbiol. Biotechnol.* **5,** 41–47 (1995).

Chou, M.-S., and Huang, J.-J., *J. Environ. Eng.* **123,** 569–576 (1997).

Cobb, G. D., and Bouwer, E. J., *Environ. Sci. Technol.* **25,** 1068–1074 (1991).

Compeau, G. C., Mahaffey, W. D., and Patras, L., *in* "Environmental Biotechnology for Waste Treatment" (G. S. Sayler, R. Fox, and J. W. Blackburn, eds.), pp. 91–109. Plenum Press, New York, 1991.

Dasu, B. N., and Sublette, K. L., *Biotechnol. Bioeng.* **34,** 405–409 (1989).

de Best, J. H., Jongema, H., Weijling, A., Doddema, H. J., Janssen, D. B., and Harder, W., *Appl. Microbiol. Biotechnol.* **48,** 417–423 (1997).

Diks, R. M. M., and Ottengraf, S. P. P., *Bioprocess Eng.* **6,** 93–99 (1991).

Duba, A. G., Jackson, K. J., Jovanovich, M. C., Knapp, R. B., Shah, N. N., and Taylor, R. T., *Environ. Sci. Technol.* **30,** 1982–1989 (1996).

Ehrhardt, H. M., and Rehm, H.-J., *Appl. Microbiol. Biotechnol.* **30,** 312–317 (1989).

Ensley, B. D., and Kurisko, P. R., *Appl. Environ. Microbiol.* **60,** 285–290 (1994).

Ergas, S. J., Kinney, K., Fuller, M. E., and Scow, K. M., *Biotechnol. Bioeng.* **44,** 1048–1054 (1994).

Ergas, S. J., and McGrath, M. S., *J. Environ. Eng.* **123,** 593–598 (1997).

Ermisch, O., and Rehm, H.-J., *DECHEMA Biotechnol. Conf.* **3,** 905–908 (1989).

Ernst, C., and Rehm, H.-J., *Appl. Microbiol. Biotechnol.* **43,** 150–155 (1995).

Evans, B. S., Dudley, C. A., and Klasson, K. T., *Appl. Biochem. Biotechnol.* **57/58,** 885–894 (1996).

Fathepure, B. Z., and Vogel, T. M., *Appl. Environ. Microbiol.* **57,** 3418–3422 (1991).

Feakin, S. J., Gubbins, B., McGhee, I., Shaw, L. J., and Burns, R. G., *Water Res.* **29,** 1681–1688 (1995).

Feidieker, D., Kämpfer, P., and Dott, W., *J. Contam. Hydrol.* **19,** 145–169 (1995).

Feng, Y., Racke, K. D., and Bollag, J.-M., *Appl. Microbiol. Biotechnol.* **47,** 73–77 (1997).

Field, J. A., Stams, A. J. M., Kato, M., and Schraa, G., *Antonie van Leeuwenhoek* **67,** 47–77 (1995).

Flathman, P. E., Jerger, D. E., and Woodhull, P. M., *Environ. Prog.* **11,** 202–209 (1992).

Freitas dos Santos, L. M., and Livingston, A. G., *Water Res.* **29,** 179–194 (1995).

Frye, R. J., Welsh, D., Berry, T. M., Stevenson, B. A., and McCallum, T., *Soil Biol. Biochem.* **24,** 607–612 (1992).

Gerritse, J., Renard, V., Visser, J., and Gottschal, J. C., *Appl. Microbiol. Biotechnol.* **43,** 920–928 (1995).

Guillet, J. E., Regulski, T. W., and McAnney, T. B., *Environ. Sci. Technol.* **8,** 923–925 (1974).
Hallas, L. E., Adams, W. J., and Heitkamp, M. A., *Appl. Environ. Microbiol.* **58,** 1215–1219 (1992).
Hapeman, C. J., Karns, J. S., and Shelton, D. R., *J. Agric. Food Chem.* **43,** 1383–1391 (1995).
Hartmans, S., de Bont, J. A. M., Tramper, J., and Luyben, K. C. A. M., *Biotechnol. Lett.* **7,** 383–388 (1985).
Heitkamp, M. A., Camel, V., Reuter, T. J., and Adams, W. J., *Appl. Environ. Microbiol.* **56,** 2967–2973 (1990).
Hirai, M., Ohtake, M., and Shoda, M., *J. Ferment. Bioeng.* **70,** 334–339 (1990).
Hirl, P. J., and Irvine, R. L., *in* "In Situ and On-Site Bioremediation, Fourth Symposium, Vol. 3, pp. 87–92. Battelle Press, Columbus, OH, 1997.
Honeycutt, R., Ballantine, L., LeBaron, H., Paulson, D., Seim, V., Ganz, C., and Milad, G., *in* "Treatment and Disposal of Pesticide Wastes" (R. F. Krueger and J. N. Seiber, eds.), pp. 343–352. American Chemical Society, Washington, DC, 1984.
Hopkins, G. D., and McCarty, P. L., *Environ. Sci. Technol.* **29,** 1628–1637 (1995).
Hopkins, G. D., Semprini, L., and McCarty, P. L., *Appl. Environ. Microbiol.* **59,** 2277–2285 (1993).
Hwang, H.-M., Hodson, R. E., and Lee, R. F., *Environ. Sci. Technol.* **20,** 1002–1007 (1986).
Hwang, J. S., and Chang, H. N., *Biotechnol. Bioeng.* **34,** 380–386 (1989).
Järvinen, K. T., Melin, E. S., and Puhakka, J. A., *Environ. Sci. Technol.* **28,** 2387–2392 (1994).
Jutras, E. M., Smart, C. M., Rupert, R., Pepper, I. L., and Miller, R. M., *Biodegradation* **8,** 31–42 (1997).
Kaake, R. H., Roberts, D. J., Stevens, T. O., Crawford, R. L., and Crawford, D. L., *Appl. Environ. Microbiol.* **58,** 1683–1689 (1992).
Kampbell, D. H., and Wilson, J. T., *J. Hazard. Mater.* **28,** 75–80 (1991).
Kampbell, D. H., Wilson, J. T., Read, H. W., and Stocksdale, T. T., *J. Air Pollut. Control Assoc.* **37,** 1236–1240 (1987).
Katayama, A., and Matsumura, F., *Environ. Sci. Technol.* **25,** 1329–1333 (1991).
Khodadoust, A. P., Wagner, J. A., Suidan, M. T., and Brenner, R. C., *Water Res.* **31,** 1776–1786 (1997).
Kiared, K., Wu, G., Beerli, M., Rothenbühler, M., and Heitz, M., *Environ. Technol.* **18,** 55–63 (1997).
Kim, J. O., *Bioprocess Eng.* **16,** 331–337 (1997).
Koyama, O., Kamagata, Y., and Nakamura, K., *Water Res.* **28,** 895–899 (1994).
Laha, S., and Luthy, R. G., *Biotechnol. Bioeng.* **40,** 1367–1380 (1992).
Leahy, M. C., Ahrens, B. W., Blazicek, T. L., and Maybach, G. B., *in* "In Situ and On-Site Bioremediation, Fourth Symposium," Vol. 3, pp. 463–468. Battelle Press, Columbus, OH, 1997.
Lee, S. H., and Carberry, J. B., *Water Environ. Res.* **64,** 682–690 (1992).
Leeson, A., Hapeman, C. J., and Shelton, D. R., *J. Agric. Food Chem.* **41,** 983–987 (1993).
LeJeune, K. E., Mesiano, A. J., Bower, S. B., Grimsley, J. K., Wild, J. R., and Russell, A. J., *Biotechnol. Bioeng.* **54,** 105–114 (1997).
Lesley, M. P., and Rangan, C. R., *J. Soil. Contam.* **6,** 95–112 (1997).
Leson, G., and Smith, B. J., *J. Environ. Eng.* **123,** 556–562 (1997).
Lewandowski, G. A., Armenante, P. M., and Pak, D., *Water Res.* **24,** 75–82 (1990).
Lin, J.-E., and Wang, H. Y., *J. Ferment. Bioeng.* **72,** 311–314 (1991).
Litherland, S. T., and Anderson, D. W., *in* "In Situ and On-Site Bioremediation, Fourth Symposium," Vol. 5, pp. 425–430. Battelle Press, Columbus, OH, 1997.
Livingston, A. G., and Willacy, A., *Appl. Microbiol. Biotechnol.* **35,** 551–557 (1991).
Loehr, R. C., and Webster, M. T., *in* "Environmentally Acceptable Endpoints in Soil" (D. G. Linz and D. V. Nakles, eds.), pp. 137–386. American Academy of Environmental Engineers, Annapolis, MD, 1997.
Maloney, S. W., Manem, J., Mallevialle, J., and Fiessinger, F., *Environ. Sci. Technol.* **20,** 249–253 (1986).
Manilal, V. B., Haridas, A., Alexander, R., and Surender, G. D., *Water Res.* **26,** 1035–1038 (1992).
Miller, D. E., and Canter, L. W., *Environ. Prog.* **10,** 300–306 (1991).
Mohn, W. W., and Tiedje, J. M., *Microbiol. Rev.* **56,** 482–507 (1992).

Mueller, J. G., Lantz, S. E., Blattmann, B. O., and Chapman, P. J., *Environ. Sci. Technol.* **25,** 1055–1061 (1991).

Munnecke, D. M., *Biotechnol. Bioeng.* **21,** 2247–2261 (1979).

Munnecke, D. M., *in* "Microbial Degradation of Xenobiotics and Recalcitrant Compounds" (T. Leisinger, A. M. Cook, R. Hütter, and J. Nüesch, eds.), pp. 251–269. Academic Press, London, 1981.

Otte, M.-P., Gagnon, J., Comeau, Y., Matte, N., Greer, C. W., and Samson, R., *Appl. Microbiol. Biotechnol.* **40,** 926–932 (1994).

Parvatiyar, M. G., Govind, R., and Bishop, D. F., *Biotechnol. Bioeng.* **50,** 57–64 (1996).

Pedersen, A. R., and Arvin, E., *Biodegradation* **6,** 109–118 (1995).

Pfaender, F. K., and Alexander, M., *J. Agric. Food Chem.* **20,** 842–846 (1972).

Pinelli, D., Fava, F., Nocentini, M., and Pasquali, G., *J. Soil Contam.* **6,** 243–256 (1997).

Rajan, R. V., Seybold, A. L., Hickey, R. F., and Hayes, T., *in* "In Situ and On-Site Bioremediation, Fourth Symposium," Vol. 3, pp. 451–456. Battelle Press, Columbus, OH, 1997.

Rittmann, B. E., and Johnson, N. M., *Water Sci. Technol.* **21**(4/5), 209–219 (1989).

Ro, K. S., Babcock, R. W., and Stenstrom, M. K., *Water Res.* **31,** 1687–1693 (1997).

Roberts, D. J., Ahmad, F., and Pendharkar, S., *Environ. Sci. Technol.* **30,** 2021–2026 (1996).

Roberts, D. J., Kaake, R. H., Funk, S. B., Crawford, D. L., and Crawford, R. L., *Appl. Biochem. Biotechnol.* **39/40,** 781–789 (1993).

Ruggiero, P., Sarkar, J. M., and Bollag, J.-M., *Soil. Sci.* **147,** 361–370 (1989).

Ryan, J. R., Loehr, R. C., and Rucker, E., *J. Hazard. Mater.* **28,** 159–169 (1991).

Schanke, C. A., Bettermann, A. D., Graham, L. L., and Rehm, B. W., *in* "In Situ and On-Site Bioremediation, Fourth Symposium," Vol. 3, pp. 267–272. Battelle Press, Columbus, OH 1997.

Schöllhorn, A., Savary, C., Stucki, G., and Hanselmann, K. W., *Water Res.* **31,** 1275–1282 (1997).

Scott, J. P., and Ollis, D. F., *Environ. Prog.* **4,** 88–103 (1995).

Shannon, M. J. R., and Bartha, R., *Appl. Environ. Microbiol.* **54,** 1719–1723 (1988).

Shimura, M., Koana, T., Fukuda, M., and Kimbara, K., *J. Ferment. Bioeng.* **81,** 573–576 (1996).

Shoda, M., *in* "Biological Degradation of Wastes" (A. M. Martin, ed.), pp. 31–46. Elsevier Applied Science, London, 1991.

Smet, E., Chasaya, G., Van Langenhove, H., and Verstraete, W., *Appl. Microbiol. Biotechnol.* **45,** 293–298 (1996).

Sorial, G. A., Smith, F. L. Suidan, M. T., Pandit, A., Biswas, P., and Brenner, R. C., *J. Environ. Eng.* **123,** 530–537 (1997).

Spuij, F., Alphenaar, A., de Wit, H., Lubbers, R., Brink, K., Gerritse, J., Gottschal, J., and Houtman, S., *in* "In Situ and On-Site Bioremediation, Fourth Symposium," Vol. 5, pp. 431–437. Battelle Press, Columbus, OH, 1997.

Stefanoff, J. G., and Garcia, M. B., Jr., *Environ. Prog.* **14,** 104–110 (1995).

Stinson, M. K., Hahn, W., and Skovronek, H. S., *in* "Innovative Hazardous Waste Treatment Technology Series" (H. M. Freeman and P. R. Sferra, eds.), Vol. 3, pp. 163–167. Technomic Publ. Co., Lancaster, PA, 1991.

Stockinger, H., Heinzle, E., and Kut, O. M., *Environ. Sci. Technol.* **29,** 2016–2022 (1995).

Stokley, K. E., Drake, E. N., Prince, R. C., and Douglas, G. S., *in* "In Situ and On-Site Bioremediation, Fourth Symposium," Vol. 4, pp. 487–492. Battelle Press, Columbus, OH, 1997.

Stucki, G., and Thüer, M., *Appl. Microbiol. Biotechnol.* **42,** 167–172 (1994).

Stucki, G., and Thüer, M., *Environ. Sci. Technol.* **29,** 2339–2345 (1995).

Suidan, M. T., Flora, J. R. V., Boyer, T. K., Wuellner, A. M., and Narayanan, B., *Water Res.* **30,** 160–170 (1996).

Swanson, W. J., and Loehr, R. C., *J. Environ. Eng.* **123,** 538–546 (1997).

Tatsumi, K., Wada, S., and Ichikawa, H., *Biotechnol. Bioeng.* **51,** 126–130 (1996).

Taylor, B. F., Amador, J. A., and Levinson, H. S., *FEMS Microbiol. Lett.* **110,** 213–216 (1993).

Uchiyama, H., Oguri, K., Nishibayashi, M., Kokufuta, E., and Yagi, O., *J. Ferment. Bioeng.* **79,** 608–613 (1995).

Vail, R. L., *Oil Gas J.* **89**(45), 53–57 (1991).

van Groenestijn, J. W., and Hesselink, P. G. M., *Biodegradation* **4,** 283–301 (1993).

Van Hoof, P. L., and Jafvert, C. T., *Environ. Toxicol. Chem.* **15,** 1914–1924 (1996).

Wani, A. H., Branion, R. M. R., and Lau, A. K., *J. Environ. Sci. Health* **A32,** 2027–2055 (1997).

Weckhuysen, B., Vriens, L., and Verachtert, H., *Appl. Microbiol. Biotechnol.* **42,** 147–152 (1994).

Widrig, D. L., Boopathy, R., and Manning, J. F., Jr., *Environ. Toxicol. Chem.* **16,** 1141–1148 (1997).

Wilson, J. T., Fogel, S., and Roberts, P. V., *in* "Detection, Control, and Renovation of Contaminated Ground Water" (N. Dee, W. F. McTernan, and E. Kaplan, eds.), pp. 168–178. American Society of Civil Engineers, New York, 1987.

Woodward, R., and Ramsden, D., *in* "Gas, Oil, Coal and Biotechnology II" (C. Akin and J. Smith, eds.), pp. 59–66. Institute of Gas Technology, Chicago, 1990.

Wright, W. F., Schroeder, E. D., Chang, D. P. Y., and Romstad, K., *J. Environ. Eng.* **123,** 547–555 (1997).

Yocum, P. S., Irvine, R. L., and Bumpus, J. A., *Water Environ. Res.* **67,** 174–180 (1995).

Young, D. M., Kitts, C. L., Unkefer, P. J., and Ogden, K. L., *Biotechnol. Bioeng.* **56,** 258–267 (1997).

CHAPTER 18

Bioremediation of Metals and Other Inorganic Pollutants

Soils, aquatic sediments, waste streams, and bodies of water frequently are contaminated with metals, metalloids, cyanide, nitrate, or radionuclides. An individual site usually contains many of these pollutants, and often environments containing organic pollutants that are being considered for bioremediation have a number of these inorganic toxicants. These toxicants may be derived from mining operations, refining ores, sludge spread on land, fly ash from incinerators, the processing of radioactive materials, metal plating, or the manufacture of electrical equipment, paints, alloys, batteries, pesticides, or preservatives. The substances of concern include ionic lead, chromium, mercury, uranium, selenium, zinc, arsenic, cadmium, gold, silver, copper, and nickel, as well as nitrate and cyanide.

In contrast with biodegradation of organic compounds, which destroys the molecules, metals and metalloids are not destroyed by microbial processes. However, the metal- or metalloid-containing molecule or ion may be modified, immobilized, or detoxified so that a bioremediation may be feasible. Also in contrast with the biodegradation of organic compounds, vestiges of which may be lost from the polluted environment as they are converted to CO_2 or CH_4, bioremediation of most metals and metalloids does not result in volatilization. Nevertheless, microbial processes may lead to appreciable or sometimes essentially complete remediation of the contaminated environment.

The chief ways by which such remediation may be accomplished include biosorption, bioaccumulation, reduction, solubilization (commonly associated with oxidation of sulfides or ferrous iron), precipitation, and methylation. Some of the technologies are fully developed and have been and are now being used in practice.

Others are still undergoing active research and must still be considered in the exploratory phase.

BIOSORPTION AND BIOACCUMULATION

Biosorption refers to the passive uptake of metals by microbial cells. This sorption is passive in that no energy is required. Indeed, the biomass in biosorption is often nonviable, some processes relying on nonviable or deliberately killed cells, others making use of a biomass that has a high percentage of viable cells. Such binding of inorganic pollutants is being used, or is being considered for use, with metals, metalloids, and radionuclides. Biosorption is characteristically rapid, and with the appropriate type of biomass, it can remove a high percentage of individual metallic cations from waste streams.

The organisms that make up the biomass may be bacteria, filamentous fungi, yeasts, or algae. The choice of organisms is particularly important because of wide differences in their capacity for sorption or their affinity for the metal. Toxicity of the metal to microorganisms is not a disadvantage in biosorptive processes because a nonviable biomass may then be used. Some of the organisms that accumulate significant amounts of metals and radionuclides are shown in Table 18.1.

The biosorption of U, Zn, Pb, Cd, Co, Ni, Cu, Hg, Th, Zn, Cs, Au, Ag, Sn, and Mn has been investigated. These studies have shown that the extent of

Table 18.1

Representative Microorganisms That Accumulate Metals and Radionuclides[a]

Microorganism	Element	Uptake (% of dry weight)
Zoogloea sp.	Co	25
	Ni	13
Citrobacter sp.	Cd	170
	U	900
Bacillus sp.	Cu	15
	Zn	14
Chlorella vulgaris	Au	10
Rhizopus arrhizus	Pb	10
	Ag	5.4
	Hg	5.8
Aspergillus niger	Th	19

[a] From Gadd (1992).

sorption varies markedly with the metal as well as with the microorganism. For some cations and organisms, the extent of removal is very large. This is shown by representative data in Table 18.1. For other metals and organisms, little is bound. Because not all metals are sorbed by the cells or biomass of a particular organism, the binding may be quite selective. In instances in which the metal is valuable, even though a pollutant, it may be recovered from the biomass and reused.

Biosorption often results from the formation of metal–organic complexes with constituents of microbial cell walls, capsules, or extracellular polymers synthesized and excreted by the organisms. It may result from the positively charged metallic cation being retained electrostatically by negatively charged functional groups in the walls, capsules, or polymers. The mechanisms of sorption and the cell constituents involved in sorption are usually unknown, although considerable research has been devoted to the possible mechanisms. Moreover, the removal of the metallic cations from solution may simply involve their precipitation on the biomass.

A variety of approaches have been considered to attain efficient metal removal by relying on inexpensive sources of biomass to sorb the contaminants. Sewage sludge, the mass of algae or cyanobacteria in a natural or managed aquatic bloom, yeasts such as *Saccharomyces cerevisiae* from alcoholic fermentations, or microbial mass from some other industrial fermentation are usually available at low cost. The sludge in the activated-sludge process of treating sewage characteristically removes a reasonable percentage of the toxic cations in solution. Similarly, the mass of algae or cyanobacteria in a bloom, which may be fostered by deliberate additions of nutrients from treated sewage, may effectively remove considerable amounts of Cu, Zn, Cd, Pb, Ni, and Hg from solution. The biomass employed for sorption may be immobilized on some inert surface, and the immobilized cells then would remove the cations from solution (Gadd, 1992), and cells may be immobilized in a gel, such as one made of polyacrylamide, and the contaminated solution allowed to contact the gel surface (Pons *et al.*, 1993). A variety of biofilms have been devised to treat contaminated water that is brought into contact with these biofilms (Summers, 1992; Brierley *et al.*, 1989; Macaskie and Dean, 1989). In addition, floating mats composed of a mixture of algae, cyanobacteria, and nonphotosynthetic bacteria have also been found to take up metallic cations from solution (Bender and Phillips, 1994).

Several proprietary products are available for the purposes of biosorption. One product is composed principally of *Bacillus* that is treated with alkali to increase the metal sorption characteristics, and the treated cells are then immobilized in polyethyleneimine beads. A second contains nonviable algae immobolized in silica gel or some other solid matrix. When placed in fixed-bed columns, they will remove metallic cations from wastewaters. A third material is composed of *Rhizopus arrhizus* that is freeze dried and immobilized in 0.70- to 0.85-mm particles. This fungal biomass has been tested for U removal. A fourth product contains a mixture of microorganisms that are immobilized in polysulfone beads, and this preparation has been evaluated for Zn and Mn removal (Brierley, 1990).

A notable illustration of the use of biosorption is the treatment of the heavy metal-containing wastewater from an operating gold mine, which each day discharges 15 million liters. The system for removal of the toxic cations involves a series of rotating biological contactors or large disks on which a biofilm of *Pseudomonas* sp. has become attached. This biomass serves to reduce the concentration of toxic metallic cations, chiefly Cu and Fe, in solution by >95% and to provide a discharge into a stream that, following the biosorption, is capable of supporting a fishery (Whitlock, 1990).

Extensive discussions of biosorption are presented by Brierley (1990), Gadd (1992), and Macaskie and Dean (1989).

In some instances, the uptake of metallic cations is an energy-requiring process, which thus is associated with viable and actively metabolizing microorganisms. Such a process is termed *bioaccumulation*. If the metal is ultimately present within the confines of the cell and the organism is viable, then the uptake may represent bioaccumulation. Laboratory investigations have demonstrated that bacteria, fungi, and algae bring about such energy-dependent processes, but it is not certain which of the metal removals in bioremediation by living organisms do, in fact, represent biosorptive or bioaccumulative phenomena.

REDUCTION

It has long been known that microorganisms can bring about the reduction of a wide array of inorganic anions and cations. Nitrate, sulfate, and carbonate are well-known examples of nonmetallic anions that are reduced microbiologically. Early studies showed that soil bacteria and fungi are capable of reducing ionic As, V, Te, S, and Mo as well as MnO_2 and chlorate (Bautista and Alexander, 1972), and enzyme preparations prepared from a bacterium were observed to reduce As, Pb, Cu, Mo, U, Se, Bi, Te, Va, Mn, Fe, W, Ag, Au, Os, and Ru (Woolfolk and Whitely, 1962). More recent research has focused on some of these as well as other elements, such as Hg and Cr. The reductions convert a higher oxidation state of the element to a lower one, e.g., Hg(II) to Hg(O), Fe(III) to Fe(II), Se(VI) to Se(O), Mn(IV) to Mn(II), U(VI) to U(IV), or As(V) to As(III). The conversion of Se is illustrated in Fig. 18.1.

The microorganisms that carry out these reductions come from a variety of genera, and the transformations relate to the physiology of the organisms in different ways. (a) Some of the reactions are catalyzed enzymatically, but some result from the excretion by the organism of a reducing compound that abiotically converts the ion or metallic compound from a higher to a lower oxidation state. (b) Among the reactions catalyzed enzymatically, some serve a role in the physiology of the

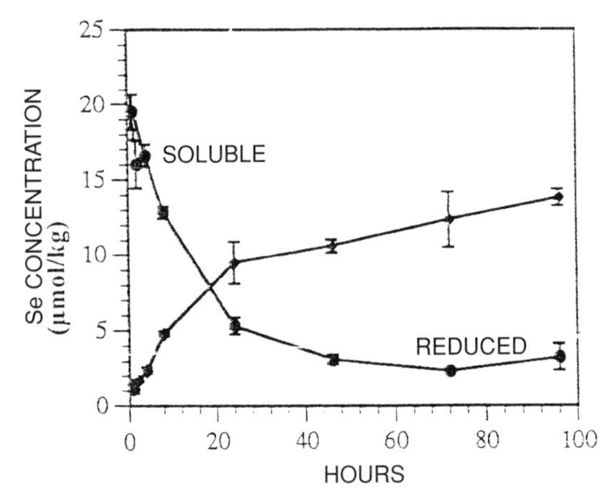

Figure 18.1 Conversion of soluble Se to reduced Se in wetland sediment. (Modified from Zhang and Moore, 1997.)

organism by acting as a terminal electron acceptor and permitting growth, much as O_2 functions for aerobes, simple organic molecules for fermentative microorganisms, sulfate for *Desulfovibrio,* nitrate for denitrifying bacteria, and CO_2 for methanogenic bacteria. Others, however, have no known role in the physiology of the responsible species and represent merely fortuitous reductions coupled to enzymatic or microbial oxidations of some substrates.

For the reductions to occur, an energy source must be available. That energy source is usually organic, but it could be inorganic. The conversions are often, but not always, inhibited by O_2. Some of the reductions occur naturally in soils, sediments, and waters, particularly when O_2 is absent, but often the conversion is so slow or so much of the oxidized form of the element remains that intervention (i.e., remediation) is required to allow for a more rapid or extensive conversion.

From the toxicological and environmental viewpoints, the reductions are important for three reasons. The reductions may change the (a) toxicity, (b) water solubility, and/or (c) mobility of the element. In other words, one oxidation state is toxic but the reduced form has less or no toxicity at environmental concentrations, one is reasonably soluble in water whereas the other is very poorly soluble, and one remains with the solids of soil or sediment and thus is nonmobile whereas the other is not retained and thus becomes mobile. An increase in water solubility and mobility can be exploited to bioremediate insoluble forms of an element in soil because the product of reduction would move out of the solids and into the water. Conversely, a decrease in solubility of an element as a result of reduction can be used to remove it from surface or groundwater. In some cases, the product of reduction is volatile, as in the reduction of nitrate to N_2 or Hg(II) to Hg(O), and the loss of the element from soil or water is (as with nitrate) or could be (as with Hg) the basis for bioremediation.

The use or possible use of reductive reactions in bioremediation is discussed in connection with individual elements.

SOLUBILIZATION/OXIDATION

Microbial oxidations may be the basis for bioremediation. The oxidative reaction that has attracted the most interest and that is in practical use is associated with the oxidations of sulfides, Fe(II), and elemental sulfur. However, it has been suggested that bacteria that oxidize arsenite to arsenate [As(III) to As(V)] may be used to enhance the removal of As from As-containing sewage because Fe(III) added to the wastewater precipitates more As(V) than As(III) (Williams and Silver, 1984).

When sulfide-containing ores become accessible to microorganisms and O_2 is present, chemoautotrophic bacteria of the genus *Thiobacillus* frequently proliferate. These bacteria use sulfides or, in the case of *Thiobacillus ferrooxidans,* both sulfides and Fe in the ores as their energy sources. The products of their activity are H_2SO_4 and Fe(III), and the quantity of H_2SO_4 may be so great in some areas that adjacent waters become highly acidic. The acidification also leads to the solubilization of K and Ca, which are not pollutants, but also Cd, Hg, Ni, Pb, Se, Ag, Al, and other elements. The acidification and uncontrolled release of metallic cations are themselves significant problems in many regions. However, when controlled, these conversions can be beneficial. The controlled or managed process is known as *bioleaching* (Ehrlich and Brierley, 1990).

Bioleaching has been used for a number of applications. For example, it has been employed to remove Cu from Cu-rich dump sites (Gokcay and Onerci, 1994) and, by the activity of *T. ferrooxidans* and *T. thiooxidans,* to solubilize Pb, Zn, U, and Cu in metal-rich deposits (Rossi and Ehrlich, 1990; McGreedy and Gould, 1990). *T. ferrooxidans* has also been the basis for removing Fe in the treatment of acid mine waters, and Fe oxidation by bacteria may contribute to Fe removal when acid waters are treated in constructed wetlands (Brierley, 1990). To remove metals from contaminated sludge, elemental S or pyrite may be deliberately added, and the S is oxidized by thiobacilli to H_2SO_4, which in turn solubilizes metallic contaminants (Jordan *et al.,* 1996; Shooner and Tyagi, 1996; Tyagi and Couillard, 1991).

PRECIPITATION

Some approaches to remediation rely on precipitation. If the metallic ion is soluble but reacts with a product of microbial metabolism to yield a water-insoluble derivative, removal of that final precipitate constitutes a remediation. Two types of precipitates are relevant, sulfides and phosphates, because microorganisms, under suitable conditions, generate H_2S from sulfate and inorganic phosphate from organic P compounds. In turn, H_2S or phosphate forms insoluble derivatives with a number of metallic ions.

The insoluble sulfide may be removed by simple sedimentation or filtration or it can be separated by centrifugation. Sulfate reduction requires anaerobiosis and a source of readily degradable organic matter, which may be decaying plant debris, animal manure, whey, or a simple compound such as methanol, ethanol, or lactic acid. The responsible microorganisms are the sulfate-reducing bacteria, typically species of *Desulfovibrio*. By such procedures, it is possible to form insoluble sulfides of Zn, Pb, Ni, Cr, Cd, Cu, Fe, Hg, and other elements.

Microbial sulfate reduction for remediation is also important in the use of constructed wetlands to remove metals from acid and coal mine drainage and other waters polluted with toxic ions and sulfate. The energy sources for the microorganisms include organic matter in the sediments, added organic substances, and/or carbonaceous materials arising from cattails or other plants growing in the wetland. Extensive removal of Zn, Cd, Cu, Pb, As, Ag, Fe, and Mn may be achieved by the appropriate usage of wetlands (Brierley, 1990; Farmer *et al.*, 1995). Precipitation as sulfides is not the only mechanism of metal removal in wetlands, and sorption, oxidation, and both abiotic and biotic processes may be involved.

Several types of reactors have been devised for the removal of metals by means of the sulfide generated by sulfate-reducing bacteria. These have been designed for the removal of Cu, Zn, and Ni from waters derived from an abandoned gold mine, and substantial removal of these cations was achieved in a pilot-scale assessment (Wildman *et al.*, 1995). A laboratory bioreactor with lactate as the energy source led to extensive removal of Cd, Pb, and Cu (Wijaya *et al.*, (1994). The sulfide formed by these anaerobes has also been used to purify metal-contaminated groundwater pumped through a biological treatment system (Barnes *et al.*, 1994).

Metal remediation as a result of the microbial formation of inorganic phosphate has also been considered. This approach relies on the cleavage of glycerol-1-phosphate by *Citrobacter* sp. (Macaskie and Dean, 1989) or tributyl phosphate by *Pseudomonas* sp. (Thomas and Macaskie, 1996), the latter being a solvent waste from U processing, and the inorganic P then precipitates U or Pb in the wastewater.

METHYLATION

Microorganisms methylate a surprisingly large number of elements, the products depending on the element accepting the methyl groups. Thus, mono-, di-, tri-, and tetramethylated products are formed. Some of the products are highly toxic, and several of the conversions are activations. However, if the product is nontoxic and volatile and if the contaminated water or environment does not contain metals that can be methylated to yield toxic derivatives, this type of transformation could be the basis for a bioremediation. Such an approach has been proposed for the removal of Se from contaminated waters.

The applicability of these various types of transformation to different elements is summarized in Table 18.2.

INDIVIDUAL POLLUTANTS

Various approaches have been proposed, devised, and sometimes practically exploited for the remediation of metals, metalloids, and other inorganic contaminants. These inorganic pollutants are discussed individually.

Chromium

Potential approaches to Cr remediation have been investigated intensively. The element in toxicologically important concentrations may be found in effluents from cooling towers or the cooling water from electricity-generating stations and in contaminated soils or groundwaters. Three approaches have been pursued: biosorption, enzymatic reduction, and abiotic reduction coupled to microbial sulfate reduction.

Chromium may exist as Cr(VI) or Cr(III). The hexavalent form is characteristic of chromates, dichromates, and chromic trioxide, and the trivalent form characterizes Cr oxides and hydroxides. Cr(VI) is toxic and highly soluble in water, whereas Cr(III) is considerably less toxic, far less water soluble, and hence less mobile if present in soil. Because of the differences in solubility of Cr(VI) and Cr(III), a reduction will result in precipitation of the element and thus diminished mobility and transport. It has long been known that bacteria in culture can reduce

Table 18.2

Processes That May Result in Metal Bioremediation

Process	Metal, metalloid, ion
Biosorption	Ag, Au, Cd, Co, Cr, Cu, Hg, Ni, Pb, Ra, U, Zn
Biological reduction	Ag, As, Au, Cr, Hg, Mo, NO_3^{-1}, Se, U
Abiotic reaction following microbial sulfate reduction	As, Cd, Cr, Cu, Hg, Pb, Zn
Methylation	Se

chromate and dichromates (Romanenko and Korenkov, 1977), and such bacteria are widespread and readily isolated from soils, sediments, and contaminated waters. Species of *Alcaligenes, Bacillus, Cornyebacterium, Enterobacter, Escherichia, Micrococcus, Pseudomonas, Vibrio,* and others have been reported to effect the reduction (Cooke *et al.*, 1995; Kvasnikov *et al.*, 1988). Both aerobes and anaerobes are active. Because of the toxicity of Cr(VI), the organisms of choice must have a high degree of tolerance, but such isolates have been obtained. Under suitable conditions, >99% of Cr(VI) may be reduced (Bhide *et al.*, 1996; Cooke *et al.*,, 1995). However, at high concentrations, the percentage of the toxic ion that is reduced diminishes appreciably (DeLeo and Ehrlich, 1994).

The microbial reduction may be accomplished in several ways. For example, Cr(VI)-contaminated soil can be amended with manure and flooded, resulting in extensive reduction of the toxic anion (Fig. 18.2), or Cr-containing water may be applied in the form of irrigation water to soil amended with organic matter, and microbial activity would then convert the hexavalent form to trivalent Cr so that the effluent from the soil contains only trace amounts of the element (Losi *et al.*, 1994b). A laboratory study suggested that Cr(VI) can be reduced in aquifer solids when benzoate was added as a source of carbon and energy (Shen *et al.*, 1996). A continuously stirred reactor supplemented with molasses as the carbon source and inoculated with *Pseudomonas mendocina* (Bhide *et al.*, 1996) and a bench-scale packed-bed reactor inoculated with *Bacillus* sp. (Chirwa and Wang, 1997) were used to reduce Cr(VI) and precipitate the trivalent form.

A second means of remediating Cr relies on the addition of sulfate and the provision of anaerobic conditions and an available C source to allow sulfate-

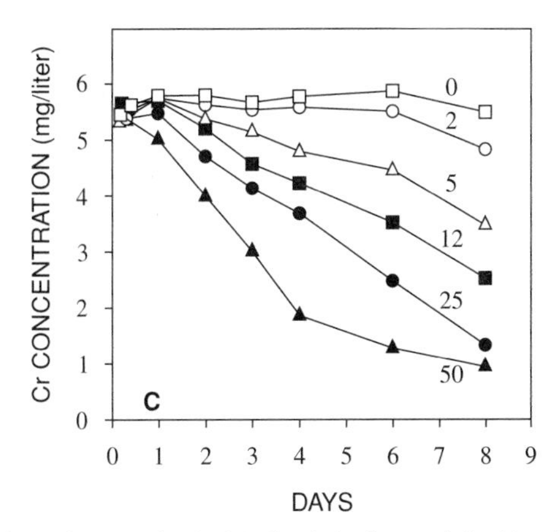

Figure 18.2 Anaerobic reduction of Cr(VI) in flooded soil amended with animal manure at 0, 2, 5, 12, 25, and 50 kg/ha. (Modified from Losi *et al.*, 1994a. Reprinted with permission of Elsevier Science.)

reducing bacteria to generate H_2S. The H_2S then reacts abiotically to reduce Cr(VI). Such an approach was used in a field test in which sulfate and fermented molasses were introduced into a shallow aquifer contaminated with Cr, the results showing remediation around the site of injection (deFilippi, 1994), and it was also investigated in a batch-scale procedure in the laboratory (Sulzbacher *et al.*, 1997).

SELENIUM

Inorganic selenium in soils, sediments, and waters is frequently present as Se(VI) in the form of selenate and Se(IV) in the form of selenite. Two dissimilar approaches have been proposed for the remediation of waters contaminated with this element; one involves a reduction and the other relies on methylation.

The ability of microorganisms to reduce oxidized Se anions has been recognized for a long time (Hall, 1926). This capacity is possessed by a broad array of aerobic bacteria, fungi, and yeasts as well as by anaerobic bacteria. Both selenate and selenite can thus be transformed, and the product is elemental Se. Because the elemental form is insoluble, the conversion may be useful in the decontamination of polluted waters, with the elemental Se being removed after the reduction by sedimentation or filtration. The conversion proceeds naturally in sediments (Zhang and Moore, 1997), but both the rate and the extent can be markedly increased to give a potentially useful technology. For example, the Se concentration in water was reduced by >95% when it was passed through a column of soil maintained under anaerobiosis; the soil in this field test was colonized with appropriate bacteria (Mattison, 1992). Another technology relies on algae grown in ponds using drainage water containing nitrate, selenate, and selenite. The algal mass and the contaminated water are introduced into a second pond that is anoxic, and here the oxidized Se is reduced to elemental Se, which is ultimately removed with the sludge composed of the microbial mass. An additional benefit is the removal of nitrate through denitrification (Gerhardt, 1994). A laboratory-scale fluidized-bed reactor containing the selenate-reducing bacterium, *Thauera selenatis,* and denitrifying bacteria has been devised to convert the anions to elemental Se. The contaminated water is pumped through a recycling sludge blanket and the reactor, and acetate is provided as a C and energy source (Lawson and Macy, 1995; Macy *et al.*, 1993). The latter approach has been evaluated in a pilot-scale system for the removal of Se from agricultural drainage water and was found to reduce the concentration of Se(VI) and Se(IV) by 98% (Cantafio *et al.*, 1996).

The methylation of selenate and selenite was also described decades ago. The original reports dealt with fungi that convert both anions to dimethyl selenide (Challenger and Bird, 1934):

$$\mathrm{Se(VI)} \rightarrow \mathrm{CH_3SeCH_3}$$
$$\mathrm{Se(IV)} \rightarrow \mathrm{CH_3SCH_3}$$

However, bacteria (Doran and Alexander, 1977) and algae (Fan *et al.*, 1997) may carry out similar transformations. The process occurs naturally in soils (Francis *et al.*, 1974) and in aquatic sediments (Chau *et al.* 1976) and leads to volatilization of the element. The volatilization of Se as a result of microbial formation of dimethyl selenide has been tested as a remediation strategy in a field study of pond sediments in a Se-contaminated reservoir containing 10–209 mg Se/kg of sediment. The most effective treatment, which involved the addition of organic wastes, increased the rate of Se volatilization over the background rates by 40-fold (Frankenberger and Karlson, 1994).

URANIUM

Strategies for U bioremediation involve biosorption, reduction, solubilization, and precipitation. For sorption, the microorganisms are retained by some solid material, and the U-containing solution is brought into contact with the immobilized biomass, which removes much of the element from the continuously flowing stream. The organisms proposed for use are the fungus *Rhizopus arrhizus,* which is converted to a particulate form (Tsezos *et al.*, 1989); residual brewery yeast retained on a membrane (Riordan, 1997); and a strain of *Pseudomonas* immobilized in polyacrylamide gel (Pons and Fuste, 1993).

U(VI) can be reduced microbiologically to U(IV) by *Clostridium* and *Desulfovibrio,* and the former bacterium may also form U(III) (Francis *et al.*, 1994; Phillips *et al.*, 1995). A process has been proposed for U bioremediation in which the element is initially extracted from the contaminated soil with bicarbonate, and the U(VI) in the extracts is then reduced by *Desulfovibrio desulfuricans* (Phillips *et al.*, 1995). The U(IV) that is generated is insoluble and thus may be removed easily from the extract.

The solubilization of U is a consequence of the action of *Thiobacillus* in the bioleaching of U from soils contaminated with nuclear wastes (Macaskie, 1991), with the autotrophic bacteria solubilizing the metal from the soil in a process analogous to their action in the bioleaching of ores containing various metals. An example of U bioremediation by precipitation is the use of a mixed culture that releases inorganic phosphate during its degradation of the tributyl phosphate used as a solvent for U extraction, and the inorganic phosphate then precipitates U as uranyl phosphate. The P salts can then be separated from the liquid phase (Thomas and Macaskie, 1996).

NITRATE

Wastewaters from municipal sewage-treatment plants frequently contain high concentrations of nitrate. Because this anion may cause eutrophication of surface waters or cause methemoglobinemia in people drinking the water, a widely used procedure for treatment of the effluent from the waste-treatment plant is to create anaerobic conditions, provide a carbon source, and allow the indigenous denitrifying bacteria to reduce the nitrate to N_2. This technology is effective and well developed for effluents following sewage treatment. It has also been used in a full-scale operation to destroy nitrate in impoundment ponds arising from the processing of U designed for nuclear weapons (Schmitt and Ballew, 1995).

Cyanide

Cyanide is common in many waste streams, and thus its potential as a substrate for microorganisms has attracted attention. Both aerobic and anaerobic microorganisms are able to destroy the cyanide, often using it as a N source for growth (Aronstein *et al.*, 1995; Nagle *et al.*, 1995). The remediation of cyanide-containing wastewater being discharged from an operating underground gold mine was accomplished by establishing a biofilm containing a strain of *Pseudomonas* on rotating biological contactors having a large surface area. The bacterium converts the cyanide to ammonium on the first two disks of the contactors, and the ammonium is oxidized to nitrate by a film of nitrifying bacteria on the last three disks (Whitlock, 1990). Because of the toxicity of cyanide, any organism selected for bioremediation must have a degree of tolerance to the inhibitor.

Arsenic

Arsenic is subject to microbial oxidation, reduction, and methylation, and it can be precipitated as a result of microbial sulfate reduction. Because methylation results in the formation of highly toxic methylated arsines, such a conversion is not an acceptable remediation strategy. Several bacterial genera are able to oxidize arsenite [As(III)] (Ilyaletdinov and Abdrashitova, 1981; Osborne and Ehrlich, 1976) to arsenate [As(V)], and such activity may be used in wastewater to form arsenate, which is precipitated more readily out of the water by treatment with ferric ions (Williams and Silver, 1984).

Bacteria and yeasts can also reduce arsenate to arsenite, sometimes using the latter as an electron acceptor for growth (Vidal and Vidal, 1980; Macy *et al.*, 1996). Such reductions occur under appropriate conditions in soils and sediments, and it has been suggested that, because arsenite is more soluble in water than arsenate, As-contaminated soils might be freed of some of the toxicant by promoting the reduction and thus increasing the leachability of the element (Dowdle *et al.*, 1996). A readily available C source would be needed to promote the transformation.

Another possible approach involves the formation of As_2S_3. This insoluble sulfate would be generated as a consequence of H_2S formation. One means could involve a bacterium such as *Desulfotomaculum auripigmentum,* which reduces both arsenate to arsenite and sulfate to H_2S, thereby leading to As_2S_3 precipitation (Newman *et al.,* 1997).

Mercury

Microorganisms from a variety of taxonomic groups have the capacity to reduce divalent mercury to yield metallic Hg:

$$Hg(II) \rightarrow Hg(O)$$

The product is volatile and thus moves into the overlying air. If these microorganisms are stimulated to greater activity, they might be the basis for a bioremediation.

Divalent mercury can also be methylated. However, the high acute toxicity of the resulting mono- and dimethyl mercury precludes consideration of Hg methylation for remediation.

Other Elements

The reaction types described above apply to a variety of other toxic metals, and appropriate strategies possibly can be devised for practical bioremediation schemes. Laboratory studies, sometimes with pure cultures and sometimes with simple experimental models for treatment systems, suggest a multitude of potentially feasible approaches. Some of the promising avenues for further exploration include the following:

1. Biosorption to remove Cd (Matis *et al.,* 1996), Ag, Cu (Simmons *et al.,* 1995), and Ra (Tsezos and Keller, 1983) from dilute aqueous solution.
2. Microbial reduction to convert the highly soluble Mo(VI) to the poorly soluble Mo(IV) (Tucker *et al.,* 1997).
3. Microbial reduction of sulfate to H_2S and abiotic precipitation of the insoluble metallic sulfides, as in the treatment of Zn from mine drainage (Farmer *et al.,* 1995) as well as Pb and other elements (Pado *et al.,* 1994).

REFERENCES

Aronstein, B. N., Patarek, J. R., Rice, L. E., and Srivastava, V. J., *in* "Bioremediation of Inorganics" (R. E. Hinchee, J. L. Means, and D. R. Burris, eds), pp. 81–87. Battelle Press, Columbus, OH, 1995.

Barnes, L. J., Scheeren, P. J. M., and Buisman, C. J. N., *in* "Emerging Technology for Bioremediation of Metals" (J. L. Means and R. E. Hinchee, eds.), pp. 38–49. Lewis Publishers, Boca Raton, FL, 1994.

Bautista, E. M., and Alexander, M., *Soil Sci. Soc. Am. Proc.* **36,** 918–920 (1972).

Bender, J., and Phillips, P., *in* "Emerging Technology for Bioremediation of Metals" (J. L. Means and R. E. Hinchee, eds.), pp. 85–98. Lewis Publishers: Boca Raton, FL, 1994.

Bhide, J. V., Dhakephalkar, P. K., and Paknikar, K. M., *Biotechnol. Lett.* **18,** 667–672 (1996).

Brierley, C. L., *Geomicrobiol. J.* **8,** 201–223 (1990).

Brierley, C. L., Brierley, J. A., and Davidson, M. S., *in* "Metal Ions and Bacteria" (T. J. Beveridge and R. J. Doyle, eds.), pp. 359–382. Wiley, New York, 1989.

Cantafio, A. W., Hagen, K. D., Lewis, G. E., Bledsoe, T. L., Nunan, K. M., and Macy, J. M., *Appl. Environ. Microbiol.* **62,** 3298–3303 (1996).

Challenger, F., and North, H. E., *J. Chem. Soc.* 68–71 (1934).

Chirwa, E. M. N., and Wang, Y.-T., *Environ. Sci. Technol.* **31,** 1446–1451 (1997).

Cooke, V. M., Hughes, M. N., and Poole, R. K., *J. Ind. Microbiol.* **14,** 323–328 (1995).

DeFilippi, L. J., *in* "Remediation of Hazardous Waste Contaminated Soils" (D. L. Wise and D. J. Trantolo, eds.), pp. 437–457. Dekker, New York, 1994.

DeLeo, P. C., and Ehrlich, H. L., *Appl. Microbiol. Biotechnol.* **40,** 756–759 (1994).

Dowdle, P. R., Laverman, A. M., and Oremland, R. S., *Appl. Environ. Microbiol.* **62,** 1664–1669 (1996).

Ehrlich, H. L., and Brierley, C. L., eds. "Microbial Mineral Recovery," McGraw-Hill, New York, 1990.

Fan, T. W.-M., Lane, A. N., and Higashi, R. M., *Environ. Sci. Technol.* **31,** 569–576 (1997).

Farmer, G. H., Updegraff, D. M., Radehaus, P. M., and Bates, E. R., *in* "Bioremediation of Inorganics" (R. E. Hinchee, J. L. Means, and D. R. Burris, eds.), pp. 17–24. Battelle Press, Columbus, OH, 1995.

Francis, A. J., Dodge, C. J., Lu, F., Halada, G. P., and Clayton, C. R., *Environ. Sci. Technol.* **28,** 636–639 (1994).

Frankenberger, W. T., Jr., and Karlson, U., *in* "Selenium in the Environment" (W. T. Frankenberger, Jr. and S. Benson, eds), pp. 369–387. Dekker, New York, 1994.

Gadd, G. M., *in* "Microbial Control of Pollution" (J. C. Fry, G. M. Gadd, R. A. Herbert, C. W. Jones, and I. A. Watson-Craik, eds.), pp. 59–88. Cambridge Univ. Press, Cambridge, UK, 1992.

Gökcay, C. F., and Önerci, S., *in* "Emerging Technology for Bioremediation of Metals" (J. L. Means and R. L. Hinchee, eds.), pp. 61–73. Lewis Publishers, Boca Raton, FL, 1994.

Hall, I. C., *J. Bacteriol.* **11,** 407–408 (1926).

Ilyaletdinov, A. N., and Abdrashitova, S. A., *Mikrobiologiya* **50,** 197–204 (1981).

Jordan, M. A., McGinness, S., and Phillips, C. V., *Miner. Eng.* **9,** 169–181 (1996).

Kvasnikov, E. I., Klyushnikova, T. M., Kasatkina, T. P., Stepanyuk, V. V., and Kuberskaya, S. L., *Mikrobiologiya* **57,** 680–685 (1988).

Lawson, S., and Macy, J. M., *Appl. Microbiol. Biotechnol.* **43,** 762–765 (1995).

Losi, M. E., Amrhein, C., and Frankenberger, W. T., Jr., *Environ. Toxicol. Chem.* **13,** 1727–1735 (1994a).

Losi, M. E., Amrhein, C., and Frankenberger, W. T., Jr., *J. Environ. Qual.* **23,** 1141–1150 (1994b).

Lundquist, T. J., Green, F. B., Tresan, R. B., Newman, R. D., Oswald, W. J., and Gerhardt, M. B., *in* "Selenium in the Environment" (W. T. Frankenberger, Jr. and S. Benson, eds.), pp. 251–278. Dekker, New York, 1994.

Macaskie, L. E., *Crit. Rev. Biotechnol.* **11,** 41–112 (1991).

Macaskie, L. E., and Dean, A. C. R., *in* "Biological Waste Treatment" (A. Mizrahi, ed.), pp. 159–201. Wiley-Liss, New York, 1989.

Macy, J. M., Lawson, S., and DeMoll-Decker, H., *Appl. Microbiol. Biotechnol.* **40,** 588–594 (1993).

Macy, J. M., Nunan, K., Hagen, K. D., Dixon, D. R., Harbour, P. J., Cahill, M., and Sly, L. I., *Int. J. System. Bacteriol.* **46,** 1153–1157 (1996).

Matis, K. A., Zouboulis, A. I., Grigoriadou, A. A., Lazaridis, N. K., and Ekateriniadou, L. V., *Appl. Microbiol. Biotechnol.* **45,** 569–573 (1996).

Mattison, P. L., "Bioremediation of Metals: Putting It to Work." Cognis, Santa Rosa, CA, 1992.
McCready, R. G. L., and Gould, W. D., *in* "Microbial Mineral Recovery" (H. L. Ehrlich and C. L. Brierley, eds.), pp. 107–125. McGraw-Hill, New York, 1990.
Nagle, N. J., Rivard, C. J., Mohagheghi, A., and Phillippidis, G., *in* "Bioremediation of Inorganics" (R. E. Hinchee, J. L. Means, and D. R. Burris, eds.), pp. 71–79. Battelle Press, Columbus, OH, 1995.
Newman, D. K., Beveridge, T. J., and Morel, F. M. M., *Appl. Environ. Microbiol.* **63,** 2022–2028 (1997).
Osborne, F. H., and Ehrlich, H. L., *J. Appl. Bacteriol.* **41,** 295–305 (1976).
Pado, R., Pawlowska-Cwiek, L., and Szwagrzyk, J., *Ekol. Pol.* **42,** 103–123 (1994).
Phillips, E. J. P., Landa, E. R., and Lovley, D. R., *J. Ind. Microbiol.* **14,** 203–207 (1995).
Pons, M. P., and Fuste, M. C., *Appl. Microbiol. Biotechnol.* **39,** 661–665 (1993).
Riordan, C., Bristard, M., Putt, R., and McHale, A. P., *Biotechnol. Lett.* **19,** 385–387 (1997).
Romanenko, V. I., and Korenkov, V. N., *Mikrobiologiya* **46,** 414–417 (1977).
Rossi, G., and Ehrlich, H. L., *in* "Microbial Mineral Recovery" (H. L. Ehrlich and C. L. Brierley, eds.), pp. 149–170. McGraw-Hill, New York, 1990.
Schmidt, G. C., and Ballew, M. B., *in* "Bioremediation of Inorganics" (R. E. Hinchee, J. L. Means, and D. R. Burris, eds.), pp. 109–116. Battelle Press, Columbus, OH, 1995.
Shen, H., Pritchard, P. H., and Sewell, G. W., *Environ. Sci. Technol.* **30,** 1667–1674 (1996).
Shooner, F., and Tyagi, R. D., *Appl. Microbiol. Biotechnol.* **45,** 440–446 (1996).
Simmons, P., Tobin, J. M., and Singleton, I., *J. Ind. Microbiol.* **14,** 240–246 (1995).
Sulzbacher, K., Ecke, H., Lagerkvist, A., and Calmano, W., *Environ. Technol.* **18,** 301–307 (1997).
Summers, A. O., *Curr. Opin. Biotechnol.* **3,** 271–276 (1992).
Thomas, R. A. P., and Macaskie, L. E., *Environ. Sci. Technol.* **30,** 2371–2375 (1996).
Tsezos, M., and Keller, D. M., *Biotechnol. Bioeng.* **25,** 201–215 (1983).
Tsezos, M., McCready, R. G. L., and Bell, J. P., *Biotechnol. Bioeng.* **34,** 10–17 (1989).
Tucker, M. D., Barton, L. L., and Thomson, B. M., *J. Environ. Qual.* **26,** 1146–1152 (1997).
Tyagi, R. D., and Couillard, D., *in* "Biological Degradation of Wastes" (A. M. Martin, ed.), pp. 307–322. Applied Science, London, 1991.
Vidal, F. V., and Vidal, V. M. V., *Mar. Biol.* **60,** 1–7 (1980).
Whitlock, J. L., *Geomicrobiol. J.* **8,** 241–249 (1990).
Wijaya, S., Henderson, W. D., Bewtra, J. K., and Biswas, N., *in* "Proc. 48th Ind. Waste Conf. Purdue, pp. 469–481, 1994.
Wildeman, T., Gusek, J., Cevaal, J., Whiting, K., and Scheuering, J., *in* "Bioremediation of Inorganics" (R. E. Hinchee, J. L. Means, and D. R. Burris, eds.), pp. 141–148. Battelle Press, Columbus, OH, 1995.
Williams, J. W., and Silver, S., *Enz. Microb. Technol.* **6,** 530–537 (1984).
Woolfolk, C. A., and Whiteley, H. R., *J. Bacteriol.* **84,** 647–658 (1962).
Zhang, Y., and Moore, J. N., *J. Environ. Qual.* **26,** 910–916 (1997).

CHAPTER 19

Recalcitrant Molecules

Many organic compounds, both low molecular weight and polymeric, persist for long periods in soils, subsoils, aquifers, surface waters, and aquatic sediments. Some of the sites containing these compounds or synthetic polymers are so rich in toxic substances that the persistence can be attributed to the inability of microorganisms to grow and bring about biodegradation in the presence of the toxins. However, many of the durable molecules are located in environments in which microorganisms are proliferating and actively metabolizing, so that the persistence cannot be ascribed to conditions inimical to microbial life. The low- and high-molecular-weight substances that thus resist biodegradation are known as *recalcitrant* molecules.

Recalcitrance is a property not commonly attributed to chemicals. The characteristic of stubbornness is more commonly associated with people, and the term usually denotes a resistance to guidance or authority. Nevertheless, even if the word stretches somewhat the limits of good English usage, the adjective "recalcitrant" aptly characterizes the resistance of the molecules to the expected omnivorous feeding habits of microbial communities in nature, and for better or worse, the term is now generally accepted.

Molecules that persist in nature are undesirable for many reasons. Some are intrinsically toxic and deleteriously affect humans, domesticated animals, agricultural crops, wildlife, fish and other aquatic animals, or microorganisms. The longer the molecule remains in nature, the greater is the exposure of susceptible individuals or populations and the greater is the risk of harmful effects. Some recalcitrant molecules are not toxic at the concentrations found in nature, but they reach hazardous levels because they are biomagnified in natural food chains. Persistent polymers are of no toxicological significance, but their obvious presence in forests or parks and on roadsides is aesthetically unpleasant. The enormous amount of nonbiodegradable packaging materials that end up in municipal wastes must ultimately be hauled away and buried, thereby not only contributing to the increasing costs of waste disposal but occupying much space in landfills. Moreover, a compound that is both mobile and persistent may be transported to previously uncontaminated areas, thereby increasing the potential exposure of susceptible individuals and populations and hence increasing the risk of harm; in contrast, a mobile compound that is readily degraded will tend to be destroyed and thus the extent of its transport is less.

Another reason for concern with persistent molecules is related to the imperfect state of knowledge of toxicology—as with all other sciences. The knowledge of toxicology is growing constantly, and what was once deemed to be safe is occasionally later found to be harmful, often as a result of assessments of previously unevaluated physiological processes. The broadening vista of toxicology has shown that compounds previously believed to be innocuous were in fact harmful to humans, animals, or plants. When this is found to be true, government regulatory agencies typically ban the chemicals. Such bans quickly result in a rapid reduction in exposure of susceptible organisms if the compounds are readily biodegraded. However, the cessation of production and use of the newly recognized toxicant does not lead to a quick diminution in exposure if the molecule is recalcitrant; it has already been released in nature, and a biological means for its rapid destruction does not exist. This problem is well illustrated by certain chlorinated hydrocarbon insecticides (such as DDT, dieldrin, heptachlor, and chlordane) that were widely used two, three, or four decades ago; although no longer manufactured or used, they remain in many soils and continue to pose problems of contamination.

A recalcitrant molecule may be wholly resistant to microbial modification. Some indeed are. Other recalcitrant molecules are destroyed, either by growth-linked processes or cometabolically, but the rates in nature are so slow that detectable levels are observed for years after the first introduction of the compound to the affected environment.

EXAMPLES OF RECALCITRANCE

PCBs are one of the more prominent groups of persistent compounds. At one time, they were widely used in transformers and capacitors, as hydraulic and heat-transfer fluids, and as solvents and plasticizers. The PCBs that are highly chlorinated are notably persistent, and sediments contaminated with these toxicants 20 or more years ago still show disturbingly high levels. Dibenzothiophene and alkylated dibenzothiophene from crude oil are known to persist in sea sediments for periods in excess of 10 years (Sinkkonen, 1989), a quaternary ammonium compound known as Amo-1618 persists in soil for more than 8 years (Marth and Mitchell, 1959), and tetralin and indane sulfonates, which are minor components of widely used linear ABS surfactants, are refractory to microbial attack (Field *et al.*, 1992). Also notably persistent in soil are such five-ring PAHs as perylene and 1,2,5,6-dibenzanthracene (Bossert and Bartha, 1986) and the tetrachloro through the octachloro dibenzo-*p*-dioxins and dibenzofurans (Orazio *et al.*, 1992).

Data showing the persistence of a number of compounds in soil are presented in Table 19.1. Except for the polychlorinated dibenzofurans and dibenzo-*p*-dioxins, the chlorobenzenes, and trichlorobenzoic acid, all the substances in the table are pesticides. The chemicals were still present at the sampling time shown, so that the actual longevities are longer than the times tabulated. Several of the insecticides, because of their resistance, contaminate the surfaces of vegetables grown underground, enter earthworms that are consumed by and hence affect birds, or are carried from treated fields into rivers and lakes long after their last application for insect control. The early dates of some of the references in the table attest to the public concern and clamor to have many of those insecticides removed from the market.

Among many of the less readily metabolized compounds that remain in soil years after their first introduction, a bi- or multiphasic rate of disappearance is evident. In the first few months, the degradation rate is rapid, but then it slows dramatically to a point that little is lost in succeeding years. This may reflect the sequestering of the compound in a manner that makes it less available to microorganisms, that is, the so-called "aging" of the molecule. It is now evident that many compounds are subject to a time-dependent sequestration in soil, including PAHs, chlorinated hydrocarbons, and chemicals of several structural classes (Alexander, 1995, 1997). This type of sequestration is considered in Chapter 10. For reasons that are unclear, but again possibly related to the aging effect, a readily metabolized substrate may be found in soil for unexpectedly long periods, as in the detection of approximately 0.1% of the parathion, an insecticide that is usually quickly degraded, applied to soil 16 to 20 years earlier (Stewart *et al.*, 1971).

The persistence of chemicals in waters is of special significance because of the use of surface and groundwaters for drinking and the frequently rapid and distant transport of pollutants in rivers and streams. A typical example is the finding of various chloroethenes, dichlorobenzenes, and alkylbenzenes as well as nonylphenol isomers in a groundwater contaminated originally with secondary sewage effluents, some of the contaminants undoubtedly persisting in the aquifer for more than 30 years (Barber *et al.*, 1988). The nonylphenols may have been derived from the nonylphenol polyethoxylate surfactants in some detergent preparations. Analysis of drinking water and samples from rivers also show the presence of a disturbing array of synthetic organic molecules (Meijers and van der Leer, 1976; Miller, 1973), some of which may have been recently introduced but many of which are known to be recalcitrant. Tetra-, penta-, hexa-, and heptachloronaph-

Table 19.1

Persistence of Several Compounds in Soil

Compound	Persistence (years)[a]	Reference
Azinphosmethyl	8	Staiff *et al.* (1975)
BHC	16	Nash and Harris (1973)
Chlordane	16	Nash and Harris (1973)
Chlorfenvinphos	4	Chisholm (1975)
Chlorobenzenes, di- to hexa-	31	Beck *et al.* (1995)
DDD	24	Boul *et al.* (1994)
DDE	24	Boul *et al.* (1994)
DDT	24	Boul *et al.* (1994)
Dicamba	4	Burnside *et al.* (1971)
Dieldrin	21	Martin *et al.* (1993)
EDB	19	Steinberg *et al.* (1987)
Endrin	16	Nash and Harris (1973)
Heptachlor	16	Nash and Harris (1973)
Isodrin	16	Nash and Harris (1973)
Lindane	21	Martin *et al.* (1993)
Mirex	12	Carlson *et al.* (1976)
Monuron	3	Birk (1955)
Paraquat	6	Fryer *et al.* (1975)
Picloram	5	Burnside *et al.* (1971)
Polychlorinated dibenzofurans	18	McLachlan *et al.* (1996)
Polychlorinated dibenzo-*p*-dioxins	18	McLachlan *et al.* (1996)
Simazine	20	Scribner *et al.* (1992)
Tordon	5	Burnside *et al.* (1971)
Toxaphene	16	Nash and Harris (1973)
2,3,6-Trichlorobenzoic acid	4	Burnside *et al.* (1965)
Trifluralin	3	Golab *et al.* (1979)

[a] Compound still present in soil at the time indicated.

thalenes have also been detected in river sediments, and these industrial chemicals also appear to be long lived (Falandysz *et al.*, 1996).

Molecular recalcitrance is not restricted to anthropogenic chemicals because natural organic materials often remain undecomposed not merely for years or decades but occasionally for millenia or even longer. The persistence of a few of these paleobiochemicals is depicted in Fig. 19.1. These ancient residues are found

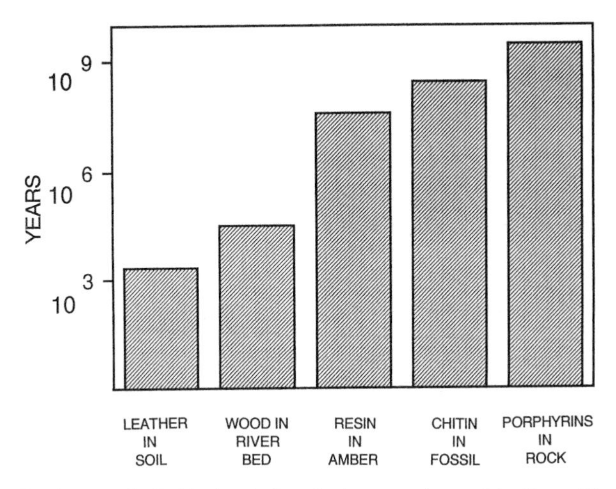

Figure 19.1 Persistence of paleobiochemicals and other natural materials. (From Alexander, 1973.)

in fossils, sedimentary rocks, and coal—where their presence is not surprising—but also in lake and marine sediments, where their occurrence is anomalous. On the basis of radiocarbon dating techniques, even part of the humus fraction of soil appears to persist for hundreds to thousands of years, and the organic matter in some peat deposits has withstood microbial destruction for tens of thousands of years. Even long-chain *n*-alkanes, which are usually readily destroyed by aerobic microorganisms, have been found to persist in soil (Bol *et al.,* 1996) and marine sediments (Eglinton *et al.,* 1997) for thousands of years. Soils and peats support active microbial communities, so that the persistence of organic fractions or discrete organic molecules is noteworthy. Even resting structures of various fungi persist in viable form for many years—evidence of the resistance to biodegradation of the organic components of the outer surfaces of the sclerotia, chlamydospores, and other resting structures (Alexander, 1973).

Thus, despite the remarkable catabolic versatility of microorganisms, they are not omnivorous. Not all organic molecules are catabolized, at least not at reasonable rates. For one reason or another, microbial communities in nature are, unfortunately, not infallible.

PERSISTENT PRODUCTS

Products generated as bacteria and fungi transform organic molecules are sometimes persistent. The original substrate in some cases may itself be long lived, but instances are known in which the original substrate is quickly destroyed yet yields a metabolite that is not. The products thus are found long after all traces of the parent compound have disappeared.

The longevity of products is well illustrated by some of the early chlorinated insecticides. DDT, for example, is itself not quickly degraded microbiologically, but it is slowly converted to a number of products, including DDE (Boul *et al.*, 1994; Spencer *et al.*, 1996), which remain in treated soil for years after the last pesticide application. The chlorinated insecticides aldrin and heptachlor are converted microbiologically in soil to their corresponding epoxides, which are known as dieldrin and heptachlor epoxide and are also insecticidal, and the products remain when little or none of the parents can be detected in treated fields (Wilkinson *et al.*, 1964). A location outside of Denver where aldrin was manufactured still contains dieldrin more than 20 years after industrial disposal of the chemical ceased.

Among the pesticides, 2,6-dichlorobenzamide formed from the herbicide dichlobenil (Verloop and Nimmo, 1970), phorate sulfone generated from the insecticide phorate (Lichtenstein *et al.*, 1973), and a keto derivative produced from the nematicide avermectin (Gullo *et al.*, 1983) are present in soil longer than their precursors. Many other examples exist among the pesticides deliberately applied to or inadvertently reaching soil. Analogous formation of long-lived products from less persistent or readily biodegradable parents occurs in surface, ground, and wastewaters. For example, nonylphenol polyethoxylate surfactants are metabolized during sewage treatment, but they are transformed to (nonylphenoxy)acetic acid and [(nonylphenoxy)ethoxy]acetic acid, which persist in the river waters that receive effluent from the sewage-treatment plants (Ahel *et al.*, 1987). Similarly, at a waste-disposal site in Ontario into which dimethylamine was introduced, this readily degradable amine was converted to the carcinogen *N*-nitrosodimethylamine, which was still detected in adjacent groundwater more than 20 years after the last addition of dimethylamine to the waste-disposal pit (Fig. 19.2). Although this potent carcinogen can be metabolized by high cell densities of microorganisms under laboratory conditions, the same rapid conversion clearly does not occur in at least some natural environments.

SYNTHETIC POLYMERS

Industrial countries use vast quantities of synthetic polymers. These polymers find uses as packaging materials, fabrics for clothing and carpets, insulation, and bedding, and large amounts are used in the construction of buildings and the manufacture of automobiles. An appreciation of the enormous quantities involved can be gained by an examination of the amounts produced in 1 year by the United States alone (Table 19.2).

In contrast to many other recalcitrant materials, the resistance of synthetic polymers is not a toxicological issue. Nevertheless, their durability in nature is the

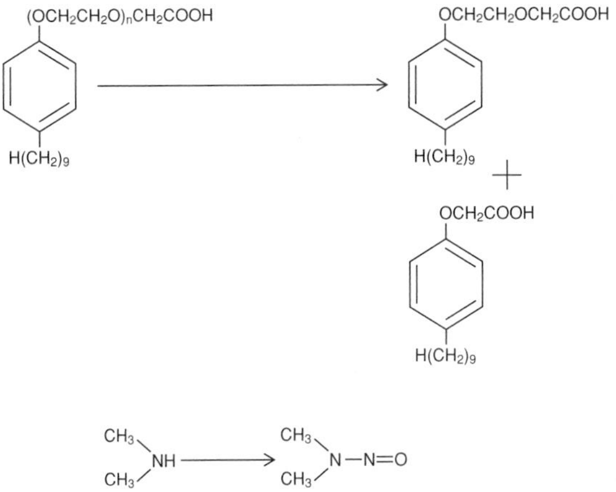

Figure 19.2 Conversion of the readily degradable nonylphenoxycarboxylic acids and dimethylamine to persistent products, namely, (nonylphenoxy)acetic and [(nonylphenoxy)ethoxy]acetic acids (top) and *N*-nitrosodimethylamine (bottom).

basis for concern. One reason is the large contribution they make to municipal solid wastes; in the United States, plastics account for 8% of the total weight and approximately 20% of the volume of municipal solid wastes (Palmisano and Pettigrew, 1992). This enormous tonnage represents much of the volume and occupies much of the space in landfills and, once buried in a landfill, such polymers

Table 19.2

Production of Synthetic Polymers in the United States in 1996[a]

Polymer	kg × 10^9
Plastics	
Polyethylene	12.03
Poly(vinyl chloride)	6.00
Polypropylene	5.44
Polystyrene	2.75
Polyamides	0.51
Synthetic fibers	
Polyester	1.74
Nylon	1.27
Olefin	1.09
Acrylic	0.21

[a] Anonymous (1997).

do not undergo reduction in weight or volume because of biodegradation. Prominent in such wastes are polyethylene, poly(vinyl chloride), and polystyrene, which are major plastics used in packaging materials. A second basis for the public outcry is aesthetic: packaging materials that end up as litter in forests, parks, and roadsides detract from human enjoyment of the surroundings. Moreover, plastic particles composed of polyethylene, polystyrene, and polypropylene have been found in many locations in the ocean, sometimes even in remote areas (Colton *et al.*, 1974; Morris, 1980), and these could cause blockages in the intestines of small fish. The plastic particles may come from solid wastes discarded by ships at sea or may originate from plastic-producing factories or processing facilities adjacent to coastal areas, rivers, or estuaries and be transported to remote locations because they are not destroyed microbiologically.

The synthetic polymers that represent the most commonly used plastics and fibers are wholly resistant to biodegradation and thus remain for as yet undetermined periods of time even in environments with highly diverse and physiologically active communities of aerobic or anaerobic microorganisms. Some of these refractory polymers are listed in Table 19.3. The recalcitrant polymers not only do not serve as C sources for any bacterium or fungus, but they are also not subject to cometabolism. The frequent reports that fungi or bacteria degrade these polymers have been discounted, and the spurious findings are attributed to microbial growth on substances added during processing, impurities, substances coating the surfaces

Table 19.3

Polymers Resistant to Microbial Degradation[a]

Acetate rayon (Estron)	Polymonochlorotrifluorethylene
Acrylonitrile-vinyl chloride (Dynel)	Polystyrene
Carboxymethyl cellulose (high degree of substitution)	Polytetrafluoroethylene (Teflon)
Cellulose acetate (fully acetylated)	Polyurethane (polyether linked)
Cellulose acetate-butyrate	Poly(vinyl butyral)
Nylon	Poly(vinyl chloride)
Phenol-formaldehyde	Poly(vinyl chloride)-acetate
Polyacrylonitrile (Orlon)	Poly(vinylidene chloride)
Polydichlorostyrene	Resorcinol-formaldehyde
Polyethylene (high molecular weight)	Silicone resins
Poly(ethylene glycol) terephthalate (Dacron)	Vinylidene chloride-vinyl chloride copolymer (Saran)
Polyisobutylene (high molecular weight)	Zein formaldehyde (Vicara)
Poly(methyl methacrylate)	

[a] From Alexander (1973).

of the polymers, or low-molecular-weight compounds added as plasticizers, lubricants, or stabilizers.

Nevertheless, a few synthetic polymers of commercial importance are biodegradable. These include high-molecular-weight polyethylene glycols (Obradors and Aguilar, 1991), poly(vinyl acetate) (Garcia Trejo, 1988), poly(vinyl alcohol) (Sakai *et al.*, 1988), polyester polyurethane (Kay *et al.*, 1991), some polycaprolactones (Oda *et al.*, 1995; Cameron *et al.*, 1988; Tanaka *et al.*, 1976), polyesters (Sasikala and Ramana, 1996), and a poly(lactic acid)-glycolic acid copolymer (Torres *et al.*, 1996). Also biodegradable are bacterial polymers that may have industrial importance, for example, poly(3-hydroxybutyrate) (Cain, 1992) and copolyesters of 3- and 4-hydroxybutyrates (Kunioka *et al.*, 1989). Among a number of the nonbiodegradable high-molecular-weight polymers, the lower-molecular-weight counterparts are susceptible to microbial decomposition, but these smaller molecules are usually not of practical importance.

The desire of society for biodegradable plastics has prompted some companies to label their products as being biodegradable. In fact, however, the synthetic polymer in most of these products is not biodegradable. The recalcitrant polymer is formulated together with starch or gelatin. Once the starch or gelatin is destroyed, the product is converted to a fragmented material, but the original synthetic polymer is essentially unaltered. The physical structure of the product that is marketed is thereby altered, but the chemical integrity of the synthetic polymer is not modified.

MECHANISMS OF RECALCITRANCE

Because of the continued presence in nature of persistent compounds, both simple and polymeric, and the need for replacements having the same utility, it is important to establish why such substances are resistant to biodegradation, that is, to establish the mechanisms of recalcitrance. Modern society requires packaging materials, fabrics, pesticides, and other chemicals, but future materials should not create the environmental problems that the existing or earlier products have made.

The mechanisms of recalcitrance may be linked to structural features of the molecules of concern, physiological limitations of living organisms, or properties of the environment in which the compounds are found, and a consideration of the reasons for persistence thus must include assessments of the contributions of chemical, microbiological, and environmental factors.

The compounds that are long lived in nature can be separated into several categories. (a) Molecules that appear to be wholly resistant to microbial attack and that are not metabolized under any conditions, at least based on present knowledge. (b) Compounds that are always slowly metabolized in nature. These may be rapidly destroyed by high cell densities of bacteria or large fungal biomasses in culture, but analogous rapid turnover is not characteristic of the natural habitats of the microorganisms. (c) Chemicals that are destroyed quickly in some environments or under certain circumstances but persist in other environments or circumstances. Many examples of each category are known.

A number of conditions must be satisfied for biodegradation to occur. (a) An enzyme must exist that can catalyze the transformation. Obviously coupled with this condition is the existence of an organism containing that enzyme. (b) The organism—presumably a bacterium or fungus—must be present in the same environment as the compound. (c) The molecule must be in a form that is available for microbial utilization. (d) If the enzyme is intracellular, as is the case for most low-molecular-weight substrates, the substrate must pass through the cell surface. (e) Should the enzyme be inducible, conditions must allow for induction to occur. (f) Environmental conditions must be suitable for microbial metabolism and, because the biomass in nature of organisms active in degrading many synthetic compounds is small, often for proliferation (Alexander, 1973). These conditions were first presented in Chapter 1. A consideration of these six requisites suggests several reasons for persistence. Some of these explanations pertain to the truly recalcitrant molecules, which are always long lived. Others pertain only to substrates that are often persistent but that are also, in one environment or another, quickly transformed.

Nonexistence of an Active Organism

Given the millions of compounds that have now been described, it is plausible that biochemical evolution has not resulted in the appearance of an enzyme capable of catalyzing a modification in many of these synthetic novelties. Enzymes are reasonably specific for the molecules on which they act, and despite the hundreds of millions of years of biochemical evolution, only a limited number of catabolic pathways have appeared. The enzymes important in catabolism are obviously critical for the provision of energy and building blocks for microorganisms, but these enzymes function on the substrates that microorganisms have encountered during their evolution and not necessarily on each and every novelty created in the laboratories of organic chemists. The nonabsolute specificity of enzymes is probably the basis for the transformations of many of the novel molecules created in recent times, but it is wishful thinking to expect that every new compound that has been synthesized has an enzyme able to catalyze its alteration. If no enzyme for a specific compound exists, an organism able to modify that molecule also will not exist.

Impermeability of the Cell

The enzymes responsible for catalyzing the biodegradation of many compounds are solely intracellular, and if a potential substrate does not cross the cell membrane and penetrate to the site in the organism where the enzyme is found, no reaction will occur. The initial stages in the metabolism of some molecules, especially those of high molecular weight, are catalyzed by extracellular enzymes, and these molecules can thus be transformed. However, if no extracellular enzyme exists to generate products to which the cell is permeable, the parent nonpenetrating molecule will be recalcitrant. It is thus possible that an intracellular enzyme of broad specificity never combines with a potential substrate simply because of the permeability barrier at the cell surface. Molecular weight is not the sole determinant of permeation through the cell membrane, and molecular shape and other properties of the chemical may prevent its transport into the organism. Impermeability of the cell may explain why high-molecular-weight polyethylenes are resistant to biodegradation whereas the low-molecular-weight polyethylenes are metabolized (Potts *et al.*, 1972): the enzymes initiating the catabolism of molecules composed of chains of $-CH_2-$ (which are called alkanes among the simpler ones and polyethylenes among the larger chains) are entirely intracellular.

Inaccessibility of Site in Molecule Potentially Acted on Enzymatically

A particular part of the enzyme (known as the active site) must combine with the substrate for a reaction to occur, and the specific site in the molecule where the reaction is to take place must be accessible. If that site is inaccessible, then no conversion will occur. Some compounds, such as alkanes or long-chain fatty acids, are acted on at the terminal ends of the molecules, but the terminal ends may be inaccessible; such inaccessibility may be a consequence of the folding or coiling of the ends of large molecules, a possible reason why some synthetic polymers may resist microbial degradation. Other compounds may be protected because they contain a substituent that sterically prevents the enzyme from combining with a molecule that otherwise might be a suitable substrate. Large molecules also have extensive cross-linkages, and these may mask the site on the potential substrate with which the enzyme must bind.

Lack of Induction of Requisite Enzymes

Enzymes involved in a particular physiological process may be active in the organism regardless of the presence of the substrate (constitutive enzymes) or they may only be formed and active when the substrate or possibly a closely related molecule is present (inducible enzymes). The initial phase of many biodegradative reaction sequences entails the activity of inducible enzymes, but the conditions may not be suitable for induction. The common inducer is the substrate itself. However, the concentration of the substrate or another inducer in the aqueous phase may be too low to result in enzyme formation, which may occur because most of the compound is sorbed or present in a NAPL, leaving a concentration in the water that is too low to promote enzyme formation.

Cometabolism

Should the population or biomass of organisms active on a substrate be small and the substrate be one that is only acted on by cometabolism, that population or biomass will not increase in size and the initially slow conversion will be maintained. This contrasts with growth-linked biodegradation, in which the slow initial stage is replaced by a period of rapid conversion. Cometabolism may be rapid in the laboratory if the organisms are grown on another substrate, but such high cell densities do not characterize natural ecosystems although they may be found in bioreactors or result from particular remediation technologies in the field.

Environmental Factors

Chemicals may persist not because they are intrinsically refractory but because of some factor in the environment that prevents rapid destruction. These are not properly refractory molecules. If growth is necessary for significant biodegradation, all nutrients must be present and available; these nutrients include inorganic ions, water, O_2 or another electron acceptor, and—for cometabolizing species—an organic compound to serve as a C and energy source. The environment must not contain toxins at levels sufficiently high to prevent growth or activity. The inhibitors may be organic, metallic cations, or salts at high concentration, but sometimes the lack of activity is the result of a pH that is too high or too low.

A well-known anomaly in the biodegradation of natural products is the resistance of peats and mucks to biodegradation. These soil materials often contain 30 to 90% organic matter and may exist in nature for thousands of years provided that they are waterlogged and hence anaerobic. Once O_2 is introduced, which takes place when the excess water is removed, decomposition proceeds at reasonable rates. Experiments involving the addition of cellulose, which is a good substrate for both aerobes and anaerobic bacteria, show that this polysaccharide is not destroyed if placed below the surface of soil that has been flooded for several weeks (Fig. 19.3). This inhibition of cellulose degradation appears to be a result of two changes, the fall in pH and the appearance of acetic, butyric, and possibly other organic acids (Kilham and Alexander, 1984). Although moderately low pH per se often does not abolish the metabolism of a polysaccharide like cellulose and low levels of fatty acids at high pH are not inhibitory, the combination of low pH and fatty acids results in antimicrobial potency. The reason is probably that the proton-

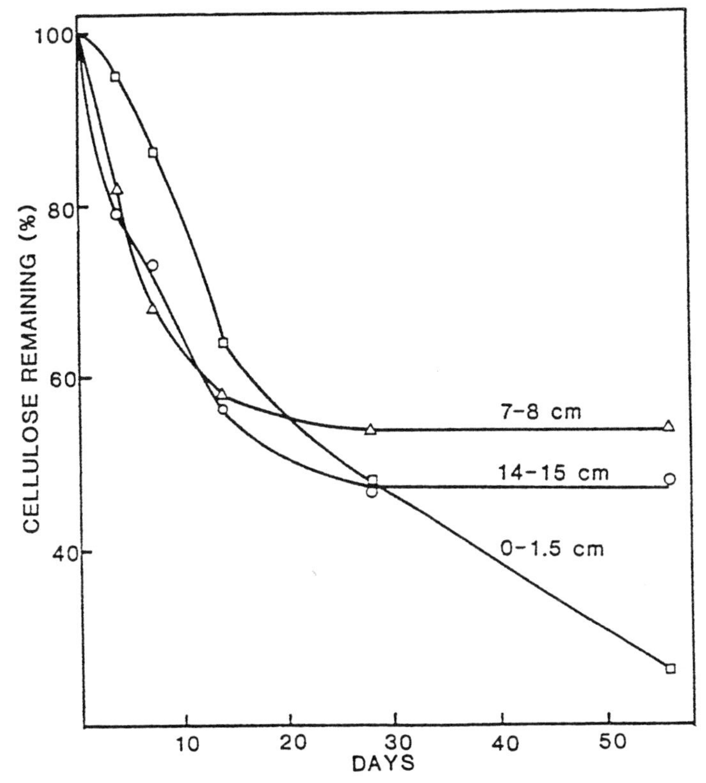

Figure 19.3 Decomposition of cellulose added at three depths to a soil flooded at time of addition of the polysaccharide. (Reprinted with permission from Kilham, and Alexander, 1984. © Williams & Wilkins.)

ated molecules (which appear in increasing concentration as the pH falls) are toxic but the unprotonated anions are not.

The poor availability or the total lack of bioavailability of a compound may be a major determinant of persistence. The molecule may not be available because of its sorption to particulate matter in the environment, its presence in a NAPL, or its sequestration at microenvironmental sites not accessible to bacteria or fungi. The longevity of aged chemicals bears witness to the problems that microorganisms encounter with substrates that become less available.

A compound present at concentrations below the threshold for growth will also persist. Subthreshold concentrations appear to be characteristic of some pollutants that are found entirely in the aqueous phase of some environments. However, the equilibrium concentration in water of a chemical that is extensively sorbed or extensively partitioned into a NAPL may also be so low as to preclude growth of species able to metabolize that substrate.

REFERENCES

Ahel, M., Conrad, T., and Giger, W., *Environ. Sci. Technol.* **21,** 697–703 (1987).

Alexander, M., *Biotechnol. Bioeng.* **15,** 611–647 (1973).

Alexander, M., *Environ. Sci. Technol.* **29,** 2713–2717 (1995).

Alexander, M., *in* "Environmentally Acceptable Endpoints in Soil" (D. G. Linz and D. V. Nakles, eds.), pp. 43–136. American Academy of Environmental Engineers, Annapolis, MD, 1997.

Anonymous, *Chem. Eng. News* **75**(25), 38–79 (1997).

Barber, L. B., II, Thurman, E. M., Schroeder, M. P., and LeBlanc, D. R., *Environ. Sci. Technol.* **22,** 205–211 (1988).

Beck, A. J., Alcock, R. E., Wilson, S. C., Wang, M.-J., Wild, S. R., Stewart, A. P., and Jones, K. C., *Adv. Agron.* **55,** 345–391 (1995).

Birk, L. A., *Can. J. Agric. Sci.* **35,** 377–387 (1955).

Bol, R., Huang, Y. Meridith, J. A., Eglington, G., Harkness, D. D., and Ineson, P., *Eur. J. Soil Sci.* **47,** 215–222 (1996).

Bossert, I. D., and Bartha, R., *Bull. Environ. Contam. Toxicol.* **37,** 490–495 (1986).

Boul, H. L., Garnham, M. L., Hucker, D., Baird, D., and Aislable, J., *Environ. Sci. Technol.* **28,** 1397–1402 (1994).

Burnside, O. C., Wicks, G. A., and Fenster, C. R., *Weeds* **13,** 277–278 (1965).

Burnside, O. C., Wicks, G. A., and Fenster, C. R., *Weed Sci.* **19,** 323–325 (1971).

Cain, R. B., *in* "Microbial Control of Pollution" (J. C. Fry, J. M. Gadd, R. A. Herbert, C. W. Jones, and I. A. Watson-Craik, eds.), pp. 293–338. Cambridge Univ. Press, Cambridge, UK, 1992.

Cameron, J. A., Bunch, C. L., and Huang, S. J., *in* "Biodeterioration" (D. R. Houghton, R. N. Smith, and H. O. Wiggins, eds.), pp. 553–561. Elsevier Applied Science, London, 1988.

Carlson, D. A., Konyha, K. D., Wheeler, W. B., Marshall, G. P., and Zaylskie, R. G., *Science* **194,** 939–941 (1976).

Chisholm, D., *Can. J. Soil Sci.* **55,** 177–180 (1975).

Colton, J. B., Jr., Knapp, F. D., and Burns, B. R., *Science* **185,** 491–497 (1974).

Cook, B. D., Bloom, P. R., and Halbach, T. R., *J. Environ. Qual.* **26,** 618–625 (1997).

Eglinton, T. I., Benitez-Nelson, B. C., Pearson, A., McNichol, A. P., Bauer, J. E., and Druffel, E. R. M., *Science* **277,** 796–799 (1997).

Falandysz, J., Strandberg, L., Bergqvist, P.-A., Kulp, S. E., Strandberg, B., and Rappe, C., *Environ. Sci. Technol.* **30,** 3266–3274 (1996).

Field, J. A., Leenheer, J. A., Thorn, K. A., Barber, L. B., II, Rostad, C., Macalady, D. L., and Daniel, S. R., *J. Contam. Hydrol.* **9,** 55–72 (1992).

Fields, R. D., and Rodriguez, F., *in* "Proceedings of The Third International Biodegration Symposium" (J. M. Sharpley and A. M. Kaplan, eds.), pp. 775–784. Applied Science Publishers, London, 1976.

Fryer, J. D., Hance, R. J., and Ludwig, J. W., *Weed Res.* **15,** 189–194 (1975).

Garcia Trejo, A., *Ecotoxicol. Environ. Saf.* **16,** 25–35 (1988).

Gilmore, D. F., Antoun, S., Lenz, R. W., Goodwin, S., Austin, R., and Fuller, C., *J. Ind. Microbiol.* **10,** 199–206 (1992).

Golab, T., Althaus, W. A., and Wooten, H. L., *J. Agric. Food Chem.* **27,** 163–179 (1979).

Gullo, V. P., Kempf, A. J., MacConnell, J. G., Mrozik, H., Arison, B., and Putter, I., *Pestic. Sci.* **14,** 153–157 (1983).

Hagenmaier, H., She, J., and Lindig, C., *Chemosphere* **25,** 1449–1456 (1992).

Kay, M. J., Morton, L. H. G., and Prince, E. L., *Int. Biodeterior. Bull.* **27,** 205–222 (1991).
Kilham, O. W., and Alexander, M., *Soil Sci.* **137,** 419–427 (1984).
Kuhr, R. J., Davis, A. C., and Taschenberg, E. F., *Bull. Environ. Contam. Toxicol.* **8,** 329–333 (1972).
Kunioka, M., Kawaguchi, Y., and Doi, Y., *Appl. Microbiol. Biotechnol.* **30,** 569–573 (1989).
Lichtenstein, E. P., Fuhremann, T. W., Schulz, K. R., Llang, T. T., *J. Econ. Entomol.* **66,** 863–866 (1973).
Marth, P. C., and Mitchell, J. W., *Plant Physiol.* **34,** Suppl., X (1959).
Martin, A., Bakker, H., and Schreuder, R. H., *Bull. Environ. Contam. Toxicol.* **51,** 178–184 (1993).
McLachlan, M. S., Sewart, A. P., Bacon, J. R., and Jones, K. C., *Environ. Sci. Technol.* **30,** 2567–2571 (1996).
Meijers, A. P., and van der Leer, R. C., *Water Res.* **10,** 597–604 (1976).
Miller, S. S., *Environ. Sci. Technol.* **7,** 14–15 (1973).
Morris, R. J., *Mar. Pollut. Bull.* **11,** 164–166 (1980).
Nash, R. G., and Harris, W. G., *J. Environ. Qual.* **2,** 269–273 (1973).
Obradors, N., and Aguilar, J., *Appl. Environ. Microbiol.* **57,** 2383–2388 (1991).
Oda, Y., Asari, H., Urakami, T., and Tonomura, K., *J. Ferment. Bioeng.* **80,** 265–269 (1995).
Orazio, C. E., Kapila, S., Puri, R. K., and Yanders, Y. F., *Chemosphere* **25,** 1469–1474 (1992).
Palmisano, A. C., and Pettigrew, C. A., *Bioscience* **42,** 680–685 (1992).
Potts, J. E., Clendinning, R. A., Ackart, W. B., and Niegisch, W. D., *Polym. Prep.* **13,** 629–634 (1972).
Reisch, M. S., *Chem. Eng. News* **71**(15), 13–16 (1993).
Sakai, K., Hamada, N., and Watanabe, Y., *Kagaku to Kogyo (Osaka)* **62**(2), 39–47 (1988); *Chem. Abstr.* **109,** 3485 (1988).
Sasikala, C., and Ramana, C. V., *Adv. Appl. Microbiol.* **42,** 97–218 (1996).
Scribner, S. L., Benzing, T. R., Sun, S., and Boyd, S. A., *J. Environ. Qual.* **21,** 115–120 (1992).
Sinkkonen, S., *Chemosphere* **18,** 2093–2100 (1989).
Spencer, W. F., Singh, G., Taylor, C. D., LeMert, R. A., Cliath, M. M., and Farmer, W. J., *J. Environ. Qual.* **25,** 815–821 (1996).
Staiff, D. C., Comer, S. W., Armstrong, J. F., and Wolfe, H. R., *Bull. Environ. Contam. Toxicol.* **13,** 362–368 (1975).
Steinberg, S. M., Pignatello, J. J., and Sawhney, B. L., *Environ. Sci. Technol.* **21,** 1201–1208 (1987).
Stewart, D. K. R., Chisholm, D., and Ragab, M. T. H., *Nature (London)* **229,** 47 (1971).
Tanaka, H., Tonomura, K., and Kamibayashi, A., *Biseibutsu Kogyo Gijutsu Kenkyusho Kenkyu Hokoku* **48,** 75–79 (1976), *Chem. Abstr.* **87,** 65110 (1977).
Torres, A., Li, S. M., Roussos, S., and Vert, M., *Appl. Environ. Microbiol.* **62,** 2393–2397 (1996).
Verloop, A., and Nimmo, W. B., *Weed Res.* **10,** 65–70 (1970).
Wilkinson, A. T. S., Finlayson, D. G., and Morley, H. V., *Science* **143,** 681–682 (1964).

CHAPTER 20

Formation and Biodegradation of Air Pollutants

A largely unappreciated role of microorganisms is related to their role in the synthesis and also in the destruction of air pollutants. This topic is often not included in considerations of biodegradation, but it is as appropriate for inclusion as are the organic toxicants. Moreover, by producing volatile substances harmful to humans, animals, plants, or the biosphere in general, microorganisms are functioning in a manner analogous to the conversions more commonly considered as activations.

The air pollutants that microorganisms form or destroy may be important in the *troposphere,* the *stratosphere,* or both. The troposphere extends upward from the earth's surface to an altitude of 8 to 12 km and is the region in which temperature decreases with increasing altitude. Above this region of the atmosphere is the stratosphere, in which temperature rises with increasing altitude. The zone at the top of the troposphere is known as the *tropopause.*

The air pollutants formed or degraded by microorganisms may have one or more effects. (a) Some may contribute to global warming, a change that could have disastrous consequences. The phenomenon is known as the *greenhouse effect.* (b) Some, such as N_2O (nitrous oxide), may reach the stratosphere and there participate in a series of reactions that result in the destruction of stratospheric ozone (O_3). Such a loss of O_3 will have a number of deleterious effects, in part because O_3 reduces the quantity of ultraviolet (UV) light that reaches the surface of the earth. The resulting increase in UV light will lead to an increase in skin cancer in humans and thus, as discussed later, microbial activity in soil is a major, albeit indirect, cause of skin cancer. An increase in UV flux to the earth's surface will likely also have harmful effects on plant life. (c) Some will increase the levels of O_3 in the troposphere, a change that has a number of undesirable effects. Volatile hydrocarbons, CO, and NO_x are implicated in such processes. NO_x represents the sum of NO (nitric oxide) and NO_2 (nitrogen dioxide). (d) Several of the gases (NO_x, some hydrocarbons, and CO) interact to form photochemical smog, which is a major pollutant affecting humans, especially in urban areas. (e) At a local level, some are obnoxious because of their foul odors. (f) Some, such as NO_x, are harmful to plants, which are often injured even at the relatively low concentrations found in the atmosphere. (g) A few participate in chemical reactions that, in turn, have major consequences. For example, a microbial product that affects the concentration of hydroxyl radicals (OH) is of great importance because OH affects the behavior and chemistry of other atmospheric constituents, which in turn may influence climate or have direct effects on living things.

A major reason for concern with atmospheric changes is the so-called greenhouse effect. The earth receives an enormous amount of energy from the sun. That energy chiefly comes from radiation in the visible region of the electromagnetic spectrum, which is not effectively absorbed in the atmosphere. Because the earth is approximately at thermal equilibrium, almost the same amount of energy is radiated from the earth. However, the radiation from the earth's surface is in the infrared region of the spectrum. In contrast with the incoming radiation, the outgoing infrared radiation is efficiently absorbed in the atmosphere. The absorption of this infrared radiation in the atmosphere results in the earth's surface being at an average temperature ca. 30°C higher than it would be if no atmosphere was present. Because the atmosphere acts to trap heat in this fashion much as a greenhouse exposed to the sunlight, the atmospheric phenomenon is known as the greenhouse effect.

Some components of the atmosphere are especially effective in absorbing the outgoing infrared radiation: CO_2, CH_4, N_2O, O_3, and chlorofluorocarbons. These are therefore called greenhouse gases. Of concern in the present context are CO_2, CH_4, and N_2O due to their microbial role in their formation. If the concentrations of these gases increase, the earth's temperature will increase. Human activities have stimulated microbial processes that have led to greater quantities of CH_4, N_2O, and other gases in the atmosphere.

Many effects are anticipated if the surface temperature of the earth increases by 1–2°C. (a) The level of the oceans will rise as a result of thermal expansion of the seas, the melting of glaciers, and possibly by some melting of polar ice caps. An enormous number of people would be affected because they live in low-lying areas, which would be under water. (b) Crop production might be seriously impacted, and farming might have to be conducted in areas more distant from the equator. (c) Natural ecosystems could be seriously impacted. Species may be lost from specific regions, and forests may no longer have the same species composition or might be displaced by grasslands. (d) Many wetlands may become flooded and be lost. Several estimates have been made of the relative contribution of individual gases to the greenhouse effect. These are based on the increases in those compounds in the atmosphere. Although the values vary somewhat, it is generally accepted that the relative contributions of CO_2, CH_4, and N_2O are in the vicinity of 60, 15, and 5%, respectively (Rodhe, 1990).

Atmospheric scientists distinguish between *sources* and *sinks* of individual compounds. Microorganisms are the sources of various atmospheric components: the sole source, a major source, or a significant contributor. In a number of cases, they serve as sinks, i.e., they are responsible for the removal of the component. They do this by metabolizing and biodegrading the compound as it makes contact with the soils or oceans, and thus the interaction between the air and the soils or seas is a controlling factor in the relative role of microorganisms as sinks. If the concentration in the atmosphere is not changing, the sums of the sources and sinks are equal. If the magnitude of the sources is larger than the sinks, the concentration in the atmosphere will increase. If the global emissions are smaller than the sinks, that concentration will fall. The soils and oceans, which are the environments from which microorganisms make their chief contributions to atmospheric chemistry, are often simultaneously both a source and a sink for specific compounds, depending on the environmental conditions for microbial activity, the concentration of the gas in the immediately overlying air, and other factors.

In considering various atmospheric constituents, it is important to distinguish among large urban centers, sites of intense industrial activity, rural areas, and larger regions or even the entire globe. A population or industrial center may be a major source of industrial pollution, but the sink may be sufficiently large that an effect in the larger region cannot be detected. Alternatively, because of the large surface area of the land mass and seas, the impact of microbial activity in soils and oceans may mask the deleterious changes associated with human activities.

Many genera of bacteria, fungi, and algae have been described that produce or consume the compounds known to be important as atmospheric pollutants or causes of atmospheric change. These are not necessarily the organisms that function in nature. The laboratory isolates may not be abundant in nature or they may be affected by environmental factors in manners that result in no appreciable gas formation or destruction. In regard to the species that degrade these gases, the organisms studied in the laboratory are usually provided the gases as substrates in concentrations far, far higher than are found in nature, and it is clear that the organisms functioning at ambient concentrations are often, or possibly always, different from those commonly used for laboratory evaluations. For these reasons, considerable uncertainty exists about the genera or even types of microorganisms important in the biodegradation or formation of these atmospheric constituents.

CARBON DIOXIDE

The major reason for concern with CO_2 is its role as a greenhouse gas. Although it is not as efficient as other gases on a molecular basis in absorbing infrared radiation and thus contributing to the greenhouse effect, its increase in concentration in recent years has been so great compared to other compounds that it is the main gas of significance to climatic change. Moreover, it is expected that its contribution to global warming will continue into the next century.

A global budget for CO_2 is presented in Table 20.1. Microorganisms clearly represent major sources of this product inasmuch as much of the CO_2 originating in soil comes from the microbial degradation of soil organic matter and plant residues, and a large part of that emanating from marine waters comes from algal respiration as well as biodegradation by heterotrophic species. It is worth noting that more than 10-fold more CO_2 is produced microbiologically than arises from the combustion of coal and other fossil fuels. However, because fossil fuel combustion is an additional input to what was presumably an equilibrium based on natural sources and sinks being equal, it is that anthropogenic contribution which represents the perturbation of great climatic importance. The key role of microorganisms as sinks for CO_2 is evident in Table 20.1 from the quantity of CO_2 fixed by marine organisms.

That the CO_2 concentration is increasing is evident from Fig. 20.1. In the last two centuries, the concentrations have increased from approximately 280 to 355 ppmv (parts per million on a volume basis, or μliter/liter). Data for the atmospheres in years back come from analyses of air bubbles in the ice caps of Greenland and the Antarctic, which have been isolated from the ambient air since the bubbles were formed. The obvious rise in amounts in the atmosphere parallels and is presumably attributable to the human burning of fossil fuels, deforestation, and alterations in land use. The changes in land use have led to a massive destruction by microorganisms of organic matter in soils.

To appreciate the magnitude of the release of CO_2 by microbial degradation of organic matter in soil, it is necessary to consider that soils contain (using the same units as in Table 20.1) $1,600,000 \times 10^9$ kg of C. Calculations suggest that

Table 20.1

Global CO_2 Budget[a]

Contribution	kg C $\times$ 10^9/year
Sources	
Oceans	105,000
Land (microbial degradation)	68,000
Fossil fuel combustion	6,000
Sinks	
Oceans, gross (algal photosynthesis)	92,000–108,000
Land, net[b]	60,000

[a] From Schlessinger (1995) and Wuebbles and Edmonds (1991).
[b] Difference between CO_2 fixed and respired by higher plants.

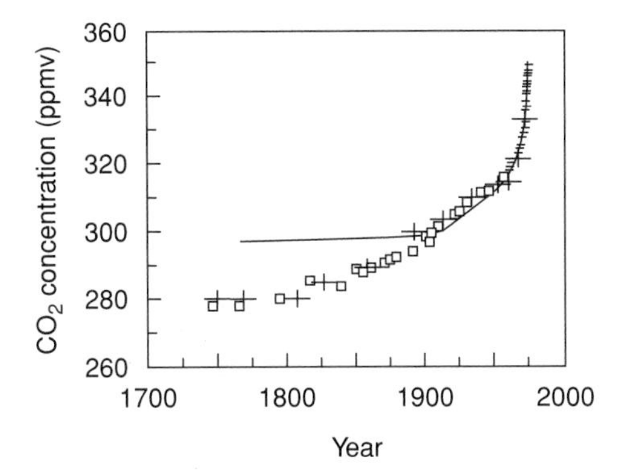

Figure 20.1 Change in CO_2 concentration with time. (From Kammen and Marino, 1993. Reprinted with permission from Elsevier Science.)

76,500 kg of that C is mineralized each year, with much of that CO_2 coming from microbial processes and some coming from root respiration (Raich and Potter, 1995). At equilibrium, an equal amount of CO_2 would be returned to soil, nearly all by the photosynthesis of higher plants. However, more intensive farming and the mechanization of agriculture late in the 18th century were accompanied by the conversion of virgin land to agriculture. For reasons not fully established, the cultivation of virgin land is accompanied by a massive disappearance of the organic matter of soils, a loss that may range from 20 to 40% or sometimes more. The decline is initially rapid, but the rate of disappearance subsequently diminishes, often after some 20 years of cultivation. The cause of the disappearance: microbial degradation of the previously unavailable organic matter. In the period between 1850 and 1980, the carbon content of soils presumably fell by about $40{,}000 \times 10^9$ kg. In addition, the changing patterns of land use associated with more intensive and more extensive agriculture resulted in the microbial decomposition of the plant material that was previously present on the land and the deliberate burning of much of the plant biomass, a contribution of about $80{,}000 \times 10^9$ kg of C. Much of this CO_2 from changes in land use arose in the temperate zone in the period after 1850, but the contribution from tropical regions has been appreciable since 1950 (Houghton, 1995). Not all of the CO_2 would result in an atmospheric increase because CO_2 dissolves in large amounts in marine waters and the biomass in some environments may increase.

It is thus evident that microbial degradation is a major contributor to the CO_2 of the atmosphere and microorganisms must be considered as a significant source of the greenhouse effect. However, their role prior to 1850 was one that had been balanced by the rate of return of C from CO_2 to the biosphere by natural processes. The current increase in CO_2 in the atmosphere is, to a significant degree, the result of combustion of fossil fuels and, because modifications in land use still release some 1100–3300 kg of additional C to the atmosphere each year (Houghton, 1995), changes in the use of tropical lands.

METHANE

Methane is significant for several reasons. After CO_2, it is the second most important greenhouse gas. On a molecular basis, it is 25–30 times more effective in absorbing infrared and thus more effective as a greenhouse gas, but the larger amounts of CO_2 generated make CH_4 second to CO_2. It reacts in the troposphere and lower stratosphere to produce O_3, whereas it destroys O_3 in the upper stratosphere. An increase in the CH_4 concentration in the atmosphere leads to a decrease in OH radicals, which are major participants in a number of processes important to atmospheric chemistry. In addition, CH_4 breaks down in the troposphere to form CO, which itself is an important pollutant.

Well over half of the CH_4 reaching the atmosphere is the product of microorganisms (Table 20.2). Because of the uncertainties in the magnitude of the emissions, it is not possible to state what percentage is microbial, but the values presented clearly show that possibly three-fourths of the CH_4 originates as an end product of microbial metabolism in ruminants, waterlogged soils planted to rice, swamps, other wetlands, and landfills where anaerobes degrade plant residues and other organic materials. Although the gas is thus largely a result of these microorganisms, much of that activity, and particularly the increase in recent years, is related to human actions.

Analyses of samples of ice cores and the current atmosphere have shown the dramatic changes in CH_4 concentrations. The atmosphere in about 1500 A.D. contained approximately 0.7 ppmv of CH_4 and possibly some 10×10^9 Kg was formed each year by anaerobes. In contrast, the modern atmosphere contains 1.7 ppmv (Kammen and Marino, 1993), and data in Table 20.2 suggest annual microbial emissions of approximately 350×10^9 kg. The rate of emission has increased markedly in more recent times, as shown in Fig. 20.2. A large part of this increase is attributable to ever larger areas of land planted to rice, continuing increases in numbers of cattle, and more landfills (all linked with microbial activity to generate CH_4), as well as other causes. The rise in emissions, ultimately arising from human activities, was evident at the beginning of the 20th century, which is parallel to the 2.7-fold rise in the numbers of cattle and doubling of the area planted to rice in that time (Khalil and Rasmussen, 1993). The increase became particularly marked after the 1940s, as shown in Fig. 20.2, but that rate appears to be slowing more recently, possibly because of the less rapid increase in animal populations and in lands used to cultivate rice (Khalil and Rasmussen, 1993).

Table 20.2

Global CH_4 Budget[a]

Contribution	kg × 10^9/year		
	Ref. 1	Ref. 2	Others
Sources			
Microbial			
Ruminants	80 ± 20	80–100	
Wetlands	130 ± 70	120–200	109[b]
Rice paddies	95	70–170	
Termites	10 ± 5	25–150	
Landfill	50 ± 25		
Oceans		1–20	
Tundra		1–5	
Other		23–80	
Nonmicrobial			
Biomass burning	30 ± 15	10–40	
Gas and oil production and transport	70 ± 40	10–20	
Coal mining	35 ± 10	10–35	
Solid waste	20	5–70	
Sinks			
Atmospheric reactions	450 ± 10		
Soil, microbial degradation	30 ± 25		29[c]

[a] From Lelieveld *et al.* (1993) and Topp and Pattey (1997) (Ref. 1) and from Tyler (1991) (Ref. 2).
[b] Bartlett and Harriss (1993).
[c] Dorr *et al.* (1993).

The microorganisms, environmental conditions, and biochemistry of methane formation have been studied intensively. In the environments of concern, be they flooded rice fields, wetlands, landfills, or the intestinal tracts of cattle or termites, certain complex materials initially undergo an anaerobic fermentation. The substrates are chiefly cellulose, hemicelluloses, other polysaccharides, and various other organic substances. These substrates largely come from plant residues in the soil, crops that are consumed by the animals, or organic materials that are placed in landfills. The initial organisms are mainly or exclusively bacteria, none of which produce CH_4. However, they degrade the complex materials to form CO_2, H_2, organic acids, and other simple organic molecules. Some of the organic products are further metabolized to yield acetic acid and more CO_2 and H_2. These three compounds are the chief, and often possibly the sole, substrates for bacteria that actually make the CH_4. These bacteria, known as *methanogens,* carry out one or both of the following two reactions under strictly anaerobic conditions:

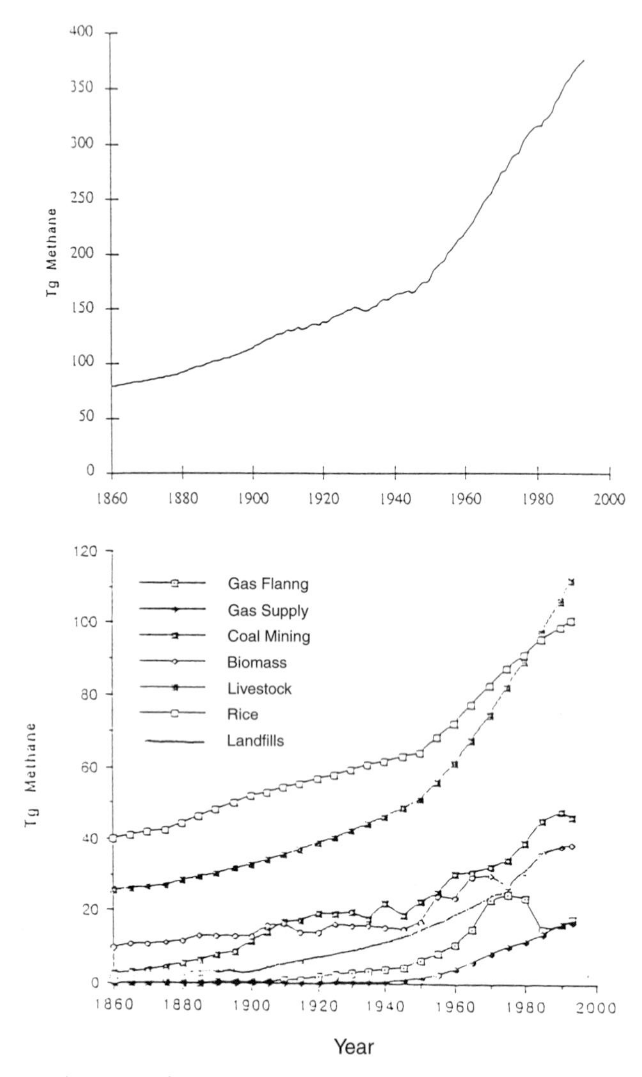

Figure 20.2 Total emissions of methane attributable to anthropogenic activity: (top) total and (bottom) according to activity. (From Stern and Kaufmann, 1996. Reprinted with permission from Elsevier Science.)

$$CH_3COOH \rightarrow CH_4 + CO_2$$
$$CO_2 + 4H_2 \rightarrow CH_4 + 2H_2O$$

The energy source for bacteria carrying out the second reaction is H_2, and thus the bacteria are chemoautotrophs. For them, CO_2 is the terminal electron acceptor.

In addition to CO_2 plus H_2 and acetic acid, methanogens can ferment formic acid, methanol, methylamine, and a few other simple compounds to form CH_4 (Topp and Pattey, 1997), but these are probably not as important in nature as acetic acid, CO_2, and H_2.

The rates of CH_4 emission from rice paddies vary from 5 to 90 g per m^2 during the growing season (Cao *et al.*, 1996). Similar processes occur in anaerobic areas of the oceans, the chief sites being in the coastal shelf and estuaries (Bange *et al.*, 1994). The activity is also prominent in termites, which feed on wood or soil organic matter or live symbiotically with a fungus, and the organic matter consumed by the termite is converted by anaerobic bacteria in its gut to give considerable amounts of CH_4 from the many new termite mounds that have appeared as a result of land clearing. Although estimates of the global contributions of termites differ greatly (Sanderson, 1996), all suggest that these arthropods are significant sources.

On a global basis, the amount of CH_4 oxidized microbiologically is in the vicinity of 10% of the total quantity destroyed (Table 20.2). A number of different methane-oxidizing bacteria, which are called *methanotrophs*, have been isolated in pure culture and studied in artificial circumstances. They carry out the following oxidation:

$$CH_4 + 2O_2 \rightarrow CO_2 + 2H_2O$$

From the reaction, it is clear that they are aerobes, and as a rule they use CH_4 but few or no other carbon sources for growth. Other bacteria are able to cometabolize but not grow on CH_4. It appears that the activity in nature is the result of methanotrophic organisms rather than other groups of microorganisms (Bender and Conrad, 1994). Nevertheless, the specific organisms degrading CH_4 at the concentration ranges found in nature are unknown as indicated by differences in the affinities (K_m) for the gas of the organisms in soil and those studied in culture (Conrad, 1996). In addition to their role in destroying CH_4 in the atmosphere, methanotrophs are probably significant in reducing the quantity emitted from paddy fields and other wetlands; they do this by living in the thin aerobic zone immediately above the underlying anaerobic soil and oxidizing the CH_4 moving upward from the anoxic regions. Chapter 7 considers thresholds for nongaseous substrates, but the biodegradation of gaseous substrates may also have a threshold; thus, at very low concentrations, CH_4 oxidation stops (Conrad, 1994). Uncharacterized anaerobes are also capable of oxidizing CH_4 and may play a role in CH_4 turnover in marine sediments (Conrad, 1995).

Surprisingly, the rates of biodegradation of CH_4 are quite similar in soils at different latitudes and in different ecosystems. Moreover, temperature often does not have the marked effect on this process that it has on other microbial transformations. The reason appears to be that the potential of the indigenous methylotroph populations of aerated soils to oxidize CH_4 is greater than the rate of diffusion of the gas from the overlying atmosphere to the sites in the soil at which the organisms function (Bartlett and Harriss, 1993; Striegl, 1993).

NITROUS OXIDE

Nitrous oxide in the atmosphere is of practical concern for two reasons. First, it is a good absorber of infrared radiation and thus is a greenhouse gas. On a molecular basis, it is more than 100 times as effective in this regard as CO_2. It is estimated by various investigators to contribute 2–6% of the potential greenhouse effect. Second, it is essentially inert and thus is not destroyed readily in the troposphere, but it slowly diffuses from ground level through the troposphere and into the stratosphere. In the stratosphere, it is converted to nitric oxide (NO), and the NO generated from the N_2O interacts with and destroys O_3 in the stratosphere. Because NO consumed in the destruction of O_3 is regenerated, the process is essentially catalytic.

Ozone is formed by the action of sunlight on O_2 to give atomic oxygen (O), and this O reacts quickly to form O_3. In the stratosphere, O_3 is an effective absorber of ultraviolet light and thus reduces the intensity of ultraviolet light that would otherwise reach the surface of the earth. If the O_3 concentration in the stratosphere is reduced, as by the action of the NO formed from N_2O (as well as by chlorine oxides formed from chlorofluorocarbons), more ultraviolet light will penetrate the atmosphere and reach the earth, leading to an increase in skin cancer and possible eye injury in humans and possible detrimental effects on crop production and phytoplankton growth.

The dominant role of microorganisms as a source of N_2O is immediately evident from a global budget for this compound (Table 20.3). Although the magnitude of the various sources is the subject of disagreement, there is no disagreement that microbial processes in soils and the oceans contribute more than two-thirds and possibly more than 90% of the N_2O that enters the atmosphere. Nevertheless, the relative contributions of agricultural lands that are fertilized, grasslands, forests, and the oceans are far from resolved.

Despite the uncertainties on the magnitude of the various sources, microorganisms thereby are, indirectly to be sure, major contributors to skin cancer.

Analysis of ice cores, measurements of levels in the atmosphere, and modeling have shown that the N_2O concentration of the atmosphere before the industrial revolution (some 150 years ago) was approximately 285 ppbv, but it has been steadily increasing thereafter. The increase has been particularly marked in recent years (Fig. 20.3), and the current concentration is ca. 310 ppbv. At the present time, the concentration is rising by 0.2–0.3% per year (Chameides and Perdue, 1997). The increase appears to be attributable to changing land use, the larger amounts of chemical fertilizers utilized for food production, the growing quantities

Table 20.3

Global N_2O Budget

Contribution	kg N × 10^9/year		
	McElroy and Wofsy (1986)	Davidson (1991)	Other references
Sources			
Microbial			
Ocean	2	2	7–10[a]
Fertilized soil	0.8	0.5	
Grasslands	0.1	0.1	0.16[b]
Forests	7.5–7.9	4.4–5.6	
Cultivated soil			4.0 ± 0.8[c]
Fertilizer			1.19[d]
Manure			1.06[d]
Nonmicrobial			
Biomass burning	0.7	0.1–0.3	0.54[d]
Fossil fuel combustion	4	<0.1	0.75[d]
Industrial			0.50[d]
Sinks			
Stratospheric reactions	10.6		10.0 ± 3.0[c]
Accumulation in stratosphere	3.5		3.8 ± 0.8[c]

[a] Bange *et al.* (1996).
[b] Mosier *et al.* (1996).
[c] Beauchamp (1997).
[d] Kroeze (1994).

of manure arising from the larger numbers of cattle, and the increases in the combustion of fossil fuel (Kroeze, 1994); N_2O from the first three of these four sources comes from microbial transformations.

The ultimate source of the N appearing as the N_2O emitted from soils and waters is (a) plant or planktonic remains, soil or oceanic organic matter, or animal wastes that are being decomposed and mineralized to yield ammonium and (b) urea, ammonium salts, or other chemical fertilizers added to farm land. Urea is also hydrolyzed microbiologically to yield ammonium. In some instances, N_2O may originate from nitrate fertilizers. The ammonium generated from organic sources or urea, or that added as fertilizers, is oxidized chiefly to nitrate under aerobic conditions in soils and waters. This conversion, known as *nitrification,* can be carried out either by chemoautotrophic bacteria or by heterotrophic fungi and bacteria. The chemoautotrophs, which have been studied extensively, get their energy by oxidizing either ammonium to nitrite or nitrite to nitrate, so the conversion is a two-step process, each of which requires O_2:

$$NH_4^+ \rightarrow NO_2^- \rightarrow NO_3^-$$

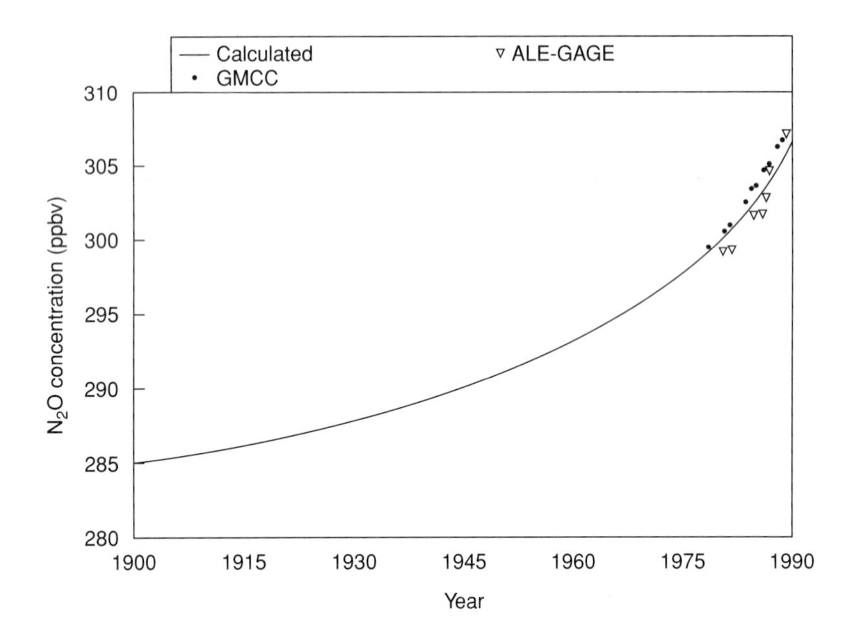

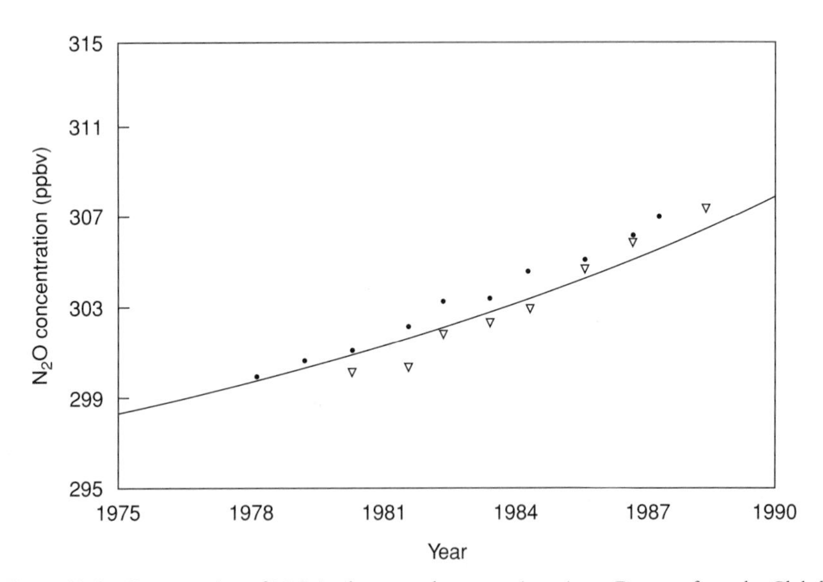

Figure 20.3 Concentration of N_2O in the atmosphere at various times. Data are from the Global Monitoring for Climate Change (GMCC) project, the Atmospheric Lifetime Experiment, and the Global Atmospheric Gases Experiment (ALE-GAGE) and calculations. (From Kroeze, 1994. Reprinted with permission from Elsevier Science.)

In contrast, heterotrophs get no energy from the oxidation and require an organic substrate as their energy source. Although several fungi and heterotrophic bacteria have been shown to nitrify in culture, which of these genera are important in nature is as yet unknown. Nitrite, organic N compounds, or both may be intermediates in heterotrophic nitrification. The relative importance of autotrophic and heterotrophic organisms to nitrification in nature remains the subject of dispute, but evidence suggests that autotrophs dominate in soils at pH values near neutrality and that heterotrophs are the chief agents of nitrate formation in acid environments in which nitrification is occurring.

Under anaerobic conditions, or in aerobic soils with anaerobic microsites, a number of bacterial genera use nitrate as their electron acceptor and reduce the nitrate mainly to N_2:

$$NO_3^- \rightarrow N_2$$

The denitrifying bacteria are abundant and can be very active, but they rely first on aerobic conditions to generate the nitrate and then on anaerobiosis for the denitrification step. Nearly all are heterotrophs. Although a few bacteria can denitrify when O_2 is available, evidence for their importance in nature is scanty.

Both denitrification and chemoautotrophic nitrification frequently are not quantitative, and not all of the N in the nitrate reduced is converted to N_2 and not all of the N from the ammonium oxidized is transformed to nitrate. In both processes, N_2O and usually smaller amounts of NO may be released. The biochemical pathways of both chemoautotrophic nitrification and denitrification, or reactions involving products formed abiotically from these intermediates, show that N_2O and NO may be formed. However, as in the pathways of the breakdown of organic compounds, an intermediate or product generated within the cell of a bacterium or fungus may be further metabolized so that none appears in the surrounding environment. Nevertheless, studies of individual microorganisms, samples of soils and waters tested in the laboratory, and field evaluations confirm that N_2O and NO are evolved.

The biochemical mechanism of denitrification involves a stepwise conversion of nitrate to N_2:

$$NO_3^- \rightarrow NO_2^- \rightarrow NO? \rightarrow N_2O \rightarrow N_2$$

Nitric oxide (NO) may itself not be the intermediate, but rather it may be in equilibrium with the actual intermediate. Some denitrifying bacteria release nitrite, NO, and N_2O into their culture medium, but others release two, one, or none of these intermediates (Firestone and Davidson, 1989; Hutchinson, 1995). In addition, some fungi form N_2O aerobically from nitrate, but no evidence exists that they are important in nature. It is also possible that N_2O or NO may be formed abiotically in nature by a reaction involving one or more of the intermediates or products of denitrification, especially in reactions involving nitrite. Sometimes nitrogen dioxide (NO_2) is found to be emitted from soils, but it is unlikely that this is itself a biochemical product, and it may originate in a reaction involving one of the actual microbial products or be the product of a simple nonbiological reaction between NO and O_2 dissolved in soil water or O_2 in gas-filled pores:

$$NO + 1/2\ O_2 \rightarrow NO_2$$

The processes associated with NO emission are considered below.

Controversy exists on the relative contributions of denitrification and nitrification to N_2O evolution in nature. Some scientists favor one process, some the other. When a soil containing nitrate is flooded or is otherwise under anaerobiosis or in anoxic waters having nitrate, denitrification is the source of N_2O. However, in soils or waters that are aerobic, N_2O (and NO) may originate during the oxidation of ammonium, but it is also possible that the nitrate formed in nitrification diffuses to an anaerobic microenvironment, where the nitrate is reduced and gives rise to N_2O (and NO). What is not controversial is that the N_2O has a microbial origin.

On a global basis, microorganisms do not appear to be a major sink for N_2O, and the chief way that it is destroyed is by reactions in the stratosphere. Nevertheless, because they have the ability to metabolize the compound, they probably are, to some degree, means for its destruction. Indeed, as the equation for the biochemical mechanism shows, N_2O is a precursor of N_2, and because the dominant product of denitrification is N_2, enormous amounts of the N_2O formed in denitrification must be destroyed in anaerobic microsites before it leaves the soil to enter the overlying air. This destruction is a reductive, anaerobic transformation. Evidence exists that samples of soil and bacteria in culture can also degrade N_2O under aerobic conditions (Vedenina and Zavarzin, 1977, 1979), so that consideration also needs to be given to an oxidative mechanism for destruction of this gas.

NITRIC OXIDE

An evaluation of the importance of NO must link this gas with NO_2 because, as discussed earlier, NO is oxidized in the atmosphere to nitrogen dioxide (NO_2). A mixture of the two is commonly designated NO_x. Among the reasons for concern with NO_x are the following. (a) It participates in reactions in the troposphere that result in O_3 formation. In a series of steps involving that O_3 with NO_x, hydrocarbons, and sunlight, photochemical smog is formed. (b) In photochemical smog, the O_3 that is formed affects lung functions, aggravates asthma, and causes other harmful effects on human health. Ozone is also toxic to plants and has been implicated in damage to trees and forest decline. Photochemical smog also causes eye irritation. (c) Both NO and NO_2 are directly phytotoxic. As little as 0.60 ppmv of NO or 0.25 ppmv of NO_2 injures plants. The effects of concentrations below 1.0 ppmv include reduction of growth, diminution in photosynthesis, and premature senescence and wilting of leaves (National Academy of Sciences, 1977a). (d) NO_x combines with water vapor in the troposphere and is thereby converted to nitric acid, which is a major component of acid rain. Acid rain appears downwind from emissions of NO_x (and SO_2), and precipitation of that rain rich in nitric acid, sulfuric acid, or both may have a devastating effect on poorly buffered lakes as well as on forest ecosystems.

A global budget for NO in the atmosphere is shown in Table 20.4. Although the amounts released microbiologically are less than those emanating from the sum of biomass burning, fossil fuel combustion, lightning, and oxidation of ammonia in the atmosphere, the quantities are still substantial. Moreover, on a regional basis in areas where burning and fuel combustion are minor, as in many forests and grasslands, the relative microbial contribution would be prominent. The two sinks are wet and dry deposition, wet deposition referring to the return to the earth of pollutants in rainfall and dry deposition designating the direct absorption of pollutants by soils, oceans, inland waters, and plants.

The microbiology of NO formation is discussed in the preceding section.

The net flux of NO from soil to the atmosphere represents a balance between its formation and destruction, and the magnitude of that destruction may be large so that little in fact escapes from the soils or waters in which it is formed. Denitrifying bacteria can reduce NO to N_2 inasmuch as the former compound is an intermediate

Table 20.4

Global NO Budget

	kg N × 10^9/year		
Contribution	Logan (1983)	Conrad (1990)	Wuebbles and Edmonds (1991)
Sources			
Microbial			
Soil	8 (4–6)	1–20	1–15
Ocean[a]	<1		<1
Nonmicrobial			
Biomass burning	12 (4–24)	2–40	4–24
Fossil fuel combustion	21 (14–28)	8–28	15–25
Lightning	8 (2–20)	1–20	8
Ammonia oxidation	1–10	1–10	
Sinks			
Wet deposition	12–42		
Dry deposition	12–22		
Total			25–85

[a] Microbial or photochemical.

or is in equilibrium with an intermediate in the pathway of denitrification. Therefore, although estimates of global sinks do not typically show an appreciable role for microorganisms, their actual role must be very large because they destroy NO and ultimately convert it to N_2 (enormous quantities of which are emitted to the atmosphere). This represents an anaerobic sink.

In addition, NO is destroyed in soils by abiotic reactions and can be metabolized aerobically. Thus, even when denitrification is not a significant sink, NO may be metabolized in soils under oxic conditions and be converted to nitrate. This conversion is microbial because it is eliminated by autoclaving the soil. Heterotrophic bacteria are able to bring about this oxidation (Baumgärtner *et al.*, 1996; Dunfield and Knowles, 1997; Rudolph *et al.*, 1996). Nevertheless, the role of this aerobic process in nature remains uncertain.

CARBON MONOXIDE

Many undesirable effects are related to increases in CO. It is highly toxic to humans, and low levels are known to cause serious harm. It is a major sink for atmospheric OH, and the destruction of OH by rising concentrations of CO causes important changes in the troposphere—the declining OH increasing the concentration and lifetime of trace gases that react with OH. Carbon monoxide is also important because increasing concentrations lead to rising levels of O_3 in the troposphere.

The global budget for CO shows the importance of microbial activity as both a source and a sink (Table 20.5). Approximately one-fourth of the CO is estimated to originate, indirectly at least, from microbial transformations, chiefly in soil. The actual source is largely not CO itself but rather is the CH_4 generated by anaerobes, the CH_4 moving to the atmosphere where it is abiotically oxidized to CO in a reaction with OH. In addition, although the estimates for the size of the microbial sink differ greatly, the microbial communities of soil obviously represent a large sink.

The quantitative contribution of oceans as a global source and the role of microorganisms in the oceans are uncertain. Values as low as 9 (Swinnerton *et al.*, 1970) to as high as 220×10^9 kg CO/year (National Academy of Sciences, 1977b) have been proposed. Both photooxidation of dissolved organic matter and microbial processes have been advanced as possible marine sources (Erickson, 1989). Nevertheless, a number of investigations (see later) clearly demonstrate the microbial potential.

The concentration in the atmosphere ranges from 50 to 170 ppbv, but the values may reach 220 ppbv. The concentration is increasing at a rate of about 1% per year, and even at locations distant from urban pollution, the troposphere is becoming richer in CO (Khalil and Rasmussen, 1990).

Table 20.5

Global CO Budget

Contribution	kg × 10^9/year	
	Khalil and Rasmussen (1990)	Seiler (1974)
Sources		
Microbial		
Oxidation of methane[a]	600	
Oceans (microbial?)[b]	40	
Nonmicrobial		
Biomass burning	680	
Fossil fuel combustion	500	
Oxidation of nonmethane hydrocarbons[c]	690	
Plants	100	
Sinks		
Microbial degradation in soil	250	450[d]
Stratospheric reactions		110

[a] Atmospheric oxidation of methane that is chiefly produced by microorganisms in the biosphere.
[b] Relative contributions of microorganisms and photochemical reactions are uncertain.
[c] Some of these hydrocarbons are produced by microorganisms.
[d] Other values: 7–10 (Potter *et al.*, 1996) and 410 (Bartholemew and Alexander, 1981).

Because the numerical values point to the abiotic oxidation of CH_4 in the atmosphere as the dominant, albeit indirect, microbial source, an assessment of the microbiology of CO formation must first consider the discussion earlier of CH_4 production by anaerobes. In addition, both soils and the oceans produce CO, but in both environments, that production may result from the abiotic decomposition of organic matter under the influence of sunlight. The fact that CO production from soil is not eliminated by inhibitors or autoclaving suggests the absence of a biological contribution (Conrad and Seiler, 1980), even though a number of bacteria (Junge *et al.*, 1971), algae (Troxler and Dokos, 1973), and fungi (Brown and Brown, 1981) in culture and uncharacterized anaerobes in a bioreactor (Bae and McCarty, 1993) generate the gas.

A direct biological role in the CO cycle (as contrast with CO made abiotically from CH_4) is evident from the global budget and tests of environmental samples. Many genera of bacteria can use CO as a source of C and energy. Included are both aerobes and anaerobes, and they are found among species of *Pseudomonas, Streptomyces, Bacillus, Nocardia, Mycobacterium,* and *Clostridium*. Some use CO as their energy but not carbon source and thus function as chemoautotrophs. In the process carried out by organisms using CO as C or energy source, CO is converted to CO_2. However, some anaerobes using CO may also form acetic acid from the

CO, and methanogens may also convert it to CH_4. Several genera of fungi and green algae are also capable of oxidizing CO to CO_2. Unfortunately, the investigations of these microorganisms have always involved CO concentrations much higher than ever occur in nature. The microorganisms in soil that remove CO from the gas phase do not use CO as a carbon source. If they did, ^{14}CO would be incorporated into the biomass of soil, but this does not occur. They also are not autotrophs, which would use CO as an energy source; if autotrophs were implicated in removing ambient levels of CO, they would fix CO_2, yet $^{14}CO_2$ provided to soils in which ^{12}CO was being oxidized is not incorporated into the soil biomass. These experiments can be summarized by two equations. If growth-linked heterotrophy were involved, then ^{14}C-biomass would be formed in these test conditions:

$$^{14}CO + 1/2\ O_2 \rightarrow {}^{14}CO_2 + {}^{14}C\text{-biomass}$$

If autotrophs were responsible, then ^{14}C-biomass would appear when soil is incubated with $^{14}CO_2$ (in the presence of ^{12}CO) since autotrophs use CO_2 as their source of carbon:

$$^{12}CO + {}^{14}CO_2 + 1/2\ O_2 \rightarrow {}^{12}CO_2 + {}^{14}C\text{-biomass}$$

Radiolabeled biomass does not appear, however. By a process of excluding heterotrophs using CO as a carbon source and autotrophs using it as an energy source, the conclusion appears to be that the responsible microorganisms act by cometabolizing CO in a simple oxidation (Bartholomew and Alexander, 1979):

$$CO + 1/2\ O_2 \rightarrow CO_2$$

The process is largely microbial and not abiotic since appreciable amounts of CO are not oxidized in soil sterilized by autoclaving or γ-irradiation (Bartholomew and Alexander, 1979). Nevertheless, the identities of the cometabolizing species have yet to be established.

NONMETHANE HYDROCARBONS

Several aliphatic hydrocarbons in addition to CH_4 and a number of aromatic hydrocarbons contribute to air pollution. As indicated in Table 20.5, the largest source of CO (an estimated 690×10^9 kg/year) arises from the oxidation of nonmethane hydrocarbons in the air. Some of these hydrocarbons are reactants in steps that create O_3 in the air. One, ethylene, at very low concentrations, has a major influence on plant growth and affects a number of processes associated with plant development.

Particular attention has been paid to ethylene because of its potential impact on plants. As shown in Table 20.6, the chief sources of this alkene are biological chiefly from soils, and a large part of that source is microbial. Some is also released

Table 20.6

Global Ethylene Budget[a]

Contribution	kg × 10^9/year
Sources	
Biological	
Soils	23.3
Oceans	2.9[b]
Nonbiological	
Biomass burning	7.1
Fossil fuel combustion	2.0
Sinks	
Reaction with OH	51.1
Reaction with O_3	4.5
To stratosphere	1.7

[a] From Sawada and Totsuka (1986).
[b] Plass-Dülmer *et al.* (1995) estimate the emission to be 0.84 × 10^9 kg/year.

from marine waters and from plants. Several C_3 hydrocarbons have been found in the air over the western Atlantic Ocean (Khalil and Rasmussen, 1988), and ethylene, ethane, and acetylene have been detected in air over the Pacific Ocean at a considerable distance from any terrestrial sources (Robinson *et al.*, 1973). Acetylene has also been found in the Indian and Pacific Oceans at concentrations greater than would be expected if the sea water and atmosphere were at equilibrium, indicating that it is produced in marine waters and then emitted to the atmosphere. From the values obtained, it was estimated that the global release of acetylene from marine environments is 0.2–1.4 × 10^9 kg/year (Kanakidou *et al.*, 1988). Sea water also contains propylene, 1-butene, and ethane (Plass-Dülmer *et al.*, 1995), some of which probably escapes to the air. It is likely that much or most of these gaseous hydrocarbons are produced in sea water as a result of microbial activity, although some may be the result of photochemical destruction of algal excretions (Wilson *et al.*, 1970). Acetylene, ethane, propane, and *n*- and isobutane have also been detected in the air at nonmarine sites remote from urban and industrial sources, so biological formation in terrestrial environments occurs (Ferman, 1981).

The formation of ethylene in soils and by pure cultures of many bacteria has been well documented. It is generally produced in higher yield in waterlogged than in unflooded soils, and bacteria and fungi of many genera possess this activity in pure culture (Primrose, 1976; Primrose and Dilworth, 1976). Because less is formed in autoclaved than in unheated soil (Smith and Cook, 1974), the activity is microbial, but the responsible groups in nature remain unknown.

Ethane, propane, propylene, butane, and butene are also formed in soils, especially when flooded or waterlogged, but the amounts are small (Van Cleemput and El-Sebaay, 1985). Although the genesis of these alkanes and alkenes is also probably microbial, the types of organisms causing the evolution are also undefined.

Some plants volatilize large amounts of a number of compounds, each of which may be a substrate for microorganisms in soils or adjacent waters. A variety of trees release isoprene and terpenes to the air, and the total of nonmethane hydrocarbons in rural air may be approximately 100 ppb C (Rasmussen, 1981). Not only trees but agronomic and horticultural crops emit isoprene and several terpenes, including α- and β-pinene, myrcene, limonene, and β-pheliandrene (Tingey, 1981). The annual global emissions of these hydrocarbons have been calculated to be 250×10^9 kg isoprene and 147×10^9 kg terpenes as well as 42×10^9 kg aromatics and 52×10^9 kg paraffins (Muller, 1992). Terpenes (chiefly limonene, camphene, sabinene, 3-carene, and α- and β-pinene) are also emitted from core samples from forest floors (Hanson and Hoffman, 1994) so a microbial role may also be possible.

Also found in areas remote from urban and industrial pollution are benzene, toluene, and xylenes, albeit at concentrations far lower than in urban locations (Ferman, 1981).

It is uncertain which of these terpenes and simple aromatic compounds are formed by microorganisms in natural habitats or what is the relative microbial contribution to the emissions of alkanes and alkenes. Nevertheless, there is little doubt that they will be biodegraded when the air containing them diffuses to the soil or ocean surface. For example, ethylene disappears when incubated together with soil, and the activity is abolished if the soil is sterilized. Calculations from these laboratory studies suggest an enormous capacity to destroy this pollutant (Abeles *et al.,* 1971; Smith *et al.,* 1973). However, those activities are not considered in the global budgets that have been prepared. Bacteria are also able to use acetylene as a carbon source (Schink, 1985), and the microflora of soil similarly can metabolize benzene in the gas phase (McFarlane *et al.,* 1981). Indeed, it is probable that each of the volatile aliphatic and aromatic hydrocarbons and terpenes will be biodegraded as molecules in the air come into contact with soils or water, provided there is no threshold.

ALKYL HALIDES

Several halogenated compounds that are biologically formed are important air pollutants. Bromine compounds are prominent in this regard because of their effectiveness, on a molecular basis, in destroying stratospheric O_3. Several brominated compounds have been found in the atmosphere, but methyl bromide (CH_3Br) is a particularly large source of atmospheric bromine. Brominated molecules have a major role in destroying O_3 in the Arctic and Antarctic through a photochemical reaction, but they also make a contribution to stratospheric O_3 depletion in nonpolar regions. Attention as an O_3 depleter is chiefly focused on CH_3Br, but bromoform ($CHBr_3$) is also important.

Microorganisms are important in the global cycling of CH_3Br in two ways (Table 20.7). They are a source in oceans and a sink in soil. Note that the values in Table 20.7 are in kg $\times$ 10^6/year in contrast to data in the previous tables, which are given as kg $\times$ 10^9/year. Considerable uncertainty exists about whether oceans are a net source or a net sink because oceans, through their algal and phytoplankton activities, release brominated organic compounds when the responsible organisms are present. Nevertheless, CH_3Br that dissolves in the oceans as it enters from the air is broken down abiotically, and that may be the dominant conversion in waters in which algae are not abundant.

Marine algae are also able to generate $CHBr_3$, and algae associated with polar ice appear to be major sources, both in the Arctic and in Antarctica. In these polar regions, the algae form brominated products that are important in O_3 destruction. Sturges *et al.* (1992) estimated that ice algae in the Arctic and Antarctica emit 10–150 $\times$ 10^6 kg organobromine gases per year. The annual global release of $CHBr_3$ is estimated to be 1000 $\times$ 10^6 kg from all oceanic sources and 10 $\times$ 10^6 kg from marine macrolgae (Wever, 1991).

The sizable emission of $CHBr_3$ in polar regions is not surprising because the phytoplankton biomass is often large, as in the spring in the Arctic Ocean, and these growths appear to contribute to the high concentrations in the Arctic air.

Table 20.7

Global CH_3Br Budget[a]

Contribution	kg $\times$ 10^9/year
Sources	
Oceans	300[a]
	26–100[b]
Fumigant	50–80[a]
	16–47[b]
Biomass burning	10–50[c]
Automobile exhaust	0.5–22[b]
Sinks	
Microbial degradation in soil	42 ± 32[d]
Oceans	50[c]

[a] Wever (1991).
[b] Ristaino and Thomas (1997).
[c] Oremland (1996).
[d] Shorter *et al.* (1995).

Algae that release $CHBr_3$, dibromomethane (CH_2Br_2), and sometimes CH_3Br are not restricted to polar regions and may also be present in temperate and tropical oceans. The responsible species are planktonic microalgae as well as some larger algae represented among the seaweeds. The waters become supersaturated with these brominated products, and they then clearly are sources to the air (Wever, 1993; Manley and Dastoor, 1987).

In soil, CH_3Br is biodegraded with the formation of CO_2 and bromide. Little activity occurs in the absence of O_2. Because an antifungal compound has no effect but an antibacterial antibiotic inhibits the oxidation, the conversion appears to be bacterial. Although methanotrophs are implicated, other as yet unknown aerobic bacteria are also important. The degradation can also occur in the water column of lakes (Oremland, 1995; Shorter *et al.*, 1995; Rice *et al.*, 1996).

Methyl chloride (CH_3Cl) is also generated biologically in the seas. The annual emission is estimated to be 2.5 to 5 $\times$ 10^9 kg (Wever, 1991). It can be generated by phytoplankton, possibly by the algal populations (Tait and Moore, 1995), as well as by macroalgae (Manley and Dastoor, 1987). However, there may also be a terrestrial source because some soil fungi may synthesize the compound (Harper and Hamilton, 1988; Wever, 1993). Several macroalgae are capable of forming chloroform ($CHCl_3$), which may be released to the air (Nightingale *et al.*, 1995).

The oceans likewise appear to be a source of methyl iodide (CH_3I), and emission rates of 0.3 to 2.0 $\times$ 10^9 kg/year have been estimated (Wever, 1991). Among the marine organisms able to synthesize CH_3I are phytoplankton species (Manley and de la Cuesta, 1997) and macroalgae (Manley and Dastoor, 1987).

ODORS

In contrast to the global or regional concerns with the previously mentioned pollutants are the highly local concerns with the odorous compounds produced by microorganisms. The latter are not issues for society at large, but they are particularly important for the people who, because they are near the source of these noxious compounds, are impacted.

The malodorous products may have several origins. In some instances, the source is an improperly operated sewage-treatment facility or stabilization pond. Off odors may result from large masses of plant or animal matter left to decay. In recent years, a major focus of attention has been the products of anaerobic microbial decomposition of manure coming from high densities of domesticated animals. For efficient agriculture, reduced costs, and optimal use of equipment, large numbers of beef and dairy cattle, poultry, sheep, and pigs are concentrated in small areas, and it is not uncommon to have thousands of cattle, sheep, or hogs—or sometimes tens of thousands—in a confined area, and the numbers of chickens in restricted areas often exceed 100,000. This results in large masses of manure. For example, a feedlot that contains 10,000 cattle or a farm that has 100,000 chickens or 1000 pigs will have wastes of 280, 14, and 3 tons each day (Munn and Phillips, 1975). The problem with these odors becomes acute if these feedlot or farm operations are near areas of habitation or places used for recreation.

The problem arises because decomposition of the large masses of these wastes consumes all the available O_2, and the further degradation is then anaerobic. Few offensive odors appear if aerobiosis is maintained, but this is often difficult or expensive to achieve. When the wastes become anaerobic, the odor threshold is soon reached, and these thresholds are, for some of the malodorous compounds, very low.

Often the major emission is NH_3, particularly if the source is animal wastes. The NH_3 is derived from urea or a range of other nitrogenous constituents, either in manure or in urine. These are mineralized readily, the pH rises, and the NH_3 formed at high pH from the ammonium–NH_3 equilibrium is volatilized (Luebs *et al.*, 1973; Denmead *et al.*, 1974). However, a variety of other compounds, most organic but some inorganic (e.g., H_2S), may be formed. Many are nitrogenous or sulfur containing. They include the following (Mosier *et al.*, 1973; Roustan *et al.*, 1977; Smith *et al.*, 1977; Young *et al.*, 1971):

(a) Organic acids: Butyric
(b) Nitrogenous: Tri-, di-, and monomethylamine, ethylamine, other alkylamines, skatole, and indole
(c) Sulfur containing: H_2S, dimethyl sulfide, dimethyl disulfide, and methane, ethane, and propane thiols (synonyms: methyl, ethyl, and *n*-propyl mercaptans)

Biofilters are used to degrade some of these off odors. That technology is reviewed in Chapter 17.

REFERENCES

Abeles, F. B., Craker, L. E., Forrence, L. E., and Leather, G. R., *Science* **173,** 914–916 (1971).

Bae, J., and McCarty, P. L., *Water Environ. Res.* **65,** 890–898 (1993).

Bange, H. W., Bartell, U. H., Rapsomanikis, S., and Andreae, M. O., *Global Biogeochem. Cycles* **8,** 465–480 (1994).

Bange, H. W., Rapsomanikis, S., and Andreae, M. O., *Global Biogeochem. Cycles* **10,** 197–207 (1996).

Bartholomew, G. W., and Alexander, M., *Appl. Environ. Microbiol.* **37,** 932–937 (1979).

Bartholomew, G. W., and Alexander, M., *Science* **212,** 1389–1391 (1981).

Bartlett, K. B., and Harriss, R. C., *Chemosphere* **26,** 261–320 (1993).

Baumgärtner, M., Koschorrek, M., and Conrad, R., *FEMS Microbiol. Ecol.* **19,** 165–170 (1996).

Beauchamp, E. G., *Can. J. Soil Sci.* **77,** 113–123 (1997).

Bender, M., and Conrad, R., *Biogeochemistry* **27,** 97–112 (1994).

Brown, S. K., and Brown, L. R., *Dev. Ind. Microbiol.* **22,** 725–731 (1981).

Cao, M., Gregson, K., Marshall, S., Dent, J. B., and Heal, O. W., *Chemosphere* **33,** 879–897 (1996).

Chameides, W. L., and Perdue, E. M., "Biogeochemical Cycles." Oxford University Press, New York, 1997.

Conrad, R., *in* "Denitrification in Soil and Sediment" (N. P. Revsbech and J. Sørensen, eds.), pp. 105–128. Plenum Press, New York, 1990.

Conrad, R., *Biogeochemistry* **27,** 155–170 (1994).

Conrad, R., *Adv. Microb. Ecol.* **14,** 207–250 (1995).

Conrad, R., *Microbiol. Rev.* **60,** 609–640 (1996).

Conrad, R., and Seiler, W., *Appl. Environ. Microbiol.* **40,** 437–445 (1980).

Davidson, E. A., *in* "Microbial Production and Consumption of Greenhouse Gases: Methane, Nitrogen Oxides, and Halomethanes" (J. E. Rogers and W. B. Whitman, eds.), pp. 219–235. American Society for Microbiology, Washington, DC, 1991.

Deanmead, O. T., Simpson, J. R., and Freney, J. R., *Science* **185,** 609–610 (1974).

Dörr, H., Katruff, L., and Levin, I., *Chemosphere* **26,** 697–713 (1993).

Dunfield, P. F., and Knowles, R., *Biol. Fertil. Soils* **24,** 294–300 (1997).

Erickson, D. J., III, *Global Biogeochem. Cycles* **3,** 305–314 (1989).

Ferman, M. A., *in* "Atmospheric Biogenic Hydrocarbons" (J. J. Bufalini and R. R. Arnts, eds.), Vol. 2, pp. 51–75. Ann Arbor Science, Ann Arbor, MI, 1981.

Firestone, M. K., and Davidson, E. A., *in* "Exchange of Trace Gases Between Terrestrial Ecosystems and the Atmosphere" (M. O. Andrea and D. S. Schimel, eds.), pp. 7–22. Wiley, New York, 1989.

Hanson, P. J., and Hoffman, W. A., *Soil Sci. Soc. Am. J.* **58,** 552–555 (1994).

Harper, D. P., and Hamilton, J. T. G., *J. Gen. Microbiol.* **134,** 2831–2839 (1988).

Houghton, R. A., *in* "Soils and Global Change" (R. Lal, J. Kimble, E. Levine, and B. A. Stewart, eds.), pp. 45–65. Lewis, Boca Raton, FL, 1995.

Hutchinson, G. L., *in* "Soils and Global Change" (R. Lal, J. Kimble, E. Levine, and B. A. Stewart, eds.), pp. 219–236. Lewis, Boca Raton, FL, 1995.

Junge, C., Seiler, W., Bock, R., Greese, K. D., and Radler, F., *Naturwissenschaften* **58,** 362–363 (1971).

Kammen, D. M., and Marino, B. D., *Chemosphere* **26,** 69–86 (1993).

Kanakidou, M., Bonsang, B., Le Roulley, J. C., Lambert, G., Martin, D., and Sennequier, G., *Nature (London)* **333,** 51–52 (1988).

Khalil, M. A. K., and Rasmussen, R. A., *Global Biogeochem. Cycles* **2,** 63–71 (1988).

Khalil, M. A. K., and Rasmussen, R. A., *Chemosphere* **20,** 227–242 (1990).

Khalil, M. A. K., and Rasmussen, R. A., *Chemosphere* **26,** 803–814 (1993).

Kroeze, C., *Sci. Total Environ.* **143,** 193–209 (1994).

Lelieveld, J., Crutzen, P. J., and Bruhl, C., *Chemosphere* **26,** 739–768 (1993).

Logan, J. A., *J. Geophys. Res.* **88,** 10785–10807 (1983).

Luebs, R. E., Davis, K. R., and Laag, A. E., *J. Environ. Qual.* **2,** 137–141 (1973).

Manley, S. L., and Dastoor, M. N., *Limnol. Oceanogr.* **32,** 709–715 (1987).

Manley, S. L., and de la Cuesta, J. L., *Limnol. Oceanogr.* **42,** 142–147 (1997).

McElroy, M. B., and Wofsy, S. C., *in* "Tropical Rain Forests and the World Atmosphere" (G. T. Prance, ed.), pp. 33–60. Westview, Boulder, CO, 1986.

McFarlane, J. C., Cross, A., Frank, C., and Rogers, R. D., *Environ. Monit. Assess.* **1,** 75–81 (1981).

Mosier, A. R., Andre, C. E., and Viets, F. G., Jr., *Environ. Sci. Technol.* **7,** 642–644 (1973).

Mosier, A. R., Parton, W. J., Valentine, D. W., Ojima, D. S., Schimel, D. S., and Delgado, J. A., *Global Biogeochem. Cycles* **10,** 387–399 (1996).

Müller, J.-F., *J. Geophys. Res.* **97,** 3787–3804 (1992).

Munn, R. E., and Phillips, M. L., *in* Progress in Biometeorology (S. W. Tromp, ed.), Vol. 1, pp. 311–331. Swets and Zeitlinger, Amsterdam, 1975.

National Academy of Sciences, "Nitrogen Oxides." National Academy of Sciences, Washington, DC, 1977a.

National Academy of Sciences, "Carbon Monoxide." National Academy of Sciences, Washington, DC, 1977b.

Nightingale, P. D., Malin, G., and Liss, P. S., *Limnol. Oceanogr.* **40,** 680–689 (1995).

Oremland, R. S., *in* "Microbial Growth on C_1 Compounds" (M. E. Lidstrom and F. R. Tabita, eds.), pp. 310–317. Kluwer, Dordrecht, The Netherlands, 1996.

Plass-Dülmer, C., Koppmann, R., Ratte, M., and Rudolph, J., *Global Biogeochem. Cycles* **9,** 79–100 (1995).

Potter, C. S., Klooster, S. A., and Chatfield, R. B., *Chemosphere* **33,** 1175–1193 (1996).

Primrose, S. B., *J. Gen. Microbiol.* **97,** 343–346 (1976).

Primrose, S. B., and Dilworth, M. J., *J. Gen. Microbiol.* **93,** 177–181 (1976).

Raich, J. W., and Potter, C. S., *Global Biogeochem. Cycles* **9,** 23–36 (1995).

Rasmussen, R. A., *in* "Atmospheric Biogenic Hydrocarbons" (J. J. Bufalini and R. R. Arnts, eds.), Vol. 1, pp. 3–14. Ann Arbor Science, Ann Arbor, MI, 1981.

Rice, P. J., Anderson, T. A., Cink, J. H., and Coats, J. R., *Environ. Toxicol. Chem.* **15,** 1723–1729 (1996).

Ristaino, J. B., and Thomas, W., *Plant Dis.* **81,** 964–977 (1997).

Robinson, E., Rasmussen, R. A., Westberg, H. H., and Holdren, M. W., *J. Geophys. Res.* **78,** 5345–5351 (1973).

Rodhe, H., *Science* **248,** 1217–1219 (1990).

Roustan, J. L., Aumaitre, A., and Salmon-Legagneur, E., *Agric. Environ.* **3,** 147–157 (1977).

Rudolph, J., Koschorreck, M., and Conrad, R., *Soil Biol. Biochem.* **28,** 1389–1396 (1996).

Sanderson, M. G., *Global Biogeochem. Cycles* **10,** 543–557 (1996).

Sawada, S., and Totsuka, T., *Atmos. Environ.* **20,** 821–832 (1986).

Schink, B., *Arch. Microbiol.* **142,** 295–301 (1985).

Schlesinger, W. H., *in* "Soils and Global Change" (R. Lal, J. Kimble, E. Levine, and B. A. Stewart, eds.), pp. 9–25. Lewis, Boca Raton, FL, 1995.

Seiler, W., *Tellus* **26,** 116–135 (1974).

Shorter, J. H., Kolb, C. E., Crill, P. M., Kerwin, R. A., Talbot, R. W., Hines, M. E., and Harriss, R. C., *Nature (London)* **377,** 717–719 (1995).

Smith, A. M., and Cook, R. J., *Nature (London)* **252,** 703–705 (1974).

Smith, K. A. Bremner, J. M., and Tabatabai, M. A., *Soil Sci.* **116,** 313–319 (1973).

Smith, M. S., Francis, A. J., and Duxbury, J. M., *Environ. Sci. Technol.* **11,** 51–55 (1977).

Stern, D. I., and Kaufmann, R. K., *Chemosphere* **33,** 159–176 (1996).

Striegl, R. G., *Chemosphere* **26,** 715–720 (1993).

Sturges, W. T., Cota, G. F., and Buckley, P. T., *Nature (London)* **358,** 660–662 (1992).

Swinnerton, J. W., Linnenbom, V. J., and Lamontagne, R. A., *Science* **167,** 984–986 (1970).

Tait, V. K., and Moore, R. M., *Limnol. Oceanogr.* **40,** 189–195 (1995).

Tingey, D. T., *in* "Atmospheric Biogenic Hydrocarbons" (J. J. Bufalini and R. R. Arnts, eds.), Vol. 1. pp. 53–79. Ann Arbor Science, Ann Arbor, MI, 1981.

Topp, E., and Pattey, E., *Can. J. Soil Sci.* **77,** 167–178 (1997).

Troxler, R. F., and Dokos, J. M., *Plant Physiol.* **51,** 72–75 (1973).

Tyler, S. C., *in* "Microbial Production and Consumption of Greenhouse Gases: Methane, Nitrogen Oxides, and Halomethanes" (J. E. Rogers and W. B. Whitman, eds.), pp. 7–38. American Society for Microbiology, Washington, DC, 1991.

Van Cleemput, O., and El-Sebaay, A. S., *Adv. Agron.* **38,** 159–181 (1985).

Vedenina, I. Ya., and Zavarzin, G. A., *Mikrobiologiya* **46,** 898–903 (1977).

Vedenina, I. Ya., and Zavarzin, G. A., *Mikrobiologiya* **48,** 581–585 (1979).

Wever, R., *in* "Microbial Production and Consumption of Greenhouse Gases: Methane, Nitrogen Oxides and Halomethanes" (J. E. Rogers and W. B. Whitman, eds.), pp. 277–285. American Society for Microbiology, Washington, DC, 1991.

Wever, R., *in* "Microbial Growth on C_1 Compounds" (J. C. Murrell and D. P. Kelly, eds.), pp. 35–45. Intercept, Andover, UK, 1993.

Wilson, D. F., Swinnerton, J. W., and Lamontagne, R. A., *Science* **168,** 1577–1579 (1970).

Wuebbles, D. J., and Edmonds, J., "Primer on Greenhouse Gases." Lewis, Chelsea, MI, 1991.

Young, R. J., Dondero, N. C., Ludington, D. C., and Loehr, R. C., *in* "Identification and Measurement of Environmental Pollutants" (B. Westley, ed.), pp. 98–104. National Research Council, Ottawa, Canada, 1971.

APPENDIX

Abbreviations, Acronyms, and Structures

ABS	Alkylbenzene sulfonate
Alachlor	2-Chloro-2′,6′-diethyl-*N*-(methoxymethyl)acetanilide
Aldicarb	2-Methyl-2-(methylthio)propionaldehyde *O*-(methylcarbamoyl)oxime
Aldrin	1,2,3,4,10,10-Hexachloro-1,4,4a,5,8,8a-hexahydro-*exo*-1,4-*endo*-5,8-dimethanonaphthalene
Ametryn	*N*-Ethyl-*N*′-(1-methylethyl)-6-(methylthio)-1,3,5-triazine-2,4-diamine
Amitrole	3-Amino-1,2,4-triazole
Asulam	Methyl(4-aminobenzenesulfonyl)carbamate
Atrazine	2-Chloro-4-ethylamino-6-isopropylamino-1,3,5-triazine
Azinphosmethyl	*O,O*-Dimethyl *S*-[4-oxo-1,2,3-benzotriazin-3(4*H*)yl]methyl phosphorodithioate
Barban	4-Chloro-2-butynyl 3-chlorocarbanilate
Benomyl	1-(Butylcarbamoyl)-2-benzimidazole carbamic acid, methyl ester
Benthiocarb	*S*-4-Chlorobenzyl *N,N*-diethylthiocarbamate

Benzoylprop-ethyl	Ethyl *N*-benzoyl-*N*-(3,4-dichlorophenyl)-DL-alaninate
BHC	Hexachlorocyclohexane
Bifenox	Methyl 5-(2,4-dichlorophenoxy)-2-nitrobenzoate
Bromoxynil	3,5-Dibromo-4-hydroxybenzonitrile
BTEX	Benzene, toluene, ethylbenzene, xylene
Butralin	4-(1,1-Dimethylethyl)-*N*-(1-methylpropyl)-2,6-dinitrobenzenamine
Buturon	3-(4-Chlorophenyl)-1-methyl-1-(1-methyl-2-propynyl)urea
Butylate	*S*-Ethyl diisobutylthiocarbamate
Captan	*N*-Trichloromethylthio-3a,4,7,7a-tetrahydrophthalimide
Carbofuran	2,4-Dihydro-2,2-dimethylbenzofuran-7-yl-methyl carbamate
Carboxin	5,6-Dihydro-2-methyl-1,4-oxathiin-3-carboxanilide
Chlomethoxynil	2,4-Dichlorophenyl-3′-methoxy-4′-nitrophenyl ether
Chloransulam	*N*-(2-Carbomethoxy-6-chlorophenyl)-5-ethoxy-7-fluoro[1,2,4]triazolo[1,5,*c*]pyrimidine-2-sulfonamide
Chlordane	1,2,4,5,6,7,8,8-Octachloro-2,3,3a,4,7,7a-hexahydro-4,7-methanoindane
Chlorfenvinphos	2-Chloro-1-(2,4-dichlorophenyl)vinyl diethyl phosphate
Chlornitrofen	2,4,6-Trichlorophenyl 4′-nitrophenyl ether
Chlorobenzilate	Ethyl 4,4′-dichlorobenzilate
Chloroneb	1,4-Dichloro-2,5-dimethoxybenzene
Chlorothalonil	Tetrachloroisophthalonitrile
Chlorotoluron	*N*-(3-Chloro-4-methylphenyl)-*N*′-dimethylurea
Chlorpropham	Isopropyl-*N*-3-chlorophenylcarbamate
Chlorpyrifos	*O,O*-Diethyl *O*-(3,5,6-trichloro-2-pyridyl) phosphorothioate

Chlorsulfuron	2-Chloro-*N*-([(4-methoxy-6-methyl-1,3,5-triazin-2-yl)amino]carbonyl) benzenesulfonamide
CIPC	See Chlorpropham
CMC	Critical micelle concentration
Cyanazine	2-Chloro-4-(1-cyano-1-methylethylamino)-6-ethylamino-*s*-triazine
Cyanox	*O,O*-Dimethyl *O*-(4-cyanophenyl) phosphorothioate
2,4-D	2,4-Dichlorophenoxyacetic acid
Dalapon	2,2-Dichloropropionic acid
Dasanit	*O,O*-Diethyl *O*-[4-(methylsulfinyl)phenyl] phosphorothioate
4-(2,4-DB)	4-(2,4-Dichlorophenoxy)butyric acid
DBP	4,4′-Dichlorobenzophenone
DCNA	2,6-Dichloro-4-nitroaniline
DDD	1,1-Dichloro-2,2-bis(*p*-chlorophenyl)ethane
DDE	1,1-Dichloro-2,2-bis(*p*-chlorophenyl)ethylene
DDT	1,1,1-Trichloro-2,2-bis(*p*-chlorophenyl)ethane
DEHP	Di(2-ethylhexyl) phthalate
Denmert	*S-n*-Butyl *S′-p-tert*-butylbenzyl *N*-3-pyridyldithiocarbonimidate
Diazinon	*O,O*-Diethyl *O*-(2-isopropyl-4-methyl-6-pyrimidinyl)phosphorothioate
Dicamba	3,6-Dichloro-2-methoxybenzoic acid
Dichlobenil	2,6-Dichlorobenzonitrile
Dichlorfop-methyl	Methyl 2-[4-(2,4-dichlorophenoxy)phenoxy]propionate
Dichlorvos	2,2-Dichlorovinyl dimethyl phosphate
Dicryl	*N*-(3,4-Dichlorophenyl)methacrylamide
Dieldrin	1,2,3,4,10,10-Hexachloro-6,7-epoxy-1,4-4a,5,6,7,8,8a-*endo,exo*-1,4:5,8-dimethanonaphthalene
Diethofencarb	Isopropyl 3,4-diethoxycarbanilate
Dietholate	*O,O*-Diethyl *O*-phenyl phosphorothioate

Dimethoate	*O,O*-Dimethyl *S*-(*N*-methylacetamide) phosphorodithioate
Dinitramine	N^3,N^3-Diethyl-2,4-dinitro-6-trifluoromethyl-*m*-phenylenediamine
Dinoseb	2-*sec*-Butyl-4,6-dinitrophenol
Diphenamid	*N,N*-Dimethyl-2,2-diphenylacetamide
Diquat	1,1′-Ethylene-2,2′-dipyridylium
Disulfoton	*O,O*-Diethyl *S*-(2-ethylthioethyl)phosphorodithioate
Diuron	3-(3,4-Dichlorophenyl)-1,1-dimethylurea
DNAPL	Dense nonaqueous-phase liquid
DNOC	4,6-Dinitro-*o*-cresol
DOC	Dissolved organic C
Dursban	See chlorpyrifos
Dyfonate	*O*-Ethyl *S*-phenyl ethylphosphonodithioate
EDB	1,2-Dibromoethane (syn: ethylene dibromide)
Endosulfan	6,7,8,9,10,10-Hexachloro-1,5,5a,6,9,9a-hexahydro-6,9-methano-2,4,3-benzodioxathiapin-3-oxide
Endothal	3,6-*endo*-Oxohexahydrophthalate
Endrin	1,2,3,4,10,10-Hexachloro-6,7-epoxy-1,4,4a,5,6,7,8,8a-octahydro-*endo,endo*-1,4:5,8-dimethanonaphthalene
Enthoprop	*O*-Ethyl *S,S*-dipropyl phosphorodithioate
EPTC	*S*-Ethyl dipropyldithiocarbamate
Ethiofencarb	α-Ethylthio-*o*-tolyl methylcarbamate
Fenitrothion	*O,O*-Dimethyl *O*-(4-nitro-*m*-tolyl)phosphorothioate
Fensulfothion	Diethyl 4-(methylsulfinyl)phenyl phosphorothionate
Fenthion	*O,O*-Dimethyl *O*-(3-methyl-4-methylthiophenyl) phosphorothioate
Flamprop-methyl	Methyl *N*-benzoyl-*N*-(3-chloro-4-fluorophenyl)-2-aminopropionate

Fluchloralin *N*-(2-Chloroethyl)-α,α,α-trifluoro-2,6-dinitro-*N*-propyl-*p*-toluidine

Fluometuron *N*′-(3-Trifluoromethylphenyl)-*N,N*-dimethylurea

Fluvalinate α-Cyano-3-phenoxybenzyl 2-[2-chloro-4-(trifluoromethyl)anilino]-3-methylbutanoate

Fonofos *O*-Ethyl-*S*-phenyl ethylphophonodithioate

Glufosinate 2-Amino-4-(methylphosphinyl)butanoic acid

Glyphosate *N*-Phosphonomethylglycine

Guthion *O,O*-Dimethyl *S*[(4-oxo-1,2,3-benzotriazin-3(4*H*)-yl)methyl] phosphorodithioate

Heptachlor 1,4,5,6,7,8,8-Heptachloro-3a,4,7,7a-tetrahydro-4,7-methanoindene

Hexazinone 3-Cyclohexyl-6-(dimethylamino)-1-methyl-1,3,5-triazine-2,4(1*H*, 3*H*)-dione

Hinosan *O*-Ethyl *S,S*-diphenyl phosphorodithiolate

HMX Octahydro-1,3,5,7-tetranitro-1,3,5,7-tetraazocine

Imugam *N*-[2,2,2-Trichloro-1-(3,4-dichloroanilino)ethyl]formamide

Ipazine 2-Chloro-4-diethylamino-6-isopropylamino-1,3,5-triazine

IPC Isopropyl *N*-phenylcarbamate

Iprodione 3-(3,5-Dichlorophenyl)-*N*-1-methylethyl-2,4-dioxo-1-imidazolidinecarboxamide

Isodrin An isomer of aldrin

Isofenphos 2-[(Ethyoxy[(1-methylethyl)amino]phosphinothioyl)oxy]-benzoic acid 1-methylethyl ester

Isoproturon *N*-(4-Isopropylphenyl)-*N*′,*N*′-dimethylurea

Karsil *N*-(3,4-Dichlorophenyl)-2-methylpentaneamide

Kepone Decachloro-octahydro-1,3,4-methano-2*H*-cyclobuta(cd)pentalene-2-one

Kitazin P *S*-Benzyl *O,O*-diisopropyl phosphorothiolate

Lindane γ-1,2,3,4,5,6-Hexachlorocyclohexane

Linuron 3-(3,4-Dichlorophenyl)-1-methoxy-1-methylurea

Malathion	*O,O*-Dimethyl *S*-(1,2-bis-carbethoxy)ethyl phosphorodithioate
MBC	Methyl benzimidazol-2-yl carbamate
MBOCA	4,4′-Methylene-bis(2-chloroaniline)
MCPA	2-Methyl-4-chlorophenoxyacetic acid
MCPB	4-(4-Chloro-2-methylphenoxy)butyric acid
Mecoprop	2-(2-Methyl-4-chlorophenoxy)propionic acid
Metamitron	4-Amino-3-methyl-6-phenyl-1,2,4-triazin-5(4*H*)one
Methyl parathion	*O,O*-Dimethyl *O*-(*p*-nitrophenyl)phosphorothioate
Metolachlor	2-Chloro-*N*-(2-ethyl-6-methylphenyl)-*N*-(2-methoxy-1-methylethyl)acetamide
Metribuzin	4-Amino-6-(1,1-dimethylethyl)-3-(methylthio)-1,2,4-triazin-5(4*H*)one
Mirex	Dodecachlorooctahydro-1,3,4-metheno-2*H*-cyclobuta(cd)pentalene
Monocrotophos	Dimethyl-(E)-1-methyl-2-methylcarbamoylvinyl phosphate
Monolinuron	3-(4-Chlorophenyl)-1-methoxy-1-methylurea
Monuron	3-(4-Chlorophenyl)-1,1-dimethylurea
NAPL	Nonaqueous-phase liquid
Napropamide	*N,N*-Diethyl-2-(1-naphthalenyloxy)propionamide
Nitrofen	2,4-Dichlorophenyl *p*-nitrophenyl ether
NTA	Nitrilotriacetic acid
Orbencarb	*S*-2-Chlorobenzyl diethylthiocarbamate
Ordram	*S*-Ethyl hexahydro-1*H*-azepine-1-carbothioate
Oryzalin	3,5-Dinitro-N^4,N^4-dipropylsulfanilamide
PAH	Polycyclic aromatic hydrocarbon
Paraoxon	Diethyl *p*-nitrophenyl phosphate
Paraquat	1,1′-Dimethyl-4,4′-bipyridinium
Parathion	*O,O*-Diethyl *O*-(*p*-nitrophenyl)phosphorothioate
PCB	Polychlorinated biphenyl

PCE Perchloroethylene (syn: Tetrachloroethylene)

PCNB Pentachloronitrobenzene

PCP Pentachlorophenol

Phorate *O,O*-Diethyl *S*-(ethylthio)methyl phosphorodithioate

Picloram 4-Amino-3,5,6-trichlopicolinic acid

Pirimiphos-methyl *O*-(2-Dimethylamino-6-methylpyrimidin-4-yl) *O,O*-dimethyl phosphorothioate

Profluralin *N*-(Cyclopropylmethyl)-α,α,α-trifluoro-2,6-dinitro-*N*-propyl-*p*-toluidine

Propachlor 2-Chloro-*N*-isopropylacetanilide

Propanil *N*-(3,4-Dichlorophenyl)propionamide

Pyrazon 5-Amino-4-chloro-2-phenylpyridazine-3(2*H*)one

RDX Hexahydro-1,3,5-trinitro-1,3,5-triazine

SBR Structure–biodegradability relationship

Sevin 1-Naphthyl *N*-methylcarbamate

Silvex 2-(2,4,5-Trichlorophenoxy)propionic acid

Simazine 2-Chloro-4,6-bis-ethylamino-1,3,5-triazine

Sucralose 4-Chloro-4-deoxy-α,D-galactopyranosyl-1,6-dichloro-1,6-dideoxy-β,D-fructofuranoside

Swep Methyl-*N*-(3,4-dichlorophenyl) carbamate

2,4,5-T 2,4,5-Trichlorophenoxyacetic acid

TCA Trichloroacetic acid

TCDD 2,3,7,8-Tetrachlorodibenzo-*p*-dioxin

TCE Trichloroethylene

Terbutryn 2-Ethylamino-4-(*t*-butylamino)-6-methylthio-*s*-triazine

Tetrachlorvinphos 2-Chloro-1-(2′,4,′,5′-trichlorophenyl)vinyl dimethyl phosphate

Thiram Tetramethylthiuram disulfide

TNT 2,4,6-Trinitrotoluene

Tordon 4-Amino-3,5,6-trichloropicolinic acid

Toxaphene Chlorinated camphene

Triadimefon	1-(4-Chlorophenoxy)-3,3-dimethyl-1-(1,2,4-triazol-1-yl)butan-2-one
Trichlorfon	2,2,2-Trichloro-1-hydroxyethyl phosphonate
Trichloronat	*O*-Ethyl *O*-2,4,5-trichlorophenyl ethylphosphonothionate
Triclopyr	3,5,6-Trichloro-2-pyridinyloxyacetic acid
Trifluralin	α,α-α-Trifluoro-2,6-dinitro-*N,N*-dipropyl-*p*-toluidine
Trithion	*O,O*-Diethyl *S*-(4-chlorophenylthio)methyl phosphorodithoate
Vernolate	*S*-Propyl dipropylthiocarbamate
Vinclozolin	3-(3,5-Dichlorophenyl)-5-ethenyl-5-methyl-2,4-oxazolidinedione
Zinophos	*O,O*-Diethyl 2-pyrazinyl phosphorothionate

INDEX

D

E

V

W

X

Z